Philosophy of Computer Science

Philosophy of Computer Science

An Introduction to the Issues and the Literature

William J. Rapaport
Department of Computer Science and Engineering
Department of Philosophy, Department of Linguistics
and Center for Cognitive Science
University at Buffalo, The State University of New York
Buffalo, NY

Registered Office
John Wiley & Sons, Inc., 111 River Street, Hoboken, NJ 07030, USA

For details of our global editorial offices, customer services, and more information about Wiley products visit us at www.wiley.com.

Wiley also publishes its books in a variety of electronic formats and by print-on-demand. Some content that appears in standard print versions of this book may not be available in other formats.

Library of Congress Cataloging-in-Publication Data

Names: Rapaport, William J., author.
Title: Philosophy of computer science : an introduction to the issues and
    the literature / William J. Rapaport.
Description: Hoboken, NJ : Wiley-Blackwell, 2023. | Includes
    bibliographical references and index.
Identifiers: LCCN 2022039093 (print) | LCCN 2022039094 (ebook) | ISBN
    9781119891901 (paperback) | ISBN 9781119891918 (adobe pdf) | ISBN
    9781119891925 (epub)
Subjects: LCSH: Computer science–Philosophy.
Classification: LCC QA76.167 .R37 2023 (print) | LCC QA76.167 (ebook) |
    DDC 004–dc23/eng/20220824
LC record available at https://lccn.loc.gov/2022039093
LC ebook record available at https://lccn.loc.gov/2022039094

Cover design: Wiley
Cover image: © 3Dsculptor/Shutterstock

Set in 9.5/12.5pt STIXTwoText by Straive, Chennai, India

This book is dedicated to my family:

Mary, Michael, Sheryl, Makayla, Laura, William, Allyson, Lexie, Rob, and Robert.

Contents

List of Figures

If you begin with Computer Science, you will end with Philosophy.[1]

---

1 "Clicking on the first link in the main text of an English Wikipedia article, and then repeating the process for subsequent articles, usually leads to the Philosophy article. In February 2016, this was true for 97% of all articles in Wikipedia, an increase from 94.52% in 2011" ("Wikipedia:Getting to Philosophy," http://en.wikipedia.org/wiki/Wikipedia:Getting_to_Philosophy). On 9 August 2021, if you began with "Computer Science," you would end with "Philosophy" in 11 links: computer science → algorithm → mathematics → quantity → counting → number → mathematical object → concept → abstraction → rule of inference → philosophy of logic → philosophy.

Preface

This is a university-level introduction to the philosophy of computer science based on a course that I created at the University at Buffalo in 2004 and taught from 2004 to 2010 (I retired in 2012). At the time I created the course, there were few other such courses and virtually no *textbooks* (only a few monographs and anthologies). Although there are now more such courses, there are only a very few *introductory* textbooks in the area. My retirement project was to turn my lecture notes into a book that could be used as an introduction to the issues and serve as a guide to the original literature; this book is the result.

The course is described in Rapaport 2005c. The syllabus, readings, assignments, and website for the last version of the course are online at http://www.cse.buffalo.edu/~rapaport/584/. The Online Resources contain suggested further readings, in-class exercises (arguments for analysis, in addition to the questions at the ends of some of the chapters), term-paper suggestions, a sample final exam, advice to the instructor on peer-editing for the exercises, and a philosophy of grading.

Many of the books and articles I discuss are available on the Web. Rather than giving Web addresses (URLs) for them, I urge interested readers to try a Google (or other) search for the documents. Books and journal articles can often be found either by visiting the author's website (e.g. most of my papers are at https://cse.buffalo.edu/~rapaport/papers.html) or by using a search string consisting of the last name(s) of the author(s) followed by the title of the document enclosed in quotation marks (For example, to find Rapaport 2005c, search for "rapaport "philosophy of computer science""). URLs that I give for Web-only items (or other hard-to-find items) were accurate at the time of writing. Some, however, will change or disappear. Documents that have disappeared can sometimes be found at the Internet Archive's Wayback Machine (https://archive .org/web/). Some documents with no public URLs may eventually gain them. And, of course, readers should search the Internet or Wikipedia for any unfamiliar term or concept.

Sidebars: Sprinkled throughout the book are sidebars in boxes, like this one. Some are **Digressions** that clarify or elaborate on various aspects of the text. Some are suggestions for **Further Reading**. Others are **Questions** for the reader to consider at that point in the text. Additional suggested readings, along with student assignments and an instructor's manual, are in the Online Resources.

Figure 1 CALVIN AND HOBBES ©2015 Watterson. Reprinted with permission of ANDREWS MCMEEL SYNDICATION. All rights reserved.

Acknowledgments

For comments on, suggestions for, or corrections to earlier versions, my thanks go especially to

Peter Boltuc, Jonathan Bona, Selmer Bringsjord, Jin-Yi Cai, Timothy Daly, Edgar Daylight, Peter Denning, Eric Dietrich, William D. Duncan, J. Michael Dunn, Frank Fedele, Albert Goldfain, James Graham Maw, Carl Hewitt, Robin K. Hill, Johan Lammens, Cliff Landesman, Nelson Pole, Thomas M. Powers, Michael I. Rapaport, Stuart C. Shapiro, Aaron Sloman, Mark Staples, Matti Tedre, and Victoria G. Traube;

as well as to

Russ Abbott, Khaled Alshammari, S.V. Anbazhagan, S. Champailler, Arnaud Debec, Roger Derham, Gabriel Dulac-Arnold, Mike Ferguson, Pablo Godoy, David Miguel Gray, Nurbay Irmak, Patrick McComb, Cristina Murta, Alexander Oblovatnyi, oleg@okmij.org, Andres Rosa, Richard M. Rubin, Seth David Schoen, Stephen Selesnick, Dean Waters, Nick Wiggershaus, and Sen Zhang;

and

the University at Buffalo Department of Computer Science Information Technology staff for help with LaTex; and my editors at Wiley: Will Croft, Rosie Hayden, and Tiffany Taylor.

About the Companion Website

This book is accompanied by a companion website:
https://cse.buffalo.edu/~rapaport/OR/

This website includes:
- An annotated list of further readings for each chapter
- Sample "position paper" assignments for argument analysis
- Sample term-paper topics
- A sample final exam
- An instructor's manual, with information on:
  - o how to use the position-paper assignments
  - o how to grade, including:
    - ■ a "triage philosophy of grading"
    - ■ suggested analyses and grading rubrics for the position papers
  - o a discussion of William Perry's scheme of cognitive development and its application to the final exam.

Part I

Philosophy and Computer Science

Part I is an introduction to both philosophy and the philosophy of computer science.

Philosophy of Computer Science: An Introduction to the Issues and the Literature, First Edition. William J. Rapaport.
© 2023 John Wiley & Sons, Inc. Published 2023 by John Wiley & Sons, Inc.

1

An Introduction to the Philosophy of Computer Science

Philosophy is often thought of as an activity, which may have considerable theoretical interest, but which is of little practical importance. Such a view of philosophy is … profoundly mistaken. … [P]hilosophical ideas and some kind of philosophical orientation are necessary for many quite practical activities. … [L]ooking at the general question of how far philosophy has influenced the development of computer science[, m]y own view is that the influence of philosophy on computer science has been very great.
—Donald Gillies (2002)

Who would have guessed that the arcane research done by the small set of mathematicians and philosophers working on formal logic a century ago would lead to the development of computing, and ultimately to completely new industries, and to the reconfiguring of work and life across the globe?
—Onora O'Neill (2013, p. 8)

There is no such thing as philosophy-free science, just science that has been conducted without any consideration of its underlying philosophical assumptions.
—Daniel C. Dennett (2013a, p. 20)

1.1   What This Book Is About

My mind does not simply receive impressions. It talks back to the authors, even the wisest of them, a response I'm sure they would warmly welcome. It is not possible, after all, to accept passively everything even the greatest minds have proposed. One naturally has profound respect for … [the] heroes of the pantheon of Western culture; but each made statements flatly contradicted by views of the others. So I see the literary and philosophical tradition of our culture not so much as a storehouse of facts and ideas but rather as a hopefully endless Great Debate at which one may be not only a privileged listener but even a modest participant.
—Steve Allen (1989, p. 2), as cited in Madigan, 2014, p. 46.

As [the logician] Harvey Friedman has suggested, every morning one should wake up and reflect on the conceptual and foundational significance of one's work.
—Robert Soare (1999, p. 25)

Philosophy of Computer Science: An Introduction to the Issues and the Literature, First Edition. William J. Rapaport.
© 2023 John Wiley & Sons, Inc. Published 2023 by John Wiley & Sons, Inc.

This book looks at some of the central issues in the philosophy of computer science. It is not designed to answer all (or even any) of the philosophical questions that can be raised about the nature of computing, computers, and computer science. Rather, it is designed to "bring you up to speed" on a conversation about these issues – to give you some background knowledge – so that you can read the literature for yourself and perhaps become part of the conversation by contributing your own views.

This book is intended for readers who might know some philosophy but no computer science, readers who might know some computer science but no philosophy, and even readers who know little or nothing about either! So, although most of the book will be concerned with *computer science*, we will begin by asking, **what is philosophy?**

Then, in Part II, we will begin our inquiry into the philosophy of computer science by asking, **what is computer science?** To answer this, we will need to consider a series of questions, each of which leads to another: is computer science a science, a branch of engineering, some combination of them, or something else altogether? And to answer those questions, we will need to ask, **what is science?** and **what is engineering?**

We next ask, **what does computer science study?** *Computers?* If so, then **what is a computer?** Or does it study *computation?* If so, then **what is computation?** Computations are said to be *algorithms*, so **what is an algorithm?** And **what is the Turing Machine model of algorithmic computation?**

In Part III, we will explore the **Church-Turing Computability Thesis.** This is the proposal that our intuitive notion of computation is completely captured by the formal notion of Turing Machine computation. But some have claimed that there are ordinary procedures (such as recipes) that are not computable by Turing Machines and that hence refute the Computability Thesis. So, **what is a procedure?** (And, for that matter, what is a recipe?) Others have claimed that the intuitive notion of computation goes beyond Turing Machine computation; so, **what is such "hypercomputation"?**

In Part IV, we explore the nature of computer programs. Computations are expressed in computer programs, which are executed by computers, so **what is a computer program**? Are computer programs "implementations" of algorithms? If so, then **what is an implementation?** Programs typically have real-world effects, so **how are programs and computation related to the world?** Some programs, especially in the sciences, are designed to model or simulate or explain some real-world phenomenon, so **can programs be considered (scientific) theories?** Programs are usually considered "software," and computers are usually considered "hardware," but **what is the difference between software and hardware?** Computer programs are notorious for having "bugs," which are often only found by running the program, so **can computer programs be logically verified** before running them?

Finally, in Part V, we look at two topics. The first is the **philosophy of artificial intelligence (AI)**: what is AI? What is the relation of computation to cognition? Can computers think? Alan Turing, one of the creators of the field of computation, suggested that the best way to deal with that question was by using what is now called the Turing Test. The Chinese Room Argument is a thought experiment devised by the philosopher John Searle, which (arguably) shows that the Turing Test won't work.

The other topic is **computer ethics**. We'll look at two questions that were not much discussed at the turn of the century but are now at the forefront of computational ethical debates: (1) should we trust decisions made by computers? (Moor, 1979) – a question made urgent by the advent of automated vehicles and by "deep learning" algorithms that might be biased; and (2) should we build "intelligent" computers? Do we have moral obligations toward robots? Can or should they have moral obligations toward us?

> **Computer Science Students Take Note:** Along the way, we will look at how philosophers reason and evaluate logical arguments. ACM/IEEE Computer Science Curricula 2020 (CC2020) covers precisely these sorts of argument-analysis techniques under the headings of Discrete Structures and Analytical and Critical Thinking. Many other CC2020 topics also overlap those in the philosophy of computer science. See https://www.acm.org/binaries/content/assets/education/curricula-recommendations/cc2020.pdf.

1.2   What This Book Is *Not* About

Have I left anything out? Yes! This book is *not* an attempt to be an encyclopedic, up-to-the-minute survey of every important issue in the philosophy of computer science. Rather, the goal is to give you the background to enable you to fruitfully explore those issues and to join in the conversation.

The questions raised earlier and discussed in this book certainly do not exhaust the philosophy of computer science. They are merely a series of questions that arise naturally from our first question: what is computer science? But there are many other issues in the philosophy of computer science. Some are included in a topic sometimes called *philosophy of computing*. Here are some examples: consider the ubiquity of computing – your smartphone is a computer; your car has a computer in it; even some refrigerators and toasters contain computers. Perhaps someday your bedroom wall will contain (or even be) a computer! How will our notion of computing change because of this ubiquity? Will this be a good or bad thing? Another topic is the role of the Internet. For instance, Tim Berners-Lee, who created the World Wide Web, has argued that "Web science" should be its own discipline (Berners-Lee et al., 2006; Lohr, 2006). And there are many issues surrounding the social implications of computers in general and social media on the Internet (and the World Wide Web) in particular.

Other issues in the philosophy of computer science more properly fall under the heading of the *philosophy of AI*. As noted, we will look at *some* of these in this book, but there are many others that we won't cover, even though the philosophy of AI is a proper subset of the philosophy of computer science.

Another active field of investigation is the *philosophy of information*. As we'll see in Section 3.8, computer science is sometimes defined as the study of how to process information, so the philosophy of information is clearly a close cousin of the philosophy of computer science. But I don't think either is included in the other; they merely have a non-empty intersection. If this is a topic you wish to explore, take a look at some of the books and essays cited at the end of Section 3.8.

And we will not discuss (except in passing; see, for example, Section 9.6.1) *analog* computation. If you're interested in this, see the Online Resources for suggested readings.

Finally, there are a number of philosophers and computer scientists who have discussed topics related to what I am calling the philosophy of computer science whom we will not deal with at all (such as the philosophers Martin Heidegger and Hubert L. Dreyfus (Dreyfus and Dreyfus, 1980; Dreyfus, 2001) and the computer scientist Terry Winograd (Winograd and Flores, 1987). An Internet search (e.g. "Heidegger "computer science"") will help you track down information on these thinkers and others not mentioned in this book. (One philosopher of computer science [personal communication] calls them the "Dark Side philosophers" because they tend not to be sympathetic to computational views of the world!)

But I think the earlier questions will keep us busy for a while as well as prepare you for examining some of these other issues. Think of this book as an extended "infomercial" to bring you up to speed on the computer-science–related aspects of a philosophical conversation that has been going on for over 2500 years, to enable you to join in the conversation.

Let's begin …

Further Reading: In 2006, responding to a talk that I gave on the philosophy of computer science, Selmer Bringsjord (a philosopher and cognitive scientist who has written extensively on the philosophy of computer science) said that philosophy of computer science was in its infancy. This may have been true at the time as a discipline so called, but there have been philosophical investigations of computer science and computing since at least Turing, 1936 (which we'll examine in detail in Chapter 8), and the philosopher James H. Moor's work goes back to the 1970s (we'll discuss some of his writings in Chapters 12 and 17.

In an early undergraduate computer science textbook, my former colleague Tony Ralston (1971, Section 1.2D, pp. 6–7) discussed "the philosophical impact of computers": he said that questions about such things as the nature of thinking, intelligence, emotions, intuition, creativity, consciousness, the relation of mind to brain, and free will and determinism "are serious questions, that the advent of computers has, philosophically speaking, reopened some of these questions and thrown new light on others, and finally, that the philosophical significance of these questions provides a worthy motivation for the study of computer science."

On social implications, see, especially, Weizenbaum, 1976 and Simon, 1977, the penultimate section of which ("Man's View of Man") can be viewed as a response to Weizenbaum. See also Dembart, 1977 for a summary and general discussion. For a discussion of the social implications of the use of computers and the Internet, be sure to read E.M. Forster's classic short story "The Machine Stops" (Forster, 1909), which predicted the Internet and email! (You can easily find versions of it online.)

See the Online Resources for more on the philosophy of computer science.

2

Philosophy: A Personal View

[T]here are those who have knowledge and those who have understanding. The first requires memory, the second philosophy. … Philosophy cannot be taught. Philosophy is the union of all acquired knowledge and the genius that applies it …
—Alexandre Dumas (1844, *The Count of Monte Cristo*, Ch. 17, pp. 168–169)

Philosophy is the microscope of thought.
—Victor Hugo (1862, *Les Misérables*, Vol. 5, Book Two, Ch. II, p. 1262)

Philosophy … works against confusion.
—John Cleese (2012), "[Twenty-First] Century," https://www.apaonline.org/resource/resmgr/John_Cleese_statements/19_Century.mp3

Consider majoring in philosophy. I did. … [I]t taught me how to break apart arguments, how to ask the right questions.
—NPR reporter Scott Simon, quoted in Keith 2014

To the person with the right turn of mind, … all thought becomes philosophy.
—Eric Schwitzgebel (2012)

Philosophy can be any damn thing you want!
—John Kearns (personal communication, 7 November 2013)

2.1   Introduction

[W]e're all doing philosophy all the time. We can't escape the question of what matters and why: the way we're living is itself our implicit answer to that question. A large part of a philosophical training is to make those implicit answers explicit, and then to examine them rigorously.
—David Egan (2019)

"What is philosophy?" is a question that is not a proper part of the philosophy of computer science. But because many readers may not be familiar with philosophy, I want to begin our

Philosophy of Computer Science: An Introduction to the Issues and the Literature, First Edition. William J. Rapaport.
© 2023 John Wiley & Sons, Inc. Published 2023 by John Wiley & Sons, Inc.

exploration with a brief introduction to how I think of philosophy and how I would like non-philosophical readers who are primarily interested in computer science to think of it. So, in this chapter, I will give you *my* definition of 'philosophy' and examine the principal methodology of philosophy: the evaluation of logical arguments.

Note on Quotation Marks: Many philosophers have adopted a convention that *single quotes* are used to form the name of a word or expression. So, when I write this:

'philosophy'

I am not talking about philosophy! Rather, I am talking about the 10-letter word spelled p-h-i-l-o-s-o-p-h-y. This use of single quotes enables us to distinguish between a *thing* that we are talking about and the *name* or *description* that we use to talk about the thing. This is the difference between Paris (the capital of France) and 'Paris' (a five-letter word). The technical term for this is the 'use-mention distinction' (http://en.wikipedia.org/wiki/Use-mention_distinction): we *use* 'Paris' to *mention* Paris. It is also the difference between a *number* (a thing that mathematicians talk about) and a *numeral* (a word or symbol that we use to talk about numbers).

I will use *double quotes* (1) when I am directly quoting someone, (2) as "scare quotes" to indicate that I am using an expression in a special or perhaps unusual way (as I just did), and (3) to indicate the *meaning* of a word or other expression (as in, 'bachelor' means "marriageable male") (Cole, 1999).

However, in both cases, some publishers (including the present one) follow a (slightly illogical) style according to which some punctuation (usually periods and commas), whether part of the quoted material or not, must appear *inside* the quotation marks. I will leave it as an exercise for the reader to determine which punctuation marks that appear inside quotation marks logically belong there! (As a warm-up exercise, is this sentence,

Here is a 5-character word: 'Paris.'

which obeys the publisher's style, true?)

2.2 A Definition of 'Philosophy'

When 'philosophy' is used informally, in everyday conversation, it can mean an "outlook," as when someone asks you what your "philosophy of life" is. The word 'philosophical' can also mean something like "calm," as when we say that someone takes bad news "very philosophically" (i.e. very calmly). Traditionally, philosophy is the study of "Big Questions" (Section 2.7) such as metaphysics (what exists?), epistemology (how can we know what exists?), and ethics (what is "good"?).

In this chapter, I want to explicate the *technical* sense of *modern, analytic, Western* philosophy – a kind of philosophy that has been done since at least the time of Socrates. 'Modern philosophy' is itself a technical term that usually refers to the kind of philosophy that has been done since the time of René Descartes (1596–1650, about 400 years ago) (Nagel, 2016). It is "analytic" in the sense that it is primarily concerned with the logical analysis of concepts (rather than literary, poetic, or speculative approaches). And it is "Western" in the sense that it has been done by philosophers working primarily in Europe (especially in Great Britain) and North America – although, of course, there are very many philosophers who do analytic philosophy in other areas of the world (and there are many other kinds of philosophy; see Adamson 2019).

Western philosophy began in ancient Greece. Socrates (470–399 BCE,[1] i.e. around 2500 years ago) was opposed to the Sophists, a group of teachers who can be caricatured as an ancient Greek version of "ambulance-chasing" lawyers, "purveyors of rhetorical tricks" (McGinn, 2012b). For a fee, the Sophists were willing to teach anything (whether it was true or not) to anyone, or to argue anyone's cause (whether their cause was just or not).

Like the Sophists, Socrates also wanted to teach and argue, but only to seek wisdom: truth in any field. In fact, the word 'philosophy' comes from Greek roots meaning "love of [*philo*] wisdom [*sophia*]." The reason Socrates only *sought* wisdom rather than claiming that he *had* it (as the Sophists did) was that he believed he *didn't* have it: he claimed that he *knew* he didn't know anything (and that, therefore, he was actually wiser than those who claimed that they *did* know things but who really didn't). As the contemporary philosopher Kwame Anthony Appiah said, in reply to the question "How do you think Socrates would conduct himself at a panel discussion in Manhattan in 2019?":

> You wouldn't be able to get him to make an opening statement, because he would say, "I don't know anything." But as soon as anybody started saying anything, he'd be asking you to make your arguments clearer – he'd be challenging your assumptions. He'd want us to see that the standard stories we tell ourselves aren't good enough. (Libbey and Appiah, 2019)

Socrates's student Plato (430–347 BCE), in his dialogue *Apology*, describes Socrates as playing the role of a "gadfly," constantly questioning (and annoying!) people about the justifications for, and consistency among, their beliefs, in an effort to find out the truth for himself from those who considered themselves to be wise (but who really weren't).

Plato defined 'philosopher' (and, by extension, 'philosophy') in Book V of his *Republic* (line 475c):

> The one who feels no distaste in sampling *every study*, and who attacks the task of learning gladly and cannot get enough of it, we shall justly pronounce the lover of wisdom, the philosopher. (Plato, 1961b, p. 714, my emphasis)

Adapting this, I define 'philosophy' as **the *personal search* for *truth*, in *any* field, by *rational* means.** This raises several questions:

1. What is "truth"?
2. Why is philosophy only the *search* for truth? (Can't the search be successful?)
3. What counts as being "rational"?
4. Why only "personal"? (Why not "universal"?)
5. What does 'any field' mean? (Is philosophy really the study of anything and everything?)

The rest of this chapter explores these questions.[2]

2.3   What Is Truth?

The study of the nature of truth is another "Big Question" of philosophy. I cannot hope to do justice to it here, but two theories of truth will prove useful to keep in mind on our journey through the philosophy of computer science: the *correspondence* theory of truth and the *coherence* theory of truth.

---

1 'BCE' is the abbreviation for 'before the common era'; i.e. BCE years are the "negative" years before the year 1, which is known as the year 1 CE (for "common era").
2 See the Online Resources for further reading on the nature of philosophy.

2.3.1 Correspondence Theories of Truth

According to the *Oxford English Dictionary* (*OED*; http://www.oed.com/view/Entry/206884), 'true' originally meant "faithful." Faithfulness requires *two things A* and *B* such that *A* is faithful to *B*. On a correspondence theory, truth is faithfulness of a *representation A* of some part of reality to the *reality B* that it is a representation of. On the one hand, there are *beliefs* (or propositions, or sentences); on the other hand, there is "reality": a belief (or a proposition, or a sentence) is true if and only if ("iff") it corresponds to reality, i.e. iff it is faithful to, or "matches," or accurately represents or describes reality.

> **Terminological Digression:** A "belief," as I am using that term here, is a mental entity, "implemented" (in humans) by certain neuron firings. A "sentence" is a grammatical string of words in some language. And a "proposition" is the meaning of a sentence. These are all rough-and-ready characterizations; each of these terms has been the subject of much philosophical analysis. For further discussion, see Schwitzgebel 2021 on belief, https://en.wikipedia.org/wiki/Sentence-(linguistics) on sentences, and McGrath and Frank 2020 on propositions.

To take a classic example, the three-word English sentence 'Snow is white.' is true iff the stuff in the real world that precipitates in certain winter weather (i.e. snow) has the same color as milk (i.e. iff it is white). Put somewhat paradoxically (but correctly – recall the use-mention distinction!), 'Snow is white.' is true iff snow is white.

How do we *determine* whether a sentence (or a belief, or a proposition) is true? Using a correspondence theory, in principle, we would have to compare the parts of the sentence (its words plus its grammatical structure, and maybe even the context in which it is thought, uttered, or written) with parts of reality, to see if they correspond. But how do we access "reality"? How can we do the "pattern matching" between our beliefs and reality? One answer is by sense perception (perhaps together with our beliefs about what we perceive). But sense perception is notoriously unreliable (think about optical illusions). And one of the issues in deciding whether our *beliefs* are true is deciding whether our *perceptions* are accurate (i.e. whether *they* match reality).

So we seem to be back to square one, which gives rise to coherence theories.

2.3.2 Coherence Theories of Truth

According to a coherence theory of truth, a set of propositions (or beliefs, or sentences) is true iff (1) they are mutually consistent, and (2) they are supported by, or consistent with, all available evidence. That is, they "cohere" with each other and with all evidence. Note that observation statements (i.e. descriptions of what we observe in the world around us) are among the claims that must be mutually consistent, so this is *not* (necessarily) a "pie-in-the-sky" theory that doesn't have to relate to the way things really are. It just says that we don't have to have independent access to "reality" in order to determine truth.

2.3.3 Correspondence vs. Coherence

Which theory is correct? Well, for one thing, there are more than two theories: there are several versions of each kind of theory, and there are other theories of truth that don't fall under either category. The most important of the other theories is the "pragmatic" theory of truth (see Glanzberg 2021, Section 3; Misak and Talisse 2019). Here is one version:

[T]he pragmatic theory of truth … is that a proposition is true if and only [if] it is useful [i.e. "pragmatic," or practical] to believe that proposition. (McGinn, 2015a, p. 148)

Fortunately, the answer to which kind of theory is correct (i.e. which kind of theory is – if you will excuse the expression – *true*) is beyond our present scope! But note that the propositions that a correspondence theory says are true must be mutually consistent (if "reality" is consistent!), and they must be supported by all available evidence; i.e. *a correspondence theory must "cohere"*. Moreover, if you include both propositions and "reality" in one large, highly interconnected network (as we will consider in Sections 16.10.4 and 18.8.3), that network must also "cohere," so the propositions that are true according to *a coherence theory of truth should "correspond to"* (i.e. cohere with) *reality*.

Let's return to the question raised in Section 2.3.1: how can we *decide* whether a statement is true? One way we can determine its truth is *syntactically* (i.e. in terms of its grammatical structure only, not in terms of what it means), by trying to *prove* it from axioms via rules of inference. It is important to keep in mind that when you prove a statement this way, you are not proving that it is true! You are simply proving that it follows logically from certain other statements: i.e. that it "coheres" in a certain way with those statements. But if the starting statements – the axioms – are true (note that I said "*if* they are true"; I haven't told you how to determine *their* truth value yet), *and* if the rules of inference "preserve truth," then the statement you prove by means of them – the "theorem" – will also be true.

Another way we can determine whether a statement is true is *semantically*: i.e. in terms of what it *means*. We can use truth tables to determine that axioms are true. This, by the way, is the only way to determine whether an *axiom* is true, since, by definition, an axiom cannot be inferred from any *other* statements. If it could be so inferred, then it would be those other statements that would be the real axioms.

But to determine the truth of a statement semantically is also to use syntax (i.e. symbol manipulation): we semantically determine the truth value of a complex proposition by symbol manipulation (via truth tables) of its atomic constituents. (For more on syntax and semantics, see Section 18.8.3.) How do we determine the truth value of an atomic proposition? By seeing if it corresponds to reality. But how do we do that? By comparing the proposition with reality: i.e. by seeing if the proposition coheres with reality.[3]

Digression: What Is a Theorem? When you studied geometry, you may have studied a version of Euclid's original presentation of geometry via a modern interpretation as an axiomatic system. Most branches of mathematics (and, according to some philosophers, most branches of science) can be formulated axiomatically. One begins with a set of "axioms": statements that are assumed to be true (or are considered so obviously true that their truth can be taken for granted). Then there are "rules of inference" that tell you how to logically infer other statements from the axioms in such a way that the inference procedure is "truth preserving": if the axioms are true (which they are, by assumption), then whatever logically follows from them according to the rules of inference is also true. (Truth is "preserved" throughout the inference.) Such statements are called 'theorems.'

Do truth and proof coincide? A logical system for which they do is said to be (semantically) "complete": all truths are theorems, and all theorems are true. Two such systems are

---

(Continued)

---

3 See the Online Resources for further reading on theories of truth.

(Continued)

propositional logic and first-order logic. Propositional logic is the logic of sentences, treating them "atomically" as simply being either true or false and not having any "parts." First-order predicate logic can be thought of as a kind of "sub-atomic" logic, treating sentences as being composed of terms standing in relations. (See Rapaport, 1992a,b.) However, if you add axioms for arithmetic to first-order logic, the resulting system is *not* complete; see the Digression on Gödel's Incompleteness Theorem. (See Sections 2.9, 6.5, 7.4.3.2, 13.2.2, 15.1, and 15.2.1 for more details.)

There are also second-order logics, modal logics, relevance logics, and many more (not to mention varieties of each). Is one of them the "right" logic? Tharp 1975 asks that question, which can be expressed as a "thesis" analogous to the Church-Turing Computability Thesis: where the Computability Thesis asks if the formal theory of Turing Machine computability entirely captures the informal, pre-theoretic notion of computability, Tharp asks if there is a formal logic that entirely captures the informal, pre-theoretic notion of logic. We'll return to some of these issues in Chapter 11.

Digression: Gödel's Incompleteness Theorem: Can *any* proposition (or its negation) be proved? Given a proposition P, we know that either P is true or else P is false (i.e. that $\neg P$ is true). So, whichever one is true should be provable. Is it? Not necessarily!

First, there are propositions whose truth value we don't know *yet*. For one example, no one knows (yet) if Goldbach's Conjecture is true. Goldbach's Conjecture says that all positive even integers are the sum of two primes; for example, $28 = 5 + 23$. For another example, no one knows (yet) if the Twin Prime Conjecture is true. The Twin Prime Conjecture says that there are an infinite number of "twin" primes": i.e. primes m, n such that $n = m + 2$; e.g. 2 and 3, 3 and 5, 5 and 7, 9 and 11, 11 and 13, etc.

Second – and much more astounding than our mere inability so far to prove or disprove any of these conjectures – there are propositions that are *known to be true* but that we can prove that *we cannot prove*! This is the essence of Gödel's Incompleteness Theorem. Stated informally, it asks us to consider proposition G, which is a slight variation on the Liar Paradox (i.e. the proposition "This proposition is false": if it's false, then it's true; if it's true, then it's false):

(G) This proposition (G) is true but unprovable.

We can assume that G is either true or false. So, suppose it is false. Then it was wrong when it said that it was *unprovable*; so, it *is* provable. But any provable proposition has to be *true* (because valid proofs are truth-preserving). That's a contradiction, so our assumption that it was false was wrong: it *isn't* false. But if it isn't false, then it must be true. But if it's true, then – as it says – it's unprovable. End of story; no paradox!

So, G (more precisely, its formal counterpart) is an example of a true proposition that cannot be proved. Moreover, the logician Kurt Gödel showed that some such propositions are true in the mathematical system consisting of first-order predicate logic plus Peano's axioms for the natural numbers (see Section 7.6.1); i.e. they are true propositions of arithmetic! For more information on Gödel and his proof, see Gödel 1931; Nagel et al. 2001; Hofstadter, 1979; Franzén 2005; Goldstein 2006.

We'll return to this question, also known as the "Decision Problem," beginning in Section 6.5.

2.4   Searching for the Truth

Thinking is, or ought to be, a coolness and a calmness …
—Herman Melville (1851, *Moby-Dick*, Ch. 135, p. 419)

Thinking is the hardest work there is, which is the probable reason why so few engage in it.
—Henry Ford (1928, p. 481)

Thinking does not guarantee that you will not make mistakes. But not thinking guarantees that you will.
—Leslie Lamport (2015, p. 41)

Let's turn to the second question: why is philosophy only the *search* for truth? Can't we *find* the truth? Perhaps not.

2.4.1   Searching vs. Finding

How does one go about *searching* for the truth, for answering questions? There are basically two complementary methods: (1) thinking hard and (2) empirical investigation. We'll look at the second of these in Section 2.5. First, let's focus on thinking hard.

Some have claimed that philosophy is just thinking really hard about things (Popova, 2012). Such hard thinking requires "rethinking explicitly what we already believe implicitly" (Baars, 1997, p. 187). In other words, it's more than merely expressing one's opinion. It's also different from empirical investigation:

> Philosophy is thinking hard about the most difficult problems that there are. And you might think scientists do that too, but there's a certain kind of question whose difficulty can't be resolved by getting more empirical evidence. It requires an untangling of presuppositions: figuring out that our thinking is being driven by ideas we didn't even realize that we had. And that's what philosophy is. (David Papineau, quoted in Edmonds and Warburton 2010, p. xx)

But we may not be able to *find* the truth, either by thinking hard or by empirical investigation. The philosopher Colin McGinn (1989, 1993) discusses the possibility that limitations of our (present) cognitive abilities may make it as impossible for us to understand the truth about certain things (such as the mind-body problem or the nature of consciousness) as an ant's cognitive limitations make it impossible for it to understand calculus. But we may not *have* to find the truth. G.E. Lessing (1778, my italics)[4] said,

> The true value of a man [sic] is not determined by his *possession*, supposed or real, of Truth, but rather by his sincere *exertion* to get to the Truth. It is not *possession* of the Truth, but rather the pursuit of Truth by which he extends his powers …

Digression: '[sic]': The annotation '[sic]' (which is Latin for "thus" or "so") is used when an apparent error or odd usage of a word or phrase is to be blamed on the original author and not on the person (in this case, me!) who is quoting the author. For example, here I want to indicate that it is Lessing who said "the true value of a *man*," where I would have said "the true value of a *person*."

---

4 Famously paraphrased by Albert Einstein (1940, p. 492); see O'Toole, 2021b.

In a similar vein, the mathematician Carl Friedrich Gauss (1808) said, "It is not knowledge, but the act of learning, not possession but the act of getting there, which grants the greatest enjoyment."

2.4.2 Asking "Why?"

> Questions, questions. That's the trouble with philosophy: you try and fix a problem to make your theory work, and a whole host of others then come along that you have to fix as well.
> —Helen Beebee (2017)

One reason the search for truth will never end (which is different from saying that it will not succeed) is that you can always ask "Why?"; i.e. you can always continue inquiring. This is

> the way philosophy – and philosophers – are[:] Questions beget questions, and those questions beget another whole generation of questions. It's questions all the way down. (Cathcart and Klein, 2007, p. 4)

You can even ask why "Why?" is the most important question (Everett, 2012, p. 38)! "The main concern of philosophy is to question and understand very common ideas that all of us use every day without thinking about them" (Nagel, 1987, p. 5). This is the reason, perhaps, that the questions children often ask (especially, "Why?") are often deeply philosophical.

The physicist John Wheeler pointed out that the more questions you *answer*, the more questions you can *ask*: "We live on an island surrounded by a sea of ignorance. As our island of knowledge grows, so does the shore of our ignorance" (https://en.wikiquote.org/wiki/John_Archibald_ Wheeler). And "Philosophy patrols the border [e.g. the shore], trying to understand how we got there and to conceptualize our next move" (Soames, 2016). The US economist and social philosopher Thorstein Veblen said, "The outcome of any serious research can only be to make two questions grow where only one grew before" (Veblen, 1908, p. 396).

Asking "Why?" is the principal part of philosophy's "general role of critically evaluating beliefs" (Colburn, 2000, p. 6) and "refusing to accept any platitudes or accepted wisdom without examining it" (Donna Dickenson, in Popova 2012). As the humorist George Carlin put it,

> [It's] not important to get children to read. Children who wanna read are gonna read. Kids who want to learn to read [are] going to learn to read. *[It's] much more important to teach children to QUESTION what they read. Children should be taught to question everything.* (https://georgecarlin.net/bogus/question.html)

Whenever you have a question, either because you do not understand something or because you are surprised by it or unsure of it, you should begin to think carefully about it. And one of the best ways to do this is to ask "Why?": *Why* did the author say that? *Why* does the author believe it? Why should *I* believe it? We can call this "looking backward" toward reasons. And a related set of questions are these: What are its *implications*? What *else* must be true if that were true? And should *I* believe those implications? Call this "looking forward" to consequences. Because we can always ask these backward- and forward-looking questions, we can understand why …

> … we should never rest assured that our view, no matter how well argued and reasoned, amounts to the final word on any matter. (Goldstein, 2014, p. 396)

> This is why philosophy must be argumentative. … Only in this way can intuitions that have their source in societal or personal idiosyncrasies be exposed and questioned. (Goldstein, 2014, p. 39)

The arguments are argued over, typically, by challenging their assumptions. It is rare that a philosophical argument will be found to be invalid (i.e. *logically* incorrect).[5] The most interesting arguments are valid ones, so that the only concern is over the truth of their "premises": the reasons for the conclusion. An argument that is found to be invalid is usually a source of disappointment – unless the invalidity points to a missing premise or reveals a flaw in the very nature of logic itself (an even rarer, but not unknown, occurrence).

2.4.3   Can There Be Progress in Philosophy?

> **Philosophy,** *n*. A route of many roads leading from nowhere to nothing.
> —Ambrose Bierce (1906, p. 157)

If the philosophical search for truth is a never-ending process, can we ever make any progress in philosophy? Mathematics and science, for example, are disciplines that not only search for the truth but seem to find it; they seem to make progress in the sense that we know more mathematics and more science now than we did in the past. We have well-confirmed scientific theories, and we have well-established mathematical proofs of theorems. But philosophy doesn't seem to be able to empirically confirm its theories or prove any theorems. Are the problems that philosophers investigate unsolvable?

I think there can be, and is, progress in philosophy. Solutions to problems in philosophy may not be as neat as they seem to be in mathematics, but in fact, they're not even that neat in mathematics! This is because solutions to problems are always *conditional*; they are based on certain *assumptions*. Most mathematical theorems are expressed as conditional statements: *If* certain assumptions are made, or *if* certain conditions are satisfied, *then* such-and-such will be the case. In mathematics, those assumptions include axioms, but axioms can be challenged and modified: consider the history of non-Euclidean geometry, which began by challenging and modifying the Euclidean axiom known as the Parallel Postulate.

Digression: Parallel Postulate: One version of the Parallel Postulate is this: For any line L, and for any point P not on L, there is only one line L' such that (1) P is on L', and (2) L' is parallel to L. For some of the history of non-Euclidean geometries, see http://mathworld.wolfram.com/ParallelPostulate.html and http://en.wikipedia.org/wiki/Parallel_postulate.

So, solutions are really parts of larger theories, which include the assumptions that the solution depends on, as well as other principles that follow from the solution. Progress can be made in philosophy (as in other disciplines) not only by following out the implications of your beliefs ("forward-looking" progress) but also by becoming aware of the assumptions that underlie your beliefs ("backward-looking" progress) (Rapaport 1982; Goldstein 2014, p. 38).

Recall Plato's view of the philosopher as a "gadfly" who investigates the foundations of, or reasons for, beliefs, always "spurring" people to ask "What *is X*?" and "Why?" This got him in trouble: his claims to be ignorant were thought (probably correctly) to be somewhat disingenuous. As a result, he was tried, condemned to death, and executed. (For the details, read Plato's *Apology*. On the "gadfly-spur" metaphors, see Marshall 2017.) One moral is that philosophy can be dangerous:

> And what is it, according to Plato, that philosophy is supposed to do? Nothing less than to render violence to our sense of ourselves and our world, our sense of ourselves in the world. (Goldstein, 2014, p. 40)

---

5 See Sections 2.5.1, 2.9, and 15.2.1.

Philosophers are the hazmat handlers of the intellectual world. It is we who stare into the abyss, frequently going down into it to great depths. This isn't a job for people who scare easily or even have a tendency to get nervous. (Eric Dietrich, personal communication, 5 October 2006)

It is violent to have one's assumptions challenged:

[P]hilosophy is difficult because the questions are hard, and the answers are not obvious. We can only arrive at satisfactory answers by thinking as rigorously as we can with the strongest logical and analytical tools at our disposal.

… I want … [my students] to care more about things like truth, clear and rigorous thinking, and distinguishing the truly valuable from the specious.

The way to accomplish these goals is not by indoctrination. Indoctrination teaches you what to think; education teaches you how to think. Further, the only way to teach people how to think is to challenge them with new and often unsettling ideas and arguments.

… Some people fear that raising such questions and prompting students to think about them is a dangerous thing. They are right. As Socrates noted, once you start asking questions and arguing out the answers, you must follow the argument wherever it leads, and it might lead to answers that disturb people or contradict their ideology. (K.M. Parsons 2015)

So, the whole point of Western philosophy since Socrates has been to make progress by getting people to think about their beliefs, to question and challenge them. It is not (necessarily) to come up with *answers* to difficult questions.[6]

2.4.4 Skepticism

If you can always ask "Why?" – if you can challenge any claims – then you can be skeptical about everything. Does philosophy lead to skepticism?[7]

Skepticism is often denigrated as being irrational. But there are advantages to always asking questions and being skeptical: "A skeptical approach to life leads to advances in all areas of the human condition; while a willingness to accept that which does not fit into the laws of our world represents a departure from the search for knowledge" (Dunning, 2007). Being skeptical doesn't necessarily mean refraining from having any opinions or beliefs. But it does mean being willing to question anything and everything that you read or hear (or think!). (Including questioning why we should question everything! See https://ubraga.com/index.php/2021/01/28/question-everything/.)

Why would you want to do this? So that you can find *reasons* for (or against) believing what you read or hear (or think)! And why is it important to have these reasons? For one thing, it can make you feel more confident about your beliefs and the beliefs of others. For another, it can help you explain your beliefs to others – not necessarily to convince them that they should believe what you believe but to help them understand why *you* believe what you do.

The heart of philosophy is not (necessarily) coming up with answers but challenging assumptions and forcing you to think about alternatives (Popper, 1978, Section 4, p. 148). My father's favorite admonition was "Never make assumptions." That is, never *assume* that something is the case or that someone is going to do something; rather, try to find out if it *is* the case, or *ask* the person. In other words, **challenge all assumptions**. Philosophers, as James Baldwin (1962) said about artists, "cannot and must not take anything for granted but must drive to the heart of every answer and expose the question the answer hides."

---

6 See the Online Resources for further reading on progress in philosophy and in science.
7 See http://www.askphilosophers.org/question/5572.

2.5   What Is "Rational"?

Rational, *adj.* Devoid of all delusions save those of observation, experience and reflection.
—Ambrose Bierce (1906, p. 170)

Active, persistent, and careful consideration of any belief or supposed form of knowledge in the light of the grounds that support it, and the further conclusions to which it tends, constitutes reflective thought.
—John Dewey (1910, p. 6)

Our third question concerns the nature of rationality. *Mere* statements (i.e. opinions) *by themselves* are **neither** rational **nor** irrational. Rather, it is **arguments** – reasoned or *supported* statements – that are capable of being rational. As the American philosopher John Dewey suggested, it's not enough to merely think something; you must also consider *reasons for* believing it (looking "backward"), and you must also consider the *consequences of* believing it (looking "forward"): Thus, being rational requires *logic*.

But there are lots of different (kinds of) logics, so there are lots of different kinds of logical rationality. And there is another kind of rationality, which depends on logics of various kinds but goes beyond them in at least one way: empirical, or scientific, rationality. Let's begin with these two kinds of rationality.

2.5.1   Logical Rationality

Philosophy: the ungainly attempt to tackle questions that come naturally to children, using methods that come naturally to lawyers.
—David Hills (2007, http://www.stanford.edu/~dhills/cv.html)

Deductive Logical Rationality   "Deductive" logic is one kind of logical rationality. Reasons $P_1, \ldots, P_n$ deductively support (or "yield," or "entail," or "imply") a conclusion C iff C *must* be true *if* all of the P_i are true. The technical term for this is 'validity': a deductive argument is said to be **valid** iff it is impossible for the conclusion to be false while all of the premises are true. This can be said in a variety of ways: a deductive argument is valid iff, *whenever* all of its premises are true, its conclusion *cannot be false.* Or: a deductive argument is valid iff, *whenever* all of its premises are true, its conclusion *must also be true.* Or: a deductive argument is valid iff the rules of inference that lead from its premises to its conclusion *preserve truth.*

For example, the rule of inference called "Modus Ponens" says that, from P and 'if P, then C,' you may deductively infer C. Using the symbols '$\rightarrow$' to represent "if… then" and '$\vdash_D$' to represent this truth-preserving relation between premises and a conclusion that is *deductively* supported by them, the logical notation for Modus Ponens is

$$P, \ (P \rightarrow C) \vdash_D C$$

For example, let $P =$ "Today is Wednesday." and let $C =$ "We are studying philosophy." So the inference becomes: "Today is Wednesday. If today is Wednesday, then we are studying philosophy. Therefore (deductively), we are studying philosophy." (For more on Modus Ponens, see Section 2.9.4.)

There are three somewhat surprising things about validity (or deductive rationality) that must be pointed out:

1. **Any or all of the premises of a valid argument can be false!**

   In the characterization of validity, note that the conditional terms 'if' and 'whenever' allow for the possibility that one or more premises are false. So, any or all of the premises of a deductively valid argument can be false, as long as *if* they *were* to be true, then the conclusion *would also have to be* true.

2. **The conclusion of a valid argument can be false!**

   How can a "truth preserving" rule lead to a false conclusion? By the principal – familiar to computer programmers – known as "garbage in, garbage out": *if* one of the P_i *is* false, even truth-preserving rules of inference can lead to a false C.

   The conclusion of a valid argument is only true *relative to the truth of its premises* (Hempel, 1945, p. 9). What this means is that you can have a situation in which a sentence is, let's say, "absolutely" or "independently" false (or you disagree with it), but it could also be true *relative* to some premises.

   The premises provide a background "context" in which to evaluate the conclusion. In other words, C *would be* true *if* all of the P_i *were* true. But sometimes the world might *not* make the premises true. And then we can't say anything about the truth of the conclusion.

   So, when can we be sure that the conclusion C of a valid argument is "really" true (and not just "relatively" true)? The answer is that C is true iff (1) all of the P_i *are* true, *and* (2) the rules of inference that lead from the P_i to C "preserve" truth. Such a deductive argument is said to be "sound": i.e. it is valid *and* all of its premises are, in fact, true.

3. **The premises of a valid argument can be *irrelevant* to its conclusion!**

   But that's not a good idea, because it wouldn't be a *convincing* argument. The classic example is that anything follows deductively from a contradiction: from the two contradictory propositions '$2 + 2 = 4$' and '$2 + 2 \neq 4$,' it can be deductively inferred that the philosopher Bertrand Russell (a noted atheist) is the Pope.

Proof: Let P and $\neg P$ (i.e. "not-P") be the two premises, let C be the conclusion, and let '$\vee$' represent inclusive disjunction ("or"). From P, we can deductively infer $(P \vee C)$ by the truth-preserving rule of Addition ("$\vee$-introduction"). Then, from $(P \vee C)$ and $\neg P$, we can deductively infer C, by the truth-preserving rule of Disjunctive Syllogism ("$\vee$-elimination"). So, in the "Pope Russell" argument, from '$2 + 2 = 4$,' we can infer that *either* $2 + 2 = 4$ *or* Russell is the Pope (or both); i.e. we can infer that at least one of those two propositions is true. But we have also assumed that one of them is false: $2 + 2 \neq 4$. So it must be the other one that is true. Therefore, Russell must be the Pope! (But remember point 2: it doesn't follow from this argument that Russell *is* the Pope. All that follows is that Russell *would be* the Pope (and so would you!) *if* $2 + 2$ both does and does not equal 4.) "Relevance" logics are one way of dealing with this problem; see Anderson and Belnap 1975; Anderson et al. 1992. For applications of relevance logic to AI, see Shapiro and Wand 1976; Martins and Shapiro 1988.

Inductive Logical Rationality "Inductive" logic is a second kind of logical rationality. In one kind of inductive logic, $P_1, \ldots, P_n \vdash_I C$ iff C is *probably* true *if* all of the P_i *are* true. Suppose you have an urn containing over a million ping-pong balls, and suppose you remove one of them at random and observe that it is red. (We can write 'Red(ball$_1$)' to mean "ball number 1 is red.") What do you think the chances are that the next ball will also be red? They are probably not very high. But suppose that the second ball that you examine is also red. And the third. … And the 999,999th. Now how likely do you think it is that the next ball will also be red? The chances are probably very high, so:

$$\text{Red(ball}_1), \dots , \text{Red(ball}_{999,999}) \vdash_I \text{Red(ball}_{1,000,000}).$$

Unlike deductive inferences, however, inductive ones do not *guarantee* the truth of their conclusion. Although it is not likely, it is quite possible that the millionth ping-pong ball will be, say, the only blue one in the urn. (For other kinds of inductive inferences, see Hawthorne 2021.)

Abductive Logical Rationality A third kind of logical rationality, "abductive" logic, is sometimes also known as "inference to the best explanation" or "circumstantial evidence." From observation O made at time t_1, and from a theory T that deductively or inductively entails O, one can *abductively* infer that T *must have been* the case at earlier time t_0. In other words, T is an *explanation* of why you have observed O. Of course, it is not necessarily a good, much less the best, explanation, but the more observations T explains, the better a theory it is. (But what is a "theory"? We'll delve into that in Section 4.6. For now, you can think of a theory as just a set of statements that describe, explain, or predict some phenomenon.) Like inductive inferences, abductive ones are not deductively valid and do not guarantee the truth of their conclusion. In fact, they have the following *invalid*(!) form, called the Fallacy of Affirming the Consequent:

$$O, (T \to O) \nvdash_D T.$$

Digression and Further Reading: O is the "consequent" of the conditional statement $(T \to O)$. "Affirming" O as a premise thus "affirms the consequent." (We will come back to this in Section 4.8.1.) But if O is true and T is false, then both premises are true, yet the conclusion (T) is not.

For the origin of the term 'abduction' in the writings of the American philosopher Charles Sanders Peirce (pronounced like 'purse'), see http://commens.org/dictionary/term/abduction. For more on abduction, see Douven 2021; Hobbs et al. 1993.

Non-Monotonic Logical Rationality In AI, a fourth kind of logical rationality is "non-monotonic" reasoning, which is arguably more "psychologically real" than the others. In *monotonic* logics (such as deductive logics), once you have proven that a conclusion C follows from a premise P, you can be assured that it will always so follow. But in *non*-monotonic logic, you might infer conclusion C from premise P at time t_0, but at later time t_1, you might learn additional information that entails $\neg C$. In that case, you must revise your beliefs. For example, you might believe that birds fly and that Tweety is a bird, from which you might conclude that Tweety flies. But if you then learn that Tweety is a penguin, you will need to revise your beliefs.

Further Reading: A great deal of work on non-monotonic logics has been done by researchers in the branch of AI called "knowledge representation"; see Ginsberg 1987; Strasser and Antonelli 2019; and the bibliography at http://www.cse.buffalo.edu/~rapaport/663/F08/nonmono.html.

2.5.2 Scientific Rationality

If philosophy is a search for truth by rational means, what is the difference between philosophy and science? Is philosophy worth doing? Or can science answer all of our questions? Or perhaps science *is* philosophy! After all, science is also a search for truth.

I would say that science *is* philosophy, as long as experiments and empirical methods are considered "rational" and yield truth. Physics and psychology, in fact, used to be branches of philosophy: the full title of Isaac Newton's *Principia* – the book that founded modern physics – was "Mathematical Principles of *Natural Philosophy*" (italics added), not "Mathematical Principles of *Physics*," and psychology split off from philosophy only at the turn of the twentieth century. The philosophers Aristotle (384–322 BCE, around 2400 years ago) and Kant (1724–1804, around 250 years ago) wrote physics books. The physicists Einstein and Mach wrote philosophy.

But scientific methodology is *not* (entirely) deductive; more often, it is inductive or abductive. Thus, it yields conclusions that may be only highly likely and are often the best we can get. So, if experiments *don't* count as being rational, and only deductive logic counts, then science is *not* philosophy. And science is also not philosophy, if philosophy is considered to be the search for *universal* or *necessary* truths: i.e. things that would be true no matter what results science came up with or what fundamental assumptions we made.

Turning the tables, we can ask whether philosophy is a science! Could (should?) philosophy be more scientific (i.e. experimental) than it is? McGinn (2012a) takes philosophy to be a science ("a systematically organized body of knowledge"), in particular what he dubs 'ontical science': "the subject consists of the search for the essences of things by means of a priori methods" (McGinn, 2012b). He argues that philosophy is a science just like physics or mathematics; it is the logical science of concepts (McGinn, 2015b, pp. 87–88). In a similar vein, the philosopher Timothy Williamson (2020, pp. 4, 82) says, "Like mathematics, philosophy is a non-natural science. ... As a systematic, methodical form of inquiry, philosophy is a science but not a natural science." (In Chapter 3, we will see that a similar claim has been made about computer science.)

There is a relatively recent movement (with some older antecedents) to have philosophers do scientific (mostly psychological) experiments in order to find out, among other things, what "ordinary" people (for example, people who are not professional philosophers) believe about certain philosophical topics.

But there is another way that philosophy can be part of a scientific worldview. The "philosophy naturalized" movement in contemporary philosophy (championed by the philosopher Willard Van Orman Quine) sees philosophy as being on a continuum with science, being aware of, and making philosophical use of, scientific results. Rather than being a passive, "armchair" discipline that merely analyzes what others say and do, philosophy can – and probably should – be a more active discipline, even helping to contribute to science (and other disciplines that it thinks about). It has certainly contributed to computer science:

> In early 20th-century logic, a question arose that was both mathematical and philosophical: what does it mean to have a 'definite method' for solving a mathematical problem without need of creativity? To answer the question, Alan Turing devised an abstract theory of imaginary *universal computing machines*. Later, in an attempt to break German codes during World War II, he actually built such a machine. Its success helped defeat Nazism. That was the origin of modern computers, which have transformed our world. It is hard, or impossible, to predict in advance what effect a philosophical idea will have on history. (Williamson, 2020, p. 93)

Philosophers can also be more "practical" in the public sphere: "The philosophers have only *interpreted* the world in various ways; the point is to *change* it" (Marx, 1845). (For more on this, see Section 5.6.)[8]

2.5.3   Computational Rationality

All of the above kinds of rationality seem to have one thing in common: they are all "declarative." That is, they are all concerned with statements (or propositions) that are true or false. But the philosopher Gilbert Ryle (1945, esp. p. 9) has argued that there is another kind of rationality, one that is "procedural" in nature: it has been summarized as "knowing *how*" to do something, rather than "knowing *that*" something is the case. Given the procedural nature of computer programs, this suggests that another kind of rationality might be "computational":

> a third, modern way of testing and establishing scientific truth – in addition to theory and experiment – is via simulations, the use of (often large) computers to mimic nature. It is a synthetic universe in a computer. … If all of the relevant physical laws are faithfully captured [in the computer program] then one ends up with an emulation – a perfect, *The Matrix*-like replication of the physical world in virtual reality. (Heng, 2014, p. 174)

One consideration that this raises is whether this is really a third way or just a version of logical rationality, perhaps extended to include computation as a kind of "logic."[9]

A Look Ahead: We will explore this kind of rationality in more detail in Sections 3.6.1 and 3.16.3. We'll discuss computational simulations in Section 14.2, and we'll return to *The Matrix* in Section 19.7.

2.5.4   Is It Always Rational to Be Rational?

> Rationality is one of humanity's superpowers. How do we keep from misusing it?
> —Joshua Rothman (2021, p. 25, col. 3)

Is there anything to be said in favor of *not* being rational?

Suppose you are having trouble deciding between two apparently equal choices. In the problem from medieval philosophy known as "Buridan's Ass" (Zupko, 2018), a donkey placed equidistant between two equally tempting bales of hay died of starvation because it couldn't decide between the two of them. My favorite way out of such a quandary is to imagine tossing a coin and seeing how you feel about how it lands: if it lands heads up, say, but you get a sinking feeling when you see that (because you would rather that it had landed *tails* up), then you know what you would have preferred, even if you had "rationally" decided that both choices were perfectly equally balanced.[10]

Further Reading: See Andrew N. Carpenter's response to the question "To what extent do philosophers/does philosophy allow for instinct, or gut feelings?" (http://www.askphilosophers .org/question/2992).

---

8 See the Online Resources for further reading on experimental philosophy, naturalistic philosophy, and science vs. philosophy.
9 See the Online Resources for further reading on knowing how vs. knowing that.
10 See the Online Resources for further reading on the limits of rationality.

2.6 Philosophy as a Personal Search

> … I'm not trying to change anyone's mind on this question. I gave that up long ago. I'm simply trying to say what I think is true.
> —Galen Strawson (2012, p. 146)

> [M]y purpose is to put my own intellectual home in order ….
> —Hilary Putnam (2015)

> "The philosophy of every thinker is the more or less unconscious autobiography of its author," Nietzsche observed …
> —Clancy Martin (2015)

So far, we have seen why philosophy is the search for truth by rational means. Why do I say that it is a "personal" search? The philosopher Hector-Neri Castañeda said that philosophy should be done "in the first person, for the first person" (Rapaport, 2005a). So, philosophy is whatever *I* am interested in, as long as I study it in a rational manner and aim at truth.

There is another way in which philosophy must be a personal search for truth. As one introductory book puts it, "the object here is not to give answers … but to introduce you to the problems in a very preliminary way *so that you can worry about them yourself*" (Nagel, 1987, pp. 6–7, my italics). The point is not to hope that someone else will tell you the answers to your questions. That would be nice, of course; but why should you believe them? The point, rather, is for you to figure out answers for yourself.

It may be objected that your first-person view on some topic, no matter how well thought out, is, after all, just your view. "Such an analysis can be of only parochial interest" (Strevens, 2019) or might be seriously misleading (Dennett, 2017, pp. 364–370). Another philosopher, Hilary Kornblith, agrees:

> I believe that the first-person perspective is just one perspective among many, and it is wholly undeserving of the special place which these philosophers would give it. More than this, this perspective is one which fundamentally distorts our view of crucial features of our epistemic situation. Far from lauding the first-person perspective, we should seek to overcome its defects. (Kornblith, 2013, p. 126)

But recall another important feature of philosophy: it is a conversation. And if you want to contribute to that conversation, you will have to take others' views into account, and you will have to allow others to make you think harder about your own views.

Psychologist William G. Perry, Jr. (1970; 1981) called the desire for an "Authority" to give you the right answers to all questions the "Dualistic" stance toward knowledge. But the Dualist soon realizes that not all questions have answers that everyone agrees with, and some questions don't seem to have answers at all (at least, not yet).

Rather than stagnating in a middle stance of "Multiplism" (a position that says because not all questions have answers, multiple opinions – proposed answers – are all equally good), a further stance is that of "Contextual Relativism": all proposed answers or opinions can (should!) be considered – and evaluated! – *relative to* and *in the context of* assumptions, reasons, or evidence that can support them. This is where philosophy as a rational search for truth comes in.

And the personal nature of philosophy appears when you "Commit" to one of these answers and *you* become responsible for defending *your* commitment against "Challenges." (This, of course, is

[just] more thinking and analysis – more philosophizing.) That commitment, however, is a personal one: the computer scientist Richard W. Hamming's warning about science and mathematics also holds for philosophy: "we do not appeal to authority, but rather *you are responsible for what you believe*" (Hamming, 1998, p. 650).

It is in this way that philosophy is done "in the first person, for the first person," as Castañeda said.

Further Reading: For more on Perry's theory, see https://cse.buffalo.edu/~rapaport/perry-positions.html. See also the answer to a question about deciding which of your own opinions to really believe, at http://www.askphilosophers.org/question/5563.

2.7 Philosophies of Anything and Everything

One of the things about philosophy is that you don't have to give up on any other field. Whatever field there is, there's a corresponding field of philosophy. … All the things I wanted to know about I could still study within a philosophical framework.
—Rebecca Newberger Goldstein, cited in Reese 2014b

[He] is a *philosopher*, so he's interested in everything …
—David Chalmers (describing the philosopher Andy Clark), as cited in Cane 2014

It is not really possible to regret being a philosopher if you have a theoretical (rather than practical or experiential) orientation to the world, because there are no boundaries to the theoretical scope of philosophy. For all X, there is a philosophy of X, which involves the theoretical investigation into the nature of X. There is philosophy of mind, philosophy of literature, of sport, of race, of ethics, of mathematics, of science in general, of specific sciences such as physics, chemistry and biology; there is logic and ethics and aesthetics and philosophy of history and history of philosophy. I can read Plato and Aristotle and Galileo and Newton and Leibniz and Darwin and Einstein and John Bell and just be doing my job. I could get fed up with all that and read Eco and Foucault and Aristophanes and Shakespeare for a change and still do perfectly good philosophy.
—Tim Maudlin, cited in Horgan 2018

Our final question concerns the scope of philosophy. Philosophy studies things that are *not* studied by any *single* discipline, the "Big Questions": What is truth? What is beauty? What is good (or just, or moral, or right)? What is the meaning of life? What is the nature of mind? Or, as the philosopher Jim Holt put it: "Broadly speaking, philosophy has three concerns: how the world hangs together, how our beliefs can be justified, and how to live" (Holt, 2009). The first of these is metaphysics, the second is epistemology, and the third is ethics.

Metaphysics tries to "understand the nature of reality in the broadest sense: what kinds of things and facts ultimately constitute everything there is" (Nagel, 2016, p. 77). It tries to answer the question "What is there?" (and also the question "Why is there anything at all?").

Here are some computationally relevant metaphysical issues: Do computer programs that deal with, say, student records model students? Or are they just dealing with 0s and 1s? (We'll discuss this in Section 13.2.3.) And, on perhaps a more fanciful level, could a computer program model students so well that the "virtual" students in the program believed that they were real? (If this sounds like the film *The Matrix*, see Section 19.7.)

Ontology is the branch of metaphysics concerned with the objects and kinds of objects that exist according to one's metaphysical (or even physical) theory, their properties, and their relations to each other (such as whether some of them are "sub-kinds" of others, inheriting their properties and relations from their "super-kinds"). Ontology is studied both by philosophers and by computer scientists. In software engineering, "object-oriented" programming languages are more focused on the kinds of objects that a program must deal with than with the instructions that describe their behavior. In AI, ontology is a branch of knowledge representation that tries to categorize the objects that a knowledge-representation theory is concerned with.[11]

Epistemology is the study of knowledge and belief:

> Epistemology is concerned with the question of how, since we live, so to speak, inside our heads, we acquire knowledge of what there is outside our heads. (Simon, 1996a, p. 162)

How do we *know* what there is? How do we know that there is anything? What *is* knowledge (and belief)? Are there other kinds of knowledge, such as knowing *how* to do something (see Section 3.16.3)? Can a computer (or a robot) be said to have beliefs or knowledge? The branch of AI called "knowledge representation" applies philosophical results in epistemology to issues in AI and computer science in general, and it has contributed many results to philosophy as well.[12]

Ethics tries to answer "What is good?" and "What ought we to do?" We'll look at some ethical issues arising from computer science in Chapters 17 and 19.

But the main branches of philosophy go beyond these "big three":

Aesthetics (or the **philosophy of art**) tries to answer "What is beauty?" and "What is art?" On whether computer programs, like mathematical theorems or proofs, can be "beautiful," see Section 3.16.1.

Logic is the study of good reasoning: What is truth? What is rationality? Which arguments are good ones? Can logic be computationally automated? (Recall our discussion in Section 2.5.)

And of central interest for the philosophy of computer science, there are numerous "philosophies of":

Philosophy of language tries to answer "What is language?" and "What is meaning?" It has large overlaps with linguistics and with cognitive science (including AI and computational linguistics).

Philosophy of mind tries to answer "What is 'the' mind?" and "How is it related to the brain?" In Section 12.6, we'll say more about this "mind-body problem" and its relation to the software-hardware distinction. The philosophy of mind also investigates whether computers can think (or be said to think), and it has close ties with cognitive science and AI, issues that we will take up in Chapter 18.

Philosophy of science tries to answer "What is science?" "What is a *scientific* theory?" and "What is a *scientific* explanation?" The philosophy of computer science is part of the philosophy of science. (We will look at the philosophy of science in Chapter 4.)

In general, **for any X, there can be a philosophy of X**: the philosophical investigation of the fundamental assumptions, methods, and goals of X (including metaphysical, epistemological, and ethical issues). X can be anything: biology, education, history, law, physics, psychology, religion, etc., including, of course, AI and computer science. *The possibility of a philosophy of X for any X is the main reason philosophy is the rational search for truth in **any** field.* Philosophy in general, and especially the philosophy of X, are "meta-disciplines": in a discipline X, you think about X

---

11 See the Online Resources for further reading on existence and ontology.
12 See the Online Resources for further reading on knowledge representation.

(in the discipline of mathematics, you think about mathematics); but in the philosophy of X, you think about *thinking about X*. Even those subjects that might be purely philosophical (metaphysics, epistemology, and ethics) have strong links to disciplines like physics, psychology, and political science, among others.

X, by the way, could also be … philosophy! The philosophy of philosophy, also known as "metaphilosophy," is exemplified by this very chapter, which is an investigation into what philosophy is and how it can be done. Some people might think the philosophy of philosophy is the height of "gazing at your navel," but it's really what's involved when you think about thinking – and after all, isn't AI just *computational* thinking about thinking?

Are there any topics that philosophy *doesn't* touch on? I'm sure there are some topics that philosophy *hasn't* touched on. But I'm equally sure there are no topics that philosophy *couldn't* touch on.[13]

2.8 Philosophy and Computer Science

[I]f there remain any philosophers who are not familiar with some of the main developments in artificial intelligence, it will be fair to accuse them of professional incompetence, and that to teach courses in philosophy of mind, epistemology, aesthetics, philosophy of science, philosophy of language, ethics, metaphysics, and other main areas of philosophy, without discussing the relevant aspects of artificial intelligence will be as irresponsible as giving a degree course in physics which includes no quantum theory.
—Aaron Sloman (1978, Section 1.2, p. 3)

The previous passage is what convinced me to take some time off from being a philosopher to learn some AI. Philosophy and computer science overlap not only in some topics of common interest (logic, philosophy of mind, philosophy of language, etc.) but also in methodology: the ability to find counterexamples; refining problems into smaller, more manageable ones; seeing implications; methods of formal logic; etc. (For more philosophical influences on AI, see McCarthy 1959; McCarthy and Hayes 1969.)

In the next chapter, we'll begin our philosophical investigation into computer science.

2.9 Appendix: Argument Analysis and Evaluation

2.9.1 Introduction

In Section 2.2, I said that the methodology of philosophy involved "rational means" for seeking truth; and in Section 2.5, we looked at different kinds of rational methods. In this appendix, we'll look more closely at one of those methods – argument analysis and evaluation – which you will be able to practice if you do the exercises in the Online Resources. Perhaps more importantly for some readers, argument analysis is a topic in Computing Curricula 2020 (https://www.acm.org/binaries/content/assets/education/curricula-recommendations/cc2020.pdf).

Unless you are gullible – willing to believe everything you read or anything that an authority figure tells you – you should want to know *why* you should believe something that you read or something that you are told. Let's consider how you might go about doing this.

---

13 See the Online Resources for further reading on philosophy in general.

2.9.2 A Question-Answer Game

Consider two players, Q and A, in a question-answer game:

Step 1. Q asks whether C is true.

Step 2. A responds: "C, because P_1 and P_2."

Player A has given an *argument* for conclusion C with reasons (also called '*premises*') P_1 and P_2. Note, by the way, that this use of the word 'argument' has nothing directly to do with the kind of fighting argument that you might have with your roommate; rather, it's more like the *legal* arguments that lawyers present to a jury. For the sake of simplicity, I'm assuming that A gives only two reasons for believing C. In a real case, there might be only one reason (for example, Fred is a computer scientist; therefore, someone is a computer scientist), or there might be more than two reasons (for examples, see any of the arguments for analysis and evaluation in the Online Resources).

Step 3. To be rational, Q should *analyze* or "verify" A's arguments. Q can do this by asking three questions:

> (a) Do I *believe* P_1? (That is, do I *agree* with it?)
> (b) Do I believe P_2? (That is, do I agree with it?)
> (c) Does C follow *validly* from P_1 and P_2?

Strictly speaking, when you're analyzing an argument, you need to say, for each premise, whether it is or is not *true*. But sometimes you don't know; after all, truth is not a matter of logic, but of correspondence with reality (Section 2.3.1): a sentence is true if and only if it correctly describes some part of the world. Whether or not you *know* the truth-value of a statement, you usually have some idea of whether you *believe* it. Because you can't always or easily tell whether a sentence *is* true, we can relax this a bit and say that sentences can be such that either you *agree* with them or you don't. So, when analyzing an argument, you can say either "This statement *is* true (or false)" or (more cautiously) "I *think* that this statement is true (or false)," or "I *believe* (or don't believe) this statement," or "I *agree* (or don't agree) with it." (Of course, you should also say *why* you do or don't agree!) Steps 3a and 3b are "*recursive*" (see Section 2.9.4): that is, for each reason P_i, Q could play another instance of the game, asking A (or someone else!) whether P_i is true. A (or the other person) could then give an argument for conclusion P_i with new premises P_3 and P_4. Clearly, this process could continue. (Recall Section 2.4.2!)

To ask whether C follows "validly" from the premises is to assume that A's argument is a *deductive* one. For the sake of simplicity, all (or at least most) of the arguments in the Online Resources are deductive. But in real life, most arguments are not completely deductive, or not deductive at all. So, more generally, in Step 3c, Q should ask whether C follows *rationally* from the premises: if it does not follow deductively, does it follow inductively? Abductively? And so on.

Unlike Steps 3a and 3b for considering the truth value of the premises, Step 3c – determining whether the *relation between* the premises of an argument and its conclusion is a rational one – is *not* similarly recursive, on pain of infinite regress.

Further Reading: The classical source of this observation is due to Lewis Carroll (of *Alice in Wonderland* fame). Carroll was a logician by profession and wrote a classic philosophy essay on this topic, involving Achilles and the Tortoise (Carroll, 1895).

Finally, it should be pointed out that the order of doing these steps is irrelevant. Q could *first* analyze the validity (or rationality) of the argument and *then* analyze the truth value of the premises (i.e. decide whether to agree with them), rather than the other way round.

Step 4. Having *analyzed* A's argument, Q now has to *evaluate* it, by reasoning in one of the following ways;

- **If** I agree with P_1,
  and if I agree with P_2,
  and if C follows validly (or rationally) from P_1 and P_2,
  then I logically must agree with C (i.e. I ought to believe C).
    - But what if I really *don't* agree with C?
      In that case, I must reconsider my having agreed with P_1, or with P_2, or with the logic of the inference from P_1 & P_2 to C.
- **If** I agree with P_1,
  and if I agree with P_2,
  but the argument is *invalid*,
  then is there a *missing premise* – an extra reason – that would validate the argument
      and that I would agree with? (See Section 2.9.3.)
    - **If** so, then I can accept C,
      else I should not yet reject C,
          but I do need a new argument for C
              (i.e. a new set of reasons for believing C).
- **If** I *disagree* with P_1 or with P_2 (or both),
  then – even if C follows validly from them –
      this argument is not a reason *for me* to believe C
      so, I need a new argument for C.
  (Recall our discussion of "first-person philosophy" in Section 2.6.)
    - There is one other option for Q in this case: Q might want to go back and reconsider the premises. Maybe Q was too hasty in rejecting them.
- What if Q cannot find a good argument for believing C? Then it might be time to consider whether C is false. In that case, Q needs to find an argument for C's negation: not-C (sometimes symbolized '$\neg C$').

This process of argument analysis and evaluation is summarized in the flowchart in Figure 2.1.

2.9.3 Missing Premises

One of the trickiest parts of argument analysis can be identifying missing premises. Often this is tricky because the missing premise seems so "obvious" that you're not even aware that it's missing. But equally often, it's the missing premise that can make or break an argument.

Here's an example from the "Textual Entailment Challenge," a competition for computational-linguistics researchers interested in knowledge representation and information extraction (Dagan et al., 2006; Bar-Haim et al., 2006; Giampiccolo et al., 2007). (We'll see real-life examples in Sections 3.5, 3.15.2, and 5.5.2.) In a typical challenge, a system is given one or two premises and a conclusion, and then it is asked to determine whether the conclusion follows from the premise. And "follows" is taken fairly liberally to include all kinds of non-deductive inference.

Here is an example:

Premise 1 (P):
Bountiful arrived after war's end, sailing into San Francisco Bay 21 August 1945.
Premise 2:
Bountiful was then assigned as hospital ship at Yokosuka, Japan, departing San Francisco 1 November 1945.
Conclusion (C): Bountiful reached San Francisco in August 1945.

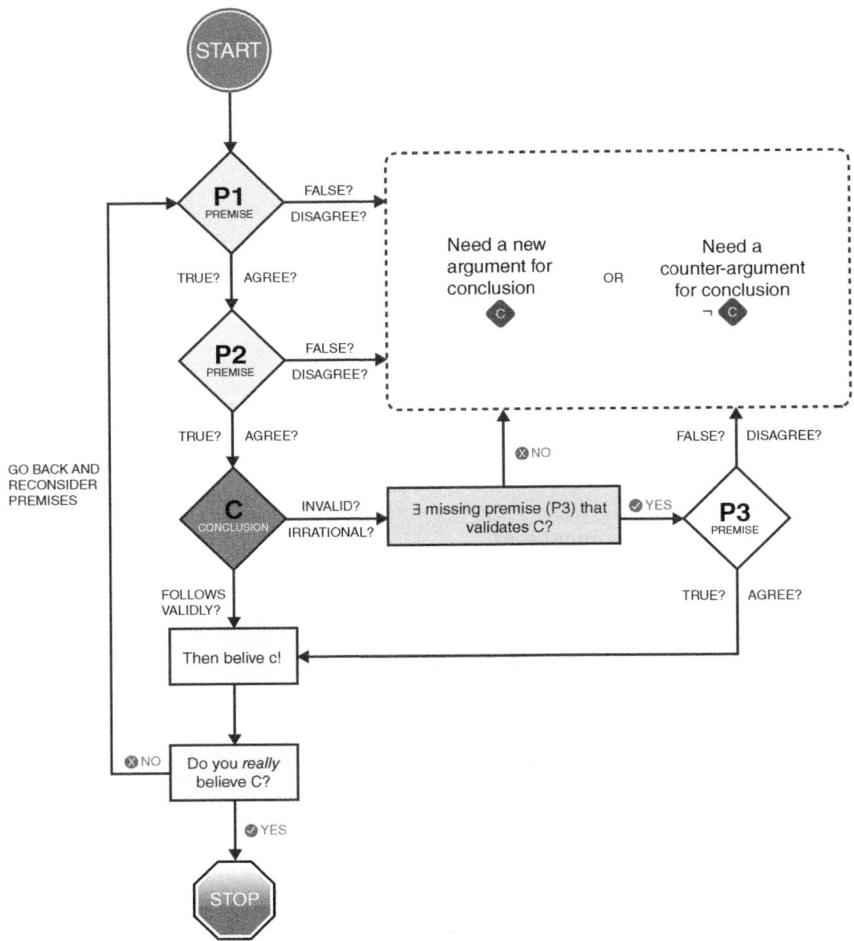

Figure 2.1 How to evaluate an argument from premises P_1 and P_2 to conclusion C. The symbol 'Ǝ' should be read: "Does there exist." Source: Flowchart by Tom Fadial.

The idea is that the two premises might be sentences from a news article, and the conclusion is something that a typical reader of the article might be expected to understand from reading it.

I hope you can agree that this conclusion does, indeed, follow from these premises. In fact, it follows from Premise 1 alone. In this case, Premise 2 is a "distractor." But what logical rule of inference allows us to infer C from P?

- P talks of "arrival" and "sailing into," but C talks only of "reaching."
- P talks of "San Francisco Bay," but C talks only of "San Francisco."

There are no *logical* rules that connect these concepts. Most people, I suspect, would think no such rules would be needed. After all, isn't it "obvious" that, if you arrive somewhere, then you have reached it? And isn't it "obvious" that San Francisco Bay must be in San Francisco?

Well, maybe. But whereas *people* might know these things, *computers won't*, unless we tell them. In other words, computers need some lexical information and some simple geographical information. (Instead of *telling* the computer these additional facts, we might tell the computer how to *find* them.) So, we need to supply some extra premises that link P with C more closely. These are the "missing premises." The argument from P to C is called an 'enthymeme,' because the missing premises are "in" (Greek 'en-') the arguer's "mind" (Greek 'thymos').

We might flesh out the argument as follows (there are other ways to do it):

> (P) Bountiful arrived after war's end, sailing into San Francisco Bay 21 August 1945.
> (P_a) If something sails into a place, then it arrives at that place.
> (C_1) ∴ Bountiful arrived at San Francisco Bay 21 August 1945.

In this first step, I've added a missing premise, P_a, and derived an intermediate conclusion C_1. Hopefully, you agree that C_1 follows validly (or at least logically in some way, i.e. rationally) from P and P_a.

We have no way of knowing whether P is true and must, for the sake of the argument, simply assume that it is true. (Well, we could look it up, I suppose; but we're not investigating whether the argument is "sound" (see Section 2.9.4), only if it is "valid": does C follow from P?)

P_a, on the other hand, doesn't have to be accepted; after all, *we* are imposing it on the (unknown) author of the argument. So, we had better impose something that is likely to be true. P_a is offered as part of the meaning of "sail into." I won't defend its truth any further here, but if you think it's *not* true, then you should either reject the argument or find a better missing premise.

We might have chosen another missing premise:

> (P_b) If something arrives in a place named 'X Bay,'
> then it arrives at a place named 'X.'
> (C_2) ∴ Bountiful arrived at San Francisco 21 August 1945.

C_2 will follow from C_1 and P_b, but is P_b true? Can you think of any bays named 'X Bay' that are *not* located in a place named 'X'? If you can, then we can't use P_b. Let's assume the worst: then we'll need something more specific, such as:

> (P_c) If something arrives in San Francisco Bay,
> then it arrives at San Francisco.

C_2 will follow from C_1 and P_c, and we can easily check the likely truth of P_c by looking at a map.

So far, so good. We've now got Bountiful *arriving* at San Francisco 21 August 1945. But what we need is Bountiful "*reaching*" San Francisco "*in*" August 1945. So let's add

> (P_d) If something arrives somewhere, then it reaches that place.

Again, this is proposed as an explication of part of the meaning of 'arrive,' and, in particular, of that part of its meaning that connects it to C.

From P_d and C_2, we can infer

> (C_3) Bountiful reached San Francisco 21 August 1945.

Are we done? Does $C_3 = C$? Look at them:

> (C_3) Bountiful reached San Francisco 21 August 1945.
> (C) Bountiful reached San Francisco in August 1945.

Think like a computer! $C_3 \neq C$. But does C_3 *imply* C? It will, if we supply one more missing premise:

> (P_e) If something occurs (*on*) DATE MONTH YEAR,
>   then it occurs *in* MONTH YEAR.

And that's true by virtue of the way (some) people talk. So, from P_e and C_3, we can infer C.

The simple argument we started with, ignoring its irrelevant premise, becomes this rather more elaborate one:

> (P) Bountiful arrived after war's end, sailing into San Francisco Bay
>   21 August 1945.
> (P_a) If something sails into a place, then it arrives at that place.
> (C_1) ∴ Bountiful arrived at San Francisco Bay 21 August 1945.
> (P_b) If something arrives in a place named 'X Bay,'
>   then it arrives at a place named 'X.'
>   (or (P_c) If something arrives in San Francisco Bay,
>   then it arrives at San Francisco.)
> (C_2) ∴ Bountiful arrived at San Francisco 21 August 1945.
> (P_d) If something arrives somewhere, then it reaches that place.
> (C_3) ∴ Bountiful reached San Francisco 21 August 1945.
> (P_e) If something occurs (on) DATE MONTH YEAR,
>   then it occurs *in* MONTH YEAR.
> (C) ∴ Bountiful reached San Francisco in August 1945.

2.9.4 When Is an Argument a "Good" Argument?

As we have seen, Q needs to do two things to analyze and evaluate an argument:

1. decide whether the premises are true (i.e. decide whether to agree with, or believe, the premises), and
2. decide whether the inference (i.e. the reasoning) from the premises to the conclusion is a valid one.

That is, there are two separate conditions for the "goodness" of an argument:

1. Factual goodness: are the premises true? (Or do you believe them?)
2. Logical goodness: is the inference valid? (Or at least rational in some way?)

Factual goodness – truth – is beyond the scope of logic, although it is definitely not beyond the scope of deciding whether to accept the conclusion of an argument. As we saw in Section 2.3, there are several ways of defining 'truth' and determining whether a premise is true. Two of the most obvious (although not the simplest to apply!) are (1) constructing a (good!) argument for a premise whose truth value is in question and (2) making an empirical investigation to determine its truth value (for instance, performing some scientific experiments or doing some kind of scholarly research).

Logical goodness (for deductive arguments) is called 'validity.' I will define this in a moment. But for now, note that these two conditions must both obtain for an argument to be "really good"; a "really good" argument is said to be "sound":

An argument is **sound** if and only if it is both valid and "factually good": i.e. if and only if *both* it is valid *and* all of its premises really are true.

Just to drive this point home, if the premises and conclusion of an argument are all true (or if you believe all of them), that by itself does not make the argument sound ("really good"). For one thing, your belief in the truth of the premises might be mistaken. But more importantly, the argument might not be valid.

And if an argument *is* valid – even if you have doubts about some of the premises – that by itself does not make the argument *sound* ("really good"). All of its premises also need to be true; i.e. it needs to be factually good.

So, what does it mean for a (deductive) argument to be "valid"?

An argument is **valid** if and only if it is necessarily "truth-preserving."

Here's another way to put it:

An argument is valid
if and only if
whenever all of its premises are *true*, then its conclusion *must also* be *true*.

And here's still another way to say the same thing:

An argument is valid
if and only if
it is *impossible* that:
all of its premises are simultaneously *true* while its conclusion is *false*.

Note that this has nothing to do with whether any of the premises actually *are* true or false; it's a "what if" kind of situation. Validity only requires that, *if* the premises *were to be* true, then the conclusion would *preserve that truth* – it would "inherit" that truth from the premises – and so it would also (have to) be true.

So you can have an argument with false premises and a false conclusion that is *invalid*, and you can have one with false premises and a false conclusion that *is* valid. Here's a valid one:

All cats are fish.
All fish can fly.
∴ All cats can fly.

In this case, everything's false, but the argument is *valid* because it has the form

All *P*s are *Q*s.
All *Q*s are *R*s.
∴ All *P*s are *R*s.

and there's no way for a *P* to be a *Q*, and a *Q* to be an *R*, without having the *P* be an *R*. That is, it's impossible that the premises are true while the conclusion is false.

Here's an invalid one, also with false premises and conclusion:

All cats are fish.
All cats can fly.
∴ All fish can fly.

Again, everything's false. However, the argument is *invalid* because it has the form

> All *P*s are *Q*s.
> All *P*s are *R*s.
> ∴ All *Q*s are *R*s.

and arguments of this form can have true premises with false conclusions. Here is an example:

> All cats are mammals.
> All cats purr.
> ∴ All mammals purr.

So, it's *possible* for an argument of this form to have true premises and a false conclusion; hence, it's *not* valid.

To repeat: validity has nothing to do with the *actual* truth or falsity of the premises or conclusion. It only has to do with the *relationship* of the conclusion to the premises.

Recall that an argument is *sound* iff it is valid *and* all of its premises *are* true. Therefore, an argument is *unsound* iff *either* it is *invalid or* at least one premise is false (*or* both). An *unsound* argument *can* be valid!

One more point: an argument with *inconsistent* premises (i.e. premises that contradict each other) is always valid(!) because it's impossible for it to have all true premises with a false conclusion, and that's because it's impossible for it to have all true premises, period. Of course, such an argument cannot be sound. (The argument that Bertrand Russell is the Pope that we saw in Section 2.5.1 is an example of this.)

All of this is fine as far as it goes, but it isn't very helpful in deciding whether an argument really is valid. How can you tell if an argument is truth-preserving? There is a simple, recursive definition, but to state it, we'll need to be a bit more precise in how we define an argument.

Definition 1 An **argument from propositions** $P_1, \ldots, P_n$ **to conclusion** C is$_{def}$[14] a *sequence* of propositions $\langle P_1, \ldots, P_n, C \rangle$, where C is *alleged* to follow logically from the P_i.

Definition 2 An argument $\langle P_1, \ldots, P_n, C \rangle$ is **valid** if and only if each proposition P_i and conclusion C is either:

> (a) *a tautology*
> (b) *a premise*
> (c) or follows validly from previous propositions in the sequence by one or more truth-preserving "rules of inference."

This needs some commenting!

(a) First, a **tautology** is a proposition that *must always be true*. How can that be? Most tautologies are (uninformative) "logical truths" such as 'Either *P* or not-*P*' or 'If *P*, then *P*.' Note that if *P* is true (or if you believe *P*), then 'Either *P* or not-*P*' has to be true (or you are logically obligated to also believe 'Either *P* or not-*P*'); and if *P* is false, then not-*P* is true, so 'Either *P* or not-*P*' still has to be true (or you are logically obligated to also believe 'Either *P* or not-*P*'). So, in either case, the disjunction has to be true (or you are logically obligated to believe it). Similar considerations hold

---

14 This symbol means "is by definition."

for 'If *P*, then *P*.' Sometimes statements of mathematics are also considered tautologies. (Whether they are "informative" or not is an interesting philosophical puzzle; see Wittgenstein, 1921.)

(b) Second, a **premise** is one of the initial reasons given for *C* or one of the missing premises added later. Premises, of course, need not be true, but when evaluating an argument for validity, we must *assume* that they are true "for the sake of the argument." Of course, if a premise is false, then the argument is unsound.

Third, clause (c) of Definition 2 might look circular, but it isn't; rather, it's recursive. A "recursive" definition begins with "base" cases that give explicit examples of the concept being defined and then "recursive" cases that define new occurrences of the concept in terms of previously defined ones. In fact, this entire definition is recursive. The base cases of the recursion are the first two clauses: tautologies must be true, and premises are assumed to be true. The recursive case consists of "rules of inference," which are argument forms that are clearly valid (truth-preserving) when analyzed by means of truth tables. (We'll say a lot more about recursion in Chapter 7.)

So, what are these "primitive" valid argument forms known as 'rules of inference'? The most famous is called 'Modus Ponens' (or '→-elimination'):

> From *P*
> and 'If *P*, then *C*,'
> you may validly infer *C*.

Why may you validly infer *C*? Consider the truth table for 'If *P*, then *C*':

P	*C*	If *P*, then *C*
true	true	true
true	**false**	**false**
false	true	true
false	false	true

This says that the conditional proposition "If *P*, then *C*" is false in only one circumstance (the boldfaced line of the truth table): when its "antecedent" (*P*) is true and its "consequent" (*C*) is false. In all other circumstances, the conditional proposition is true. So, if the antecedent of a conditional is true, and the conditional itself is true, then its consequent must also be true. (Look at the first line of the truth table.) Modus Ponens preserves truth.

Another important rule of inference is called 'Universal Instantiation' (or 'Universal-Quantifier Elimination'):

> From 'For all *x*, *F(x)*' (i.e. for all *x*, *x* has property *F*),
> you may validly infer *F(a)*, for any individual *a* in the "domain of discourse" (i.e. in the set of things that you are talking about).

A truth-table analysis won't help here because this is a rule of inference for "first-order predicate logic," not for "propositional logic." The formal definition of truth for first-order predicate logic is beyond our scope, but it should be pretty obvious that if it is true that *everything* in the domain of discourse has some property *F*, then it must also be true that any *particular thing* in the domain (say, *a*) has that property. (For more rules of inference and for the formal definition of truth in first-order predicate logic, see any good introductory logic text or the Further Reading on the correspondence theory of truth, in Section 2.3.1.)

There are, however, a few terminological points to keep in mind:

- *Sentences* can only be *true* or *false*
(or you can agree or disagree with them).
- *Arguments* (which are sequences of sentences) can be *valid* or *invalid*, and they can be *sound* or *unsound*.
- *Conclusions* of arguments (which are sentences) can *follow validly* or *not follow validly* from the premises of an argument.

Therefore:

- *Sentences* (including premises and conclusions) *cannot* be valid, invalid, sound, or unsound (because they are not arguments).
- *Arguments* *cannot* be true or false (because they are not sentences).

2.9.5   Examples of Good and Bad Arguments

There is only one way to have a sound argument: it must be valid and have only true premises. But there are lots of ways to have *invalid* arguments! More importantly, it is possible to have an *invalid* argument whose conclusion is *true*! Here's an example:

All birds fly.	(true)
Tweety the canary flies.	(true)
Therefore, Tweety is a bird.	(true)

This is invalid, even though both of the premises as well as the conclusion are true (but see Section 18.3.3!): it is invalid because an argument with the same *form* can have true premises and a *false* conclusion. Here is the form of that argument:

$$\forall x[B(x) \rightarrow F(x)]$$
$$C(a) \wedge F(a)$$
$$\therefore B(a)$$

In English, this argument's form is

> For all x, if x has property B, then x has property F.
> a has property C, and a has property F.
> $\therefore a$ has property B.

That is,

> For all x, if x is a bird, then x flies.
> Tweety is a canary, and Tweety flies.
> Therefore, Tweety is a bird.

Here's a counterexample – i.e. an argument with this form that has true premises but a false conclusion:

All birds fly.	(true)
Bob the bat flies.	(true)
Therefore, Bob is a bird.	(false)

Just having a true conclusion doesn't make an argument valid. And such an argument doesn't *prove* its conclusion (even though the conclusion *is* true).

At the end of this chapter is a collection of valid (V), invalid (I), sound (S), and unsound (U) arguments with different combinations of true (T) and false (F) premises and conclusions. Make sure you understand why each is valid, invalid, sound, or unsound.

2.9.6 Summary

To *analyze* an argument, you must identify its premises and conclusion and supply any missing premises to help make it valid. To *evaluate* the argument, you should then determine whether it is valid (i.e. truth preserving) *and* decide whether you agree with its premises.

If you agree with the premises of a valid argument, then you are logically obligated to believe its conclusion. If you don't believe its conclusion, even after your analysis and evaluation, then you need to revisit both your evaluation of its validity (maybe you erred in determining its validity) as well as your agreement with its premises: if you really *disagree* with the conclusion of a *valid* argument, then you must (logically) *disagree* with at least one of its premises.

You should be sure to use the technical terms correctly: you need to distinguish between *premises* – which can be *true* or *false* (but cannot be "valid," "invalid," "sound," or "unsound") – and *arguments* – which can be *valid* (if the argument's conclusion must be true whenever its premises are true), *invalid* (i.e. not valid; the argument's conclusion could be false even if its premises are true), *sound* (if the argument is valid *and* all of its premises are true), or *unsound* (i.e. not sound: either invalid or else valid-with-at-least-one-false-premise) (but cannot be "true" or "false").

And you should avoid using such non-technical (hence ambiguous) terms as 'correct,' 'incorrect,' 'right,' or 'wrong.' You also have to be careful about calling a *conclusion* "valid," because that's ambiguous between meaning you think it's true (and are misusing the word 'valid') and meaning you think it follows validly from the premises.

Perhaps most importantly, keep in mind that often the point of argumentation is not to convince someone of your position (something that rarely happens, no matter how rational we think we are) but to help the other person think through the issues.

A	(1)	All pianists are musicians.	T		
	(2)	Lang Lang is a pianist.	T	V	S
	(3)	∴ Lang Lang is a musician.	T		
B	(1)	All pianists are musicians.	T		
	(2)	Lang Lang is a musician.	T	I	U
	(3)	∴ Lang Lang is a pianist.	T		
C	(1)	All musicians are pianists.	F		
	(2)	The violinist Itzhak Perlman is a musician.	T	V	U
	(3)	∴ Itzhak Perlman is a pianist.	F		
D	(1)	All musicians are pianists.	F		
	(2)	Itzhak Perlman is a violinist.	T	I	U
	(3)	∴ Itzhak Perlman is a pianist.	F		

E	(1)	All cats are dogs.	F		
	(2)	All dogs are mammals.	T	V	U
	(3)	∴ All cats are mammals.	T		
F	(1)	All cats are dogs.	F		
	(2)	All cats are mammals.	T	I	U
	(3)	∴ All dogs are mammals.	T		
G	(1)	All cats are dogs.	F		
	(2)	Snoopy is a cat.	F	V	U
	(3)	∴ Snoopy is a dog.	T		
H	(1)	All cats are birds.	F		
	(2)	Snoopy is a cat.	F	I	U
	(3)	∴ Snoopy is a dog.	T		
I	(1)	All cats are birds.	F		
	(2)	All birds are dogs.	F	V	U
	(3)	∴ All cats are dogs.	F		
J	(1)	All cats are birds.	F		
	(2)	All dogs are birds.	F	I	U
	(3)	∴ All cats are dogs.	F		
K	(1)	All cats are mammals.	T		
	(2)	All dogs are mammals.	T	I	U
	(3)	∴ All cats are dogs.	F		

Part II

Computer Science, Computers, and Computation

Part II begins our exploration of the philosophy of computer science by looking at **what computer science is** (Chapter 3). For computer science to be considered either as a science or as a branch of engineering, we need to know **what science is** (Chapter 4) and **what engineering is** (Chapter 5). If computer science is a study of computers, then we need to know **what a computer is** (Chapters 6 and 9). And if computer science is a study of computation, then we need to know **what an algorithm is** (Chapters 7 and 8).

Philosophy of Computer Science: An Introduction to the Issues and the Literature, First Edition. William J. Rapaport.
© 2023 John Wiley & Sons, Inc. Published 2023 by John Wiley & Sons, Inc.

3

What Is Computer Science?[1]

Thanks to those of you who [gave their own] faculty introductions [to the new graduate students]. For those who [weren't able to attend], I described your work and courses myself, and then explained via the Reductionist Thesis how *it all comes down to strings and Turing machines operating on them.*
—Kenneth Regan, email to University at Buffalo Computer Science & Engineering faculty (27 August 2004); italics added.

The Holy Grail of computer science is to capture the messy complexity of the natural world and express it algorithmically.
—Teresa Marrin Nakra, quoted in Davidson, 2006, p. 66.

3.1 Introduction

To *define* any scholarly discipline is a formidable task.
—Anthony Ralston (1971, p. 1)

The fundamental question of this book is, **what is computer science?** Almost all of the other questions we will be considering flow from this one. (Is it a *science*? Is it the science *of computers*? What is science? What is a computer? And so on.) In this chapter, we will look at several definitions of 'computer science.' Each definition raises issues that we will examine in more detail later, so a final answer (if there is one!) will have to await the end of the book (Chapter 20). Before we try to answer our "fundamental" question, it's worth asking some preliminary ones: What should this discipline be called? Why should we even bother seeking a definition? What does it mean to give a definition or to ask what something is?[2]

3.2 Naming the Discipline

When our discipline was newborn, there was the usual perplexity as to its proper name.
—Frederick P. Brooks (1996, p. 61)

1 An earlier version of this chapter appears as Rapaport, 2017c.
2 Philosopher and logician J. Michael Dunn suggested to me that we might also ask, what does it mean to ask what does it mean to ask what something is?

Philosophy of Computer Science: An Introduction to the Issues and the Literature, First Edition. William J. Rapaport.
© 2023 John Wiley & Sons, Inc. Published 2023 by John Wiley & Sons, Inc.

Should we call the discipline 'computer science' (which seems to assume that it is the *science* of a certain kind of *machine*), or 'computer *engineering*' (which seems to assume that it is *not* a science but a branch of engineering), or '*computing* science' (which seems to assume that it is the science of what those machines *do*), or 'informatics' (a name more common in Europe), or something else altogether?

Until we have an answer to our question, think of the subject as being called by a 15-letter word 'computerscience' that may have as little to do with computers or science as 'cattle' has to do with cats. To save space and to suppress presuppositions, I'll often just refer to it as "CS."[3]

3.3   Why Ask What CS Is?

Let's now turn to the question of why we might *want* a definition. There are at least two kinds of motivations, academic ones and philosophical ones. And among the academic motivations, there are political, pedagogical, and publicity ones.

Academic Politics

Here is an academic political reason for asking what CS is: *where should a CS department be administratively housed?* Intellectually, this might not matter: after all, a small school might not even have academic departments, merely teachers of various subjects. But deciding where to place a CS department can have political repercussions:

> In a purely intellectual sense such jurisdictional questions are sterile and a waste of time. On the other hand, they have great importance within the framework of institutionalized science – e.g., the organization of universities and of the granting arms of foundations and the Federal Government. (Forsythe, 1967b, p. 455)

Sometimes a department is housed in a particular school or college[4] only because it is hoped that it will get better treatment there (more funding, more resources), or only because it is forced to be there by the administration. It may have very little, if any, academic or intellectual reason for being housed where it is. Some possible locations for CS include

- A college or school of **arts and sciences**, which typically includes other departments in the humanities, social sciences, and natural sciences
- A college or school of **engineering**, which typically includes disciplines such as chemical engineering, electrical engineering, mechanical engineering, etc.
- A college or school of **informatics**, which might also include disciplines such as communications, library science, etc.

Another possibility is that CS should not be (merely) a department but an entire school or college itself, with its own dean, and perhaps with its own departments.

---

3 See the Online Resources for further reading on CS's names.
4 In the US, colleges and universities are usually administratively divided into smaller units, variously known as 'schools,' 'colleges,' 'faculties,' 'divisions,' etc., each typically headed by a "dean" and divided into still smaller units, called 'departments.'

Academic Pedagogy

Perhaps a more important academic purpose for asking what CS is concerns pedagogy: *what should be taught in an introductory CS course?*

Should it be a **programming** course? That is, is CS the study of programming? Or, worse, should students be led to *think* that's what it is? I don't know any computer scientists who think CS is *just* the study of programming (Denning et al., 2017), but the typical introductory course tends to lead students (and the general public) to think so.

Should it be a **computer literacy** course? That is, is CS all about how to *use* computers?

Should it be a course in the **mathematical theory of computation**? That is, is CS the study of computation?

Should it be a course that introduces students to several different branches of CS, including, perhaps, some of its history?

And so on.

Academic Publicity

A related part of the academic purpose for asking the question concerns *publicity for prospective students and the general public*: How should a CS department characterize itself so as to attract good students? How should the discipline of CS characterize itself so as to encourage primary- or secondary-school students to consider it as something to study in college or to consider it as an occupation? (For more motivations along these lines, see Denning, 2013, p. 35.) How should the discipline characterize itself so as to attract more women and minorities to the field? How should it characterize itself to the public at large so that ordinary citizens might have a better understanding of CS?

Philosophical Motivations

Perhaps the academic (and especially political) motivations for asking what CS is are ultimately of little more than practical interest. But deep intellectual or philosophical issues underlie those questions, and this will be the focus of our investigation: **What is CS "really"?** Is it like some other academic discipline? For instance, is it like physics, or mathematics, or engineering? Or is it "*sui generis*"?

Digression: 'sui generis': '*Sui generis*' is a Latin phrase meaning "own kind." As an analogy, a poodle and a pit bull are both kinds of dogs. But a wolf is not a dog; it is its own kind of animal ("*sui generis*"). Some biologists believe that dogs are actually a kind of wolf; others believe that dogs are *sui generis*.

To illustrate this difference, consider two very different comments by two Turing Award-winning computer scientists (as cited in Gal-Ezer and Harel, 1998, p. 79). Marvin Minsky, a co-founder of artificial intelligence, once said

> Computer science has such *intimate relations* with so many other subjects that *it is hard to see it as a thing in itself*. (Minsky, 1979, my italics; cf. Forsythe, 1967a, p. 6)

On the other hand, Juris Hartmanis, a founder of computational complexity theory, has said

> Computer science *differs* from the known sciences so deeply that it has to be viewed as *a new species among the sciences*. (Hartmanis, 1993, p. 1; my italics; cf. Hartmanis, 1995a, p. 10)[5]

So, is CS like something "old," or is it something "new"? But we have yet another preliminary question to consider …

The Turing Award: The A.M. Turing Award, given annually by the Association for Computing Machinery, is considered the "Nobel Prize" of computer science. See http://amturing.acm.org/ and Vardi, 2017.

3.4   What Does It Mean to Ask What Something Is?

> It does not make much difference how you divide the Sciences, for they are one continuous body, like the ocean.
> —Gottfried Wilhelm Leibniz (1685, p. 220)

> We will not try to give a few-line definition of computer science since no such definition can capture the richness of this new and dynamic intellectual process, *nor can this be done very well for any other science*.
> —Juris Hartmanis (1993, p. 5; my italics)

3.4.1   Determining Boundaries

> When sharp formulations are offered for concepts that had been vague, they sometimes result in bizarre rulings along the edges, bizarre but harmless.
> —Willard van Orman Quine (1987, p. 217)

> *We should quell our desire to draw lines.* We don't need to draw lines.
> —Daniel C. Dennett (2013a, p. 241)

A fundamental principle should be kept in mind whenever you ask what something is, or what kind of thing something is: **There are no sharp boundaries in nature**; there are only continua. A "continuum" (plural = 'continua') is like a line with no gaps in it and hence no natural places to divide it up. The real-number line is the best example. Another is the color spectrum: although we can identify the colors red, orange, yellow, green, blue, and so on, there are no sharp (or non-arbitrary) boundaries where red ends and orange begins; worse, one culture's "blue" might be another's "green" (Berlin and Kay, 1969; Grey, 2016). A third example is the assignment of letter grades to numerical scores: there is often no natural (or non-arbitrary) reason why a score of (say) 75 should be assigned a letter grade of (say) 'B−' while a 74 is a 'C+' (Rapaport, 2011).

---

5 Hartmanis, 1995a covers much of the same ground, and in many of the same words, as Hartmanis, 1993, but is more easily accessible, having been published in a major journal that is widely available online, rather than in harder-to-find conference proceedings. Moreover, the issue of *ACM Computing Surveys* containing Hartmanis, 1995a also contains commentaries (including, especially, Denning, 1995, Loui, 1995, Plaice, 1995, Stewart, 1995, Wulf, 1995) and a reply by the author (Hartmanis, 1995b).

An apparent counterexample to the lack of sharp boundaries in nature might be biological species: dogs are clearly different from cats, and there are no "intermediary" animals – ones that are not clearly either dogs or else cats. But both dogs and cats evolved from earlier carnivores (it is thought that both evolved from a common ancestor some 42 million years ago). If we traveled back in time, we would not be able to say whether one of those ancestors was a cat or a dog; in fact, the question wouldn't even make sense.[6]

Moreover, although logicians and mathematicians like to define categories in terms of "necessary and sufficient conditions" for membership, this only works for abstract, formal categories. For example, we can define a sphere of radius r and center c as the set of *all* and *only* those points that are r units distant from c.

Necessary and Sufficient Conditions: "All" such points is the "sufficient condition" for being a sphere; "only" such points is the "necessary condition": S is a sphere of radius r at center c **if and only if** $S = \{p : p$ is a point that is r units distant from $c\}$. That is, p is r units from c **only if** p is a point on S (i.e. **if** p is r units from c, **then** p is a point on S). So, being a point that is r units from c **is a sufficient condition for** being on S. And **if** p is a point on S, **then** p is r units from c. So, being a point that is r units from c **is a necessary condition for** being on S.

However, as philosophers and cognitive scientists have pointed out, non-abstract, non-formal ("real") categories usually don't have such precise, defining characteristics. The most famous example is the philosopher Ludwig Wittgenstein's unmet challenge to give necessary and sufficient defining characteristics for something being a game (Wittgenstein, 1958, Section 66ff). Instead, he suggested that games (such as solitaire, basketball, chess, etc.) all share a "family resemblance": the members of a family don't necessarily all have the same features in common (having blue eyes, being tall, etc.) but instead resemble each other (mother and son, but not father and son, might have blue eyes; father and son, but not mother and son, might both be tall, and so on). And the psychologist Eleanor Rosch has pointed out that even precisely definable, mathematical categories can have "blurry" edges: most people consider 3 to be a "better" example of a prime number than, say, 251, or a robin to be a "better" example of a bird than an ostrich is.

In his dialogue *Phaedrus*, Plato suggested that a good definition should "carve nature at its joints" (Plato, 1961a, lines 265e–266a). But if "nature" is a continuum, then there are no "joints." Hence, we do not "carve nature at *its* joints"; rather, we "carve nature" at "joints" that are usually of our *own* devising: **we impose our own categories on nature.**

But I would not be a good philosopher if I did not immediately point out that, just as Plato's claim is controversial, so is this counter-claim! After all, isn't the point of science to describe and explain a reality that exists independently of us and our concepts and categories – i.e. independently of the "joints" that *we* "carve" *into* nature? (We'll return to the topic of the goal of science in Section 4.4.) And aren't there "natural kinds"? Dogs and cats, after all, do seem to be kinds of things that are there in nature, independently of us, no matter how hard it might be to *define* them.

Is CS similar to such a "natural kind"? Here, I think the answer is that it pretty clearly is not. There would be no academic discipline of CS without humans, and there probably wouldn't even be any computers without us (although we'll see some reasons to think otherwise in Section 9.7).[7]

Let's consider a few examples of familiar terms whose definitions are controversial.

---

6 See https://www.quora.com/Do-cats-and-dogs-have-a-common-ancestor-If-so-do-we-know-when-they-started-to-split-off and the *Wikipedia* article "Carnivora" (http://en.wikipedia.org/wiki/Carnivora).
7 See the Online Resources for further reading on Wittgenstein and on categorization.

'Planet'

Consider the case of poor Pluto – not Mickey Mouse's dog, but the satellite of the Sun: it used to be considered a planet, but now it's not, because it's too small (Lemonick, 2015). I don't mean it is now *not a planet* because of its size. Rather, I mean now it is *no longer considered* a planet because of its size. Moreover, if it were to continue being categorized as a planet, then we would have to count as planets many other small bodies that orbit the Sun, eventually having to consider all (non-human-made) objects in orbit around the Sun as planets, which almost makes the term useless, because it would no longer single out some things (but not others) as being of special interest. To make matters even worse, the Moon was once considered a planet! When it was realized that it did not orbit the Sun directly, it was "demoted." But curiously, under a proposed new definition of 'planet' (as having an "orbit-clearing mass"), it might turn out to be (considered) a planet once more! (Battersby, 2015).

Note that, in either case, the *universe* has not changed; only our *descriptions* of it have:

> Exact definitions are undoubtedly necessary but are rarely perfect reflections of reality. Any classification or categorization of reality imposes arbitrary separations on spectra of experience or objects. (Craver, 2007)

So, depending on how we define 'planet,' either something that we have always considered to *be* one (Pluto) might turn out *not* to be one, or something that we have (usually) *not* considered one (the Moon) might turn out to *be* one! Typically, when trying to define or "formalize" an informal notion, one finds that one has *excluded* some "old" things (i.e. things that were informally considered to fall under the notion), and one finds that one has *included* some "new" things (i.e. things that one hadn't previously considered to fall under the notion). Philosopher Ned Block has called the former kind of position "chauvinism" and the latter position "liberalism" (Block, 1978, pp. 263, 265–266, 277). When this happens, we can then either reformulate the definition or bite the bullet about the inclusions and exclusions. One attitude toward exclusions is often that of sour grapes: our intuitions were wrong; those things really weren't Xs after all. The attitude toward inclusions is sometimes "Wow! That's right! Those things really *are* Xs!" Alternatively, a proposed definition or formalization might be *rejected* because of its chauvinism or liberalism.

'Computation'

The next two cases will be briefer because we will discuss them in more detail later in the book. The first is the very notion of 'computation' itself: according to the Church-Turing Computability Thesis, a mathematical function is computable if and only if it is computable by a Turing Machine. This is not necessarily either a definition of 'computable' or a mathematical theorem; arguably, it is a suggestion about what the *informal* notion of "computability" should mean. But some philosophers and computer scientists believe there are functions that *are* informally computable but *not* computable by a Turing Machine. (We'll discuss these in Chapter 11.)

Terminological Digression: If you don't yet know what these terms are, be patient; we will begin discussing them in Chapter 7.

Should 'machine' be capitalized in 'Turing Machine'? Not capitalizing it suggests that a Turing Machine is a machine of some kind. But machines are typically physical objects, whereas a Turing Machine is an abstract mathematical notion. Capitalizing 'machine' turns it into a proper name, allowing us to ask whether a Turing Machine is or is not a machine without begging any questions. So I will capitalize 'Turing Machine' except in direct quotations.

'Thinking'

Our final case is the term 'thinking': if thinking is categorized as any process of the kind that cognitive scientists study – including such things as believing, consciousness, emotion, language, learning, memory, perception, planning, problem-solving, reasoning, representation, sensation, etc. (Rapaport, 2012b, p. 34) – then it is (perhaps!) capable of being carried out by a computer. Some philosophers and computer scientists accept this way of thinking about thinking and therefore believe that computers will eventually be able to think (even if they do not yet do so). Others believe that if computers can be said to think when you accept this categorization, then there must be something wrong with the categorization. (We'll explore this topic in Chapter 18.)

> **Other Examples:** Another example is 'life' (Machery, 2012; Allen, 2017, p. 4239). We'll come back to this in Sections 10.1 and 18.9. Angere, 2017 is another case study, which shows how even 'square' and 'circle' may have counterintuitive definitions, allowing for the (mathematical) existence of square circles (or round squares)!

3.4.2 Extensional and Intensional Definition

An "extensional" definition of a term t is given by presenting the *set* of items that are considered ts. For example, we once might have said that x is a planet (of the Sun) iff $x \in$ {Mercury, Venus, Earth, Mars, Jupiter, Saturn, Uranus, Neptune, Pluto}. Now, however, we say that x is a planet (of the Sun) iff $x \in$ {Mercury, Venus, Earth, Mars, Jupiter, Saturn, Uranus, Neptune}. Note that these two extensional definitions of 'planet' are *different*. For another example, the (current as of March 2022) extensional definition of 'US President' is {Washington, Adams, Jefferson, …, Obama, Trump, Biden}.

An "intensional" definition can be given in terms of necessary and sufficient conditions or in terms of a family resemblance. For example, an intensional definition of 'US President' might be given by citing Article II of the US Constitution: roughly, x is US President iff x has been vested with the executive power of the United States. Note that this intensional definition holds even if an extensional definition changes (such as the extensional definition in the previous paragraph of 'US President,' which changes roughly every four or eight years).

Two concepts can be said to be "extensionally equivalent" if exactly the same sets of things fall under each concept. Importantly, two extensionally equivalent concepts can be (and usually are) "intensionally distinct"; i.e. they really are different concepts. Here's a mundane example: large supermarkets these days not only carry groceries but also pharmaceuticals, greeting cards, hardware, etc. Large drugstores these days carry not only pharmaceuticals but also groceries, greeting cards, hardware, etc. And large "general stores" also carry pretty much the same mix of products. We tend to think of Walgreens as a drugstore, Wegmans as a supermarket, and Walmart as a general store: they are "intensionally distinct." But because they sell the same mix of products, they are "extensionally equivalent."

Here is an important example from computability theory (which we'll look at in detail in Chapter 7): recursive function theory and the theory of Turing Machines are extensionally equivalent but intensionally distinct. They are extensionally equivalent because it is mathematically provable that all functions that are recursive are Turing computable and vice versa. But they are intensionally distinct because the former is concerned with a certain way of defining mathematical functions, while the latter is concerned with algorithms and computation. From the point of view of what facts can be proved about functions, it doesn't matter which formalism is used because they are extensionally equivalent. Their intensional distinctness comes into play when one formalism might be easier or more illuminating to use in a given situation. We'll come back to this in Section 3.7. (For more on extensions and intensions, see Rapaport, 2012a.)

An Extensional Definition of CS

To the extent that it is we who impose *our* categories on nature, there may be no good answer to the question "What is CS?" beyond something like "computer science is that which is taught by computer science departments" (Abrahams, 1987, p. 472). Perhaps intended more seriously, the computer scientist Peter J. Denning (2000, p. 1) defines "The discipline of computer science ... [as] the body of knowledge and practices used by computing professionals in their work." But then we can ask, *what is it that computer scientists do?* Of course, one can beg that last question – i.e. argue in a circle – by saying that computer scientists do computer science! Turing Award winner Richard W. Hamming (1968, p. 4) suggests something like this, citing the (humorous) "definition" of mathematics as "what mathematicians do," but he goes on to point out that "there is often no clear, sharp definition of ... [a] field."

Begging the Question: 'To beg the question' is a slightly archaic term of art in philosophy and debating. The phrase does *not* mean "to *ask* a question" – i.e. to "beg" in the sense of "to raise or invite" a question. In debating, a "question" is the *topic* being debated. 'To beg the question' means: "to *request* (i.e. "to beg") that the *topic being debated* (i.e. the "question") be *granted as an assumption* in the debate." That is, it means "to assume as a premise ("to beg") the conclusion ("the question") that you are arguing for." A modern synonym for 'beg the question' is 'argue in a circle.'

Nevertheless, it's worth looking briefly at what computer scientists do. It has been said that CS is "a sort of spectrum ... with 'science' on the one end and 'engineering' on the other" (Parlante, 2005, p. 24), perhaps something like this:

> abstract, mathematical theory of computations
> abstract, mathematical theory of computational complexity
> abstract, mathematical theory of program development
> software engineering
> ...
> operating systems
> ...
> AI
> ...
> computer architecture
> ...
> VLSI
> networks
> social uses of computing, etc.

But this is less than satisfactory as a *definition*.

Intensional Definitions of CS

As with most non-mathematical concepts, there are probably no necessary and sufficient conditions for being CS. At best, the various branches of the discipline share only a family resemblance. We can try to give an intensional definition by splitting the question of what CS is into two parts:

1. What is its **object**? (*What* does it study or investigate?)
2. What is its **methodology**? (*How* does it study those objects?)

We'll begin with the second.

Is the *methodology* of CS the same as that of some other discipline? Or does it have its own distinctive methodology? If the latter, is its methodology not only unique but also something brand new? Methodologically, CS has been said to be (among many other things)

- An *art form* (Knuth, 1974a, p. 670, has said that programs can be beautiful)
- An *art and science*:

> Science is knowledge which we understand so well that we can teach it to a computer; and if we don't fully understand something, it is an art to deal with it. … [T]he process of going from an art to a science means that we learn how to automate something. (Knuth, 1974a, p. 668)

- A *liberal art* along the lines of the classical liberal arts of logic, mathematics, or astronomy (Perlis, 1962; Lindell, 2001, p. 210)
- A branch of *mathematics* (Dijkstra, 1974; Davis, 1978),
- A *natural science* (McCarthy, 1963; Newell et al., 1967; Shapiro, 2001)
- An *empirical study* of the artificial (Simon, 1996b)
- A combination of *science and engineering* (Hartmanis, 1993, 1995a; Loui, 1995)
- Just *engineering* (Brooks, 1996)
- Or – generically – a "study"

But a study (or a science, or an engineering, or an art, or …) *of what*? Is its *object* the same as that of some other discipline? Does it study exactly what science, or engineering, or mathematics, or – for that matter – psychology or philosophy studies studies? Or does it have its own, distinctive object of study (computers? algorithms? information?)? Or does it study something that has never been studied before? The logician Jon Barwise (1989a) suggested that we can understand what CS is in terms of what it "traffics" in. So here's an alphabetical list of some of the *objects* that it traffics in:

> algorithms, automation, complexity, computers, information, intelligence, numbers (and other mathematical objects), problem-solving, procedures, processes, programming, symbol strings

This is, of course, only a very partial list. Just as, for any X, there can be a philosophy of X (see Section 2.7), one can use computational methods to study pretty much any X. (For some examples, see Reese, 2014a.)

Perhaps advertising blurbs for CS like the ones you find at university websites should not be taken too seriously. But the authors of several published essays that try to define 'computer science' – all of whom are well-respected computer scientists – presumably put a lot of thought into them. They *are* worth taking seriously, which is the main purpose of this chapter.

3.5 CS as the Science of Computers

The first such answer that we will look at comes from three Turing Award winners: Allen Newell, Alan Perlis, and Herbert Simon. Here is their definition, presented as the conclusion of an argument:

Wherever there are phenomena, there can be a science to describe and explain those phenomena. … There are computers. Ergo,[8] **computer science is the study of computers**. (Newell et al., 1967, p. 1373, my boldface)

This argument is actually missing two premises (recall Section 2.9.3). Their two explicit premises only imply that there *can* be a science of computers. They do not, by themselves, imply that there *is* such a science *or* that that science is CS rather than some other discipline. The missing premises are

(A) There *is* a science of computers.

(B) There is no *other* discipline that is the science of computers besides CS.

3.5.1 Objection: Computers Are Not Natural

Newell, Perlis, and Simon's first premise is that, for any phenomenon[9] *p*, there can be a science of *p*. An objection to this that they consider is that this premise holds not for *any* phenomenon but only when *p* is a *natural* phenomenon. For example, the computer engineer Michael C. Loui (1987, p. 175) notes that there are toasters but no *science* of toasters. The objection goes on to point out that computers (like toasters) aren't natural; they are *artifacts*. So, it doesn't follow that there can be (much less that there *is*) a *science* of computers. It might still be the case that there is some other kind of discipline that studies computers (and toasters!), such as engineering.

The computer scientist Bruce W. Arden (1980, p. 6) also argues that neither mathematics nor CS is a science, because its object is not natural phenomena: the object of mathematics is "human-produced systems … concerned with the development of deductive structures," and the object of CS is "man-made" [sic]. But what is CS's object? Computers? Yes, they're clearly human-made, and this leads us back to Newell, Perlis, and Simon's arguments. Or is it algorithms? Perhaps they're also human-made in whatever sense mathematical structures are. But in Section 3.10, we'll look at a claim that algorithms are a special case of a *natural* entity ("procedures").

The computer historian Michael S. Mahoney (2011, pp. 159–161) rejects this objection that "the computer is an artifact, not a natural phenomenon" because he thinks there is no sharp dividing line "between nature and artifact." He gives two reasons: (1) We *use* artifacts to study nature – "We know about nature through the models we build of it." (2) "Artifacts work by the laws of nature, and by working reveal those laws." In other words, artifacts are *part* of nature. The philosopher Timothy Williamson (2007, p. 43) makes a similar point about scientific instruments: "The scientific investigation of [a] physical quantity widens to include the scientific investigation of its interaction with our experimental equipment. After all, our apparatus is part of the same natural world as the primary topic of our inquiry." The same could be said about computers and computation: we use computers and computational methods to study both computers and computation. (Mahoney even goes on to suggest that nature itself might be ultimately computational. We will explore that idea in Section 9.7.2.)

Newell, Perlis, and Simon's reply to the objection is to deny the premise that the phenomenon that a science studies must be natural. They point out that there *are* sciences of artifacts; e.g. botanists study hybrid corn.[10] In fact, in 1969, Simon published the first edition of his book *The Sciences of the Artificial* (Simon, 1996b). And the computer scientist Donald Knuth (2001, p. 167) has called CS "an *unnatural* science [because] [c]omputer science deals with artificial things, not bound by the constraints of nature."

---

8 'Ergo' is Latin for "therefore."

9 By the way, 'phenomen**on**' is the correct singular term. If you have two or more of them, you have two or more phenomen**a**.

10 Curiously, they say that it is *zoologists* who study hybrid corn!

The objector might respond that the fact that Simon had to write an entire book to argue that there could be sciences of artifacts shows that the objection – that science only studies natural phenomena – is not *obviously* false. Moreover, botanists study mostly *natural* plants: hybrid corn is not only not studied by all botanists, it is certainly not the only thing that botanists study (botany is not defined as the science of hybrid corn). Are there any *natural* phenomena that computer scientists study? As I have already hinted, we will see a positive answer to this question in Section 3.10.

But let's not be unfair. There certainly are sciences that study artifacts *in addition* to natural phenomena: ornithologists study both birds (which are natural) and their nests (which are artifacts); apiologists study both bees (natural) and their hives (artifacts). On the other hand, one might argue (a) that beehives and birds' nests are not *human*-made phenomena and (b) that 'artifact' should be used to refer not to any *manufactured* thing (as opposed to living things), but only to things that are manufactured by *humans*: i.e. to things that are not "found in nature," so to speak. The obvious objection to this claim is that it unreasonably singles out humans as being apart from nature (Figure 3.1).[11]

3.5.2 Objection: Computers Are Tools, Not Phenomena

A related objection has to do with the observation that it is wrong to define a subject by its *tools*:

> The debate over the appropriate place of computing in grade schools and high schools echoes the debate in universities decades ago, when computers and software were initially seen as mere plumbing. And certainly not something worthy of study in its own right. A department of computer science? Why not a department of slide rules? (Lohr, 2008)

And as Hammond (2003) notes, "Theoretical Computer Science doesn't even use computers, just pencil and paper." Newell et al. (1967, p. 1373) also say that astronomy is the science of stars. And, of course, telescopes are used to study the stars. But as the computer scientist Edsger W. Dijkstra is alleged to have said, "Computer Science is no more about computers than astronomy is about telescopes" (O'Toole, 2021a). Dijkstra (1986) also said that calling the discipline 'computer science' "is like referring to surgery as 'knife science.'" This may be true, but the problem, of course, is that the closest term that computer scientists have corresponding to 'surgery' is probably 'computing,' and defining 'computer science' as the science of computing may be legitimate but not very clarifying (at least, not without a further description of computing, preferably not in terms of computers!). Newell, Perlis, and Simon address this in their Objection 4: "The computer is such a novel and complex instrument that its behavior is subsumed under no other science" (Newell et al., 1967, p. 1374). This is also support for missing premise B.

But it is also wrong to define a subject without saying what its tools enable. Even if what Newell, Perlis, and Simon say about the novelty of computers is true, it can be argued that a new tool can open up a new science or, at least, a new scientific paradigm (see Section 4.9): "Paradigm shifts have often been preceded by 'a technological or conceptual invention that gave us a novel ability to see things that could not be seen before' " (Mertens, 2004, p. 196, quoting Robertson, 2003). Horsman et al. (2017) discuss this at length, commenting that, "With all due respect to Dijkstra, ... computer science *is* as much about computers as astronomy is about telescopes" (p. 31, my italics). In fact, there once *was* a science that only studied such a tool – microscopes![12] It is worth a short digression to look at "microscopy," a science that no longer exists![13]

---

11 See the Online Resources for further reading on artifacts.
12 Another "science" of an artifact might be bicycle science (Wilson and Papadopoulos, 2004). But it's really not clear if this is a science or a branch of engineering.
13 Thanks to Stuart C. Shapiro for suggesting this. A more complete story is told in the Online Resources.

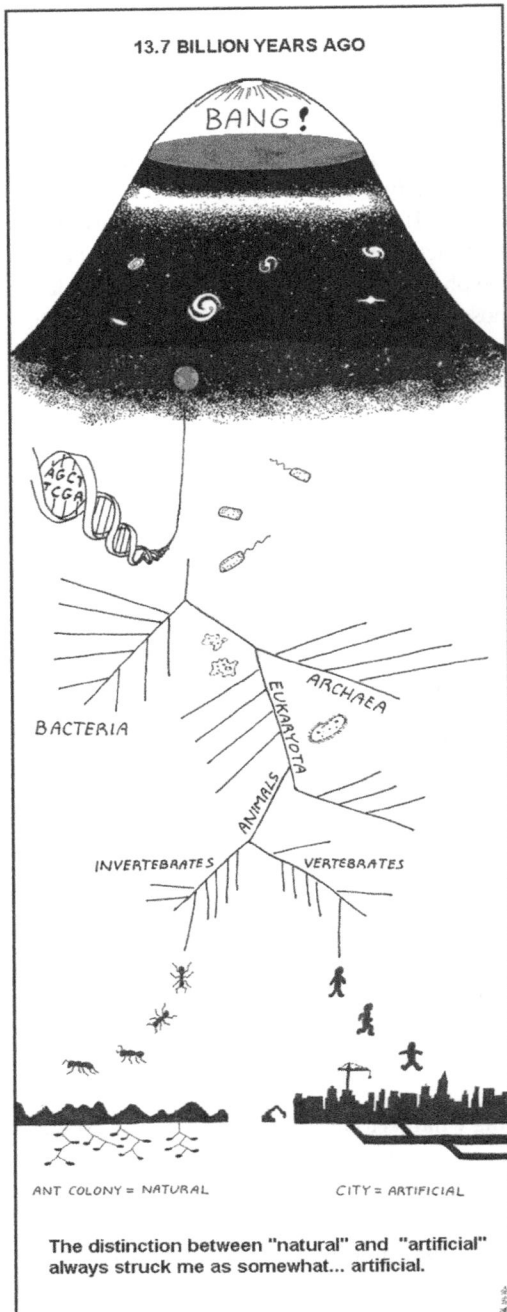

Figure 3.1 Artificial vs. Natural. Source: Abstruse Goose, Artificial. Retrieved from https://abstrusegoose.com/215.

3.5.3 Digression: The Once-upon-a-Time Science of Microscopy

> … Marcello Malpighi (1628–1694), was a great scientist whose work had no dogmatic unity.[14] He was one of the first of a new breed of explorers who defined their mission neither by the doctrine of their master nor by the subject that they studied. They were no longer 'Aristotelians' or 'Galenists.' Their eponym, their mechanical godparent, was some device that extended their senses and widened their vistas. What gave his researches coherence was a new instrument. Malpighi was to be a 'microscopist,' and his science was 'microscopy' …. His scientific career was held together not by what he was trying to confirm or to prove, but by the vehicle which carried him on his voyages of observation.
> —Daniel Boorstin (1983, p. 376)

In a similar fashion, surely[15] computers are "device[s] that [have] extended [our] senses and widened [our] vistas," and the science of computer scientists is, well, *computer* science. After all, one of the two principal professional associations is the Association for Computing *Machinery* (ACM). What "holds" computer scientists "together … [is] the vehicle which carrie[s them] on [their] voyages of observation."

But this is not necessarily a positive analogy.

> The applications of computers to a discipline should be considered properly a part of the natural evolution of the discipline. … The mass spectrometer has permitted significant advances in chemistry, but there is no 'mass spectrometry science' devoted to the study of this instrument. (Loui, 1987, p. 177)

Similarly, the microscope has permitted significant advances in biology (and many other disciplines), but arguably, *microscopy no longer exists as an independent science devoted to the study of that instrument or the things studied with it.* Could the same thing happen to *computer* science that happened to *microscope* science? If so, what would fall under the heading of the things that can be studied with computers? A dean who oversaw the Department of Computer Science at my university once predicted that the same thing would happen to our department: the computer-theory researchers would move into the mathematics department; the AI researchers would find homes in psychology, linguistics, or philosophy; those who built new kinds of computers would move (back) into electrical engineering; and so on. This hasn't happened yet (although McBride, 2007 suggests that it is already happening, while Mander, 2007 disagrees). Nor do I foresee it happening in the near future, if at all. After all, as the computer scientist George Forsythe pointed out, in order to teach "nontechnical students" about computers and computational thinking, "specialists in other technical fields" about how to use computers as a tool (alongside "mathematics, English, statistics"), and "computer science specialists" about how to "lead the future development of the subject,"

> The first major step … is to create a department of computer science … Without a department, a university may well acquire a number of computer scientists, but they will be scattered and relatively ineffective in dealing with computer science as a whole. (Forsythe, 1967a, p. 5)

But the breakup of CS into component disciplines is something to ponder.

---

14 That is, Malpighi did not study any *single*, natural phenomenon; rather, he studied all phenomena that are only visible with a microscope.
15 'Surely,' by the way, is a word that any philosopher should surely(?) take with a grain of salt! (Dennett, 2013a, Ch. 10).

3.5.4 Objection: CS Is Just a Branch of …

The microscopy story is, in fact, close to an objection that Newell, Perlis, and Simon consider to missing premise B: perhaps the science of computers is not CS but some other subject – electrical engineering, or mathematics, or even psychology. For example, the computer historian Paul Ceruzzi comes close to saying that CS is identical to electrical (more precisely, electronic) engineering: it emerged from electrical engineering (Ceruzzi, 1988, p. 257), and it "came to define the daily activities of electrical engineers" (Ceruzzi, 1988, p. 258).

One problem with trying to conclude from this that CS is (nothing but) electrical engineering is that there are now other technologies that are beginning to come into use, such as quantum computing and DNA computing.[16] Assuming that those methods achieve some success, it becomes clear that (and how) CS goes beyond any particular *implementation* technique or technology and becomes a more *abstract* discipline in its own right. (On abstraction and implementation, see Chapter 13.) And Ceruzzi himself declares, "The two did not become synonymous" (Ceruzzi, 1988, p. 273).

Newell, Perlis, and Simon reply that, although CS does intersect electrical engineering, mathematics, psychology, etc., there is no other *single* discipline that subsumes *all* computer-related phenomena. (This is missing premise B.) This, however, assumes that CS *is* a single discipline, a cohesive whole. Is it? I began my professional university career in a philosophy department; although certain branches of philosophy (ethics and history of philosophy, for instance) were not my specialty, I was expected, and was able, to participate in philosophical discussions on these topics and even to teach them. But my colleagues in CS often do not, and are not expected to, understand the details of those branches of CS that are far removed from their own. As a computer scientist specializing in AI, I had far more in common with colleagues in the philosophy, psychology, and linguistics departments than I did with my computer-science colleagues down the hall who specialized in, say, computer networks or computer security. (And this is not just an autobiographical confession on my part; my colleagues in computer networks and computer security would be the first to agree that they have more in common with some of their former colleagues in electrical engineering than they do with me.) So, perhaps CS is *not* a coherent whole. (See the Question for the Reader in Section 20.2.)

3.5.5 Objection: What about Algorithms?

The most interesting – and telling – objection to Newell, Perlis, and Simon's view is that CS is really the study not (just) of *computers* but (also) of *algorithms*: very roughly, the programs and rules that tell computers what to do. (We'll devote a great deal of time, beginning with Chapter 7, to looking at what algorithms and programs are; at this point, I will just assume that you already have an idea of what they are and won't try to define them further.) For example, Bajcsy et al. (1992, p. 1, my italics) explicitly mention "the (*incorrect*) assumption that … [CS] is based *solely* on the study of a *device* …."

What is interesting about this objection is how Newell, Perlis, and Simon respond: they agree with it! They now say,

> In the definition [of CS as the science of computers], '*computers*' *means … the hardware, their programs or algorithms, and all that goes along with them.* **Computer science is the study of the phenomena surrounding computers**." (Newell et al., 1967, p. 1374, my italics and my boldface)

---

16 See the Online Resources for further reading on quantum and DNA computing.

In the end, they even allow that the study of computers may also be an engineering discipline (Newell et al., 1967, p. 1374). So, they ultimately water down their definition to something like this: *computer science is the science and engineering of computers, algorithms, and other related phenomena.*

Readers would be forgiven if they objected that the authors have changed their definition! But instead of making that objection, let's turn to an interestingly different, yet similar, definition due to another celebrated computer scientist.

3.6 CS Studies Algorithms

Turing Award-winner Donald Knuth (Roberts, 2018) gave an apparently different answer to the question of what CS is:

> [C]omputer science is … **the study of *algorithms***. (Knuth, 1974b, p. 323; my boldface, Knuth's italics)

3.6.1 Only Algorithms?

Knuth cited, approvingly, a statement by Forsythe (1968) that the central question of CS is, "What can be automated?" Presumably, a process can be automated – i.e. done automatically, by a machine, without human intervention – if it can be expressed as an algorithm. (We'll return to this in Section 20.3.1; for a book-length discussion, see Arden, 1980.) Knuth (1974b, p. 324) even noted that the name 'comput*ing* science' might be better than 'comput*er* science,' because the former sounds like the discipline is the science of what you *do* with computers as opposed to the science of the *tools* themselves.

Knuth pointed out that

> a person does not really understand something until he [sic] teaches it to someone else. Actually a person does not *really* understand something until he can teach it to a *computer*, i.e. express it as an algorithm. (Knuth, 1974b, p. 327)

The celebrated cellist Janos Starker once said something similar: "When you have to explain what you are doing, you discover what you are really doing" (Fox, 2013). Expressing something as an algorithm requires "real" understanding because every step must be spelled out in excruciating detail:

> It is a commonplace that a computer can do anything for which precise and unambiguous instructions can be given. (Mahoney, 2011, p. 80)

That is, a computer can do anything for which an *algorithm* can be given. After all, isn't an algorithm merely "precise and unambiguous instructions"? Thought of this way, the comment is almost trivial. But consider that to give such instructions (to give an algorithm) is to be able to explicitly *teach* the computer (or the executor, more generally) how to do that thing (Figure 3.2).

But there is a potential limitation to Knuth's theory that we *teach* computers how to do something – more specifically, to the theory that, insofar as CS is the study of what tasks are *computable*, it is the study of what tasks are *teachable*. The potential limitation is that teaching is *explicit* or "conscious." It is what psychologist and Nobel laureate Daniel Kahneman (2011, p. 21) has called a "System 2" task:

Figure 3.2 We're awesome at teaching. Source: XKCD, Progeny. Retrieved from http://xkcd.com/894/.

> *System 2* allocates attention to the effortful mental activities that demand it, including complex computations. The operations of System 2 are often associated with the subjective experience of agency, choice, and concentration.

But there is another algorithmic way of getting a computer to do something: by *training* it, either via a connectionist, neural-network algorithm or via a statistical, machine-learning algorithm. 'Learning,' in this sense of 'machine learning,' is different from being explicitly *taught*. Such training is *implicit* or "*un*conscious." It is "System 1" thinking:

> *System 1* operates automatically and quickly, with little or no effort and no sense of voluntary control. (Kahneman, 2011, p. 20)

We, as external, third-person observers, don't necessarily consciously or explicitly know how to do a System-1 task. Knowing *how* is not necessarily the same as knowing *that* (Section 2.5.3). An example might help to explain the difference: consider tic-tac-toe. A computer (or a human player) might be programmed – i.e. explicitly "taught" – to play winning tic-tac-toe by using a "conscious" or "System 2" algorithm that it explicitly follows. Most older children and adults have been taught a version of this algorithm (Zobrist, 2000):[17]

> For player X to win or draw (i.e. to not lose), **do**: **begin**
>
>> **if** there are 2 Xs in a row, **then** make 3 Xs in a row
>> **else if** there are 2 Os in a row, then block with an X
>> **else if** 2 rows intersect with an empty square
>>> such that each row contains 1 X, no Os,
>>> **then** place X at the intersection
>> **else if** 2 rows intersect with an empty square
>>> such that each row contains 1 O, no Xs,
>>> **then** place X at the intersection

---

17 This "algorithm" is written in informal pseudocode. Terms in **boldface** are control structures. Note its form: to accomplish goal *G*, do algorithm *A*. We'll return to this in Section 16.5.

       else if there is a vacant corner square, **then** put X there
       else place X on any vacant square.

  end

Alternatively, a computer can be programmed to *learn how* to play winning tic-tac-toe in a "System 1" manner without expressing (or being able to express) that strategy "consciously" (i.e. in a "System 2" manner). Such a learning mechanism can be found in Michie, 1974. Briefly, the computer is "rewarded" for each random move that leads to a win or draw, and such moves are thus caused to be made more frequently in future games.[18]

An algorithm in the form of a System-1–style artificial neural network is akin to building into the computer the ability to do that thing. Such a computer might not be able to tell us *how* it was doing it; it would not necessarily have any "conscious" access to its algorithm. An algorithm in the form of an explicit machine-learning program that would enable the computer to learn how to do that thing is somewhere in the middle. It would be "conscious" of its ability to learn but not necessarily of how to do the thing; it might not necessarily be able to teach someone or something else how to do it, unless it could observe itself doing it and develop a theory of how to do it (which theory would be expressed in a System-2–style, explicit algorithm). (We'll return to these issues in Sections 3.11 and 3.16.3; in Section 12.4.4, when we discuss the difference between following rules and behaving in accordance with them; and again in Section 17.2, when we discuss whether computers can make decisions.)

Let's say for now that something is computable just in case "precise and unambiguous instructions can be given" for it. (We'll be more precise and unambiguous(!) in Chapter 7.) So, the question becomes, what tasks are amenable to "precise and unambiguous instructions"? Presumably chess is computable in this sense because there are explicit rules for how to play chess. (Playing *winning* chess is a different matter!) But vision would seem *not* to be thus computable. After all, one cannot give "precise and unambiguous instructions" that would enable someone to see. Yet there *are* computer-vision systems (see http://aitopics.org/topic/vision for an overview), so vision *does* seem to be computable in a different sense: a behavior is computable if it can be *described* in terms of such instructions. The entity that exhibits that behavior naturally might not *use*, or be able to use, those instructions in order to behave that way. But we might be able to give those instructions to another system that *could* use them to exhibit that behavior. For instance, the human brain might not literally compute in the sense of executing an algorithm in order to see, but a computer using that algorithm might be able to exhibit visual behavior. (Whether it "sees," phenomenologically, is a philosophical question! See Section 19.2.) Similarly, the solar system might not be executing Kepler's laws, but an artificial solar system might. (We'll look into this issue in Section 9.7.2.)

3.6.2 Or Computers, Too?

Knuth goes on to point out, however, that you need computers in order to properly study algorithms because "human beings are not precise enough nor fast enough to carry out any but the simplest procedures" (Knuth, 1974b, p. 323). Indeed, he explicitly copies Newell, Perlis, and Simon's strategy, revising his initial definition to include computers – i.e. the phenomena "surrounding" algorithms:

> When I say that computer science is the study of algorithms, I am singling out only one of the "phenomena surrounding computers," so computer science actually includes more. (Knuth, 1974b, p. 324)

---

18 See the Online Resources for implementations of Michie's algorithm.

Does CS *have to* study computers? (Recall Hammond's suggestion about pencils [Section 3.5.2].) If so, does that mean CS really *is* the study of computers? Let's consider some similar questions for other disciplines. Do you *need* a compass and straightedge to study geometry, or can you study it just by proving theorems about points, lines, and angles? After all, the mathematicians David Hilbert (1899) and Oswald Veblen (1904) wrote completely axiomatic treatments of geometry without any (significant) mention of compass or straightedge. Do you *need* a microscope to study biology? I doubt that Watson and Crick used one when they discovered the structure of DNA.[19] Do you *need* a calculator (or a computer!) to study physics or mathematics (or does it just help you perform calculations more quickly and easily)? Even if you do need these tools, does that make geometry the study of compasses and straightedges, or physics and mathematics the study of calculators, or biology the study of microscopes? I think most people would say that these disciplines are not studies of those tools. On the other hand, "deep learning" algorithms do seem to need computers to determine if they will really do what they are intended to do, and do so in real time (Lewis-Kraus, 2016). (We'll return to this in Section 3.14.)

About 10 years later, Knuth (1985, pp. 170–171) backed off from the "related phenomena" definition, more emphatically defining CS as "primarily the study of algorithms," because he "think[s] of algorithms as encompassing the whole range of concepts dealing with well-defined processes, including the structure of data that is being acted upon, as well as the structure of the sequence of operations being performed," preferring the name 'algorithmics' for the discipline. Knuth also suggested that what computer scientists have in common (and what differentiates them from people in other disciplines) is that they are all "algorithmic thinkers" (Knuth, 1985, p. 172). (We will see what this means in Section 3.16.4 and Chapter 7.)

3.7 Physical Computers vs. Abstract Algorithms

So far, it may seem that we have two very different definitions of CS: *the study of computers* or *the study of algorithms*. But just as Newell, Perlis, and Simon said that CS is the study of computers *and related phenomena such as algorithms*, Knuth said that it is the study of algorithms *and related phenomena such as computers*! Stated a bit more bluntly, Newell, Perlis, and Simon's definition comes down to this: computer science is the science of *computers and algorithms*. Knuth's definition comes down to this: computer science is the study of *algorithms and computers*.

Ignoring for now the subtle difference between "science" and "study," what we have here are extensionally equivalent but intensionally distinct definitions. They may approach the discipline from different viewpoints (one from the viewpoint of a physical tool, one from the viewpoint of an abstract procedure), but the "bottom line" is the same – only the emphasis is different.

On the other hand, Arden (1980, p. 9) claims that "the *study of algorithms* and the *phenomena related to computers* are not coextensive, since there are important organizational, policy, and nondeterministic aspects of computing that do not fit the algorithmic mold." But I don't think either Newell et al. (1967) or Knuth (1974b) had those things in mind. And if "phenomena related to computers" is taken as widely as Arden does, then it encompasses pretty much everything, thus making any definition based on such a wide notion virtually useless. The classical sciences (physics, chemistry, biology, etc.) also have "important organizational, policy, and nondeterministic aspects," but those aren't used in trying to define what those sciences are about.

---

19 However, their collaborator Rosalind Franklin did use X-ray machines.

So, we now have two (only slightly different) definitions:

1. Computer science is the study of computers (and related phenomena such as the algorithms that they execute).
2. Computer science is the study of algorithms (and related phenomena such as the computers that execute them).

This strongly suggests that it would be wrong to treat CS as being primarily about algorithms *or* primarily about computers. *It is about both.* We'll see this more clearly in Chapter 6 when we trace the parallel histories of computers (as they evolved from calculating machines) and computing (as it evolved from the search for a foundation for mathematics).

But others beg to differ …

3.8 CS Studies Information

Others who have offered definitions of 'computer science' say "A plague on both your houses":[20] CS is *not* the study of computers *or* of algorithms but of **information**:

> I consider computer science, in general, to be the art and science of *representing and processing information* and, in particular, processing information with the logical engines called automatic digital computers. (Forsythe, 1967a, p. 3, my italics)

Denning (1985, p. 16, my italics) defined it as "the body of knowledge dealing with the design, analysis, implementation, efficiency, and application of *processes that transform information*" (see also Denning et al., 1989, p. 16). And Barwise (1989a, pp. 386–387) said that computers are best thought of as "information processors," rather than as numerical "calculators" or as "devices which traffic in formal strings … of meaningless symbols."

But contrary to Barwise, information processing is arguably nothing but symbol manipulation: after all, information has to be expressed in physical symbols, and symbols can be manipulated independently of their meaning (we'll go into this in more detail in Sections 16.11.2 and 18.8.3) But information processing is *interpreted* symbol manipulation. Moreover, not all symbol manipulation is necessarily information in some sense. So perhaps, although computers may be nothing but symbol manipulators, as Newell and Simon (1976) argue, it is as information processors that they have an impact. But what kind of information?

The influential theory of information due to Claude E. Shannon (1948) is purely "syntactic"; it is not concerned with the semantic meaning of the information. And the data structures textbook by Tenenbaum and Augenstein (1981, p. 6) claims that …

> … information itself has no meaning. Any meaning can be assigned to a particular bit pattern as long as it is done consistently. It is the interpretation of a bit pattern that gives it meaning.

(We'll return to their view in Section 13.2.3.)

And why constrain the algorithmic processes to only those that concern information? The algorithmic processes that undoubtedly underlie your use of Facebook on your laptop, tablet, or smartphone may not seem to be related to "information" in any technical sense. One answer might be found in an earlier (1963) statement by Forsythe:

---

20 Shakespeare, *Romeo and Juliet*, Act III, scene 1.

Machine-held strings of binary digits can simulate a great many *kinds* of things, of which numbers are just one kind. For example, they can simulate automobiles on a freeway, chess pieces, electrons in a box, musical notes, Russian words, patterns on a paper, human cells, colors, electrical circuits, and so on. (Forsythe, quoted in Knuth, 1972b, p. 722; see also Shannon, 1953, especially p. 1235; Hamming, 1980, pp. 7–8)

(This is an expression of one of the "Great Insights" of CS, which we will look at in Section 7.4.1.) What's common to all the items on Forsythe's list, encoded as (and thus simulated by) bit strings, is the information contained in them.

But here, the crucial question is, what is information? The term 'information' as many people use it informally has many meanings: it could refer to Shannon's mathematical theory of information, or Fred Dretske's (1981) or Kenneth Sayre's philosophical theories of information (1986), or several others. And if 'information' isn't intended to refer to some specific *theory*, then it seems to be merely a vague synonym for 'data' (itself a vague term!). As the philosopher Michael Rescorla observes, "Lacking clarification [of the term 'information'], the description [of "computation as 'information processing' "] is little more than an empty slogan" (Rescorla, 2020, Section 6.1). And the philosopher Gualtiero Piccinini has made the stronger claim that computation is distinct from information processing in *any* sense of 'information.' He argues, for example, that semantic information *requires* representation, but computation does *not*; so, computation is distinct from semantic information processing (Piccinini, 2015, Ch. 14, Section 3).

It is important to decide what information is, but that would take us too far afield. As I noted in Section 1.2, the philosophy of information is really a separate topic from (but closely related to!) the philosophy of computer science.[21]

Question for the Reader: Are there any kinds of algorithmic processes that manipulate something *other than* information? If there *aren't*, does that make this use of the term 'information' rather meaningless (as simply applying to everything that computers manipulate)? On the other hand, if there *are*, does that mean defining CS as the study of information is incorrect? (In Chapter 10, we'll look at some algorithms that apparently manipulate something other than information: namely, recipes that manipulate food.)

3.9   CS as a Mathematical Science

The concept of *computation* is arguably the most dramatic advance in *mathematical thinking* of the past century.
—Dennis J. Frailey (2010, p. 2, my italics)

To the extent that CS studies algorithms and an algorithm is a mathematical notion, could CS just be a branch of mathematics? Before we investigate whether CS is a mathematical science, let's ask another question: *is mathematics even a science at all?* As we saw in Section 2.5, sometimes a distinction is made between, on the one hand, *experimental* disciplines that investigate the physical world and, on the other, purely *logical* disciplines like mathematics. Let's assume, for the sake of argument, that mathematics is at least a special *kind* of science – a "formal" science – and let's consider whether CS might be more like mathematics than it is like empirical sciences.

---

21 See the Online Resources for further reading on information theory.

> **Terminological Digression: 'Formal':** This is as good a place as any to discuss the meaning of the word 'formal,' as it appears in phrases like 'formal logic' or 'formal science.' In this use, it relates to "form," "shape," or "structure" and is thus almost synonymous with 'syntactic.' It is *not* synonymous with words like 'prim' or 'methodical,' and it has nothing directly to do with concepts like "a formal dinner party." It is also not necessarily to be contrasted with 'informal' but rather to 'contentful.' See https://www.merriam-webster.com/dictionary/formal. On "formal" sciences in general, see http://en.wikipedia.org/wiki/Formal_science. We will go into much more detail on formal systems and syntax vs. semantics in Section 13.2.2.

Turing was a mathematician, and his celebrated 1936 paper that, according to many, created the field of CS was a successful attempt to solve a mathematical problem:

> Turing was born in 1912, and his undergraduate work at Cambridge during 1931–1934 was primarily mathematical. Turing machines were judged as *a mathematical interpretation of computational problem solving*; and computing was interpreted as *an entirely mathematical discipline*. (Wegner, 2010, p. 2, my italics)

And the theory of computational complexity is also clearly mathematical, as are other aspects of CS. Dijkstra (1974, p. 608) argues that "programming [i]s a mathematical activity." He doesn't explicitly say that (all) of CS is a branch of mathematics, but it is quite clear that large portions of CS – not only programming – can be considered branches of mathematics. Here is Dijkstra's argument (Dijkstra, 1974, p. 608):

1. A discipline *D* is a mathematical discipline if and only if *D*'s assertions are:
   (a) "unusually precise,"
   (b) "general in the sense that they are applicable to a large (often infinite) class of instances," and
   (c) capable of being reasoned about "with an unusually high confidence level."
2. Programming satisfies "characteristics" (1a)–(1c).
3. ∴ Programming is a mathematical discipline.

Dijkstra does not argue for premise (1). He takes the "only if" half (mathematical disciplines satisfy (1a)–(1c)) as something that "most of us can agree upon." And he implicitly justifies the "if" half (disciplines that satisfy (1a)–(1c) are mathematical) on the grounds that the objects of mathematical investigation need not be restricted to such usual suspects as sets, numbers, functions, shapes, etc., because what matters is *how* objects are studied, not *what* they are. However, the question of what mathematics is (and whether it is a science) is beyond our scope (but see Benacerraf and Putnam, 1984, Pincock, 2011, Horsten, 2019).

He argues for premise 2 on the grounds that programming clearly requires extraordinary precision, that programs can accept a wide range of inputs (and thus are general), and that contemporary program-verification techniques are based on logical reasoning. I can't imagine anyone seriously disagreeing with this! We will look into program-verification techniques in Chapter 15, so let's assume that programming satisfies 1a and 1c for now.

That leaves characteristic 1b: are programs really general in the same way that mathematical assertions are? A typical general mathematical assertion might be something like this: *for any triangle*, the sum of its angles is 180 degrees. The generality of mathematical assertions comes from their being "universally quantified" ("for any x …"). Is that the kind of generality that programs exhibit? A program (as we will see more clearly in Chapter 7) computes a (mathematical) function.

Insofar as mathematical functions are "general," so are programs. Consider a simple mathematical function: $f(x) = 2x$. If we universally quantify this,[22] we get: $\forall x[f(x) = 2x]$. This is general in the same way that our assertion about triangles was. An algorithm for computing f might look like this:[23]

> Let x be of type integer; **begin**
>   input(x);
> $f \leftarrow 2 * x$;
>   output(f)
> **end.**

The "preamble," which specifies the type of input, plays the role of the universal quantifier.[24] Thus, the *program* does seem to be general in the same way that a mathematical assertion is. So we can agree with Dijkstra about programming being mathematical.[25]

Mathematician Steven G. Krantz wrote that "Computer scientists, physicists, and engineers frequently do not realize that the technical problems with which they struggle on a daily basis are *mathematics*, pure and simple" (Krantz, 1984, p. 599). As a premise for an argument to the conclusion that CS is nothing but mathematics, this is obviously weak: after all, one could also conclude from it that physics and engineering are nothing but mathematics, a conclusion that I doubt Krantz would accept and that I am certain no physicist or engineer would accept.

Let's see if we can strengthen Krantz's premise. Suppose all the problems a discipline $D1$ (such as CS) is concerned with come from discipline $D2$ (such as mathematics). Does it follow that $D1$ is nothing but $D2$? (Does it follow that CS is nothing but mathematics?) Here's an analogy: if you want to express some literary idea, you can write a story or a poem. Does it follow that prose fiction and poetry are the same thing? Probably not; rather, prose and poetry are two different ways of solving the same problem (in our example, the problem of expressing a certain literary idea). Similarly, even if both CS and mathematics study the same problems, they do so in different ways: mathematicians prove (declarative) theorems; computer scientists express their solutions algorithmically.

So, perhaps a better contrast between CS and mathematics is that mathematics makes *declarative* assertions, whereas CS is concerned with *procedural* statements (Loui, 1987, p. 177; Knuth, 1974a, Section 3; Abelson et al., 1996). But is that distinction enough to show that CS is *not* mathematics? After all, Euclidean geometry – which is clearly mathematics – is procedural, not declarative. (We discuss this in further detail in Section 3.16.3.)

There is yet another way to think about the relationship between mathematics and CS:

> I think it is generally agreed that mathematicians have somewhat different thought processes from physicists, who have somewhat different thought processes from chemists, who have somewhat different thought processes from biologists. Similarly, the respective "mentalities" of lawyers, poets, playwrights, historians, linguists, farmers, and so on, seem to be unique. Each of these groups can probably recognize that other types of people have a different approach to knowledge; and it seems likely that a person gravitates to a particular kind of occupation according to the mode of thought that he or she grew up with, whenever a choice is possible. C.P. Snow wrote a famous book about "two cultures," scientific vs. humanistic, but in fact there seem to be many more than two. (Knuth, 1985, p. 171)

---

22 The inverted 'A' is logical notation for 'for all.'
23 The notation '$a \leftarrow b$' means "assign the value b to variable (or storage unit) a."
24 Technically, it is a "restricted" universal quantifier because it specifies the type of the variable. See, e.g. https://www.encyclopediaofmath.org/index.php/Restricted_quantifier.
25 See the Online Resources for further reading on Dijkstra.

There is a saying that, to a hammer, everything looks like a nail (http://en.wikipedia.org/wiki/ Law_of_the_instrument). This can be taken two ways: as a warning not to look at things from only one point of view, or as an observation to the effect that everyone has their own point of view. I take Knuth's remarks along the latter lines. And, of course, his eventual observation is that computer scientists look at the world algorithmically. Given the wide range of different points of view that he mentions, one conclusion could be that, just as students are encouraged to study many of those subjects so as to see the world from those points of view, so we should add algorithmic thinking – computer science – to the mix because of its unique point of view.

Knuth, on the other hand, is quite clear that he does *not* view CS as a branch of mathematics, or vice versa (Knuth, 1974b, Section 2), primarily because mathematics allows for *infinite* searches and *infinite* sets, whereas CS presumably does not. But there is no reason in principle why one couldn't write an algorithm to perform such an infinite search. The algorithm would never halt, but that is a *physical* limitation, not a theoretical one. (We'll return to infinite processes in Chapters 7 and 8.)[26]

3.10   CS as a Natural Science of *Procedures*

> So does nature compute, and does computation actually predate its invention, or rather discovery, by human beings? If it is the case, then this would actually lend credence to the claim that Computer Science is actually a science and not just and only a branch of engineering.
> —Erol Gelenbe (2011, p. 1)

Then there are those who agree that CS *is* a natural science but of neither computers, algorithms, *nor* information. Stuart C. Shapiro agrees with Newell, Perlis, and Simon that CS is a science, but he differs on what it is a science of, siding more with Knuth, but not quite:

> Computer Science is a **natural science** that studies **procedures**. (Shapiro, 2001, my boldface)

For Shapiro, CS *is* a *science*, which, like any science, has both theoreticians (who study the limitations on and kinds of, possible procedures) as well as experimentalists. And, as Newell and Simon (1976) suggest in their discussion of empirical results (see Section 3.11), there are "fundamental principles" of CS as a science. Newell and Simon cite two: (1) the Physical Symbol System Hypothesis that a computer "has the necessary and sufficient means for general intelligent action" (Newell and Simon, 1976, p. 116, col. 2) and (2) the Heuristic Search Hypothesis that computers solve problems intelligently by searching through symbol structures (Newell and Simon, 1976, p. 120, col. 2). Shapiro cites two others: (3) the Church-Turing Computability Thesis to the effect that any computation can be expressed as a Turing Machine program and (4) the Böhm-Jacopini Theorem that codifies "structured programming."

And although procedures are not natural *objects*, they are measurable natural *phenomena*, in the same way that events are not (natural) "objects" but are (natural) "phenomena." There are surely some procedures "in nature," such as a bird's procedure for building a nest. Dennett has …

> … argued that natural selection is an *algorithmic* process, a collection of sorting algorithms that are themselves *composed* of generate-and-test algorithms that exploit randomness … in the generation phase, and some sort of mindless quality-control

---

26 See the Online Resources for further reading on CS and mathematics.

testing phase, with the winners advancing in the tournament by having more offspring. (Dennett, 2017, p. 43)

And the computer scientist Peter Denning observed that "Computer science … is the science of information processes and their interactions with the world," adding that "There are many *natural* information processes" (Denning, 2005, p. 27, my emphasis). Denning (2007) cites examples of the "discovery" of "information processes in the deep structures of many fields": biology, quantum physics, economics, management science, and even the arts and humanities, concluding that "computing is *now* a natural science," not (or no longer?) "a science of the artificial." For example, there can be algorithmic (i.e. computational) theories or models of biological phenomena such as cells, plants, and evolution (see Section 16.6).[27]

For Shapiro, procedures include, but are not limited to, algorithms. Whereas algorithms are typically considered precise, to halt, and to produce correct solutions, the more general notion allows for variations on these themes:

1. Procedures (as opposed to algorithms) may be imprecise, such as in a recipe. Does *computer* science really study things like recipes? According to Shapiro (personal communication), the answer is 'yes': an education in CS should help you write a better cookbook because it will help you understand the nature of procedures better! However, Denning (2017, p. 38) says, "There is no evidence to support this claim." (We'll return to recipes in Chapter 10.)[28]
2. Procedures need not halt: a procedure might go into an infinite loop either by accident or, more importantly, on purpose, as in an operating system or a program that computes the infinite decimal expansion of π. (But see Chapter 8.)
3. Nor do they have to produce a correct solution: a chess procedure does not always play optimally. (We will return to these issues in Chapters 7 and 11.)

Moreover, Shapiro says that computer science is *not just* concerned with procedures that manipulate *abstract* information but also with procedures that are linked to sensors and effectors that allow computers to "sense and operate on *the world and objects in it*" (p. 3, my italics). The philosopher and AI researcher Aaron Sloman makes a similar point when he says that one of the "primary features" of computers (and of brains) is "Coupling to environment via physical transducers" (Sloman, 2002, Section 5, #F6, pp. 17–18). This allows for "perceptual processes that control or modify actions" and "is how internal information manipulation often leads to external behaviour." We'll return to this idea when we discuss interactive computation (Section 11.8) and the relation of computers to the world (Section 16.4.1).

Procedures are, or could be, carried out in the real world by physical agents, which could be biological, mechanical, electronic, etc. Where do computers come in? According to Shapiro, a computer is simply "a general-purpose procedure-following machine." (But does a computer "follow" a procedure or merely "execute" it? For some discussion of this, see Dennett, 2017, p. 70; we'll come back to this in Section 12.4.4.)

So, Shapiro's view seems to be a combination of Knuth's and Newell, Perlis, and Simon's: CS is the natural science of procedures and surrounding phenomena such as computers.

---

27 See the Online Resources for further reading on evolution and natural computation.
28 See the Online Resources for further reading on recipes.

3.11   CS as an *Empirical Study*

In a classic paper from 1976, Newell and Simon updated their earlier characterization. Instead of saying that CS is the *science* of computers and algorithms, they now said that it is the *"empirical"* *"study* of the phenomena surrounding computers," "not just the hardware but *the programmed, living machine"* (Newell and Simon, 1976, pp. 113, 114; my italics).

 The reason that they say that CS is not an "experimental" *science* is that it doesn't always strictly follow the scientific (or "experimental") method. (In Section 4.7, we'll talk more about what that method is. For an opposing view that CS *is* an experimental science, see Plaice, 1995.) CS is, like experimental sciences, *empirical* – because programs running on computers are *experiments*, though not necessarily like experiments in other experimental sciences. For example, often, just *one* experiment will suffice to answer a question in CS, whereas in other sciences, *numerous* experiments have to be run.

 Another difference between computer "science" and other experimental sciences is that, in CS, the chief objects of study (the computers and the programs) are not "black boxes" (Newell and Simon, 1976, p. 114); i.e. most natural phenomena are things whose internal workings we cannot see directly but must infer from experiments we perform on them. But we know exactly how and why computer programs behave as they do (they are "glass boxes," so to speak), because *we* (not nature) designed and built the computers and the programs. We can understand them in a way that we cannot understand more "natural" things. Although this is the case for "classical" computer programs, "A neural network, however, was a black box" (Lewis-Kraus, 2016, Section 4). (We'll return to this in Sections 3.14 and 17.6.2.)

 A distinction can be made between a *procedure* and a *process*: a *procedure* might be expressed in a static computer program or the static way that a computer is hardwired – a textual or physical implementation of an abstract algorithm or procedure. A *process* is a dynamic entity – the program in the "process" of actually being executed by the computer. By "programmed, living machines," Newell and Simon meant computers that are actually running programs – *not just* the static machines sitting there waiting for someone to use them, *nor* the static programs just sitting there on a piece of paper waiting for someone to load them into the computer, *nor* the algorithms just sitting there in someone's mind waiting for someone to express them in a programming language – but *"processes" that are actually running on a computer*. (We'll look at the program-process distinction in more detail in Chapter 12.)

 To study "programmed living machines," we certainly do need to study the algorithms that they are executing. After all, we need to know what they are doing; i.e. it seems to be necessary to know what algorithm a computer is executing. On the other hand, to study an algorithm, it does *not* seem to be *necessary* to have a computer around that can execute it or to study the computer that is running it. It can be helpful and valuable to study the computer and to study the algorithm actually being run on the computer, but the mathematical study of algorithms and their computational complexity doesn't *need* the computer. That is, the algorithm can be studied as a mathematical object, using only mathematical techniques, without necessarily executing it. It may be very much more convenient, and even useful, to have a computer handy, as Knuth notes, but it does not seem to be necessary. If that's so, then it would seem that *algorithms* are really the essential object of study of CS: both views require algorithms, but only one requires computers.

 Can you study computers without studying algorithms? Compare the study of computers with the study of brains and the nervous system. Although neuroscience studies both the anatomy of the brain (its static, physical structure) and its physiology (its dynamic activity), historically, at least, it has generally treated the brain as a "black box": its parts are typically named or described not in terms of what they *do* (their *function*), but in terms of *where they are located* (their *structure*).

For example, the "frontal lobe" is so-called because it is in the *front* of the brain; its *functions* include memory, planning, and motivation. The "temporal lobe" is so-called because it is near the *temples* on your head; its *functions* include processing sensory input. And the "occipital lobe" is so-called because it is near the occipital bone (itself so-called because it is "against" (*ob-*) the head (*caput*)); its *functions* include visual processing. (On the function-structure distinction, see Bechtel and Abrahamsen, 2005, Section 3.) Of course, modern neuroscience, especially modern *cognitive* neuroscience, well understands that it cannot fully understand the brain without understanding its processing (its algorithms, if indeed it executes algorithms) (Dennett, 2017, p. 341). Only recently have new maps of the brain begun to identify its regions *functionally*: i.e. in terms of what the regions do, rather than where they are located (Zimmer, 2016).

Suppose a person from the nineteenth century found what *we* know to be a laptop computer lying in the desert and tried to figure out what it was, how it worked, and what it did, with no documentation. They might identify certain physical features: a keyboard, a screen, internal wiring (if they were from the nineteenth century, they might describe these as buttons, glass, and strings), and so on. More likely, they would describe the device as we do the brain, in terms of the *locations* of the parts: an array of button-like objects on the lower half, a glass rectangle on the upper half, and so on. But without knowledge of what the entire system and each of its parts were supposed to *do* – what their *functions* were – they would be stymied. And that kind of knowledge about computers requires the study of algorithms. (We'll return to this in Section 3.14.)[29]

3.12   CS as *Engineering*

We have just looked at some reasons for classifying CS as a science of one kind or another. An alternative is that CS is *not a science* at all but a kind of *engineering*. For now, we will assume that engineering is, strictly speaking, something different from science. Again, a clearer answer to this will have to wait until Chapter 5, where we look more closely at what engineering is.

Frederick P. Brooks, Jr. – another Turing Award winner, perhaps best known as a software engineer – says that CS isn't science because, according to him, it is not concerned with the "discovery of facts and laws" (Brooks, 1996). Rather, he argues, CS is "an engineering discipline": computer scientists are "toolmakers," "concerned with *making things*" with physical tools such as computers and with abstract tools such as algorithms, programs, and software systems for others to use. CS, he says, is concerned with the usefulness and efficiency of the tools it makes; it is *not*, he says, concerned with newness for its own sake, as scientists are. Here is Brooks's argument:

1. "[A] science is concerned with the *discovery* of facts and laws."
   (Brooks, 1996, p. 61, col. 2)
2. "[T]he scientist *builds in order to study*; the engineer *studies in order to build*. (Brooks, 1996, p. 62, col. 1)
3. The purpose of engineering is to *build* things.
4. Computer scientists "are concerned with *making things*, be they computers, algorithms, or software systems" (Brooks, 1996, p. 62, col. 1).
5. ∴ "the discipline we call 'computer science' is in fact not a science but a *synthetic*, an engineering, discipline" (Brooks, 1996, p. 62, col. 1).

The accuracy of premise 1's notion of what science is will be our concern in Chapter 4. By itself, however, Brooks's first premise doesn't *necessarily* rule out CS as a science. First, computer scientists who study the mathematical theory of computation certainly seem to be studying scientific

---

29 See the Online Resources for further reading on CS's empirical nature.

laws. Second, computer scientists like Newell, Simon, and Shapiro have pointed out that Heuristic Search, the Physical Symbol System Hypothesis, the Computability Thesis, and the Böhm-Jacopini theorem certainly seem to be scientific theories, facts, or laws. And "Computer *programming* is an exact science in that all the properties of a program and all the consequences of executing it in any given environment can, in principle, be found out from the text of the program itself by means of purely deductive reasoning" (Hoare 1969, p. 576, my italics). (We'll look into this claim in more detail in Chapter 15.) So, it certainly seems that at least *part* of CS *is* a science. (We'll return to this in Section 3.15.) We'll assume the truth of the first premise for the sake of argument (revisiting it in the next chapter).

The point of the second premise seems to be to set up a distinction between science and engineering. If a scientist's goal is to discover facts and laws – i.e. to study rather than to build – then anything built by the scientist is only built for that ultimate purpose. But building is the ultimate goal of engineering, and any studying (or discovery of facts and laws) that an engineer does along the way to building something is merely done for that ultimate purpose. For science, building is a side effect of studying; for engineering, studying is a side effect of building. Both scientists and engineers, according to Brooks, build and study, but each focuses more on one than the other. (Does this remind you of the algorithms-vs.-computers dispute?) However, not every engineer agrees that the distinction is as sharp as Brooks suggests: the engineer Henry Petroski (2008) argues that all scientists are sometimes engineers, and all engineers are sometimes scientists.

Premise 2 supports premise 3, which is a missing premise that Brooks does not explicitly state. It defines engineering as a discipline whose goal is to build things: i.e. a "synthetic" – as opposed to an "analytic" – discipline. To *analyze* is to pull apart; to *synthesize* is to put together: "We speak of engineering as concerned with 'synthesis,' while science is concerned with 'analysis' " (Simon, 1996b, p. 4). "Where physical science is commonly regarded as an analytic discipline that aims to find laws that generate or explain observed phenomena, CS is predominantly (though not exclusively) synthetic, in that formalisms and algorithms are created in order to support specific desired behaviors" (Hendler et al., 2008, p. 63). Similarly, Arden (1980, pp. 6–7) argues that engineering is concerned with "implementation, rather than understanding," which "is the best distinction" between engineering and science. And implementation is surely on the "building" side of the spectrum (as we'll see in more detail in Chapter 13). Whether or not Brooks's notion of *engineering* is accurate will be our focus in Chapter 5. So, let's assume the truth of the second and third premises for the sake of argument.

Clearly, if premise 4 is true, then the conclusion will follow validly (or, at least, it will follow that computer scientists belong on the engineering side of the science–engineering, or studying–building, spectrum). So, is it the case that computer scientists are (only? principally?) concerned with building or "making things"? And, if so, what kind of things?

Interestingly, Brooks seems to suggest that computer scientists *don't* build computers, even if that's what he says in premise 4! Here's why: he says that "Even when we build a computer the computer scientist designs only the abstract properties – its architecture and implementation. Electrical, mechanical, and refrigeration engineers design the realization" (Brooks, 1996, p. 62, col. 1). I think this passage is a bit confused. (You'll understand why I say that when we look into the notion of implementation in Chapter 13. Briefly, I think the "abstract properties" *are* the design *for* the realization; the engineers *build* the realization – they don't *design* it.) But it makes an interesting point: Brooks seems to be saying that computer scientists only design *abstractions*, whereas other (real?) engineers *implement them in reality*. This is reminiscent of the distinction between the relatively abstract *specifications* for an algorithm (which typically lack detail) and its relatively concrete (and highly detailed) implementation in a computer *program* (we'll look into this in Chapter 10). Brooks (following Zemanek, 1971) calls CS "the engineering of abstract objects": if engineering is a discipline that builds, then what CS-considered-as-engineering builds is *implemented abstractions* (see Chapter 13 for further discussion).

In 1977, when he first wrote these words (see Brooks, 1996, p. 61, col. 1), very few people other than scientists, engineers, business people, and a few educational institutions had access to computing machines (typically, large mainframes or only slightly smaller "minicomputers") – certainly there were no personal computers (sometimes these used to be called "microcomputers") or laptops, tablets, or smartphones. So, for Brooks, what computer scientists build, unlike what other engineers build, are not things for direct human benefit but rather things that in turn can be used to build such directly beneficial things. Put more simply, his answer to the question "What is a computer?" seems to be: a computer is a tool (and a computer scientist, who makes such tools, is a "toolsmith") (Brooks, 1996, p. 62, col. 1).

But much of what he says *against* CS being considered a *science* smacks of a different battle, one between science and engineering, with scientists belittling engineers. Brooks takes the opposite position: "as we honor the more mathematical, abstract, and 'scientific' parts of our subject more, and the practical parts less, we misdirect young and brilliant minds away from a body of challenging and important problems that are our peculiar domain, depriving the problems of the powerful attacks they deserve" (Brooks, 1996, p. 62, col. 2).

(We'll come back to these issues in Section 5.9, question 2.)

3.13   Science *xor* Engineering?

So, is CS a science of some kind (natural or otherwise), or is it not a science at all but some kind of engineering? The term 'xor' in the title of this section refers to the "exclusive-or" of propositional logic: the title of this section means "science or engineering, *but not both*?" Here, we would be wise to listen to two skeptics about the exclusivity of this choice:

> Let's remember that there is only one nature – the division into science and engineering, and subdivision into physics, chemistry, civil and electrical, is a human imposition, not a natural one. Indeed, the division is *a human failure*; it reflects *our limited capacity to comprehend the whole*. That failure impedes our progress; it builds walls just where the most interesting nuggets of knowledge may lie. (Wulf, 1995, p. 56; my italics)

> Debates about whether [CS is] science or engineering can … be counterproductive, since we clearly are *both, neither, and more* …(Freeman, 1995, p. 27, my italics)

Let's consider Freeman's three options.

3.14   CS as "Both"

> [L]ike electricity, these phenomena [surrounding computers] belong both to engineering and to science.
> —Donald E. Knuth (1974b, p. 324)

> Computer science is both a scientific discipline and an engineering discipline. … The boundary [between "the division of computer science into theory" (i.e. science) "and practice" (i.e. engineering)] is a fuzzy one.
> —Paul Abrahams (1987, p. 472)

Could CS be *both* science *and* engineering – perhaps the *science* of computation and the *engineering* of computers – i.e. the study of the "programmed living machine"?

It certainly makes no sense to have a computer without a program: "A computer without a program is just a box with parts in it" (qFiasco, 2018, p. 38). It doesn't matter whether the program is *hardwired* (in the way that a Turing Machine is; see Section 8.12); i.e. it doesn't matter whether the computer is a *special*-purpose machine that can only do one task. Nor does it matter whether the program is a piece of *software* (like a program inscribed on a Universal Turing Machine's tape; see Section 8.13); i.e. it doesn't matter whether the computer is a *general*-purpose machine that can be loaded with different "apps" allowing the *same* machine to do many *different* things.

*Without a program, a computer wouldn't be able to **do** anything.*

But it also makes very little sense to have a program without a computer to run it on. Yes, you can study the program mathematically; e.g. you can try to verify it (see Chapter 15), and you can study its computational complexity. But what good would it be (for that matter, what fun would it be) to have, say, a program for passing the Turing Test that never had an opportunity to pass it? Hamming said,

> Without the [computing] machine almost all of what we [computer scientists] do would become idle speculation, hardly different from that of the notorious Scholastics of the Middle Ages. (Hamming, 1968, p. 5)

So, *without a computer, a **program** wouldn't be able to do anything.*
This is reminiscent of Immanuel Kant's slogan that

> Thoughts without content are empty, intuitions without concepts are blind. … The understanding can intuit nothing, the senses can think nothing. Only through their union can knowledge arise. (Kant, 1929, p. 93 (A51/B75))

Similarly, we can say, "Computers without programs are empty; programs without computers are blind. Only through the union of a computer with a program can computational processing arise."

Philosophical, Historical, and Literary Digression: In more modern terms, Kant can be understood as saying that (1) the part of the brain that *thinks* ("the understanding") doesn't *sense* ("intuit") the external world and that (2) the part of the brain (or nervous system) that *senses* cannot think. Thoughts have to be thoughts *about* something; i.e. they have to have "content." "The understanding" provides organizing principles ("concepts") in order to *think* about what is sensed. "The understanding" *by itself* doesn't sense the external world; the senses *by themselves* don't think. Only through the "union" of rational thought and empirical sensation "can knowledge arise." This was Kant's way of resolving the opposing views of the nature of knowledge due to the rationalist philosophers (Descartes, Leibiniz, and Spinoza) and the empiricist philosophers (Locke, Berkeley, and Hume). (Recall our discussion in Section 2.5, of the different kinds of "rationality.") For an informal presentation of some of Kant's ideas, see Cannon, 2013. For a more philosophically sophisticated introduction, see Rohlf, 2020. For more on what Kant meant by 'intuition,' see http://www.askphilosophers.org/question/204. We'll return to Kant in Sections 4.4.1 and 15.6.2.

A literary version of 'computers without programs are empty' is the legend of the Golem, a purely material statue that comes to life when certain Hebrew words are inscribed on it (https://www.jewishvirtuallibrary.org/the-golem). As Ted Chiang's (2002) story "Seventy-Two Letters" suggests, the linguistic text can be thought of as the computer program for a robot.

So, CS must be both a science (that studies algorithms) and an engineering discipline (that builds computers). But we need not be concerned with the two fighting words 'science' and 'engineering,' because, fortunately, there are two very convenient terms that encompass both: 'scientific' and 'STEM.' Surely both natural science and engineering, as well as "artificial science," "empirical studies," many of the social sciences, and mathematics, are all *scientific* (as opposed, say, to the arts and humanities). And, lately, both the National Science Foundation and the popular press have taken to referring to "STEM" disciplines – science, technology, engineering, and mathematics – precisely in order to have a single term to emphasize their similarities and interdependence, and to avoid having to try to spell out differences among them.[30]

Let's agree for the moment that CS might be *both* science *and* engineering (or, perhaps, can be divided into a science subdiscipline and an engineering subdiscipline). What about Freeman's other two options: *neither* and *more*? Let's begin with "more."

3.15 CS as "More"

Perhaps CS is science *together with being something else*, or perhaps CS is engineering plus something else. The computer scientist Juris Hartmanis takes the first position; the computer engineer Michael C. Loui takes the second.

3.15.1 CS as a *New Kind of Science*

> [C]omputer science differs from the known sciences so deeply that it has to be viewed as a new species among the sciences.
> —Juris Hartmanis (1993, p. 1)

Hartmanis comes down on the side of CS being a science, but it is a "new species *among the sciences*." What does it mean to be a "new species"? A clue comes in Hartmanis's next sentence:

> This view is justified by observing that theory and experiments in computer science play a different role and do not follow the classic pattern in physical sciences. (Hartmanis, 1993, p. 1)

This strongly suggests that CS is not a *physical* science (such as physics or biology), and Hartmanis confirms this suggestion on p. 5: "computer science, *though not a physical science*, is indeed a science" (my italics; see also Hartmanis, 1993, p. 6; Hartmanis, 1995a, p. 11). The non-physical sciences are typically taken to include both social sciences (such as psychology) and formal sciences (such as mathematics). This seems to put CS either in the same camp as (either) the social sciences or mathematics or in a brand-new camp of its own: i.e. *sui generis*.

Hartmanis said that he would not define CS (recall the epigraph to Section 3.4). But immediately after saying that, he seems to offer a definition:

> At the same time, it is clear that *the objects of study in computer science are information and the machines and systems which process and transmit information.* From this alone, we can

---

30 Nothing should be read into the ordering of the terms in the acronym: the original acronym was the less mellifluous 'SMET'! (See https://www.nsf.gov/pubs/1998/nsf98128/nsf98128.pdf.) And educators have been adding the arts to create 'STEAM' (http://stemtosteam.org/). On CS and art, see Section 3.16.1.

see that computer science is concerned with the abstract subject of information, which gains
reality only when it has a physical representation, and the man-made devices which process
the representations of information. The goal of computer science is to endow these information processing devices with as much intelligent behavior as possible. (Hartmanis, 1993,
p. 5, my italics; see also Hartmanis, 1995a, p. 10)

Although it may be "clear" to Hartmanis that information, an "abstract subject," is (one of) the
"objects of study in computer science," he does not share his reasons for that clarity. Since, as we
have seen, others seem to disagree that CS is the study of information, it seems a bit unfair for Hartmanis not to defend his view. But he cashes out this promissory note in Hartmanis 1995a (p. 10,
my italics), where he says that "what sets it [i.e. CS] apart from the other sciences" is that it studies
"processes [such as information processing] that are *not directly governed by physical laws*." And
why are they not so governed? Because "information and its transmission" are "abstract entities"
(Hartmanis, 1995a, p. 8). This makes CS sound very much like mathematics. That is not unreasonable, given that it was this aspect of CS that led Hartmanis to his ground-breaking work on
computational complexity, an almost purely mathematical area of CS.

But, says Hartmanis, it's not just information that is the object of study; it's also information-
processing *machines*: i.e. computers. Computers, however, don't deal directly with information,
because information is abstract (i.e. non-physical). For one thing, this suggests that insofar as CS
is a new species of *non-physical* science, it is not a species of *social* science: despite its name, the
"social" sciences deal with pretty physical things: societies, people, speech, etc. So if CS is a science
but is neither physical nor social, then perhaps it is a "formal" science like mathematics.

To say that computers don't deal directly with information but deal only with "physical ... *representations* of information" suggests that CS has a split personality: part of it deals directly with something *abstract* (information), and part of it deals directly with something *real* but that is (merely?)
a representation of that abstraction (hence dealing *indirectly* with information). Such real (physical?) representations are called "implementations." (These issues will be the topics of Chapters 13
and 16.)

Here is another reason Hartmanis thinks CS is not a physical science and probably also not a
social science:

> [C]omputer science is concentrating more on the *how* than the *what*, which is more the
> focal point of physical sciences. In general the *how* is associated with engineering, but com-
> puter science is **not** a subfield of engineering. (Hartmanis, 1993, p. 8; Hartmanis's italics,
> my boldface)

But there are two ways to understand "how": algorithms are the prime formal entities that codify
how to accomplish some goal. But as Hartmanis quickly notes, engineering is the prime discipline
that is concerned with how to do things, how to build things. The first kind of "how" is mathematical and abstract (indeed, it is *computational*! – see Sections 3.16.3 and 3.16.4); the second is
more physical. One way to see this as being consistent with Hartmanis's description of the objects
of study of CS is to say that, insofar as CS studies abstract information, it is concerned with how
to process information (i.e. it is concerned with algorithms); and insofar as CS studies computers,
it is concerned with how to process *representations* (or *implementations*) of information (i.e. it is
concerned with the physical devices).

But that latter task would seem to be part of engineering (perhaps, historically, electrical engineering; perhaps, in the future, quantum-mechanical or bioinformatic engineering; certainly computer engineering!). Then why does he say that "computer science is *not* a subfield of engineering"?

In fact, he seems to regret this strong statement, for he next says that "the engineering in our field has different characterizations than the more classical practice of engineering" (Hartmanis, 1993, p. 8): so, CS certainly *overlaps* engineering, but just as he claims that CS is a new species of science, he also claims that "it is a new form of engineering" (Hartmanis, 1993, p. 9). In fact, he calls it "[s]omewhat facetiously … the engineering of mathematics"; however, he also says that "we should not try to draw a sharp line between computer science and engineering" (Hartmanis, 1993, p. 9).

To sum up so far, Hartmanis views CS as a new species both of science and of engineering. This is due, in part, to his view that it has two different objects of study: an abstraction (namely, information) as well as its implementations (i.e. the physical representations of information, typically in strings of symbols). But isn't it also reasonable to think that, perhaps, there are really two different (albeit new) disciplines, namely, a new kind of science *and* a new kind of engineering? If so, do they interact in some way more deeply and integratively than, say, chemistry and chemical engineering, so that it makes sense to say that "they" are really a single discipline?

Hartmanis suggests two examples that show a two-way interaction between these two disciplines (or two halves of one discipline?): Alan Turing's interest in the *mathematical* nature of computation led to his development of *real* computers; and John von Neumann's interest in *building* computers led to his *theoretical* development of the structure of computer architecture (Hartmanis, 1993, p. 10). The computational logician J. Alan Robinson made similar observations:

> Turing and von Neumann not only played leading roles in the design and construction of the first working computers but were also largely responsible for laying out the general logical foundations for understanding the computation process, developing computing formalisms, and initiating the methodology of programming: in short, for founding computer science as we now know it. …
>
> Of course, no one should underestimate the enormous importance of the role of engineering in the history of the computer. Turing and von Neumann did not. They themselves had a deep and quite expert interest in the very engineering details from which they were abstracting, but they knew that the logical role of computer science is best played in a separate theater. (Robinson, 1994, pp. 5, 12)

Hartmanis explicitly says that CS *is* a science and is *not* engineering, but his comments imply that it is both. I don't think he can have it both ways.

3.15.2 CS as a *New Kind of Engineering*

> Hartmanis calls CS "a new species among the sciences" …. [I]t would be more accurate to call computer science a new species of engineering …
> —Michael C. Loui (1995, p. 31)

There are two parts to Loui's argument: CS is a *kind* of engineering, and it is a *new* kind. We have already looked at one argument (Brooks's) for the first part. (For an analysis of Loui's argument for it, see Rapaport, 2017c, Section 12.1.) Now let's look at Loui's argument for it being a *new* kind of engineering.

Here is his argument for this (Loui, 1995, p. 31, italics added):

1. "[E]ngineering disciplines have a scientific basis."
2. "The scientific fundamentals of computer science … are rooted … in mathematics."

3. "Computer science is therefore a *new* kind of engineering."

   This argument can be made valid by adding two missing premises:

   (A) Mathematics is a branch of science.
   (B) No other branch of engineering has mathematics as its basis.

We are assuming (for now) that CS *is* a *kind* of engineering. From that and premise 1, we can infer that CS (as an engineering discipline) must have a scientific basis. We need missing premise A so that we can infer that the basis of CS (which, by premise 2, is mathematics) is indeed a scientific one. What about premise B? Although mathematics can be considered essential to all (other) branches of engineering, it does appear that the scientific basis of "traditional engineering disciplines such as mechanical engineering and electrical engineering" is physics (Loui, 1987, p. 176), that the scientific basis of chemical engineering is chemistry, and so on, even though mathematics is essential to physics, chemistry, etc. Thus, CS is *mathematical engineering* – as Hartmanis suggested! – and therefore differs from all other branches of engineering.

Both Loui and Hartmanis agree that CS is a new kind of something or other; each claims that the scientific and mathematical aspects of it are central; and each claims that the engineering and machinery aspects of it are also central. But one *calls* it 'science,' while the other *calls* it 'engineering.' This is reminiscent of the dialogue between Newell, Perlis, and Simon on the one hand and Knuth on the other: extensionally equivalent but intensionally distinct. In fact, toward the end of his essay, Loui says this: "It is impossible to define a reasonable boundary between the disciplines of computer science and computer engineering. They are the same discipline" (Loui, 1987, p. 178).

3.16 CS as "Neither"

> It seems that the broad field of computing (which is probably a better name than computer science) has aspects that are similar to many other fields – mathematics, statistics, logic, natural science, social science, engineering, linguistics, philosophy, various fine arts, etc. – and also different from all other fields. Perhaps, to borrow a phrase from Dijkstra, computing is a "radical novelty." Maybe we shouldn't try to shoehorn computing into any preexisting category. It is just what it is.
> —H. Conrad Cunningham, ACM SIGCSE mailing list (4 February 2021)

Perhaps CS is something else altogether: an **art**, or the study of **complexity**, or a branch of **philosophy**, or a **way of thinking**, or **AI**, or **magic**(!).

3.16.1 CS as *Art*

Knuth titled his multi-volume classic *The Art of Computer Programming* (Knuth, 1973), defending his use of the word 'art' in the title not by saying that all of CS is an art but that 'art' can be applied to, at least, computer programming. I will let you decide whether CS is an art. We'll return to this briefly in Section 12.3.2. (Relevant readings are in the Online Resources.)

3.16.2 CS as the Study of *Complexity*

> [T]he art of programming is the art of organising complexity, of mastering multitude and avoiding its bastard chaos as effectively as possible.
> —Edsger W. Dijkstra (1972, p. 6)

It has been suggested that CS is the study of *complexity* – not just the mathematical subject of "computational complexity," which is really more a study of *efficiency* – but complexity in general and in all of nature. Ceruzzi (1988, pp. 268–270) ascribes this to the electrical engineer and MIT president Jerome Wiesner (1958). But all Wiesner says is that "Information processing systems are but one facet of … communication sciences … i.e. the study of …the problems of organized complexity' " (quoted in Ceruzzi, 1988, p. 269). Even if computer science is part of a larger discipline ("communication sciences"?) that studies complexity, it doesn't follow that CS itself *is* the study of complexity. Again, I will let the reader investigate this. (Relevant readings are in the Online Resources.)

3.16.3 CS as the *Philosophy*(!) of Procedures

At least one introductory CS text claims that CS is neither a science nor the study of computers (Abelson et al., 1996, "Preface to the First Edition"). Rather, it is what the authors call 'procedural epistemology':

> the study of the structure of knowledge from an *imperative* point of view, as opposed to the more *declarative* point of view taken by classical mathematical subjects. Mathematics provides a framework for dealing precisely with notions of "what is." Computation provides a framework for dealing precisely with notions of "how to." (Italics added.)

Epistemology, recall, is the branch of *philosophy* that studies knowledge and belief (Section 2.7).

"How to" is certainly important and interestingly distinct from "what is." But is there really a difference between "how to" and "what is"? Many imperative statements can equally well be expressed as declarative ones: consider, for example, Lisp programs, which appear to be merely declarative definitions of recursive functions. Or consider that each "$p :- q$" rule of a Prolog program can be interpreted either procedurally ("to achieve p, execute q") or declaratively ("p if q").

Or consider Euclid's *Elements*, which was originally written in "how to" form (Toussaint, 1993): to construct an equilateral triangle using only compass and straightedge, follow this algorithm (http://www.perseus.tufts.edu/hopper/text?doc=Perseus:text:1999.01.0086:book=1:type=Prop:number=1). Compare: to compute the value of this function *using only the operations of a Turing Machine*, follow this algorithm. (For further discussion of the "to accomplish goal G, do algorithm A" formula, see Section 16.5.) But today, geometry is expressed in "what is" form: the triangle that is constructed by following that algorithm is equilateral: "When Hilbert gave a modern axiomatization of geometry at the beginning of the present century, he asserted the bald existence of the line. Euclid, however, also asserted that it can be constructed" (Goodman, 1987, Section 4). (We'll return to this topic in Section 10.3.) Note that the declarative version of a geometry theorem can be considered a formal proof of the correctness of the procedural version. This is closely related to the notion of program verification, which we'll look at in Chapter 15.

Even if procedural language can be intertranslated with declarative language, the two are surely distinct. And, just as surely, CS *is* concerned with procedures! So, we need to be clearer about what we mean by 'procedure' as well as phrases like 'computational thinking' and 'algorithmic thinking.'

3.16.4 CS as *Computational Thinking*

A currently popular view is to say that CS is a "way of thinking':' that "computational," or "algorithmic," or "procedural" thinking – about anything(!) – is what makes CS unique:

CS is the new "new math," and people are beginning to realize that CS, like mathematics, is unique in the sense that many other disciplines will have to adopt that *way of thinking*. It offers a sort of conceptual framework for other disciplines, and that's fairly new. ... Any student interested in science and technology needs to learn *to think algorithmically*. That's the next big thing. (Bernard Chazelle, interviewed in Anthes, 2006, my italics)

Computer scientist Jeannette Wing's notion of "computational thinking" (Wing, 2006, echoing Papert, 1980) is thinking in such a way that a problem's solution "can effectively be carried out by an information-processing agent" (Wing, 2010; see also Guzdial, 2011). Here, it is important not to limit such "agents" to computers but to include humans (as Wing (2008a, p. 3719) admits).

Five years before Perlis (along with Newell and Simon) defined CS as the science of *computers*, he emphasized what is now called *computational thinking*:

> [T]he purpose of ... [a] first course in programming ... is not to teach people how to program a specific computer, nor is it to teach some new languages. *The purpose of a course in programming is to teach people how to construct and analyze processes.* ... The point is not to teach the students how to use [a particular programming language, such as] ALGOL, or how to program [a particular computer, such as] the 704. These are of little direct value. The point is *to make the students construct complex processes out of simpler ones* (and this is always present in programming) in the hope that the basic concepts and abilities will rub off. A properly designed programming course will develop these abilities better than any other course. (Perlis, 1962, pp. 209–210, my italics)

Some of the features of computational thinking that various people have cited include abstraction, hierarchy, modularity, problem analysis, structured programming, the syntax and semantics of symbol systems, and debugging techniques. (Note that all of these are among the methods for handling complexity!)

Denning (2009, p. 33) also recognizes the importance of computational thinking. However, he dislikes it as a *definition* of CS, primarily on the grounds that it is too narrow:

> Computation is present in nature even when scientists are not observing it or thinking about it. Computation is more fundamental than computational thinking. For this reason alone, computational thinking seems like an inadequate characterization of computer science. (Denning, 2009, p. 30)

A second reason Denning thinks defining CS as computational thinking is too narrow is that there are other equally important forms of thinking: "design thinking, logical thinking, scientific thinking, etc." (Denning et al., 2017).[31]

3.16.5 CS as *AI*

> Computation ... is the science of how machines can be made to carry out *intellectual processes*.
> —John McCarthy (1963, p. 1, my italics)

---

31 See the Online Resources for further reading on computational thinking.

The goal of computer science is to endow these information processing devices with as much *intelligent behavior* as possible.
—Juris Hartmanis (1993, p. 5, my italics) (cf. Hartmanis, 1995a, p. 10)

Computational Intelligence *is* the manifest destiny of computer science, the goal, the destination, the final frontier.
—Edward A. Feigenbaum (2003, p. 39)

These aren't exactly definitions of CS, but they could be turned into one: computer science – note: CS, *not* AI! – is the study of how to make computers "intelligent" and how to understand cognition computationally.

As we will see in more detail in Chapter 6, the history of computers supports this: it is a history that began with how to get machines to do *some* human thinking (in particular, certain mathematical calculations) and then more and more. And (as we will see in Chapter 8) the Turing Machine model of computation was motivated by how *humans* compute: Turing (1936, Section 9) analyzed how humans compute and then designed what we would now call a computer program that does the same thing. But the branch of CS that analyzes how humans perform a task and then designs computer programs to do the same thing is AI. So, the Turing Machine was the first AI program! But *defining* CS as AI is probably best understood as a special case of its fundamental task: determining what tasks are computable.

3.16.6 Is CS *Magic*?

[T]he computing scientist could not care less about the specific technology that might be used to realize machines, be it electronics, optics, pneumatics, or magic.
—Edsger W. Dijkstra (1986)

To engender empathy and create a world using only words is the closest thing we have to magic.
—Lin-Manuel Miranda (2016)[32]

The great science-fiction author Arthur C. Clarke famously said that "Any sufficiently advanced technology is indistinguishable from magic" (http://en.wikipedia.org/wiki/Clarke's_three_laws). Could it be that the advanced technology of CS is not only indistinguishable from magic but really *is* magic? Not magic as in tricks, but magic as in Merlin or Harry Potter? Brooks makes an even stronger claim than Clarke:

The programmer, like the poet, works only slightly removed from pure thought-stuff. He [sic] builds castles in the air, creating by the exertion of the imagination .... Yet the program construct, unlike the poet's words [or the magician's spells?], is real in the sense that it moves and works, producing visible outputs separate from the construct itself. ... **The magic of myth and legend has come true in our time.** *One types the correct incantation on a keyboard, and a display screen comes to life, showing things that never were nor could be.* (Brooks, 1975, pp. 7–8, my boldface).[33]

---

32 https://www.nytimes.com/2016/04/10/books/review/lin-manuel-miranda-by-the-book.html
33 For an illustration of "things that never were nor could be," see https://www.google.com/books/edition/From_Animals_to_Animats_3/kcMoUj3aIfoC?hl=en&gbpv=1&printsec=frontcover.

What is "magic"? Here's how one anthropologist defines it:

> In anthropology, magic generally means beliefs in the use of symbols to control forces in nature … (Stevens, 1996, p. 721, col. 1)

> A definition of magic can be constructed to say something like the following: Magic involves the human effort to manipulate the forces of nature directly, through symbolic communication and without spiritual assistance. (Stevens, 1996, p. 723, col. 2).[34]

Clearly, programming involves exactly that kind of use of symbols. Or, as Abelson and Sussman put it in their introductory CS text,

> A computational process is indeed much like a sorcerer's idea of a spirit. It cannot be seen or touched. It is not composed of matter at all. However, it is very real. It can perform intellectual work. It can answer questions. It can affect the world by disbursing money at a bank or by controlling a robot arm in a factory. *The programs we use to conjure processes are like a sorcerer's spells.* They are carefully composed from symbolic expressions in arcane and esoteric programming languages that prescribe the tasks we want our processes to perform. (Abelson et al., 1996, my italics) (https://web.archive.org/web/20010727165536/https://mitpress .mit.edu/sicp/full-text/book/book-Z-H-9.html)

How is magic supposed to work? Anthropologist James G. Frazer (1915) "had suggested that primitive people imagine magical impulses traveling over distance through 'a kind of invisible ether' " (Stevens, 1996, p. 722, col. 1). That sounds like a description of electrical currents running from a keyboard to a CPU, or information traveling across the Internet, or text messaging. According to another anthropologist, Bronisław Malinowski,

> The magical act involves three components: the *formula*, the *rite*, and the condition of the *performer*. The rite consists of three essential features: the dramatic expression of emotion through *gesture* and physical attitude, the use of objects and substances that are *imbued with power by spoken words*, and, most important, the *words* themselves. (Stevens, 1996, p. 722, col. 2, my italics; citing Malinowski)

A "wizard," *gesturing* with a "wand," *performs* a "spell" consisting of a *formula* expressed in the *words* of an arcane language; the spell has real-world effects, *imbuing objects with power*. We see all of this in computing: *programs* play the role of spells; the *programmer* plays the role of the wizard; *a mouse, trackpad, or touchscreen* plays the role of the wand; *programming languages* (or, in the case of Siri or Alexa, English itself) play the role of the arcane language; and *computations* are "powered" by "words" with real-world effects.

Here is another aspect of the role of symbols in magic:

> [A symbol] can take on the qualities of the thing it represents, and it can take the place of its referent; indeed, as is evident in religion and magic, *the symbol can become the thing it represents*, and in so doing, the symbol takes on the power of its referent. (Stevens, 1996, p. 724, my italics)

---

34 For more on definitions of 'magic,' see Stairs, 2014.

We see this happening in computers when we treat desktop icons (which are symbols) or the screen output of a WYSIWYG word processor (such as a page of a Microsoft Word document) as if they were the very things they represent. More significantly, we see this in the case of those computer simulations in which the simulation of something really *is* that (kind of) thing: in online banking, the computational simulation of transferring funds between accounts *is* the transferring of funds; digitized signatures on online Word or PDF documents carry legal weight. And a National Research Council report (cited by Samuelson et al., 1994, p. 2324, notes 44, 46; p. 2325, note 47) talks about user interfaces as "illusions":

> Unlike physical objects, the virtual objects created in software are not constrained to obey the laws of physics. ... In the desktop metaphor, e.g. the electronic version of file folders can expand, contract, or reorganize their contents on demand, quite unlike their physical counterparts. (Samuelson et al., 1994, p. 2334)

Isn't that magic?

However, there *is* a difference between computing and "the magic of myth and legend": the latter lacks (or at least fails to specify) any causal (i.e. physical) connection between incantation and result, whereas computation is quite clear about the connection – recall the emphasis on algorithms. Thus, although CS may have the outward appearance of magic and even accomplish (some of) the things that magic accomplishes, the way it does it is different. CS has a method; magic does not. Actually, CS has more in common with magic *tricks* than with "real" magic:

> "I'm writing a book on magic," I explain, and I'm asked, "Real magic?" By real magic people mean miracles, thaumaturgical acts, and supernatural powers. "No," I answer: "Conjuring tricks, not real magic." *Real magic, in other words, refers to the magic that is not real, while the magic that is real, that can actually be done, is not real magic.* (Lee Siegel, quoted in Dennett, 2017, p. 318, my italics)

Magic tricks require intermediary steps that accomplish the illusions of magic. In a "Rhymes with Orange" comic,[35] a student magician waves a wand in front of a math problem, and the answer magically appears; the student's teacher says, "No relying on the wand – I want to see how you arrived at the right answer." Put another way, magic does what it does magically; CS does those things computationally:

> Everything going on in the software [of a computer] has to be physically supported by something going on in the hardware. Otherwise the computer couldn't do what it does from the software perspective – **it doesn't work by magic.** But usually we don't have to know how the hardware works – only the engineer and the repairman do. We can act as though the computer just carries out the software instructions, period. **For all we care, as long as it works, it might as well** *be* **magic.** (Jackendoff, 2012, p. 99, my boldface, italics in original)

3.17 Summary

Is CS a science, a branch of engineering, or something else? Or all of the above? Does it study computers, algorithms, information, or something else? Or all of the above? In the next two chapters, we will look more deeply at science and engineering. We will then look further into the nature of computers and algorithms.

---

35 Behind a paywall at https://comicskingdom.com/rhymes-with-orange/2017-10-26.

3.18 Questions for the Reader

1. Many of the definitions of CS that you can find on various academic websites are designed with one or more of the purposes discussed in Section 3.3 in mind. Link to the websites for various CS departments (including your own school's!), and make a list of the different definitions or characterizations of CS that you find.

   Which purposes were they designed for? Do you agree with them? Do they agree with each other? Are any of them so different from others that you wonder if they are really trying to describe the same discipline?

2. In Section 3.14, I said that it makes no – or very little – sense to have a program without a computer to run it on. Some of the earliest AI programs (for playing chess) were executed by hand (Shannon, 1950, Turing, 1953; https://chessprogramming.wikispaces.com/Turochamp). And journalist Steve Lohr (2008) quotes a high-school mathematics and CS teacher as saying, "I do feel that computer science really helps students understand mathematics … And I would use computers more in math, if I had access to a computer lab." That a computer is useful, but not necessary, is demonstrated by the "Computer Science Unplugged" project (http://csunplugged .org/).

   So, *did* these programs "have a computer to run on"? Were the humans, who hand-executed them, the "computers" that these programs "ran on"? When you debug a computer program, do you do the debugging by hand?[36]

3. Forsythe (1967b, p. 454), observed that,

   > in the long run the solution of problems in field X on a computer should belong to field X, and CS should concentrate on finding and explaining the principles ["the methodology"] of problem solving [with computers].

   Should contributions made by AI researchers to philosophy or psychology be considered the results of AI? Or are they philosophical or psychological results that were only *produced* or *facilitated* by computational techniques?

4. In this chapter, we asked what CS is: Is it a science? A branch of engineering? Or something else? But we could also have responded to the question with another one: *Does it matter?* Is it the case that, in order for a discipline to be respectable, it has to be (or claim to be!) a science? Or is it the case that a discipline's usefulness is more important? (For instance, whether or not medicine is a science, perhaps what really matters is that it is a socially useful activity that draws upon scientific – and other! – sources.)[37]

   So: does it matter what CS is? And what would it mean for a discipline to be "useful"?

5. A related (but distinct) question is, what is a computer scientist? Bill Gasarch considers a number of reasons why the answer to this question is not straightforward (https://blog .computationalcomplexity.org/2018/09/what-is-physicist-mathematician.html): Does it depend on whether the person is in a CS department? Whether the person's degree is in CS? What the person's research is? For example, the computer scientist Scott Aaronson received a prize in physics, yet he insists that he is not a physicist (Aaronson, 2018).

   Read Gasarch's post, and try to offer some answers. (We'll return to this issue in Section 14.3.4.)

---

36 Thanks to Stuart C. Shapiro for this suggestion.
37 Thanks to Johan Lammens (personal communication, 2017) for the observations in this question.

4

Science

Science is the great antidote to the poison of enthusiasm and superstition.
—Adam Smith (1776, V.1.203)

The most remarkable discovery made by scientists is science itself. The discovery must be compared in importance with the invention of cave-painting and of writing. Like these earlier human creations, science is an attempt to control our surroundings by entering into them and understanding them from inside. And like them, science has surely made a critical step in human development which cannot be reversed. We cannot conceive a future society without science.
—Jacob Bronowski (1958, my italics)

[A] science is an evolving, but never finished, interpretive system. And fundamental to science … is its questioning of what it thinks it knows. … Scientific knowledge … is a system for coming to an understanding.
—Avron Barr (1985)

Science is *all about* the fact that we don't know everything.
Science is the learning process.
—Brian Dunning (2007)

[S]cience is not a collection of truths. It is a continuing exploration of mysteries.
—Freeman Dyson (2011b, p. 10)

4.1 Introduction

All these processes are very complex, and they tend to follow the rule that the more you find out about them, the more you discover that you didn't know …. That is both the joy and the frustration of science ….
—Gregory L. Murphy (2019, Section 1)

Recall from Chapter 3 that I am referring to computer science as 'CS' so as not to beg any questions about whether it is a science simply because its name suggests that it is. Nonetheless, we have seen that one answer to our principal question – what is CS? – is that it *is* a science (or that parts of it are). Some say that it is a science of computers, some that it is a science of algorithms or procedures, some that it is a science of information processing. And, of course, some say that it is not a science at all but a branch of engineering or something else entirely. In Chapter 5, we will explore what engineering is so that we will have more information to help us decide whether CS is a branch of engineering. In the present chapter, we will explore what it means to be a science, to help us decide whether CS is one (or whether parts of it are).

In keeping with the definition of philosophy as the *personal search* for truth by rational means (Section 2.6), I will provide considerations to help *you* find *and defend* an answer that *you* like. It is more important for you to determine an answer *for yourself* than it is for me to present you with *my* view; this is part of what it means to do philosophy *in* the first person *for* the first person. And it is very important for you to be able to *defend* your answer; this is part of what it means to be rational. We will follow this strategy throughout the rest of the book.

4.2 Science and Non-Science

According to the *OED* (http://www.oed.com/view/Entry/172672), 'science' derives from the Latin verb *scire*, which means "to know." ('Scientist' was coined by the philosopher William Whewell in 1834, to parallel 'artist'; http://www.oed.com/view/Entry/172698.) But of course, 'science' has come to mean much more than "knowledge."

Let's begin by contrasting 'science' with some other terms. First, of course, science is often opposed to engineering. Because this will be our focus in Chapter 5, I won't say more about it here.

Second, science is sometimes opposed to "art," not only in the sense of the fine arts (such as painting and music) but also in the sense of an informal body of experiential knowledge, or tricks of the trade: information that is the result of personal experience, perhaps unanalyzable (or, at least, unanalyzed), and creative. This is "art" in the sense of "the art of cooking." By contrast, science is formal, objective, and systematic. This contrast can be seen in the titles of two classic texts in CS: Donald Knuth's 1973 The **Art** of Computer Programming and David Gries's 1985 The **Science** of Programming. The former is a multi-volume handbook of different techniques, cataloged by type and analyzed. The latter is a compendium of formal methods for program development and verification, an application of logic to programming. (For a detailed defense of the title of Knuth's work, see Knuth, 1974a.)

Finally, science is opposed (both semantically and philosophically) to "pseudo-science": any discipline that masquerades as science but is not science. The problem of determining the dividing line between "real" science and "pseudo"-science is called the 'demarcation problem.' For example, almost everyone will agree that *astronomy* is a "real" science and that *astrology* is not. But what is the difference between "real" and "pseudo"-sciences? We will return to this in Section 4.8, because explaining the contrast between science and pseudo-science is part of the philosophical exploration of what science is.

One might think the philosophy of science would be the place to go to find out what science is, but philosophers of science these days seem to be more interested in questions such as the following (the first two of which are the closest to our question):

- What is a scientific theory?
  (Here, the emphasis is on the meaning of the term 'theory.')
- What is a scientific explanation?
  (Here, the emphasis is on the meaning of the term 'explanation.')
- What is the role of probability in science?
- What is the nature of induction? (Why) will the future resemble the past?

- What is a theoretical term? That is, what do the terms of (scientific) theories mean? Do they necessarily refer to something in the real world? For example, there used to be a scientific concept in the theory of heat called 'phlogiston,' but we no longer think this term refers to anything.
- How do scientific theories change? When they do, are their terms "commensurable" – i.e. do they mean the same thing in different theories? For example, what is the relationship between 'phlogiston' and 'heat'? Does 'atom,' as used in ancient Greek physics, or even nineteenth century physics, mean the same as 'atom' as used in twenty-first century physics?
- Are scientific theories "realistic": do they attempt to describe the world? Or are they merely "instrumental"? That is, are they just very good predicting devices that don't necessarily bear any obvious resemblance to reality, as sometimes seems to be the case with our best current theory of physics – namely, quantum mechanics?

And so on.

These are all interesting and important questions, and it is likely that a good answer to our question "What is science?" will depend on answers to many of them. If so, then a full answer will be well beyond our present scope, and the interested reader is urged to explore a good book on the philosophy of science. Here, we will only be able to consider a few of these questions.[1]

4.3   Science as Systematic Study

Sir Francis Bacon – a contemporary of Shakespeare who lived about 400 years ago (1561–1626) – devised one of the first "scientific methods." He introduced science as a *systematic study*. (So, when you read about computer scientists who call CS a "study" rather than a "science," maybe they are not trying to deny that CS is a science but are merely using a euphemism.)

> [Bacon] told us to ask questions instead of proclaiming answers, to collect evidence instead of rushing to judgment, to listen to the voice of nature rather than to the voice of ancient wisdom. (Dyson, 2011a, p. 26)

He emphasized the importance of "replicability":

> Replicability begins with the idea that science is not private; researchers who make claims must allow others to test those claims. (Wainer, 2012, p. 358)

Perhaps science is merely *any systematic* activity, as opposed to a *chaotic* one. There is a computer program called AlphaBaby, designed to protect your computer from young children who want to play on the computer but might accidentally delete all of your files while randomly hitting keys. AlphaBaby's screen is blank; when a letter or numeral key is hit, a colorful rendition of that letter or numeral appears on the screen; when any other key is hit, a geometric figure or a photograph appears. Most children hit the keys randomly ("chaotically") rather than systematically investigating which keys do what ("scientifically").

Timothy Williamson (2011) suggests something similar when he characterizes the "scientific spirit" as "emphasizing values like curiosity, honesty, accuracy, precision and rigor." And the magician and skeptical investigator known as The Amazing Randi said, "Science, after all, is simply a logical, rational and careful examination of the facts that *nature* presents to us" (quoted in Higginbotham, 2014, p. 53, my italics). Although Shapiro and Denning (Section 3.10), would be happy with

---

1 See the Online Resources for further reading on the philosophy of science.

the word 'nature' here, mathematicians who think of their discipline as a "formal" science might not be. But I think it can be eliminated without loss of meaning and still apply to both mathematics and computer "science." (For further discussion of this aspect of science in the context of whether both philosophy and CS are sciences, see Section 3.18, Question 4.)

Studying something, X, systematically includes

- finding positive and negative instances of X – things that are *are* Xs and things that are *not*;
- making changes in Xs or their environment (i.e. doing experiments);
- observing Xs and the effects of experiments performed with them;
- finding correlations (and perhaps causal relationships) among Xs, their behavior, and various aspects of their environment.

4.4 The Goals of Science

At least three different things have been identified as the goals of science: *description*, *explanation*, and *prediction*. They are not independent of each other: at the very least, you need to be able to describe things in order to explain them or predict their behavior. But they are distinct: a theory that predicts doesn't necessarily also explain (for some examples, see Piccinini, 2015, p. 94).

4.4.1 Description

Ernst Mach was a physicist and philosopher of science who lived about 130 years ago (1838–1916), at the time when the atomic theory was being developed. He was influenced by Einstein's theory of relativity and is probably most famous for having investigated the speed of sound (which is now measured in "Mach" numbers).

For Mach, the goal of science was to discover regular patterns among our sensations to enable the prediction of future sensations, and then to *describe* those patterns in an efficient manner. Scientific theories, he argued, are (merely) *shorthand – or summary – descriptions* of how the world *appears* to us.

According to one version of the philosophical position known as "physicalism," our sensory perception yields reliable (but corrigible)[2] knowledge of ordinary, medium-sized physical objects and events. For Mach, because atoms were not observable, there was no reason to think that they exist. Perhaps it seems odd to you that a physicist would be interested in our *sensations* rather than in the *world* outside of our sensations. This makes it sound as if science should be done "in the first person, for the first person," just like philosophy! That's almost correct; many philosophically oriented scientists at the turn of the last century believed that science should begin with observations, and what are observations but our sensations? Kant distinguished between what he called 'noumena' (or "things in themselves," independent of our concepts and sensations) and what he called 'phenomena' (or things as we perceive and conceive them as filtered through our conceptual apparatus; recall Section 3.14). He claimed that we could only have knowledge about phenomena, not noumena, because we could not get outside of our first-person, subjective ways of conceiving and perceiving the world. This is why some philosophers of science have argued that sciences such as quantum mechanics are purely instrumental and only concerned with prediction, rather than being realistic and concerned with the way the world "really" is. (We'll come back to this in Sections 4.4.3 and 4.5, and in Section 15.6.2 when we discuss the relation of computer programs to the world.)[3]

---

2 That is, "correctable."
3 See the Online Resources for further reading on physicalism and Kant.

4.4.2 Explanation

By contrast, the atomic theory was an attempt to *explain* why the physical world appears the way it does. Explanatory theories are not merely *descriptive* summaries of our observations but try to *account* for them, often by going beyond observations to include terms that refer to things (like atoms) that we might not be able to observe (yet). On this view, the task of science is not, in spite of Mach, merely to *describe* the complexity of the world in simple terms but to *explain* the world:

> This is the task of natural science: to show that the wonderful is not incomprehensible, *to show how it can be comprehended* … . (Simon, 1996b, p. 1, my italics)

One major theory of the nature of scientific explanation is the philosopher Carl Hempel's Deductive-Nomological Theory (Hempel, 1942, 1962). It is "deductive" because the statement that some object c has property Q is explained by showing that it can be validly deduced from two premises: that c has property P and that all Ps are Qs. And it is "nomological" because the fact that all Ps are Qs is lawlike or necessary, not accidental: anything that is a P *must* be a Q. (This blending of induction and deduction is a modern development; historically, Bacon [and other "empiricists," chiefly in Great Britain] emphasized experimental "induction and probabilism," while Descartes [and other "rationalists," chiefly on the European continent] emphasized "deduction and logical certainty" (Uglow, 2010, p. 31).)

One of the paradoxes of explanation (or the "paradox of analysis") is that by showing how something mysterious or wonderful or complicated is really just a complex structure of simpler things that are non-mysterious or mundane, we lose sight of the original thing that we were trying to understand or analyze. We will see this again in Section 18.9 when we look at Dennett's notion of Turing's "strange inversion." It is also closely related to the notion of recursion (see Section 7.4.3.2), where complex things are defined in terms of simpler ones. Herbert Simon demurs:

> … the task of natural science … [is] to show how it [the wonderful] can be comprehended – *but not to destroy wonder. For when we have explained the wonderful, unmasked the hidden pattern, a new wonder arises at how complexity was woven out of simplicity.* (Simon, 1996b, pp. 1–2, my italics)

So, for instance, the fact – if it is a fact (explored in Chapter 18) – that non-cognitive computers can exhibit (or even merely simulate) cognitive behaviors is itself something worthy of wonder and further (scientific) explanation.

4.4.3 Prediction

> … prediction is always the bottom line. It is what gives science its empirical content, its link with nature. … This is not to say that prediction is the *purpose* of science. It was once … when science was young and little; for success in prediction was … the survival value of our innate standards of subjective similarity. But prediction is only one purpose among others now. A more conspicuous purpose is technology, and an overwhelming one is satisfaction of pure intellectual curiosity – which may once have had its survival value too.
> —Willard van Orman Quine (1987, p. 162)

Mach thought that the job of science was to *describe* the world. Einstein "thought the job of physics was to give a complete and intelligible account of … [the] world" (J. Holt, 2016, p. 50) – i.e. to *explain* the world. Both scientific descriptions and explanations of phenomena enable us to make *predictions* about their future behavior. This stems, in part, from the fact that scientific descriptions must be *general* or *universal* in nature: they must hold for all times, including future times. As the philosopher Moritz Schlick put it,

> For the physicist ... the absolutely decisive and essential thing, is that the equations derived from any data **now** also hold good of *new* data. (Schlick, "Causality in Contemporary Physics" (1931), quoted in Coffa, 1991, p. 333; my boldface, Schlick's italics)

Thus, "[t]he 'essential characteristic' of a law of nature 'is the *fulfillment of predictions*' " (Coffa, 1991, p. 333, embedded quotations from Schlick).

According to Hempel (1942, Section 4), prediction and explanation are opposite sides of the same coin. As we saw in the previous section, to *explain* an event is to find (perhaps abductively) one or more "initial conditions" (usually, earlier events) and one or more general laws such that the event to be explained can be deduced from them. For Hempel, to *predict* an event is to use already-known initial conditions and general laws to deduce a future event:

> The customary distinction between explanation and prediction rests mainly on a pragmatical difference between the two: While in the case of an explanation, the final event is known to have happened, and its determining conditions have to be sought, the situation is reversed in the case of a prediction: here, the initial conditions are given, and their "effect" – which, in the typical case, has not yet taken place – is to be determined. (Hempel, 1942, p. 38)

But some scientists and philosophers hold that prediction is the *only* goal that is important, and that description and explanation are either not important or impossible to achieve. One of the main reasons for this comes from quantum mechanics. Some aspects of quantum mechanics are so counter-intuitive that they seem to fail both as descriptions of reality as we think we know it and as explanations of that reality: for example, according to quantum mechanics, objects seem to be spread out rather than located in a particular place – until we observe them; there seems to be "spooky" action at a distance (quantum entanglement); and so on. Yet quantum mechanics is the most successful scientific theory (so far) in terms of the predictions it makes. Niels Bohr (one of the founders of quantum mechanics) said "that quantum mechanics was meant to be an *instrument for predicting* our observations," neither a description of the world nor an explanation of it (J. Holt, 2016, p. 50, my italics).[4]

4.5 Instrumentalism vs. Realism

The explanation-vs.-prediction debate underlies another issue: is there a world to be described or explained? That is, do scientific theories tell us what the world is "really" like, or are they just "instruments" for helping us get around in it? Here's a simplified way of thinking about what a scientific theory is. We can begin by considering two things: the *world* and our *beliefs about* it (including our descriptions of it). Those beliefs are theories about what the world is like. Such theories are *scientific* if they can be tested by empirical or rational evidence to see if they are true (i.e. correspond to what the world is really like). A theory is *confirmed* if it can be shown that it is

---

4 See the Online Resources for further reading on quantum mechanics.

consistent (or "coheres") with the way the world really is. And a theory is *refuted* if it can be shown that it is not the way the world really is (that it does not "correspond" to reality).

A picture might help (Figure 4.1)

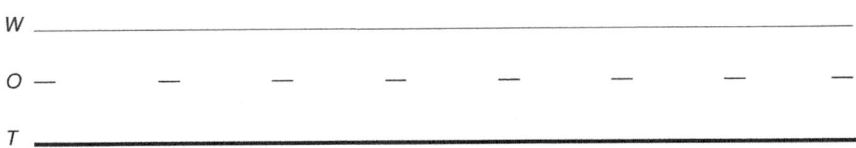

Figure 4.1 World, Observations, Theory.

Line *W* (a continuous line) is intended to represent the world, a continuum. Line *O* (a line with gaps) is intended to represent observations that we can make about the world: some parts of the world we have observed (or we can observe), and they are represented in *O* by the line segments; others we have not observed (or we cannot observe), and those are the gaps. The solid lines in *O* represent things we believe about the world; the gaps represent things we don't know (yet) about the world. Line *T* is intended to represent a scientific theory about the world (about line *W*); here, the gaps are filled in. Those fillings-in are predictions about what the world is like at those locations where we cannot observe it; they are guesses (hypotheses) about the world.

Suppose we have an explanatory scientific theory of something (e.g. atomic theory). Such theories, as we have seen, often include "unobservables" – terms referring to things that we have not (yet) observed (e.g. atoms or quarks) but whose existence would help explain things that we *have* observed. One way of looking at this is to think of an experiment as taking some input (perhaps some change deliberately made to some entity being studied) and observing what happens after the experiment is over – the output of the experiment. Between the input and the output, something happens, but we don't necessarily know what it is. It is as if what we are studying is a "black box," and all we can observe are its inputs and outputs. A scientific theory (or, for that matter, a computer algorithm!) can be viewed as an explanation of what's going on inside the black box. Can it be viewed merely as a *description* of what's going on inside? Probably not, because you can only describe what you can observe, and, by hypothesis, we can't observe what's going on inside the black box. Such a theory will usually involve *unobservables* structured in various ways.

Do the unobservables that form part of such an explanatory theory really exist? If you answer 'yes,' then you are a "realist"; otherwise, you are an "instrumentalist." A realist believes in the real existence of explanatory unobservables. An instrumentalist believes they are merely useful tools (or "instruments") for making predictions. The debate between realism and instrumentalism is as old as science itself. Galileo (1564–1642) …

> … and the Church came to an implicit understanding: if he would claim his work only as "*istoria*," and not as "*dimonstrazione*," the Inquisitors would leave him alone. The Italian words convey the same ideas as the English equivalents: a new story about the cosmos to contemplate for pleasure is fine, a demonstration of the way things work is not. You could calculate, consider, and even hypothesize with Copernicus. You just couldn't believe in him. (Adam Gopnik 2013, p. 107)

In Mach's time, it was not clear how to treat the atomic theory. Atoms were clearly of instrumental value, but there was no observable evidence of their existence. But they were so useful scientifically that it eventually became unreasonable to deny their existence, and, eventually, they were observed.

In our time, black holes moved from being "merely" theoretical entities to being considered among the denizens of the universe, despite not having been observed directly.

Quantum mechanics poses a similar problem. If the world really is as quantum mechanics says it is, then the world is really weird. But quantum mechanics is our best current theory about how the world is. So, possibly quantum mechanics is merely a useful calculating tool for scientific prediction and shouldn't be taken literally as a description of the real world.

> **Digression and a Look Ahead:** One kind of instrumentalism is related to "syntactic understanding." There are (at least) two ways to understand something: (1) you can understand something *syntactically* in terms of something else that you are more familiar with, and (2) you can understand something *semantically* in terms of itself, by being very familiar with it directly (Rapaport, 1986, 1995). The physicist Jeremy Bernstein has said that there is "a misguided but humanly understandable desire to explain quantum mechanics by something else – something more familiar. But if you believe in quantum mechanics there is nothing else" (Bernstein and Holt, 2016, p. 62). On Bernstein's instrumentalist view, quantum mechanics can only be understood in terms of itself.
>
> Syntactic understanding (perhaps of the kind that Bernstein says we have of quantum mechanics) is a "base case" of understanding. Semantic understanding is a "recursive" case of understanding. (We'll discuss recursion in Section 7.4.3.2.) However, *other* things might be understandable in terms of quantum mechanics: recent research in cognitive science suggests that quantum-mechanical methods applied at the macroscopic level might provide better explanations of certain psychological findings about human cognition than more "standard" methods (Wang et al., 2013). We'll return to these two kinds of understanding in Sections 13.2.2 and 18.8.

Can an instrumentalist theory evolve into a realist one?

> Though Galileo … wants to convince … [readers] of the importance of looking for yourself, he also wants to convince them of the importance of *not* looking for yourself. The Copernican system is counterintuitive, he admits – the Earth certainly doesn't seem to move. It takes intellectual courage to grasp the argument that it does. (Adam Gopnik, Adam, 2013, p. 107)

So, just as the Copernican theory, initially proposed merely as an instrumentalist claim, became a realist-explanatory theory, eventually the quantum-mechanical view of the world may come to be accepted as a realist description.[5]

4.6 Scientific Theories

It is important to distinguish between the everyday sense of 'theory' and the scientific sense. In the *everyday sense*, a "theory" is merely an idea; it may or may not have any evidence to support it. In this everyday sense, 'theory' can be contrasted with 'fact.' In the *scientific sense*, a "theory" is a set of statements (1) that describe, explain, or predict some phenomenon, often formalized mathematically or logically (or even computationally, as we'll see in Chapter 14), *and* (2) that are grounded in empirical or logical evidence. (The terms 'theory' and 'theorem' are etymologically

---

5 See the Online Resources for further reading on instrumentalism vs. realism.

related.) To be "scientific," a theory must be accompanied by confirming evidence, *and* (as we'll see in Section 4.8) its statements must be precise enough to be capable of being falsified. The best theory is one that is (1) consistent, (2) as complete as possible (i.e. explains as much as possible), and (3) best-supported by good evidence.

Anti-evolutionists (creationists and advocates of "intelligent design") sometimes criticize the scientific theory of evolution as "merely a theory." Anyone who does so is confusing the everyday sense (in which 'theory' is opposed to 'fact') with the scientific sense. Evolution *is* a theory in the scientific sense. (It is also a fact! More precisely, the world really is as the scientific theory of evolution describes it.) Gravity, too, is not "just a theory" (https://www.gocomics.com/nonsequitur/2014/11/07).

The scientific notion of theory comes in (at least) two varieties: syntactic and semantic. We will have a lot to say about syntax and semantics in Section 13.2.2. For now, let's just say that regarding the *syntactic* approach to scientific theories,

> a theory was conceived of as an axiomatic theory. That means, as a set of sentences, defined as the class of logical consequences of a smaller set, the axioms of that theory. (van Fraassen, 1989, p. 220)

By contrast, the *semantic* approach to scientific theories focuses on the *models* that interpret those sentences and "link their terms with their intended domain" (van Fraassen, 1989, p. 221). Just as there is a syntactic vs. a semantic view of scientific theories, there is a syntactic vs. a semantic view of computer programs (see Chapter 16). We will return to this topic in Chapter 14, where we will consider whether computer programs can be scientific theories.[6]

4.7 "The" Scientific Method

> [T]here is no such thing as *the* scientific method. Case studies of particular theories in physics, biology, etc., have convinced me that no one paradigm can fit all of the various inquiries that go under the name of 'science.'
> —Hilary Putnam (1987, p. 72)

> … "the scientific method" originated[,] not in any field or practice of science, but in the popular, professional, industrial, and commercial exploitation of its authority. This exploitation crucially involved the insistence that science held an exclusive monopoly on truth, knowledge, and authority, a monopoly for which "the scientific method" was a guarantee. … I would call it a feat of branding equal to "diamonds are forever" or "Coke is it" … This is not to deny, of course, that the sciences include procedures of observation, controlled experimentation, and analysis, and that these procedures are crucial to the progress of scientific understanding. But no list of four or five discrete steps can describe them, and they don't operate … [by] carrying the scientist inexorably toward transcendent truth.
> —Jessica Riskin (2020, pp. 49–50)

People often talk about "the scientific method." Either there isn't any such thing (as Putnam said) or there are *many* scientific methods (plural) of studying something (Kitcher, 2019): (some) biologists and astronomers use (some) different methods from (some) physicists. And disciplines besides the natural sciences (notably mathematics and engineering but also the social sciences and even many of the humanities) also use scientific methods (Blachowicz, 2016; Ellerton, 2016).

---

6 See the Online Resources for further reading on evolution and on scientific theories.

But let's look at one version of a scientific method, a version that is interesting in part because it was described by the mathematician John Kemeny, who was also a computer scientist. (He was co-inventor of the BASIC computer programming language and helped develop time sharing. He also worked with Einstein and was president of Dartmouth College.) His book *A Philosopher Looks at Science* presents the scientific method as a cyclic procedure (Kemeny, 1959, Chs. 5, 10). Because cyclic procedures are called 'loops' in computer programming, I will present Kemeny's version of the scientific method as an infinite loop (a non-halting algorithm):

Algorithm Scientific-Method[7]
begin
    while there is a new fact to observe, **do:**
        {That is, **repeat until** there are no new facts to observe.
        This will never happen, so we have *permanent inquiry*.}
    begin
    1. *observe* things & events;
        {Express these observations as descriptive statements about
        particular objects a, b, ..., such as: $Pa \rightarrow Qa$, $Pb \rightarrow Qb$, ...
        Observations may be "theory-laden," i.e. based on assumptions.}
    2. *induce* general statements;
        {Make summary descriptions, such as: $\forall x[Px \rightarrow Qx]$}
    3. *deduce* future observations;
        {Make predictions, such as: $Pc/\therefore Qc$}
    4. *verify* predictions against observations;
        {**if** Qc
            then general statement is confirmed or is consistent with theory
            else revise theory (or ...)}
    end
end.

Kemeny's version of the scientific method is a loop consisting of observations, followed by inductive inferences, followed by deductive predictions, followed by verifications. (Perhaps a better word than 'verification' is 'confirmation'; we'll discuss this in Section 4.8.) The scientist begins by making individual observations of specific objects and events and describes these in language: object a is observed to have property P, object a is observed to have property Q, object a's having property P is observed to be correlated with object a's having property Q, and so on. Next, the scientist uses inductive inference to infer from a series of events of the form Pa precedes Qa, Pb precedes Qb, etc. that whenever any object x has property P, it will also have property Q. So, the scientist who observes that object c has property P will deductively infer (i.e. will predict) that object c will also have property Q – *before observing whether it does*. The scientist will then perform an experiment to see whether Qc. If Qc is observed, then the scientist's theory that $\forall x[Px \rightarrow Qx]$ will be verified; otherwise, the theory will need to be revised in some way (see Section 2.5.1; we'll discuss this in more detail in Section 4.8). For Kemeny, an observation is explained by means of a deduction from a theory, following Hempel's deductive-nomological theory (Section 4.4.2).

Finally, according to Kemeny, **a discipline is a science if and only if it follows the scientific method.** This rules out astrology on the grounds that astrologers never verify their predictions.

---

7 This "algorithm" is written in the informal pseudocode introduced in Ch. 3. Expressions in {braces} are comments.

(Or on the grounds that their predictions are so vague that they are always trivially verified. See Section 4.8.)

However, there are at least two other views of the nature of science that generally agree on the distinctions between science as opposed to art, engineering, and pseudo-sciences such as astrology but that differ on the nature of science itself: Popper's theory of falsifiability and Kuhn's theory of scientific revolutions, to which we now turn.[8]

4.8 Falsifiability

> Science is always wrong .... Science can never solve one problem without raising ten more problems.
> —George Bernard Shaw, 1930, https://quoteinvestigator.com/2021/12/21/science-ten/

4.8.1 Science as Conjectures and Refutations

According to the philosopher Karl Popper (1902–1994), the "real" scientific method sees science as a sequence of *conjectures* and *refutations* (Popper, 1953; cf. Popper, 1959 and, on engineering, Popper, 1972):

1. Conjecture a theory (to explain some phenomenon).
2. Compare its predictions with observations
   (i.e. perform experiments to test the theory).
3. **If** an observation differs from a prediction,

   then the theory is *refuted* (or *falsified*)
   else the theory is *confirmed.*

It is important to note that 'confirmed' *does not mean "true"*! Rather, it means we have evidence that is consistent with the theory (recall the coherence theory of truth, Section 2.3.2) – i.e. the theory is *not yet falsified*! This is because there might be some *other* explanation for the predicted observation. Just because a theory T predicts that some observation O will be made, and that observation is indeed made, it does not follow that the theory is true! This is because the Fallacy of Affirming the Consequent (Section 2.5.1) –

A) $O, (T \rightarrow O) \nvdash_D T$

– is an invalid argument. If O is true, but T is false, then the second premise is still true, so we could have true premises and a false conclusion. This might also be called the fallacy of circumstantial evidence, where O is the circumstantial evidence that could support T, but there might be *another* theory that *also* predicts O and that *is* true.

So, according to Popper, **a theory or statement is scientific if and only if it is falsifiable.** By 'falsifiable,' Popper meant something like "capable of being falsified *in principle*," not "capable of being falsified with the techniques and tools that we *now* have available to us."

For Popper, falsifiability also ruled out astrology (and other superstitions), Freudian psychotherapy, and Marxist economics as candidates for scientific theories. The reason Popper claimed that astrology, etc. were only pseudo-sciences was that they cannot be falsified because they are too vague. The vaguer a statement or theory is, the harder it is to falsify. As the physicist Freeman Dyson once wrote, "Progress in science is often built on wrong theories that are later corrected. It

---

8 See the Online Resources for further reading on scientific methods.

is better to be wrong than to be vague" (Dyson, 2004, p. 16). When I was in college, one of my friends came into my dorm room, all excited about an astrology book he had found that, he claimed, was really accurate. He asked me what day I was born; I said "September 30." He flipped the pages of his book, read a horoscope to me, and asked if it was accurate. I said that it was. He then smirked and told me that he had read me a random horoscope, for April 16. The point was that the horoscope for April 16 was so vague that it also applied to someone born on September 30![9]

4.8.2   The Logic of Falsifiability

It is worthwhile to explore the logic of falsifiability a bit more. Although the Fallacy of Affirming the Consequent seems to describe what goes on, it needs to be made more detailed because it is not the case that scientists deduce predictions from *theories alone*. There are usually *background beliefs* that are independent of the theory being tested (e.g. beliefs about the accuracy of one's laboratory equipment). And one does not usually test a complete theory T but merely one new *hypothesis H* that is being considered as an addition to T. So it is not simply that argument (**A**) should have as a premise that theory T predicts observation O. Rather, theory T *conjoined with background beliefs B, conjoined with the actual hypothesis H* being tested, is supposed to logically predict that O will be observed:

$$(T \ \& \ B \ \& \ H) \to O$$

Suppose O is not observed:

$$\neg O$$

What follows from these two premises? By the rule of inference called 'Modus Tollens,' we can infer

$$\neg(T \ \& \ B \ \& \ H)$$

But from this, it follows (by De Morgan's Law) that

$$\neg T \ \lor \ \neg B \ \lor \ \neg H$$

That is, either T is false, *or B is false, or H* is false, *or any combination of them* is false. This means if you strongly believe in your theory T that seems to be inconsistent with your observation O, *you do not need to give up T*. Instead, you could give up hypothesis H, or some *part* of T, or (some part of) your background beliefs B (e.g. you could blame your measuring devices as being too inaccurate). (For a detailed example, see Horsman et al., 2014, Section 5.) As Quine (1951) pointed out, you could even give up the laws of logic if the rest of your theory has been well confirmed; this is close to the situation that obtains in contemporary quantum mechanics with the notion of "quantum logic."

However, sometimes you *should* give up an entire theory. This is what happens in the case of "scientific revolutions," such as (most famously) when Copernicus's theory that the Earth revolves around the Sun (and not vice versa) replaced the Ptolemaic theory, small revisions to which were making it overly complex without significantly improving it. (See Section 4.9.)[10]

---

9 See the Online Resources for further reading on pseudo-science, and see the "Calvin and Hobbes" comic at http://www.gocomics.com/calvinandhobbes/2012/04/20.
10 See the Online Resources for further reading on rules of inference and quantum logic.

4.8.3 Problems with Falsifiability

One problem with falsifiability is that not all alleged pseudo-sciences are vague. Is astrology really a Popperian pseudo-science? Although the popular newspaper style of astrology no doubt is (on the grounds of vagueness), "real" astrology, which might be considerably less vague, might actually turn out to be testable and, presumably, falsified, hence falsifiable. But that would make it scientific (albeit false)!

That points to another problem with falsifiability as the mark of science: Are false statements scientific? This is related to the "pessimistic meta-induction" that all statements of science are false.[11] But this isn't quite right: although it might be the case that any given statement of science that is currently held to be true may turn out to be false, it doesn't follow that all such statements *are* false or will eventually be found to be false. What does follow is that all statements of science are *provisional*:

> Newton's laws of gravity, which we all learn in school, were once thought to be complete and comprehensive. Now we know that while those laws offer an accurate understanding of how fast an apple falls from a tree or how friction helps us take a curve in the road, they are inadequate to describe the motion of subatomic particles or the flight of satellites in space. For these, we needed Einstein's new conceptions.
>
> Einstein's theories did not refute Newton's; they simply absorbed them into a more comprehensive theory of gravity and motion. Newton's theory has its place, and it offers an adequate and accurate description, albeit in a limited sphere. As Einstein once put it, "The most beautiful fate of a physical theory is to point the way to the establishment of a more inclusive theory, in which it lives as a limiting case." It is this continuously evolving nature of knowledge that makes science always provisional. (Natarajan, 2014, pp. 64–65)

4.9 Scientific Revolutions

In Section 3.5.2, we talked briefly about "scientific paradigms." Thomas Kuhn (1922–1996), a historian of science, rejected both the classic scientific method and Popper's falsifiability criterion. Kuhn (1962, Ch. 9) claimed that the history of science shows that the real scientific method works as follows:

1. There is a period of "normal" science based on a "paradigm" – roughly, a generally accepted theory. During that period of normal science, a Kemeny-like or Popper-like scientific method is in operation. Dyson (2004, p. 16) refers to the "normal" scientists as "conservatives ... who prefer to lay one brick at a time on solid ground."
2. If that paradigmatic theory is challenged often enough, there will be a "revolution," and a new theory – a new paradigm – will be established, completely replacing the old one. Dyson (2004, p. 16) refers to the "revolutionaries" as "those who build grand castles in the air."
3. A new period of normal science follows, now based on the new paradigm, and the cycle repeats.

The most celebrated example of a scientific revolution was the Copernican revolution in astronomy (Kuhn, 1957). "Normal" science at the time was based on Ptolemy's "paradigm" of an Earth-centered theory of the solar system. But this was so inaccurate that its advocates had to keep patching it up to make it consistent with observations. Copernicus's new paradigm – the heliocentric theory that we now believe – overturned Ptolemy's paradigm. Other scientific revolutions

---

11 See the Online Resources for further reading on the pessimistic meta-induction.

include those of Newton (who overthrew Aristotle's physics), Einstein (who overthrew Newton's), Darwin (whose theory of evolution further "demoted" humans from the center of the universe), Watson and Crick ("whose discovery of the … structure of DNA … changed everything" in biology (Brenner, 2012, p. 1427)), and Chomsky in linguistics (even though some linguists and cognitive scientists today think Chomsky was wrong (Boden, 2006)).[12]

4.10 Other Alternatives

> [T]raditional views about how science is carried out are often idealized or simplistic. Science proceeds in anything but a linear and logical fashion.
> —Lawrence M. Krauss (2016, p. 85)

Besides the triumvirate of Bacon's or Kemeny's scientific method, Popper's falsificationism, and Kuhn's scientific revolutions, there are other approaches to the nature of science. For instance, the philosopher of science Michael Polanyi argued in favor of science as being "socially constructed," not purely rational or formal (see Kaiser, 2012 for an overview). And the philosopher of science Paul Feyerabend critiqued the rational view of science from an "anarchic" point of view (Feyerabend, 1975; Preston, 2020). But exploration of alternatives such as these is beyond our scope.[13]

4.11 CS and Science

4.11.1 Is CS a Science?

These are only a handful among many views of what science is. Is CS a science according to any of them? This is a question that I will leave to the reader to ponder. But here are some things to consider:

Does CS follow Kemeny's scientific method? For that matter, does *any* science (like physics, chemistry, or biology) really follow it? Does *every* science follow it (what about astronomy or cosmology)? Or is it just an idealized vision of what scientists are supposed to do?

Is CS scientific in Popper's or Kuhn's sense? Are any parts of it falsifiable (Popper)? Have there been any revolutions in CS (Kuhn)? Is there even a current Kuhnian paradigm?

The philosopher Timothy R. Colburn (2000, p. 168) draws an analogy between (1) the scientific method of formulating, testing, and (dis)confirming *hypotheses* and (2) the problem-solving method of CS consisting of formulating, testing, and accepting or rejecting an *algorithm*. Besides suggesting that CS is (at least in part) scientific, this analogizes algorithms to scientific hypotheses or theories. (See Chapter 14 for further discussion.) Even if this is just an idealization, does CS even come close? What kinds of theories are there in CS? How are they tested? If CS is a science, is it "provisional"? (Nelson Pole has suggested[14] that "if there is a bug lurking in every moderately complex program, then *all* programs are provisional.") Are any computer-science theories ever refuted? Similarly, Denning (2005, p. 28) says that "The scientific paradigm … is the process of forming hypotheses and testing them through experiments; successful hypotheses become models that explain and predict phenomena in the world." He goes on to say, "Computing science follows

---

12 See the Online Resources for further reading on scientific revolutions.
13 See the Online Resources for further reading on other alternatives.
14 Private communication, 9 March 2015.

this paradigm in studying information processes." For readers who are studying CS, think about your own experiences. Do you agree with Denning that CS follows this scientific method?

Here are two other issues to think about. First, the Church-Turing Computability Thesis identifies the informal notion of computation with formal notions like the Turing Machine (as we'll see in more detail in Chapters 7 and 8). "Hypercomputation," a name given to various claims that the informal notion of computation goes beyond Turing Machine computability, could be considered an attempt to falsify the Computability Thesis (Kaznatcheev, 2014). Or the hypercomputation challenges to the Computability Thesis could be examples of Kuhnian revolutionary paradigmatic challenges to the "normal" science of CS (Cockshott and Michaelson, 2007, Section 2.5, p. 235). And Stepney et al., 2005 offer a long list of paradigms that they think can and should be challenged. Keep this in mind when you read Chapter 11 on hypercomputation.

Second, two traditions in AI have been logically oriented: knowledge-based AI (sometimes called "Good Old-Fashioned AI" (GOFAI) Haugeland, 1985); and connectionist AI, which is based on "artificial neural networks" instead of on logic. Although the former dominated AI research in the early days and, arguably, still has an important role to play (Levesque, 2017, Seabrook, 2019; B.C. Smith, 2019; Landgrebe and Smith, 2021), most AI now is based on the latter. When three connectionist researchers (Geoffrey Hinton, Yann LeCun, and Yoshua Bengio) received the Turing Award, another AI researcher (Oren Etzioni) said, "What we have seen is nothing short of a paradigm shift in the science. History turned their way, and I am in awe" (quoted in Metz, 2019a). Keep this in mind when you read Chapter 18 on AI.

4.11.2 What Kind of Science Might CS Be?

What about disciplines like mathematics? Mathematics is certainly scientific in some sense, but is it a science like physics or biology? Is CS, perhaps, more like mathematics than like these (other) sciences? This raises another question: even if CS is a science, what kind of science is it?

Hempel (1966) distinguished between *empirical* sciences and *non-empirical* sciences. The former explore, describe, explain, and predict various occurrences in the world. Such descriptions or explanations are empirical statements that need empirical (i.e. experimental) support. The empirical sciences include the *natural* sciences (physics, chemistry, biology, some parts of psychology [e.g. cognitive psychology], etc.) and the *social* sciences (other parts of psychology [e.g. clinical and social psychology], sociology, anthropology, economics, perhaps political science, perhaps history, etc.).

The non-empirical sciences are logic and mathematics. Their statements don't need empirical support. Yet they are true and confirmed by empirical evidence (although exactly how and why this is the case is still a great mystery).

Is CS an empirical science? A non-empirical science? CS arose from logic and mathematics. But it also arose from the development of the computer as a tool to solve logic and mathematics problems. (We will explore this twin history of computers and algorithms in Chapter 6.) This brings it into contact with the empirical world and empirical concerns such as space and time limitations on computational efficiency (or "complexity").

One possible way of adding CS to Hempel's taxonomy is to take a cue from the fact that psychology doesn't neatly belong to just the natural or just the social sciences. So, perhaps CS doesn't neatly belong to just the empirical or just the non-empirical sciences, but parts of it belong to each. And it might even be the case that the non-empirical aspects of CS are not simply a third kind of non-empirical science, on a par with logic and mathematics, but are themselves parts of both logic and of mathematics.

Or it might be the case that we are barking up the wrong tree altogether. What if CS isn't a science at all? This possibility is what we turn to in the next chapter.[15]

4.12 Questions to Think About

1. Hempel's empirical–non-empirical distinction may be an arbitrary division of a continuous spectrum (Section 3.4.1):

   > The history of science is partly the history of an idea that is by now so familiar that it no longer astounds: the universe, including our own existence, can be explained by the interactions of little bits of matter. *We scientists are in the business of discovering the laws that characterize this matter.* We do so, to some extent at least, by a kind of reduction. The stuff of biology, for instance, can be reduced to chemistry and the stuff of chemistry can be reduced to physics. (Orr, 2013, p. 26, my italics)

   This view, sometimes called the "unity of science" (Oppenheim and Putnam, 1958), can be extended at both ends of the spectrum that Orr mentions: at one end,

   > if physics was built on mathematics, so was chemistry built on physics, biology on chemistry, psychology on biology, and … sociology … on psychology … .
   > (Grabiner, 1988, p. 225, citing Comte, 1830, Vol. I, Ch. 2, Introduction)

   At the other end, mathematics is built on logic and set theory (Quine, 1976) (see Figure 4.2). However, not everyone thinks this chain of reductions is legitimate (Fodor, 1974).

   Does CS fit into this sequence? If it doesn't, does that mean it's not part of science? After all, it's not obvious that CS is "in the business of discovering the laws that characterize … matter." Wheeler, 1989 suggests that the universe's matter consists of information. Then, if you are also willing to say that CS is the science of information (or of information processing), you could conclude that it is a (physical) science.

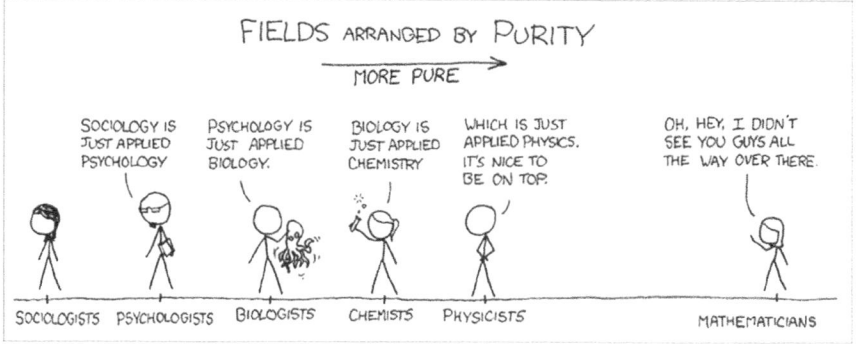

Figure 4.2 Fields arranged by purity. Source: http://xkcd.com/435/ Licensed under a Creative Commons Attribution-NonCommercial 2.5 License.

---

15 See the Online Resources for further reading on CS and science.

2. Read some of the essays cited in the Online Resources for Section 4.8.1 that have been critical of the scientific status of disciplines such as Freudian psychoanalysis, economics (Marxist or otherwise!),[16] astrology, etc., and consider whether the arguments that have been used to justify or challenge their status as a science can be applied to CS.

3. The computer scientist and philosopher Amnon H. Eden (Eden, 2007) seeks to bring clarity to the science-vs.-mathematics-vs.-engineering controversy by taking up a distinction due to the computer scientist Peter Wegner (1976) among three different "Kuhnian paradigms": a view of CS as (1) a "rationalist" or "mathematical" discipline, (2) a "technocratic" or "technological" discipline, and (3) a "scientific" discipline. (Tedre and Sutinen, 2008 also discusses these three paradigms.) Eden then argues in favor of the scientific paradigm.

   But must there be a single paradigm? Are there any disciplines with multiple paradigms? Does the existence of multiple paradigms mean there is no unitary discipline of CS? Or can all the paradigms co-exist?

4. In Section 2.5.2, we saw that McGinn (2015b) argues that philosophy is a science just like physics (an empirical science) or mathematics (a "formal" science), likening it more to the latter than the former (p. 85). To make his argument, he offers this characterization of science:

   > [W]hat distinguishes a discourse as scientific are such traits as these: rigor, clarity, literalness, organization, generality (laws or general principles), technicality, explicitness, public criteria of evaluation, refutability, hypothesis testing, expansion of common sense (with the possibility of undermining common sense), inaccessibility to the layman, theory construction, symbolic articulation, axiomatic formulation, learned journals, rigorous and lengthy education, professional societies, and a sense of apartness from naïve opinion. (McGinn, 2015b, p. 86)

   Does CS fit that characterization?

---

16 For non-Marxist economics, you might consider Rosenberg, 1994, Leiter, 2004, 2005, 2009, Chetty, 2013.

5

Engineering

[Engineering is] the art of directing the great sources of power in nature for the use and convenience of man [sic].
—Thomas Tredgold, 1828 (cited in Florman, 1994, p. 175)

Engineering … is a great profession. There is the fascination of watching a figment of the imagination emerge through the aid of science to a plan on paper. Then it moves to realization in stone or metal or energy. Then it brings jobs and homes to men [sic]. Then it elevates the standards of living and adds to the comforts of life. That is the engineer's high privilege.
—Herbert Hoover (1954),[1]
https://hooverpresidentialfoundation.org/speeches/engineering-as-a-profession/

[T]he scientist builds in order to study; the engineer studies in order to build.
—Frederick P. Brooks (1996, p. 62, col. 1)

[S]cience tries to understand the world, whereas engineering tries to change it.
—Mark Staples (2015, p. 2)[2]

5.1 Defining 'Engineering'

We began by asking what CS is, and we considered that it might be what it says it is: a science (Chapter 3). So we then looked at what science is (Chapter 4).

We also considered that CS might be a branch of engineering; so now it is time to ask what engineering is. What is the relationship of engineering to science? And what is the relationship of CS to engineering?

The philosophy of engineering is much less well developed than the philosophy of science, and, for some reason, there seem to be fewer attempts to try to define 'engineering.' For instance, if you link to various university websites for schools or departments of engineering, you will rarely find a definition.

The etymology of 'engineer' is of little help. According to the *OED* (http://www.oed.com/view/Entry/62225 and http://www.oed.com/view/Entry/62223), 'engineer' comes from 'engine' + '-or' (where '-or' means "agent"), and 'engine,' in turn, comes from the Latin *'ingenium,'* which had

---

1 Yes; the 31st President of the United States.
2 Recall that Marx said that philosophers should change the world, not merely understand it (see Section 2.5.2). Was Marx proposing a discipline of "philosophical engineering"?

Philosophy of Computer Science: An Introduction to the Issues and the Literature, First Edition. William J. Rapaport.
© 2023 John Wiley & Sons, Inc. Published 2023 by John Wiley & Sons, Inc.

multiple meanings, including "natural disposition," "mental powers," and "clever device" – none of which seems to help: the word has evolved too much for us to be able to figure out what it means from its origins.

Dictionary definitions are even less helpful than usual. Actually, dictionary definitions are rarely useful: first, different dictionaries don't always agree. Second, some are better than others. Third, dictionaries at best tell you how people *use* a term, but if people use a term "incorrectly," dictionaries are duty bound to record that.[3] Finally, dictionaries can be misleading: *Webster's Ninth New Collegiate Dictionary* (Mish, 1983, p. 259) defined 'college' as "a body of clergy living together and supported by a foundation"! This may once have been true, and may even still be true in a very limited sense of the term, but why is it listed as the *first* definition? The answer is that Merriam-Webster dictionaries list definitions in *historical order*! So, caution is always advised when citing a dictionary.

Nevertheless, it is instructive to see how that same dictionary defined 'engineering':

1. "The activities or function of an engineer ..."
2. "The application of science and mathematics ... [to make] matter and ... energy ... useful to people ..."

We'll come back to the second definition in a moment. As for the first, it cannot be understood without understanding 'engineer,' which is defined thus:

1. "A member of a military group devoted to engineering work."
2. "A person ... trained in ... engineering."[4]

Independently of the "military group" condition (see Section 5.3), both of these definitions of 'engineer' require us to already understand 'engineering'!

As we saw in Section 3.4.2, Hamming (1968, p. 4) observed that "the only generally agreed upon definition of mathematics is 'Mathematics is what mathematicians do,' which is followed by 'Mathematicians are people who do mathematics.' " So this dictionary agrees explicitly with Hamming: engineering is what engineers do; engineers are people who do engineering!

Only the second definition of 'engineering' in *Webster's Ninth* holds out some hope for independent understanding. Arguably, however, it seems to rule out by definition that CS is engineering, because it is not at all clear that computer scientists "apply science and mathematics to make matter and energy useful." Some might do that (by a stretch of meaning), but surely not all do.

According to the National Research Council's Committee on the Education and Utilization of the Engineer, engineering is, by their definition,

> Business, government, academic, or individual efforts in which knowledge of mathematical and/or natural sciences is employed in research, development, design, manufacturing, systems engineering, or technical operations with the objective of creating and/or delivering systems, products, processes, and/or services of a technical nature and content intended for use. (Florman, 1994, pp. 174–175)

Even Florman admits that this is a mouthful! Perhaps it can be simplified to something like this: efforts in which mathematics and natural science are used to produce something useful. If so, then is engineering (merely) applied science?

---

3 More precisely, if "the meaning" of a word is simply how people use it (Wittgenstein, 1958, Section 43), then there might be no such thing as an "incorrect" use. Many dictionaries take it as their task merely to record how people use a word, without taking a stand on whether any of those uses are "incorrect."
4 For the complete definitions, see Mish, 1983, p. 412 or the updated version at www.merriam-webster.com.

Michael Davis, a philosopher of engineering, points out that this definition, because of its vagueness (the overuse of 'and/or'), includes too much (such as accountants, because they use mathematics). He does say that it emphasizes three important "elements" of engineering: (1) the centrality of mathematics and science, (2) the concern with the physical world (which might, therefore, rule out software; but see Section 12.3 on that topic), and (3) the fact that "unlike science, engineering does not seek to understand the world but to remake it." But he goes on to say that "those three elements ... do not define" engineering. So, at best, they are necessary but not sufficient conditions (Davis, 1996, p. 98).

Here is another definition-by-committee (note the lists of verbs and nouns):[5]

> **Engineering** is the *knowledge* required, and the *process* applied, to conceive, design, make, build, operate, sustain, recycle or retire, something of significant technical content for a specified purpose; – a concept, a model, a product, a device, a process a system, a technology. (Malpas, 2000, p. 31, my italics)

This comes down to much the same thing as others have said: designing or building useful things. It emphasizes two aspects of this: one is that the designing or building must be *knowledge-based*. This presumably rules out designing or building that is based not on scientific knowledge but on experience alone (what Knuth might call "art"; see Section 4.2). The other aspect is that engineering is a *process*, in the sense of "knowing how" to do something (Malpas, 2000, p. 5). This has an algorithmic flair – after all, algorithms are methods of describing how to do something.

Finally, Henry Petroski (an engineer) notes that we speak of "the sciences" in the plural (as we do of "the humanities") but of engineering in the singular, "even though there are many" "engineerings" (Petroski, 2005, p. 304). So determining what engineering is may be as difficult as determining what CS is. More than for science or even CS, it seems that engineering *is* what engineers *do*. In Sections 5.3 and 5.4, we will consider a variation on this theme – that engineering is what engineers *study*; in Section 5.5, we will look at what it is that they *do*.[6]

5.2 Engineering as Science

> The scientist seeks to understand what is; the engineer seeks *to create what never was*.
> —Theodore von Kármán (cited in Petroski, 2008, my italics)[7]

Could engineering and science be the same discipline? That would certainly short-circuit the debate about whether CS is one or the other! Citing von Kármán, Petroski (2008) argued that all scientists are sometimes engineers (e.g. when they create a *new theory* that "never was") and that all engineers are sometimes scientists (e.g. when they seek to understand how an *existing* bridge works). Another engineer, Samuel C. Florman, also suggested as much (note the italicized phrase!):

> It is generally recognized ... that *engineering is* "*the* art or *science of* making practical application of the knowledge of pure sciences." ... The engineer uses the logic of science to achieve practical results. (Florman, 1994, pp. x–xi, my italics)

---

5 That is, a "klunky" definition designed to be acceptable to a variety of competing interests. The standard joke about such definitions is that a camel is a horse designed by a committee. See http://en.wikipedia.org/wiki/Design_by_committee.

6 See the Online Resources for further reading on the philosophy of engineering.

7 Recall Brooks's comment (Section 3.16.6) that computer programs "show ... things that never were."

One philosopher who has tried to explain engineering – Mario Bunge – also places it among the sciences. First, along with Kemeny (see Section 4.7), Bunge defines science as any discipline that applies the scientific method. Next, he says that there are two kinds of science: pure and applied. *Pure* sciences apply the scientific method to increasing our knowledge of reality (e.g. cell biology). *Applied* sciences apply the scientific method to enhancing our welfare and power (e.g. cancer research). Among the applied sciences are operations research (mathematics applied to management), pharmacology (chemistry applied to biology), *and engineering* (Bunge, 1974). Given this taxonomy, CS would not necessarily be a *branch* of engineering, although it might be an applied science *alongside* engineering. Yet there is a "pure" component of CS: namely, the mathematical theory of algorithms, computability, and complexity (Chapter 7).

And Quine said something that suggests engineering might be a part of science:

> I have never viewed prediction as the main purpose of science, although it was probably the survival value of the primitive precursor of science in prehistoric times. *The main purposes of science are understanding (of past as well as future),* **technology**, *and control of the environment.* (Quine, 1988, my italics and boldface)

If "technology" can be identified with engineering, then this puts engineering squarely into the science camp, rendering the science-vs.-engineering debates moot (although still not eliminating the need to ask what engineering – or technology – is). (On how technology might differ from engineering, see Section 5.5.1 and Bunge, 1974, Fiske, 1989.)

5.3 A Brief History of Engineering

> Rather than treat software engineering as a subfield of CS, I treat it as an element of the set, {Civil Engineering, Mechanical Engineering, Chemical Engineering, Electrical Engineering, … }. This is not simply a game of academic taxonomy, in which we argue about the parentage or ownership of the field; *the important issue is the content and style of the education.*
> —David Lorge Parnas (1990, p. 1, my italics)

Michael Davis (1998) offers an insight into what engineering might be. (See also Davis, 1995a, b, 1996, 2009.) He starts with a history of engineering, beginning some 400 years ago in France, where there were "engines" – i.e. machines – and "engineers" who worked with them. These "engineers" were soldiers: either those who used "engines of war" such as catapults and artillery, or those who had been carpenters and stonemasons in civilian life and who continued to ply these trades as soldiers. From this background comes the expression "railroad engineer" and such institutions as the US Army Corps of Engineers.

In 1676, the French army created a corps of engineers (separate from the infantry) who were charged with *military* construction. So, at least in seventeenth century France, an engineer was someone who did whatever it was that those soldiers did. Forty years later, in 1716, there were *civil* engineers: soldiers who built infrastructure (like bridges and roads) *for civilians.*

A distinction was drawn between engineers and architects. At that time, engineers in France were trained at the École Polytechnique ("Polytechnic School"), a university whose curriculum began with a year of science and mathematics, followed gradually by more and more applications to construction (e.g. of roads), culminating in a specialization. These engineers were concerned with reliability and other practical matters. And they were trained as army officers, and hence (presumably) more disciplined for larger projects. Architects, on the other hand, were more like artists, chiefly concerned with aesthetics.

So, at this time, engineering was the application of science "for the use and convenience of" people and for "improving the means of production" (Tredgold, as quoted in Davis, 1998, p. 15). Engineering was *not* science: engineers *used* science but didn't *create* new knowledge. Nor was engineering *applied* science: engineers were concerned with human welfare (and not even with generality and precision), whereas applied scientists are concerned with applying their scientific knowledge.

5.4 Conceptions of Engineering

Davis (2011, pp. 31–33) cites four different conceptions of engineering:

"Engineering as tending engines": This would include railroad engineers and building-superintendent engineers. Clearly, neither computer scientists nor software engineers are engineers in this sense, but neither are electrical, civil, mechanical, or chemical engineers. (Although perhaps a company's information technology [IT] staff "tend" computational "engines.")

"Engineering-as-invention-of-useful-objects": Davis criticizes this sense as both "too broad" (including architects and accountants) and "anachronistic" (applying to inventors of useful objects before 1700, which is about when the modern sense of 'engineer' came into use). Note that this seems to be the sense of engineering used by many who argue that CS *is* engineering; they view engineering as designing and building useful artifacts.

"Engineering-as-discipline": Here, the issue concerns "the body of knowledge engineers are supposed to learn," which includes "courses concerned with the material world, such as chemistry and statistics." Again, this would seem to rule out both CS and software engineering on the grounds that neither needs to know any of the "material" natural sciences like chemistry or physics (although both software engineers and computer scientists probably need some statistics) and both need "to know things other engineers do not."

"Engineering-as-profession": This is Davis's favored sense (argued for in his other writings).

Davis concludes that engineering must be defined by two things: (1) its professional *curriculum* (by its specific knowledge) and (2) a professional commitment to use that knowledge consistent with a *code of ethics*. So, rather than saying that engineering is what engineers *do*, Davis says that engineering is what engineers *learn* and how they *ought* (ethically) to use that knowledge. This, of course, raises the question, what is it that engineers learn? Mark Staples[8] observes that Davis's definition of engineering in terms of its curriculum "is circular .... How does engineering knowledge become accepted into engineering curricula?"

There is another question central to our concerns: is what engineers learn also something that computer scientists learn? Here, Davis's explicit argument against software engineering (currently) being engineering (and his implicit argument against CS (currently?) being engineering) is that, although both are professions, neither is (currently) part of the profession of engineering as it is taught and licensed in engineering schools. Even CS departments that are academically housed in engineering schools typically do not require their students to take "engineering" courses (Guzdial, 2021), their academic programs are not accredited in the same way,[9] and their graduates are not required to become "professional engineers" in any legal senses.

---

8 Personal communication, 2015.
9 Many of them are accredited, of course, but not as *engineering* curricula.

5.5 What Engineers Do

There are two very general tasks that various authors put forth as what engineers do: they *design* things, and they *build* things.

5.5.1 Engineering as Design

Petroski (2003, p. 206) says that engineering's fundamental activity is *design*. And philosopher Carl Mitcham (1994) distinguishes between the *engineer* as *designer* and the *technician* or technologist as *builder*. So, engineering is not science, because science's fundamental activity is analysis (Petroski, 2003, p. 207), whereas design and building are synthesizing activities. Mark Staples[10] points out that, contra Petroski, engineering is more than just design, because architects also design but are not engineers.

One aspect of design has been picked up by Hamming (1968). When one designs something, one has to make choices. Hamming suggests that "science is concerned with *what* is possible while engineering is concerned with *choosing*, from among the many possible ways, *one* that meets a number of often poorly stated economic and practical objectives." This fits well with much of the work – even theoretical work – that is done by computer scientists. As we saw in Sections 3.6 and 3.7, one definition of CS is that it is concerned with what can be automated (in the sense of "computed"). One way of expressing this is as follows: for what tasks can there be an algorithm that accomplishes the task? But there can be many algorithms, all of which accomplish the exact same task. How can we choose among them? We can ask which ones are more efficient: Which use less memory ("space")? Which requires fewer operations (less "time")? So, in addition to asking what can be computed, CS also asks, what can be computed *efficiently*? If that is computer *engineering*, so be it, but that would put one of the most abstract, theoretical, and mathematical branches of CS – namely, the theory of computational complexity – smack dab in the middle of computer engineering, and that doesn't seem correct.

5.5.2 Engineering as Building

We have seen that many people say that what engineers do is *build* or *create* things. For example, computer scientist Paul Abrahams (1987, p. 472) argues as follows:

1. Someone who "discover[s] how things work" is a scientist.
2. Someone who "learn[s] how to build things" is an engineer.
3. Therefore, "[c]omputer science is both a scientific discipline and an engineering discipline."

The conclusion can be made valid by adding two missing premises:

A. Computer scientists discover how things work.
B. Computer scientists learn how to build things.

Is the argument sound?

As for missing premise A, computer scientists can be said to discover how things work *algorithmically*. As for B, computer scientists can be said to build both software (e.g. computer programs) and hardware (e.g. computers).

The explicit premises *seem* to be true. But is premise 1 *really* true? Is life, or the universe, a "thing"? Do scientists really try to learn how the kinds of physical objects that engineers build

---

10 Personal communication, 2015.

work (and nothing else)? This seems overly simplistic. Nevertheless, this "analytic vs. synthetic" distinction (i.e. a distinction between *analyzing* – taking something apart – in order to learn how it works, on the one hand, and *synthesizing* – putting things together – in order to build something, on the other hand) seems to be a feature of many characterizations of science vs. engineering.

Concerning premise 2, Petroski (2005, p. 304), says, "engineering … is an activity that creates things." Note two points: first, engineering is creative; this is related to claims about engineering as designing and building things. But second, it is an activity, even grammatically: the word 'engineering' is a gerund – a word that (as Petroski says) "expresses … action." Is science also an activity? Or is engineering different from science in this respect? Insofar as science is an activity, it is an activity that produces "knowledge." Engineering is an activity that uses that scientific knowledge to design and build artifacts. Yet one way to discover how things work is to try to build them; so, is all engineering a kind of science?

5.6 The Engineering Method

Just as some people speak of a "scientific method," others have proposed an "engineering method." Presumably, just as 'science' can be defined as any discipline that follows "the scientific method," so 'engineering' can be defined as any discipline that follows "the engineering method."

In Section 4.7, we saw one view of the scientific method, according to which it is a loop that cycles through observation of facts, induction of general statements, deduction of future observations, and verification of the deduced predictions against observations before cycling back to more observations. Similarly, Robert Malpas (2000) describes the engineering method both linearly and as a cycle (Figure 5.1). It begins by inputting a set of requirements, followed by analysis, then synthesis, then evaluation and execution, and outputting a solution. The cycle comes in between the input and the output: the evaluation and execution cycle back both to the analysis and to the synthesis, as well as adding to a knowledge base that, along with a set of resources, interacts with the analysis, synthesis, and evaluation-execution.

But neither this nor the scientific method is carved in stone; both are more like guidelines or even after-the-fact descriptions of behavior rather than rules that must be slavishly followed. Are "engineering methods" significantly different from "scientific methods"? Malpas's engineering method doesn't seem so.

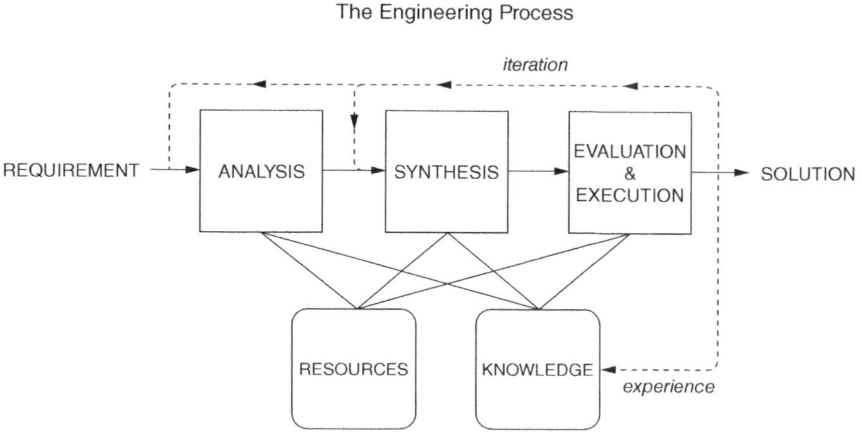

Figure 5.1 Malpas's engineering method (Malpas, 2000, p. 35). Source: Malpas, 2000 / Reproduced with permission of the Royal Academy of Engineering.

Instead of scientific methods, engineering methods, or even methods used in the humanities, the engineer Billy Vaughn Koen (2009) seeks a "universal method": heuristics. Koen (1988) defines this universal method as

> the use of engineering *heuristics* to cause the best change in a poorly understood situation within the available resources. (Koen, 1988, p. 308, my italics)

For Koen

> A heuristic is anything that provides a plausible aid or direction in the solution of a problem but is in the final analysis … incapable of justification and fallible. It is anything that is used to guide, discover and reveal a possible, but not necessarily, correct way to solve a problem. Though difficult to define, a heuristic has four characteristics that make it easy to recognize: it does not guarantee a solution; it may contradict other heuristics; it reduces the search time for solving a problem; and its acceptance depends on the immediate context instead of on an absolute standard. (Koen, 1988, p. 308).

Many other disciplines use heuristics; writers, for example, are often told to "write simply." (See Section 5.9, Question 4.) What makes a heuristic an *engineering* heuristic? According to Koen, the first two characteristics differentiate the use of heuristics from science and mathematics. So, they demarcate engineering from science and mathematics. The third and fourth characteristics make their use more practical than at least some scientific or mathematical theories.

Koen (1988, p. 309) states "that the *engineering strategy for causing desirable change in an unknown situation within the available resources* and the *use of heuristics* is an absolute identity." First, Koen is saying that what engineers do is cause changes. This does contrast with science (and math), whose goal is, presumably, to understand things, not to change them, and it is consistent with the quote from Staples cited as an epigraph to this chapter.

Second, Koen's engineering method is not as "formal" as, say, Malpas's, because it is simply the use of heuristics ("the engineering strategy" = "the use of heuristics"). But what kind of heuristics? Much of what Koen says suggests that the kind of heuristic reasoning used by engineers is akin to what Herbert Simon called "satisficing": being satisfied with having a reasonable answer to a question rather than the "correct" one. Satisficing is due to what Simon called "bounded rationality," which is necessary in practical situations, given limits ("bounds") on our time and knowledge.

Koen's notion of heuristics is akin to informal "rules of thumb." But there is a more computational notion: A *heuristic for problem p* can be defined as an *algorithm* for some other problem p', where the solution to p' is "good enough" as a solution to p (Rapaport, 1998, p. 406). Being "good enough" is, of course, a subjective notion; Oommen and Rueda (2005, p. 1) call the "good enough" solution "a *sub-optimal* solution that, hopefully, is arbitrarily close to the *optimal*." The idea is also related to Simon's notions: we might not be able to solve p because of limitations in space, time, or knowledge, but we might be able to solve p' algorithmically within the required spatio-temporal-epistemic limits. And if the *algorithmic* solution to p' gets us closer to a solution to p, then it is a *heuristic* solution to p. So, it is still an algorithm.[11]

5.7 Software Engineering

In addition to the question of whether CS is a kind of engineering, there is the question of the nature of software engineering. Computer scientists (whether or not they consider themselves to be

---

11 See the Online Resources for further reading on heuristics.

scientists or engineers) often consider software engineering as a branch of CS. Courses in software engineering are often, perhaps even usually, taught in CS departments. But is software engineering engineering?

For Davis, software engineering would be (real?) engineering if and only if there is a professional curriculum for it, along with a code of professional ethics. Interestingly, he also suggests that this might not happen until "real" engineering starts becoming more computational (Davis, 2011, p. 34).

The software engineer David Parnas has a different take on CS's relationship to engineering:

> Just as the scientific basis of electrical engineering is primarily physics, the scientific basis of software engineering is primarily computer science. This paper contrasts an education in a science with an education in an engineering discipline based on the same science. (Parnas, 1990, p. 2).

In other words, software engineering is CS engineering. There are two interesting implications of this. First, it suggests that Parnas views CS as a science because he takes it to be the scientific basis of a branch of engineering. Second, this view of things is inconsistent with the view advocated by, for instance, Loui and Hartmanis, who take *CS* (or parts of it) as being a kind of engineering whose scientific basis is primarily *mathematics*: i.e. as mathematical engineering (Section 3.15). On the other hand, one might argue that if software engineering is based on CS, which, in turn, is based on mathematics, then software engineering must ultimately be based on mathematics, too, which suggests that software engineering would be *mathematical-engineering* engineering!

And that might not be far from the truth, considering that much of *formal* software engineering is based on (discrete) mathematics and logic (such as the formal analysis of computer programs and their development (Mili et al., 1986) or the use of program-verification methods in the development of programs (Chapter 15). So, is software engineering unique in being a kind of engineering that is based on another kind of engineering rather than on a science? Or is software engineering indeed based on a science: namely, CS? Parnas quite clearly believes that CS is a science, *not* an engineering discipline. Why?

Part of the reason concerns his definition of 'engineering': "Engineers are professionals *whose education prepares them* to use mathematics, science, and the technology of the day, *to build products* that are important to the safety and well-being of the public" (Parnas, 1990, p. 2, my italics). This echoes Davis's claim about the central role of *education* in the nature of being an engineer, as well as Brooks's (and others') claim that the purpose of engineering is to *use* science to *build* humanly useful things.

To complete his argument that CS is not engineering, Parnas needs a premise that states that CS education *doesn't* prepare computer scientists to use CS to build things, or perhaps just that computer scientists don't build things. (That leaves open the possibility that CS might be a branch of mathematics or a "technology of the day," but it's pretty clear from the first quote that he thinks it is a science.) This missing premise is the gist of his entire article. But at least one part of his argument is this: proper training in software engineering – "designing, building, testing, and 'maintaining' software products" (Parnas, 1990, p. 2) – requires more than a course or two offered in a CS curriculum. Rather, it requires an "accredited professional programme … modelled on programmes in traditional engineering disciplines" (Parnas, 1990, p. 2).[12]

But we still don't have a clear statement of why he thinks CS *is* a science and is *not* engineering. As for the latter, it's not engineering because there is no "rigid accreditation process … [hence, no]

---

12 The spelling in this quote is Canadian-British spelling. 'Programme' is used in the sense of an "academic program," not in the sense of a "computer program."

well documented 'core body of knowledge' … for computer science" (Parnas, 1990, p. 2). Such accreditation might be necessary but is surely not sufficient: one might force such a core body of knowledge and such an accreditation process on, say, physics, but that wouldn't make physics an engineering discipline.

Some clarity arises here:

> It is clear that two programmes are needed [e.g. both physics and electrical engineering, or both computer science and software engineering], not because there are two areas of science involved [e.g. physics and electrical engineering], but because there are two very different career paths. *One career path is that of graduates who will be designing products for others to use. The other career path is that of graduates who will be studying the phenomena that interest both groups and extending our knowledge in this area.* (Parnas, 1990, p. 3, my italics)

So: scientists study phenomena and extend knowledge; engineers design products. So: CS studies phenomena and extends knowledge; software engineers design software products. The distinction between science and engineering, for Parnas, is that between learning and building (Parnas, 1990, p. 4). Note that Parnas agrees with Brooks about the distinction but draws the opposite conclusion: that CS is *not* engineering![13]

5.8 CS and Engineering

A science and an engineering discipline can both be about the same thing. For example, both chemists and chemical engineers study chemistry. What, then, is the common object of computer *science* and computer *engineering*? Is it computers? Algorithms? Information? Perhaps computer *science* studies algorithms and procedures, whereas computer *engineering* studies computers and computer systems. If so, then who studies the relations between these, such as "programmed living machines"? (Recall Section 3.7.)

Trying to *distinguish* between science and engineering may be the wrong approach. It is worth recalling W.A. Wulf's cautionary remarks, which we quoted in Section 3.13:

> Let's remember that *there is only one nature – the division into science and engineering … is a human imposition, not a natural one.* Indeed, the division is *a human failure*; it reflects *our limited capacity to comprehend the whole.* That failure impedes our progress; *it builds walls* just where the most interesting nuggets of knowledge may lie. (Wulf, 1995, p. 56; my italics)

Is CS a science that tries to understand the world computationally? Or is it an engineering discipline that tries to change the world by building computational artifacts? (Or both? Or neither?) No matter our answer, it has to be the science or engineering (or whatever) *of* something. We have seen at least two possibilities: it studies computers, or it studies computation (algorithms). To further explore which of these might be central to CS, let us begin by asking, "What is a computer?" Later, we will inquire into what computation is.[14]

---

13 See the Online Resources for further reading on software engineering.
14 See the Online Resources for further reading on CS and engineering.

5.9 Questions to Think About

1. Link to various engineering websites, and try to find a definition of 'engineer' or 'engineering.' Here are two good ones to begin with:
   (a) "What Is Engineering?"
       Whiting School of Engineering, Johns Hopkins University,
       http://www.jhu.edu/~virtlab/index.php
   (b) "What is engineering and what do engineers do?"
       National Academy of Engineering of the National Academies,
       http://www.nae.edu/About/FAQ/20650.aspx
2. In Section 3.12, we saw that Brooks argued that CS was not a science but a branch of engineering, in part because the purpose of engineering is to build things, and that's what computer scientists do.

   How would you evaluate his argument now that you have thought more deeply about what engineering is?
3. Loui (1987, p. 176) said that "The ultimate goal of an engineering project is a product … that benefits society," giving bridges and computer programs as sample "products." But not all computer programs benefit society – think of computer viruses. Presumably, Loui meant something like "product that *is intended to* benefit society."

   But does that mean, then, that a computer programmer who writes a spreadsheet program *is* an engineer (no matter how sloppily the programmer writes it), whereas a computer programmer who writes a computer virus is *not* an engineer (even if the program was designed according to the best software engineering principles)?
4. If the central feature of engineering is, say, the application of scientific (and mathematical) techniques for producing or building something, then arguably part of CS is engineering – especially those parts concerned with building computers and writing programs. Here's something to think about: just as (some) computer scientists write programs, journalists and novelists write essays. Moreover, they use heuristics, such as "write simply," "avoid using the passive voice," and so on. The engineer Alice W. Pawley (2009, p. 310, col. 2) makes a similar point concerning a National Academy of Engineering definition of engineers as "men and women who create new products":

   > Without knowing how the NAE defines "product," one could argue that an academic who writes a book on how food is portrayed in Victorian novels has created a product (the book) based on abstract ideas (theories about the historical display of food).

   Are journalists, novelists, and other writers therefore engineers? Their products are not typically applications of science and mathematics, so perhaps they aren't. But might they be considered, say, language engineers?
5. What phenomena does Parnas think computer scientists study?
6. Does Parnas consider electrical *engineering* to be an "area of science"?
7. Evaluate the validity and soundness of the following argument:[15]
   (a) Engineers are cognitive agents who build artifacts for some identifiable purpose.
   (b) Birds build nests for housing their young.

---

15 Thanks to Albert Goldfain for questions 7 and 8.

(c) Beavers build dams because the sound of rushing water annoys them.[16]

(d) Computer engineers build computers for computation.

(e) ∴ Birds, beavers, and computer engineers are all engineers.

8. Evaluate the validity and soundness of the following argument:

(a) Engineers are cognitive agents who build artifacts for some identifiable purpose *and* who know what that purpose is.

(b) Birds and beavers do not *know* why they build nests and dams, respectively; they are only responding to biological or evolutionary instincts.

(c) Computer engineers *do* know what the purpose of computation is.

(d) ∴ Computer engineers are engineers, but birds and beavers are not.

16 https://web.archive.org/web/20110714160115/http://naturealmanac.com/archive/beaver_dams/beaver_dams.html

6

Computers: A Brief History

Let us now return to the analogy of the theoretical computing machines ... It can be shown that a single special machine of that type can be made to do the work of all. *It could in fact be made to work as a model of any other machine.* The special machine may be called the universal machine ...
—Alan Turing (1947, my italics)

If it should turn out that the basic logics of a machine designed for the numerical solution of differential equations coincide with the logics of a machine intended to make bills for a department store, *I would regard this as the most amazing coincidence I have ever encountered.*
—Howard Aiken (1956, my italics), cited in Martin Davis, 2012[1]

There is no reason for any individual to have a computer in their home.
—Ken Olsen (1974)[2]

Many people think that computation is for figuring costs and charges in a grocery store or gas station.
—Robin K. Hill (2008)

6.1 Introduction

Let us take stock of where we are. We began by asking what CS is, and we saw that it might be a science, a branch of engineering, a combination of both, or something else. To help us answer that question, we then investigated the nature of science and of engineering.

It is now time to ask, "What is CS the science, or engineering, or study *of*?" The subject matter of CS might be *computers* (the physical objects that compute) (Newell et al., 1967). Or it might be *computing* (the algorithmic processing that computers do) (Knuth, 1974b). Or, of course, it might be something else (such as the information that gets processed; see Sections 3.8 and 3.16). In the next few chapters, we will begin to examine the first two options. In this chapter, our focus will be to seek answers to the question **what is a computer?** from the *history* of computers. In Chapter 9, we will look at some *philosophical* issues concerning the nature of computers. Chapters 7 and 8 will investigate computing.

1 However, five years *before* Aiken said this, the Lyons tea company in Great Britain became the first company to computerize its operations (Martin, 2008).
2 For the citation and history of this quote, see https://quoteinvestigator.com/2017/09/14/home-computer/. That website offers an interesting alternative interpretation: home *computers* might not be needed if there are home *terminals*: i.e. if what is now called "cloud computing" becomes ubiquitous.

Philosophy of Computer Science: An Introduction to the Issues and the Literature, First Edition. William J. Rapaport.

6.2   Would You Like to Be a Computer?

> Towards the close of the year 1850, the Author first formed the design of rectifying the Circle[3] to upwards of 300 places of decimals. He was fully aware, at that time, that the accomplishment of his purpose would add little or nothing to his fame as a Mathematician, though it might as a Computer; …
> —William Shanks (1853), as cited in B. Hayes, 2014a, p. 342

Let's begin our historical investigation with some terminology. Some 130 years ago, in the 2 May 1892 issue of *The New York Times*, the ad shown in Figure 6.1 appeared.

So, over a century ago, the answer to the question "What is a computer?" was: a human who computes! In fact, until at least the 1940s (and probably the 1950s), that was the meaning of 'computer.' When people wanted to talk about a *machine* that computed, they would use the phrase 'computing *machine*' or (later) '*electronic* (digital) computer.' (In Chapters 8 and 18, when we look at Alan Turing's foundational papers in CS and AI, this distinction will be important.) Interestingly, nowadays, when one wants to talk about a *human* who computes, we need to use the phrase '*human* computer' (Pandya, 2013). In this book, for the sake of familiarity, I will use the word 'computer' for the machine and the phrase 'human computer' for a human who computes.

Digression on Human Computers: 'Computer' in the human sense was used by some people even in the 1960s, as told in the 2016 film *Hidden Figures* (https://en.wikipedia.org/wiki/Hidden_Figures); see also Bolden, 2016, Natarajan, 2017. For a history of human computers (most of whom were women), see Lohr, 2001, Grier, 2005, Skinner, 2006, Thompson, 2019; and the website "Computer Programming Used to Be Women's Work" *Smart News Blog*, https://www.smithsonianmag.com/smart-news/computer-programming-used-to-be-womens-work-718061/. The other kind of human computers, of course, are mathematicians (of either sex):

> Historians might … wonder if mathematicians who devised algorithms were programmers …. Modern programmers would … say no because these algorithms were not encoded for a particular machine. (Denning and Martell, 2015, p. 83)

But they were! They were encoded for *human* computers! Curiously, on the very next page, Denning and Martell say exactly that:

> The women who calculated ballistic tables for the Army during World War II were also programmers, although their programs were not instructions for a machine but for themselves to operate mechanical calculators. In effect, they were human processing units.

But why should this be treated merely as a kind of metaphor? These women *were* the computers!

---

3 "Rectifying" – or "squaring" – the circle is the Euclidean-geometry problem of constructing a square with the same area as a circle using only compass and straightedge. It is logically impossible to do so. We'll return to such impossibility proofs in Section 6.5.

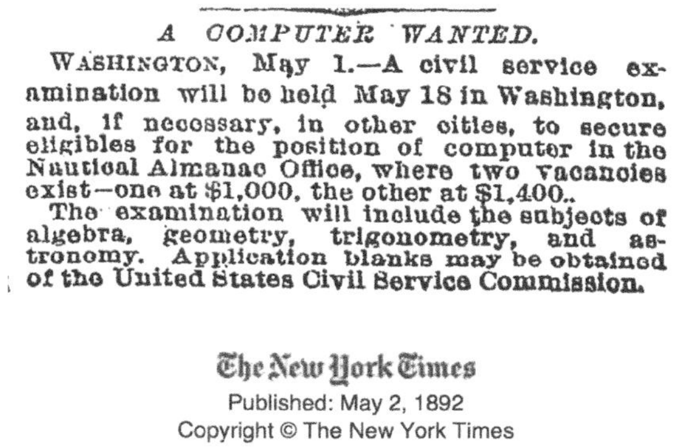

Figure 6.1   1892 computer ad. Source:
https://dirkvl.info/2011/04/05/48510398 / last
accessed 7 June 2022. Public Domain.

A COMPUTER WANTED.

WASHINGTON, May 1.—A civil service examination will be held May 18 in Washington, and, if necessary, in other cities, to secure eligibles for the position of computer in the Nautical Almanac Office, where two vacancies exist—one at $1,000, the other at $1,400..

The examination will include the subjects of algebra, geometry, trigonometry, and astronomy. Application blanks may be obtained of the United States Civil Service Commission.

The New York Times
Published: May 2, 1892
Copyright © The New York Times

6.3   Two Histories of Computers

Just as there are two views of CS as science and as engineering, there are both scientific and engineering histories of computers; they begin in parallel, intersect, and eventually converge. One of these is the *engineering* history of building a *machine that could compute* (or calculate): i.e. a machine that could duplicate – and therefore assist, replace, or even supersede – human computers. The other is the *scientific* history of providing a *logical foundation for mathematics*.

These histories were probably never purely parallel but more like a tangled web, with at least two human "bridges" connecting them: the first was a philosopher who lived about 340 years ago, and the other was a mathematician who was active much more recently (about 90 years ago) – Gottfried Wilhelm Leibniz (1646–1716) and Alan Turing (1912–1954). (There were, of course, other significant people involved in both histories, as we will see.) Moreover, both histories begin in ancient Greece, the engineering history beginning with the need for computational help for astronomical purposes (including navigation), and the mathematical-logical history beginning with Aristotle's study of logic.[4]

6.4   The Engineering History

The engineering history concerns the attempt to create machines that would do certain mathematical computations. The two main reasons for wanting to do this seem to be (1) to make life easier and less boring for humans (let a machine do the work!) and – perhaps of more importance – (2) to produce computations that are more accurate (both more precise and with fewer errors) than those that humans produce. It is worth noting that the goal of having a machine perform an intellectual task that would otherwise be done by a human is one of the motivations underlying AI. In this section, we will only sketch some of the highlights of the engineering history.

The engineering history to follow focuses only on digital computers. But "Before electronic calculators, the mechanical slide rule dominated scientific and engineering computation" (Stoll, 2006). The slide rule is an *analog* calculator![5]

6.4.1   Ancient Greece

The attempt to build calculating machines can be traced back to at least the second century BCE, when a device now known as the Antikythera Mechanism was constructed. This device was used

---

4 See the Online Resources for further reading on computer history.
5 See the Online Resources for further reading on the engineering history.

to calculate astronomical information, possibly for use in agriculture or religion. Although the Antikythera Mechanism was discovered in 1900, a full understanding of what it was and how it worked was not figured out until the 2000s (Freeth et al., 2006).[6]

6.4.2 Seventeenth Century Calculating Machines

Skipping ahead almost 2000 years to about 350 years ago, two philosopher-mathematicians are credited with more familiar-looking calculators: Blaise Pascal (1623–1662), who helped develop the theory of probability, also invented an adding (and subtracting) machine that worked by means of a series of connected dials.[7] And Leibniz (who invented calculus almost simultaneously with, but independently of, Isaac Newton) invented a machine that could add, subtract, multiply, and divide. As we'll see in Section 6.5, Leibniz also contributed to the *scientific* history of computing with an idea for something he called a "*calculus ratiocinator*" (loosely translatable as a "reasoning system").[8]

6.4.3 Babbage's Machines

> We both went to see the *thinking* machine (for such it seems) last Monday.
> —Lady Byron (Ada Lovelace's mother), writing in 1833 about Babbage's
> Difference Engine (as cited in Stein, 1984, p. 38)

Two of the most famous antecedents of the modern electronic computer were due to the English mathematician Charles Babbage, who lived about 190 years ago (1791–1871). The first machine he designed was the "Difference Engine" (1821–1832), inspired in part by a suggestion made by French mathematician Gaspard de Prony (1755–1839).

De Prony, who later headed France's civil-engineering college, needed to construct highly accurate logarithmic and trigonometric tables for large numbers and was himself inspired by Adam Smith's 1776 text on economics, *The Wealth of Nations*. Smith discussed the notion of the "division of labor": the manufacture of pins could be made more efficient by breaking the job down into smaller units, with each laborer who worked on one unit becoming an expert at that one job. De Prony, realizing that it would take him too long using "difference equations" by hand,[9] applied this division of labor to computing the log and trig tables, using two groups of human computers, each as a check on the other.

De Prony's technique is essentially what modern computer programmers call "top-down design" (Mills, 1971) and "stepwise refinement" (Wirth, 1971): to accomplish some task T, analyze it into subtasks $T_1, \ldots, T_n$, each of which should be easier to do than T. This technique can be repeated: analyze each T_i into sub-subtasks $T_{i_1}, \ldots, T_{i_m}$, and so on, until the smallest sub…subtask is so simple that it can be done without further instruction (this is an aspect of "recursion"; see Section 7.6).

It should be noted that, besides its positive effects on efficiency, the division of labor has negative ones, too: it

> would make workers as "stupid and ignorant as it is possible for a human creature to be." This was because no worker needed to know how to make a pin, only how to do his part in the process of making a pin. (Skidelsky, 2014, p. 35, quoting Adam Smith (1776, Book v, Ch. I, Part III, Art. II), https://www.marxists.org/reference/archive/smith-adam/works/wealth-of-nations/book05/ch01c-2.htm

---

6 See the Online Resources for further reading on the Antikythera Mechanism.
7 I wish I still had the plastic one that I bought in a New York City candy store in the 1950s!
8 See the Online Resources for further reading on Pascal, Leibniz, and other early calculating machines.
9 Difference equations are a discrete-mathematical counterpart to differential equations. They involve taking successive differences of sequences of numbers.

More recently, several writers have pointed out that very few of us know every detail about the facts that we know or the activities that we know how to perform (see, for example, Dennett, 2017, Ch. 15). So this negative effect might be unavoidable.

Babbage wanted a *machine* to replace de Prony's *people*; this was to be his Difference Engine. He later conceived of an "Analytical Engine" (1834–1856), which was intended to be a general-purpose problem-solver (perhaps more closely related to Leibniz's goal for his *calculus ratiocinator*). Babbage was unable to completely build either machine: the engineering methods available to him in those days were simply not up to the precision required. However, he developed techniques for what we would today call "programming" these machines, using a nineteenth century version of punched cards (based on a technique invented by Joseph Marie Jacquard for use in looms – a sequence of punched cards constituted a "program" for weaving a pattern in the cloth on the loom). Working with Babbage, Lady Ada Lovelace (1815–1852) – daughter of the poet Lord Byron – wrote a description of how to program the (yet-unbuilt) Analytical Engine; she is thus considered the first computer programmer:

> … the important difference between the two machines is that the Difference Engine followed an unvarying computational path …, while the Analytical Engine was to be truly programmable …. (Stein, 1984, p. 49)

This suggests that the relationship between the Difference Engine and the Analytical Engine was similar to that between a Turing Machine (which can only compute a single function) and a Universal Turing Machine (which can compute any function whose algorithm is stored on its tape). We'll return to these in Chapters 8 and 9.[10]

6.4.4 Electronic Computers

The modern history of electronic, digital computers is itself rather tangled and the source of many historical and legal disputes. Here is a brief survey:

1. John Atanasoff (1903–1995) and his student Clifford Berry (1918–1963), working at Iowa State University, built the ABC (Atanasoff-Berry Computer) in 1937–1942. This may have been the first electronic, digital computer, but it was not a general-purpose (programmable) computer, and it was never completed. It was, however, the subject of a patent-infringement suit, about which more in a moment.
2. Konrad Zuse (1910–1995), in Germany, developed the Z3 computer in 1941, which was programmable.
3. In 1943, the Colossus computer was developed and installed at Bletchley Park, England, for use in cryptography during World War II. Bletchley Park was where Alan Turing worked on cracking the Nazi's code-making machine, the Enigma.[11]
4. Howard Aiken (1900–1973), inspired by Babbage, built the Harvard Mark I computer in 1944; it was designed to compute differential equations. (Recall the Aiken epigraph at the beginning of this chapter.)
5. After the war, in 1945, Turing decided to try to implement his "*a*-machine" (what is now called the 'Turing Machine'; see Section 6.5, and – for more detail – Chapter 8), and developed the ACE (Automatic Computing Engine) (Copeland, 1999; Campbell-Kelly, 2012). It was also around this time that Turing started thinking about AI and neural networks.
6. John Presper Eckert (1919–1995) and his student John Mauchly (1907–1980), working at the University of Pennsylvania, built the ENIAC (Electronic Numerical Integrator And

---

10 See the Online Resources for further reading on Babbage, Lovelace, Adam Smith, and de Prony.
11 See the Online Resources for further reading on Colossus and Enigma.

Computer) in 1946, a *general-purpose* – i.e. programmable – electronic computer. In 1945, with the collaboration of the mathematician John von Neumann (1903–1957) – who outlined an architecture for computers that is still used today – they began to develop the EDVAC (Electronic Discrete Variable Automatic Computer), which used binary arithmetic (rather than decimal). Completed in 1949, it evolved into the first commercial computer: the UNIVAC (UNIVersal Automatic Computer). UNIVAC became famous for predicting, on national TV, the winner of the 1952 US presidential election. The company that made UNIVAC evolved into the Sperry Rand Corporation, which owned the patent rights. The Honeywell Corporation, a rival computer manufacturer, successfully sued Sperry Rand in 1973, on the grounds that Mauchly had visited Atanasoff in 1941 and that it was Atanasoff and Berry – not Eckert and Mauchly – who had "invented" the computer, thus vacating Sperry Rand's patent claims.[12]

6.4.5 Modern Computers

> Where a calculator like ENIAC today is equipped with 18,000 vacuum tubes and weighs 30 tons, computers in the future may have only 1000 vacuum tubes and perhaps weigh only $1\frac{1}{2}$ tons.
> —*Popular Mechanics*, March 1949 (cited in Meigs, 2012)

A few years ago, one of our daughters looked at a pile of MacBooks in our living room and asked, "Can you hand me a computer?" Early computers, however, were large, cumbersome, and expensive, so there weren't very many of them (and they couldn't have been "handed" around!):

> *There are currently over one hundred computers installed in American universities.* Probably two dozen or more will be added this year. In 1955 the number was less than twenty-five. ... [C]onsidering the costs involved in obtaining, maintaining, and expanding these machines, the universities have done very well in acquiring hardware with their limited funds. (Perlis, 1962, p. 181, my italics)

Of course, a university with, say, 5000 students now probably has at least 5000 computers – and probably double that amount if you include smartphones – not to mention the computers owned by the universities themselves! And each one of those 10,000 or more computers is at least as powerful as, if not more so than, the 100 of a half-century ago.

Although the early computers were mostly intended for military uses,

> The basic purpose [of computers at universities], at present [i.e. in 1962], is to do computations associated with and supported by university research programs, largely government financed. ... *Sad to state, **some** uses occur merely because the computer is available, and seem to have no higher purpose than that.*
> —Alan J. Perlis (1962, p. 182, my italics)

Today, I wouldn't be surprised if *most* uses (Candy Crush? Skype? Facebook? Twitter? Amazon?) of the 10,000 computers at an average contemporary university "have no higher purpose"! (At this point, you are urged to re-read the chronologically ordered epigraphs at the beginning of this chapter.)

It is also worth noting the simultaneous *decrease in size* of computers from the 1940s to now, as well as the increase in their ease of use: ENIAC needed an entire room for all the hardware and a team of experts to run it; 40 years later, a child could use a desktop personal computer; and less

---

12 See the Online Resources for further reading on Atanasoff, the ENIAC-ABC controversy, and von Neumann.

than 40 years after that, infants and toddlers could use iPads and iPhones. (For illustrations of this, see https://cse.buffalo.edu/~rapaport/111F04/summary.html.)[13]

6.5 The Scientific History

> Logic's dominant role in the invention of the modern computer is not widely appreciated. The computer as we know it today was invented by Turing in 1936, an event triggered by an important logical discovery announced by Kurt Gödel in 1930. Gödel's discovery … decisively affected the outcome of the so-called Hilbert Program. Hilbert's goal was to formalize all of mathematics and then give positive answers to three questions about the resulting formal system: is it consistent? is it complete? is it decidable? Gödel found that no sufficiently rich formal system of mathematics can be both consistent and complete. In proving this, Gödel invented, and used, a high-level symbolic programming language: the formalism of primitive recursive functions. As part of his proof, he composed an elegant modular functional program …. This computational aspect of his work … is enough to have established Gödel as the first serious programmer in the modern sense. Gödel's computational example inspired Turing … [who] disposed of the third of Hilbert's questions by showing … that the formal system of mathematics is not decidable. Although his original computer was only an abstract logical concept, … Turing became a leader in the design, construction, and operation of the first real computers.
> —J. Alan Robinson (1994, pp. 6–7)[14]

The scientific history paralleling the engineering history concerns not the construction of a physical device that could compute but the logical and mathematical analysis of computation itself.

This story begins, perhaps, with Leibniz, who not only constructed a calculating machine, as we have seen, but also wanted to develop a *"calculus ratiocinator"*: a formalism in a universally understood language (*"characteristica universalis"*) that would enable its users to precisely express any possible question and then to rationally calculate its answer. Leibniz's motto (in Latin) was: *Calculemus!* (Let us calculate!). In other words, he wanted to develop an algebra of thought.

This task was taken up around 180 years later (around 180 years ago) by the English mathematician George Boole (1815–1864), who developed an algebra of logic, which he called *The Laws of Thought* (Boole, 2009).[15] This was what is now called propositional logic. But it lacked a procedure for determining the truth value of a given (atomic) statement (recall Section 2.3.1).

Boole's work was extended by the German mathematician Gottlob Frege (1848–1925, around 130 years ago), who developed what is now called first-order logic (or the first-order predicate calculus).[16] Frege advocated a philosophy of mathematics called "logicism," which viewed mathematics as a branch of logic. Thus, to give a firm foundation for mathematics, it would be necessary to provide a system of logic that itself would need no foundation.

---

13 See the Online Resources for further reading on modern computers.

14 Roughly, a formal system is "consistent" if *no false* propositions can be proved within it, and it is "complete" if *every true* proposition *can* be proved within it. What Gödel "found" was that if arithmetic is consistent, then it is incomplete, because an arithmetical version of the English sentence "This sentence is unprovable" is true but unprovable. For more on Gödel, see the Digression in Section 2.3.3 on whether any proposition can be proved. We will discuss "primitive recursive functions" in Section 7.6. For a discussion of the relationship between Gödel's theorems and Turing Machines, see Feferman, 2011, Rapaport, 2021.

15 Interesting historical aside: Boole's great-great-grandson is Turing Award winner and AI researcher Geoffrey Hinton.

16 None of these things called 'calculus' (plural: 'calculi') are related to the differential or integral calculus. 'Calculus' just means "system for calculation."

Unfortunately, the English philosopher Bertrand Russell (1872–1970, around 100 years ago) discovered a problem while reading the manuscript of Frege's book *The Foundations of Arithmetic*. This problem, now known as Russell's Paradox, concerned the logic of sets: a set that has as members all and only those sets that do *not* have themselves as members would *both* have itself as a member *and not* have itself as a member. This inconsistency in Frege's foundation for mathematics began a crisis that, arguably, resulted in the creation of the theory of computation.

That story continues with work done by the German mathematician David Hilbert (1862–1943, around 115 years ago), who wanted to set mathematics on a rigorous, logical foundation, one that would be satisfactory to all philosophers of mathematics, including "intuitionists" and "finitists." (Intuitionists believe that mathematics is a construction of the human mind and that any mathematical claim that can only be proved by showing that its assumption leads to a contradiction should not be accepted. Finitists believe that only mathematical objects constructible in a finite number of steps should be allowed into mathematics.) It is worth quoting Hilbert at length:

> Occasionally it happens that we seek … [a] solution [to a mathematical problem] under insufficient hypotheses or in an incorrect sense, and for this reason do not succeed. **The problem then arises: to show the impossibility of the solution under the given hypotheses, or in the sense contemplated.** Such proofs of impossibility were effected by the ancients, for instance when they showed that the ratio of the hypotenuse to the side of an isosceles right triangle is irrational. In later mathematics, the question as to the impossibility of certain solutions plays a preeminent part, and we perceive in this way that old and difficult problems, such as the proof of the axiom of parallels, the squaring of the circle, or the solution of equations of the fifth degree by radicals have finally found fully satisfactory and rigorous solutions, although in another sense than that originally intended. It is probably this important fact along with other philosophical reasons that gives rise to **the conviction (which every mathematician shares, but which no one has as yet supported by a proof) that every definite mathematical problem must necessarily be susceptible of an exact settlement, either in the form of an actual answer to the question asked, or by the proof of the impossibility of its solution and therewith the necessary failure of all attempts.** Take any definite unsolved problem, … However unapproachable these problems may seem to us and however helpless we stand before them, we have, nevertheless, **the firm conviction that their solution must follow by a finite number of purely logical processes.** Is this **axiom of the solvability of every problem** a peculiarity characteristic of mathematical thought alone, or is it possibly a general law inherent in the nature of the mind, **that all questions which it asks must be answerable**? This conviction of the solvability of every mathematical problem is a powerful incentive to the worker. We hear within us the perpetual call: There is the problem. Seek its solution. **You can find it by pure reason, for in mathematics there is no *ignorabimus* ["We will not know"].** (Hilbert, 1900, pp. 444–445, my boldface)

Hilbert proposed the following "decision problem" (*Entscheidungsproblem*) for mathematics: to devise a procedure according to which it can be decided by a finite number of operations whether a given statement of first-order logic is true under all interpretations. Because this involves truth, it is a semantic notion. But because first-order logic is "complete" – the set of truths equals the set of theorems (recall Section 2.3.3 and footnote 14) – this is usually restated syntactically, in terms of provability:

> By the Entscheidungsproblem of a system of symbolic logic is here understood the problem to find an effective method by which, given any expression Q in the notation of the system, it can be determined whether or not Q is provable in the system. (Church, 1936a, p. 41, note 6)

We will return to this in Section 8.3. A mathematical statement that was decidable in this way was also said to be "effectively computable" or "effectively calculable," because one could compute, or calculate, whether or not it was a theorem in a finite number of steps. (We'll return to "effectiveness" in Section 7.3.)

Many mathematicians took up Hilbert's challenge: in the United States, Alonzo Church (1903–1995) developed the "lambda-calculus," claiming that any effectively computable mathematical function could be computed in the lambda-calculus. The Austrian (and later American) logician Kurt Gödel (1906–1978), who had previously proved the incompleteness of arithmetic (and thus became the most respected logician since Aristotle; see footnote 14), developed the notion of "recursive" functions, which was also co-extensive with effectively computable functions. Emil Post, a Polish-born American logician (1897–1954), developed "production systems," which also capture the notion of effective computability (Soare, 2009, Section 5.2, p. 380). And the Russian A.A. Markov (1903–1979) developed what are now known as Markov algorithms. (We will look in more detail at some of these systems in Chapter 7.)

But central to our story was the work of the English mathematician Alan Turing (1912–1954), who – rather than trying to develop a mathematical theory of effectively computable functions in the way that the others approached the subject – gave an analysis of *what human computers did*. Based on that analysis, he developed a formal, mathematical model of a human computer, which he called an "*a*-machine," and which we now call, in his honor, a Turing Machine. In his classic paper published in 1936, Turing presented his informal analysis of human computation, his formal definition of an *a*-machine, his claim that functions computable by *a*-machines were all and only the functions that were "effectively computable," a (negative) solution to Hilbert's Decision Problem (by showing that there was a mathematical problem that was *not* decidable computationally: namely, the Halting Problem), a demonstration that a single, "universal," Turing Machine could do the work of all other Turing Machines, *and* – as if all that were not enough – a proof that a function was computable by an *a*-machine if and only if it was computable in Church's lambda-calculus. (To fully appreciate his accomplishment, be sure to calculate how old he was in 1936!) We will look at Turing's work in much greater detail in Chapter 8.

Later, others proved that both Turing's and Church's methods were also logically equivalent to all of the others: recursive functions, production systems, Markov algorithms, etc. Because all of these theories had been proved to be logically equivalent, this finally convinced almost everyone that the notion of "effective computability" (or "algorithm") had been captured precisely. This is now known as the Church-Turing Computability Thesis (which will be the topic of Part III). Indeed, Gödel himself was not convinced until he read Turing's paper, because Turing's was the most intuitive presentation of them all. (But in Chapters 10 and 11, we will look at the arguments of those who are still not convinced.)[17]

---

17 See the Online Resources for further reading on the scientific-mathematical history.

6.6 The Histories Converge

> … it is really only in von Neumann's collaboration with the ENIAC team that two quite separate historical strands came together: the effort to achieve high-speed, high-precision, automatic calculation and the effort to design a logic machine capable of significant reasoning.
>
> The dual nature of the computer is reflected in its dual origins: hardware in the sequence of devices that stretches from the Pascaline to the ENIAC, software in the series of investigations that reaches from Leibniz's combinatorics to Turing's abstract machines. Until the two strands come together in the computer, they belong to different histories ….
> —Michael S. Mahoney (2011, p. 26)[18]

The two histories are nicely "bookended" by Leibniz and Turing: Leibniz began with constructing a mechanical arithmetic calculator and later proposed his *calculus ratiocinator*. And Turing began with his version of a *calculus ratiocinator* – the Turing Machine – and later constructed an electronic computer (Carpenter and Doran, 1977). (For interesting comments on the interrelations between the two histories, see von Neumann, 1948.)

6.7 What Is a Computer?

The twin histories suggest different answers to our question.

6.7.1 An Engineering Answer

If computers can be defined "historically," then they are

> machines which (i) perform calculations with numbers, (ii) manipulate or process data (information), and (iii) control continuous processes or discrete devices … in real time or pseudo real time. (Martin Davis, 1977, p. 1096)

Note that (ii) can be considered a generalization of (i) because numbers are a kind of data and because performing calculations with numbers is a kind of manipulation of data. And, because being continuous or being discrete pretty much exhausts all possibilities, criterion (iii) doesn't really seem to add much. So this characterization comes down to (ii) alone: *a computer is a machine that manipulates or processes data (information).*

Or does it? One possible interpretation of clause (iii) is that the *output* of a computer need not be limited to *data* but might include instructions to other "processes … or devices": i.e. real-world effects. (We'll look into this in Chapter 16.)

According to Martin Davis (1977, pp. 1096–1097), computers had evolved to have the following "key characteristics" (perhaps among others):

"Digital operation": This focuses on only the discrete aspect of (iii).

"Stored program capability": This is understood as "the notion that the instructions for the computer be written in the same form as the data being used by the computer" and is attributed to von Neumann. (We will return to this issue in Section 9.3.2.)

---

18 Mahoney, 2011, p. 88, also emphasizes the fact that these histories "converged" but were not "coincident."

"Self-regulatory or self-controlling capability": This is not merely the automaticity of any machine but seems to include the ideas of feedback and "automatic modifiable stored programs."

"Automatic operation": This is singled out from the previous characteristic because of its emphasis on operating "independently of human operators and human intervention."

"Reliance on electronics": This is admitted to be somewhat parochial in the sense that electronic computers were, in 1977, the dominant way of implementing them, but Davis recognized that other kinds of computers would eventually exist. (Recall Section 3.5.4.)

So, ignoring the last item and merging the previous two, we come down to a version of our previous characterization: *a (modern) computer is an automatic, digital, stored-program machine for manipulating data or information.*

What is the nature of the "information" that is manipulated? Davis said that it is *numbers.* But numbers are *abstract* entities not susceptible to (or capable of) *physical* manipulation. So, are computers machines that (somehow) manipulate non-physical (abstract) *numbers?* Or are they machines that manipulate physical (concrete) *numerals* – i.e. *physical symbols* that *represent* numbers? There are actually two contrasts to be made: the first contrasts *numbers* with *numerals.* The second contrasts numbers *and* numerals in particular with *symbols more generally.*

The first contrast is closely related to issues in the philosophy of mathematics. Is mathematics itself more concerned with numerals than with numbers, or the other way around? "Formalists" and "nominalists" suggest that it is only the symbols for numbers that we really deal with. "Platonists" suggest that it is numbers that are our real concern, but at least some of them admit that the only way that we can directly manipulate numbers is via numerals – although some Platonists, including Gödel, suggest that we have a kind of perceptual ability, called 'intuition,' that allows us to access numbers directly. There are also related questions about whether numbers exist and, if so, what they are. But these issues are beyond our scope. (For more on the philosophy of mathematics, see the suggested readings in Section 3.9.)

Even if humans can "intuit" numbers, computers pretty clearly have to deal with them via numerals. So, "The mathematicians and engineers then [in the 1950s] responsible for computers [who] insisted that computers only processed *numbers* – that the great thing was that instructions could be translated into numbers" (Newell, 1980, p. 137) were probably wrong.

But even if we modify such a claim so that we replace numbers by numerals, we are faced with the second contrast given previously. Do computers *only* manipulate numerals (or numbers)? What about all the things you use your personal computers for (not to mention your smartphones) – how many of them involve numerals (or numbers)? An answer to that question will depend in part on how we interpret the symbols that a computer deals with. Certainly there are ways to build computers that, apparently, can deal with more than merely numerical symbols. The Lisp machines of the 1980s are prime examples: their fundamental symbols were Lisp lists.[19] But insofar as any computer is ultimately constructed from physical switches that are in either an "on" or "off" position, we are left with a symbol system that is binary – hence numerical – in nature. Whether we consider these symbols to be numerals or not may be more a matter of taste or convenience than anything more metaphysical. We will return to some of these issues in Section 7.4.1 and Chapter 16.

6.7.2 A Scientific Answer

If the engineering history suggests that a computer is an automatic, digital, stored-program machine for manipulating data or information, what does the scientific history suggest? Is a

---

19 Lisp is a programming language whose principal data structure is a "linked list." See e.g. S.C. Shapiro, 1992b.

computer merely a physical implementation of a Turing Machine? But Turing Machines are hopelessly inefficient and cumbersome ("register" machines, another Turing-equivalent model of computation, are closer to modern computers; see Section 9.3.1). As Perlis has observed,

> What is the difference between a Turing machine and the modern computer? It's the same as that between Hillary's ascent of Everest and the establishment of a Hilton hotel on its peak. ("Epigrams in Programming," http://www.cs.yale.edu/homes/perlis-alan/quotes.html)

To clarify some of this, it will be necessary for us to look more closely at the nature of "effective computation" and "algorithms," which we will do in the next chapter. Armed with the results of that investigation, we will return to the question of what a computer is (from a philosophical point of view), in Chapter 9.

7

Algorithms and Computability

Thou must learne the Alphabet, to wit, the order of the Letters as they stand. … Nowe if the word, which thou art desirous to finde, begin with (a) then looke in the beginning of this Table, but if with (v) looke towards the end. Againe, if thy word beginne with (ca) looke in the beginning of the letter (c) but if with (cu) then looke toward the end of that letter. And so of all the rest. &c.

—Robert Cawdrey, *A Table Alphabeticall, conteyning and teaching the true writing, and understanding of hard usuall English wordes* (1604), cited in Gleick, 2008, p. 78.

This nation is built on the notion that the rules restrain our behavior ....
—*New York Times*, 2006

Algorithmic behavior existed long before there was an algorithm (Figure 7.1).
—Janice Min, quoted in Rutenberg, 2019

7.1 Introduction

[C]omputer science is not really that much about computers. What computer science is mostly about is *computation*, a certain kind of process such as sorting a list of numbers, compressing an audio file, or removing red-eye from a digital picture. The process is typically carried out by an electronic computer of course, but it might also be carried out by a person or by a mechanical device of some sort.

The hypothesis underlying AI … is that ordinary *thinking* … is also a computational process, and one that can be studied without too much regard for who or what is doing the thinking.

—Hector J. Levesque (2017, pp. ix–x)

We have been examining two questions: (1) Is CS a science (or something else)? And (2) what is its subject matter? Is the subject of CS comput*ers*: (physical) devices that compute – or is it comput*ing*: the algorithmic processes that computers do? (Or perhaps it studies something else, such as information or information processing.) What computers *are* is intimately related to what they *do*. That is why our investigation is moving back and forth between these two topics.

In the previous chapter, we began our investigation into what *computers* are by looking at their history. In this chapter and the next, we ask what *computing* is. Then we will be in a better position

Philosophy of Computer Science: An Introduction to the Issues and the Literature, First Edition. William J. Rapaport.

Figure 7.1 BABY BLUES ©2004 Baby Blues Bros LLC. Dist. By ANDREWS MCMEEL SYNDICATION. Reprinted with permission. All rights reserved.

to return to our question of what a computer is, looking at it from a philosophical, rather than a historical, point of view. And after *that*, we will return to the question of what computing is, again looking at some philosophical issues before returning to our fundamental question: what is CS?

7.2 Functions and Computation

> The question before us – what is computation? – is at least as old as computer science. It is one of those questions that will never be fully settled because new discoveries and maturing understandings constantly lead to new insights and questions about existing models. It is like the fundamental questions in other fields – for example, "what is life?" in biology and "what are the fundamental forces?" in physics – that will never be fully resolved. Engaging with the question is more valuable than finding a definitive answer.
> —Peter J. Denning (2010, p. 2)

To understand computation, we first need to understand what a *function* is. The English word 'function' has at least two very different meanings: the *ordinary, everyday meaning* is, roughly, "purpose." Instead of asking, "What is the *purpose* of this button?" we might say, "What is the *function* of this button?" To ask for the function – i.e. the purpose – of something is to ask what it does or what it is for. We'll return to this "teleological" meaning in Chapter 16. In this chapter, we will be interested in its *mathematical meaning*, as when we say that some "dependent variable" *is a function of* – i.e. depends on – some "independent variable."[1]

7.2.1 Mathematical Functions

Many introductory textbooks define a mathematical function as an "assignment" or "mapping" of *inputs* (the "independent variables") to *outputs* (also called "values" or "dependent variables"). But I have never seen a definition of 'assignment.' Such an "assignment" is not quite the same thing as an assignment of a value to a variable in a programming language or in a system of logic. A better term might be 'association': an output *is associated with* an input. A much more rigorous way of defining a function is to give a definition based on set theory, thus explicating the notion of "association." There are two ways to do this: "extensionally" and "intensionally" (recall Section 3.4.2).

---

1 See the Online Resources for the etymologies of some of these terms.

Historical Digression: The mathematical sense of the word was initiated by Leibniz and was an extension of its teleological meaning. Euler defined a *function* as what I call a 'formula':

> A function of a variable quantity is an analytical expression of some kind composed from that variable quantity and from constant numbers or magnitudes.
> —Leonhard Euler (1748, p. 7), http://www.17centurymaths.com/contents/euler/introductiontoanalysisvolone/ch1vol1.pdf

For the history of the *word*, see the *OED* entry on the noun 'function' in its mathematical sense (sense 5; http://www.oed.com/view/Entry/75476). On the history of the *concept*, see O'Connor and Robertson, 2005.

7.2.2 Functions Described Extensionally

A *binary relation* is a set of ordered pairs of elements from two sets. (The "two" sets can be the same one; you can have a binary relation among the members of a single set.) A *function described extensionally* is a set of input-output pairs such that no two of them have the same input (or first element). That is, it is a special kind of binary relation in which no two *distinct* members of the relation have the same first element (but different second elements). That is, the input (or independent variable) of a function must always have the same output (or dependent variable). As a rule of thumb, a binary relation is a function if "same input implies same output."

Because we are considering a binary relation as a set of ordered pairs, let's write each member of a binary relation from A to B as an ordered pair $\langle a, b \rangle$, where $a \in A$ and $b \in B$. In mathematical English, here is the precise way to say what a function is:[2]

> Let A, B be sets. (Possibly, $A = B$.)
> Then f **is a *function* from** A **to** $B =_{def}$

1. f is a binary relation from A to B,
   and

2. (a) **for all** $a \in A$, and
   (b) **for all** $b \in B$, and
   (c) **for all** $b' \in B$,
   > **if** $\langle a, b \rangle$ is a member of f, and
   > **if** $\langle a, b' \rangle$ is *also* a member of f,
   > > **then** $b = b'$

Mathematical Digression: In clause 2, keep in mind that b' might be the same as b! The best way to think about these sequences of "for all" (or "universally quantified") statements is this: imagine that sets are bags containing their members. (a) "For all $a \in A$" means: put your hand in bag A, remove a (randomly chosen) member, look at it to see what it is, *and return it to the bag*. (b) "For all $b \in B$" means: put your hand in bag B, remove a (randomly chosen) member, look at it to see what it is, *and return it to the bag*. Finally, (c) "For all $b' \in B$" means exactly the same thing as in case (b), which means, in turn, that b' in step (c) might be the same member of B that you removed *but then replaced* in step (b); you might simply have picked it out twice.

---

2 The notation '$=_{def}$' should be read as "means by definition."

Here are some examples of functions in this extensional sense:

1. $f = \{\langle 0,0 \rangle, \langle 1,2 \rangle, \langle 2,4 \rangle, \langle 3,6 \rangle, \dots\}$
   Using "functional" notation – where f(input) = output – this is sometimes written $f(0) = 0$, $f(1) = 2$, …
2. $g = \{\langle 0,1 \rangle, \langle 1,2 \rangle, \langle 2,3 \rangle, \dots\}$
   This is sometimes written $g(0) = 1$, $g(1) = 2$, …
3. Here is a *finite* function (i.e. a function with a finite number of members – remember: a function is a *set*, so it has members):

$$h = \{\langle \text{'yes'}, \text{print 'hello'} \rangle,$$
$$\langle \text{'no'}, \text{print 'goodbye'} \rangle,$$
$$\langle \text{input} \neq \text{'yes'} \,\&\, \text{input} \neq \text{'no'}, \text{print 'sorry'} \rangle\}$$

The idea behind h is this:

> h prints 'hello', if the input is 'yes';
> h prints 'goodbye', if the input is 'no';
> and h prints 'sorry', if the input is neither 'yes' nor 'no'.

4. Here is a *partial* function (i.e. a function that has no outputs for some possible inputs):

$$k = \{\dots, \langle -2, \tfrac{1}{-2} \rangle, \langle -1, \tfrac{1}{-1} \rangle, \langle 1, \tfrac{1}{1} \rangle, \langle 2, \tfrac{1}{2} \rangle, \dots\}$$

Here, $k(0)$ has no output; we say that $k(0)$ is *undefined*.

5. Another example of a partial function is

$$h' = \{\langle \text{'yes'}, \text{print 'hello'} \rangle,$$
$$\langle \text{'no'}, \text{print 'goodbye'} \rangle\}$$

Here, h'('yeah'), h'('nope'), and h'('computer') are all *undefined*.

A function defined extensionally *associates* or *relates* its inputs to its outputs but does not show how to *transform* an input into an output. For that, we need a "formula" or an "algorithm" (but these are not the same thing, as we will soon see).

Mathematical Digression: Here are some terms that we will need later:

1. A function is *"one-to-one"* (or "injective") $=_{def}$ if two of its *outputs* are the same, then their inputs must have been the same (or: if two inputs differ, then their outputs differ). For example, $f(n) = n + 1$ is a one-to-one function. However, $g = \{\langle a, 1 \rangle, \langle b, 1 \rangle\}$ is not a one-to-one function (it is, however, a "two-to-one" function).
2. A function is *"onto"* (or "surjective") $=_{def}$ *everything* in the set of possible outputs "came from" something in the set of possible inputs. For example, $h(n) = n$ is an onto function. However, the previous one-to-one function $f(n)$ is not an onto function if its inputs are restricted to non-negative numbers, because 0 is not the result of adding 1 to any non-negative number, so it is not in the set of actual outputs.
3. A function is a *"one-to-one correspondence"* (or "bijective") $=_{def}$ it is *both* one-to-one *and* onto. For example, the previous onto function $h(n)$ is also one-to-one.

For more formal definitions and examples, see http://www.cse.buffalo.edu/~rapaport/191/ F10/lecturenotes-20101103.html.

7.2.3 Functions Described Intensionally

> Editor: We are making this communication intentionally short to leave as much room as possible for the answers. 1. Please define "Algorithm." 2. Please define "Formula." 3. Please state the difference. T. WANGSNESS, J. FRANKLIN *TRW Systems, Redondo Beach, California* (Wangsness and Franklin, 1966).[3]

Sometimes, functions are described "intensionally" by *formulas*. But – unlike an extensional description – this is not a unique way to describe them because two different formulas can describe the same function. Here are some examples (using the same or similar function names from Section 7.2.2):

1. $f(i) = 2i$
2. $g(i) = i + 1$
3. $g'(i) = 2i - i + 7/(3 + 4)$
   Note that g and g' use two different formulas to describe the same function;
   i.e. $g = g'$, even though their formulas are different.

Question for the Reader: How would you state the fact that the two *formulas* are different? Note that you cannot do this by saying "$i + 1 \neq 2i - i + 7/(3 + 4)$."

---

4. $h(i) = \begin{cases} \text{'hello'}, & \textbf{if } i = \text{'yes'} \\ \text{'goodbye'}, & \textbf{if } i = \text{'no'} \\ \text{'sorry'}, & \textbf{otherwise} \end{cases}$

5. **if** $i \neq 0$, **then** $k(i) = \frac{1}{i}$.

6. The next function takes as input a year y and outputs an ordered pair consisting of the month m and day d that Easter falls on in year y (Stewart, 2001):
   $E = \{\langle y, \langle m, d \rangle \rangle : \langle m, d \rangle = \langle$
   $((((19 * (y \bmod 19)) + (y/100) - ((y/100)/4) - (((8 * (y/100)) + 13)/25) + 15) \bmod 30) - (((y \bmod 19) + (11 * (((19 * (y \bmod 19)) + (y/100) - ((y/100)/4) - (((8 * (y/100)) + 13)/25) + 15) \bmod 30)))/319) + (((2 * ((y/100) \bmod 4)) + (2 * ((y \bmod 100)/4)) - ((y \bmod 100) \bmod 4) - (((19 * (y \bmod 19)) + (y/100) - ((y/100)/4) - (((8 * (y/100)) + 13)/25) + 15) \bmod 30) - (((y \bmod 19) + (11 * (((19 * (y \bmod 19)) + (y/100) - ((y/100)/4) - (((8 * (y/100)) + 13)/25) + 15) \bmod 30)))/319) + 32) \bmod 7) + 90)/25,$

   $((((19 * (y \bmod 19)) + (y/100) - (y/100)/4 - (((8 * (y/100)) + 13)/25) + 15) \bmod 30) - (((y \bmod 19) + (11 * (((19 * (y \bmod 19)) + (y/100) - (y/100)/4 - (((8 * (y/100)) + 13)/25) + 15) \bmod 30)))/319) + (((2 * ((y/100) \bmod 4)) + (2 * ((y \bmod 100)/4)) - ((y \bmod 100) \bmod 4) - (((19 * (y \bmod 19)) + (y/100) - (y/100)/4 - (((8 * (y/100)) + 13)/25) + 15) \bmod 30) - (((y \bmod 19) + (11 * (((19 * (y \bmod 19)) + (y/100) - (y/100)/4 - (((8 * (y/100)) + 13)/25) + 15) \bmod 30)))/319) + 32) \bmod 7) + (((((19 * (y \bmod 19)) + (y/100) - (y/100)/4 - (((8 * (y/100)) + 13)/25) + 15) \bmod 30) - (((y \bmod 19) + (11 * (((19 * (y \bmod 19)) + (y/100) - (y/100)/4 - (((8 * (y/100)) + 13)/25) + 15) \bmod 30)))/319) + (((2 * ((y/100) \bmod 4)) + (2 * ((y \bmod 100)/4)) - ((y \bmod 100) \bmod 4) - (((19 * (y \bmod 19)) + (y/100) - (y/100)/4 - (((8 * (y/100)) + 13)/25) + 15) \bmod 30) - (((y \bmod 19) + (11 * (((19 * (y \bmod 19)) + (y/100) - (y/100)/4 - (((8 * (y/100)) + 13)/25) + 15) \bmod$

---

3 For the published answers, see the Further Reading box at the end of this section.

30)))/319) + 32) mod 7) + 90)/25 + 19) mod 32
⟩}

A function described *extensionally* is like a black box; we know the inputs and outputs but not how they are related. A function described *intensionally* via a *formula* is less opaque and gives us more understanding of the relationship between the input and the outputs.[4]

A function described intensionally via an *algorithm* – roughly, a set of instructions for computing the output of the function – gives us even more understanding, not only telling us *what* the relationship is but also giving explicit instructions on *how* to make the conversion from input to output. Although formulas may look a lot like algorithms, they are not the same thing. Consider, for example, the formula '2 + 4 ∗ 5': without an explicit statement of a rule telling you whether to multiply first or to add first, there is no way of knowing whether the number expressed by that formula is 30 or 22. Such a rule, however, would be part of an *algorithm* telling you how to calculate the value of the formula.

Or consider the formula '$2x + 1$': should you first calculate $2x$ and then add 1 to it? Or should you store 1 somewhere (say, by writing it on a piece of paper), then calculate $2x$, and finally add $2x$ to 1? And how should you calculate $2x$? Take 2, and then multiply it by x? Or take x, and then multiply it by 2? One of these might be easier to do than the other; for instance, 2×1000 might take only 1 step, whereas 1000×2 might take 999 steps. Of course, the commutative laws of addition and multiplication tell us that, in this case, as far as the output is concerned, it doesn't matter in which order you compute the value of the formula; however, one of these algorithms might be more efficient than the other. In any case, here we have a clear case of only one formula but at least two (and possibly four) distinct algorithms.

Perhaps an even clearer example is function E – the one that tells you when Easter occurs. I dare you to try to use this formula to find out when Easter will occur next year! Where would you even begin? To use it, you would need an *algorithm* such as the one at http://tinyurl.com/yb9jvbpl (or http://techsupt.winbatch.com/webcgi/webbatch.exe?techsupt/nftechsupt.web+WinBatch/How~To+Easter~finder.txt. If neither of these links works, do the following: link to http://techsupt.winbatch.com/ and then search for "Easter finder.") (A related matter is knowing whether the formula is even correct! We'll explore this issue in Chapter 15.)[5]

Some functions expressed as formulas might be seen as containing an implicit algorithm for how to compute them:

> [A] term in the series for arctan 1/5 can be written either as $(1/5)^m/m$ or as $1/(m5^m)$. Mathematically these expressions are identical, but they imply different computations. In the first case you multiply and divide long decimal fractions; in the second you build a large integer and then take its reciprocal. (B. Hayes, 2014a, p. 344)

But these formulas can only be interpreted as algorithms with additional information about the order of operations (roughly, do things in innermost parentheses first, then do exponentiations, then multiplication and division from left to right, then addition and subtraction from left to right).

Functions describable by formulas are not the only kind of functions. There are functions that lack formulas for computing them. For example, "table look-up" functions are essentially extensional functions for which the only way to identify the correct output for a given input is to look it up in a table (rather than to compute it); usually, this is the case when there is no lawlike pattern relating the inputs and the outputs. Of course, there are non-computable functions, such as the

---

4 See the Online Resources for further reading on this point.
5 I created this formula by working backward from the algorithm given in Stewart, 2001, so it's quite possible that I introduced a typographical error! Even if I didn't, I am assuming the algorithm in Stewart, 2001 is correct. And that could be a big assumption.

Halting Problem (to be discussed in Section 7.7). And there are random functions. (For more on this, see Chaitin, 2005.)

One of the central purposes – perhaps *the* central question – of CS is to figure out which functions *do* have algorithms for computing them! This includes "non-mathematical" functions, such as the (human) cognitive "functions" that take as input sensory information from the environment and produce as output (human, cognitive) behavior. To express this another way, the subfield of CS known as AI can be considered to have as its purpose figuring out which such cognitive functions are computable.

Digression: Algorithm vs. Formula – The Answer. The published answers to the question asked in the epigraph to this section are Huber, 1966, Knuth, 1966. Knuth's answer is a commentary on Huber's. Huber's answer, roughly, is that an algorithm is a set of instructions for computing the value of a function by "executing" (or carrying out, or following) the instructions, whereas a formula is an expression describing the value of a function; it can be "evaluated" (i.e. the value of the function can be determined from the formula) but not executed (because a formula does not come equipped with an algorithm for telling you *how* to evaluate it). In a sense, a formula is "static," whereas an algorithm is (potentially) "dynamic." See also B. Hayes, 2006, p. 204, cols. 2–3, and Chater and Oaksford, 2013, p. 1172. I have collected lots of examples of informal algorithms (not all of them serious ones) at http://www.cse.buffalo.edu/~rapaport/584/whatisanalg.html.

7.2.4   Function "Machines"

Sometimes, functions are characterized as "machines" that accept input into a "black box" with a "crank" that mysteriously transforms the input into an output. In Figure 7.2, f is a machine into which you put a; you then turn a crank (clockwise, let's suppose), f grinds away at the input by means of some mysterious mechanism, and finally the machine outputs b (i.e. $f(a)$). But this view of a function as being something "active" or "dynamic" that *changes* something is incorrect.[6]

Despite what you may have been told elsewhere (I was told this in high school), this "machine" is **NOT** what a function is! A function, as we saw in Section 7.2.2, is merely the *set* of input-output pairs. So, what is the machine? **It is a *computer*!** And the mysterious "gears" hidden inside the black box implement an *algorithm* that computes the function.

Figure 7.2   A function "machine" f that transforms input a into output $b = f(a)$. Source: Author's drawing.

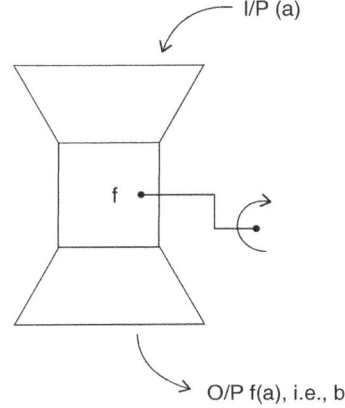

Interestingly, Gödel made this observation in the 1930s in an unpublished paper!

> [Turing] has shown that the computable functions defined in this way [i.e. in terms of Turing Machines] are exactly those for which you can construct a machine with a finite number of parts which will do the following thing. If you write down any number n_1, ..., n_r on a slip of paper and put the slip into the machine and turn the crank, then after a finite number of turns the machine will stop and the value of the function for the argument n_1, ..., n_r will be printed on the paper. (Gödel, 1938, p. 168)

So, the machine pictured in Figure 7.2 is a Turing Machine! And the problem with this machine metaphor for a function, as we will see, is that not all functions can be computed by algorithms; i.e. there are functions for which there are no such "function machines."

7.2.5 Computable Functions

We can combine the two central concepts of function and algorithm as follows:

> A function f **is computable** means, roughly, that there is an "algorithm" that *computes* f. (Cf. Church, 1936b, pp. 356, 358.)

This is only a rough definition or characterization because, for one thing, we haven't yet defined 'algorithm.' But using our informal characterization of an algorithm from Section 7.2.3, it makes sense to define a function as being computable if we can … well … compute it! So:

> A function f **is computable** iff there is an algorithm A_f such that, for all inputs i, $A_f(i) = f(i)$.

That is, a function f is computable by an algorithm A_f if both f and A_f have the same input-output "behavior" (i.e. if both define the same binary relation, or set of input-output pairs). Moreover, A_f must specify *how* f's inputs and outputs are related. A *function* only *shows* its input-output pairs but is silent about *how* they are related. A *formula* for a function does show how they are related but is silent about how to use that relationship to *transform* the input into the output. That is done by an *algorithm* for that function: a procedure, or a mechanism, or a set of intermediate steps or instructions that transforms the input into the output; it shows you explicitly how to find the output by starting with the input – how to get from the input to the output. Algorithms shouldn't be magic or merely arbitrary.[7]

It seems easy enough to give examples of algorithms for some of the functions listed earlier:

1. **Algorithm** $A_f(i)$
   begin
      input i;
      multiply i by 2;
      output result.
   end

---

7 Except possibly in the "base case," where the "algorithm" is so simple or basic that it consists merely in giving you the output directly, without any intermediate processing. (See Section 7.4.3.2 for an explanation of "base case.")

2. **Algorithm** $A_g(i)$
   begin
      input i;
      add 1 to i;
      output result.
   end
3. **Algorithm** $A_{g'}(i)$
   begin
      input i;
      multiply i by 2;
      call the result x;
      subtract i from x;
      add 3 to 4;
      call the result y;
      divide 7 by y;
      add x to y;
      output result.
   end
4. For $E(m, d)$, see the English algorithm in Stewart, 2001 or the computer program online at the URL given in Section 7.2.3. Note that even though that algorithm may not be easy to follow, it is certainly much easier than trying to compute the output of E from the formula. (For one thing, the algorithm tells you where to begin!)
5. **Algorithm** $A_k(i)$
   begin
      if $i \neq 0$
        then
          begin
            divide 1 by i;
            output result
          end.
   end

Note that Algorithm $A_k(i)$ doesn't tell you what to do if $i = 0$, because there is no "**else**"-clause. So, what would happen if you input 0? Because the algorithm is silent about what to do in this case, anything might happen! If it were implemented on a real computer, it would probably "hang" (i.e. do nothing), or crash, or go into an infinite loop.

Question for the Reader: The philosopher Richard Montague (1960, p. 433) suggested that – for a more general notion of computation than a mere Turing Machine (one that would apply to both digital and analog computation) – a computer needs an output signal that indicates when the computation is finished. As we will see in Chapter 8, in Turing's theory of computation, the machine simply halts.

How do you know that a machine has halted rather than merely being in an infinite loop? What is the difference between a program *halting* and a program *hanging*?

Good programming technique would require that the program be rewritten to make it "total" instead of "partial," perhaps with an error handler like this:

Algorithm $A'_k(i)$
begin
  if $i \neq 0$
    then
      begin
        divide 1 by i;
        output result
      end
    else output "illegal input."
end

> **Question for the Reader:** Is $A'_k(i)$ merely a different algorithm for function k, or is it really an algorithm for a *different function* (call it k')? We'll return to this puzzle in Section 15.1.

But our notion of algorithm is still a rough one. *Can* it be made more precise?

7.3 'Algorithm' Made Precise[8]

> The meaning of the word *algorithm*, like the meaning of most other words commonly used in the English language, is somewhat vague. In order to have a *theory of algorithms*, we need a mathematically precise definition of an algorithm. However, in giving such a precise definition, we run the risk of not reflecting exactly the intuitive notion behind the word.
> —Gabor T. Herman (1983, p. 57)

7.3.1 Ancient Algorithms

Before anyone attempted to define 'algorithm,' many algorithms were in use by mathematicians – e.g. ancient Babylonian procedures for finding lengths and for computing compound interest (Knuth, 1972a), Euclid's procedures for construction of geometric objects by compass and straightedge (Toussaint, 1993), and Euclid's algorithm for computing the greatest common divisor of two integers. And algorithms were also used by ordinary people – e.g. the algorithms for simple arithmetic with Hindu-Arabic numerals (Robertson, 1979). In fact, the original, eponymous use of the word referred to those arithmetic rules as devised by Abū 'Abdallāh Muḥammad ibn Mūsā Al-Khwārizmī, a Persian mathematician who lived around 1200 years ago (780–850 CE). 'Algorithm' is merely a corruption of what looks as if it might be his last name: 'Al-Khwarizmi' really just means something like "the person who comes from Khwarizm," a lake that is now known as the Aral Sea (Knuth, 1985, p. 171).[9]

7.3.2 "Effectiveness"

When David Hilbert investigated the foundations of mathematics, his followers began to try to make the notion of algorithm precise, beginning with discussions of "effectively calculable," a

---

8 This section is adapted from Rapaport, 2012b, Appendix.
9 See the Online Resources for further reading on Al-Khwarizmi.

phrase first used by Jacques Herbrand in 1931 (Gandy, 1988, p. 68) and later taken up by Alonzo Church (1936b) and his student Stephen Kleene (1952) but left largely undefined, at least in print.[10]

J. Barkley Rosser (another of Church's students) made an effort to clarify the contribution of the modifier 'effective':

> "Effective method" is here used in the rather special sense of a method each step of which is [1] *precisely predetermined* and which is [2] *certain to produce the answer* [3] in a *finite number of steps*. (Rosser, 1939, p. 55, my italics and enumeration)

But what, exactly, does 'precisely predetermined' mean? And does 'finite number of steps' mean (a) that the written statement of the algorithm has a finite number of instructions or (b) that when executing them, only a finite number of tasks must be performed? In other words, what gets counted: written steps or executed instructions? One written step – "**for** $i \leftarrow 1$ to 100 **do** $x \leftarrow x + 1$" – can result in 100 executed instructions. And one written step – "**while** true **do** $x \leftarrow x + 1$" – can even result in infinitely many executed instructions! Here is what Hilbert had to say about finiteness:

> It remains to discuss briefly what general requirements may be justly laid down for the solution of a mathematical problem. I should say first of all, this: that it shall be possible to establish the correctness of the solution by means of a finite number of steps based upon a finite number of hypotheses which are implied in the statement of the problem and which must always be exactly formulated. This requirement of logical deduction by means of a finite number of processes is simply the requirement of rigor in reasoning. (Hilbert, 1900, pp. 440–441)

7.3.3 Three Attempts at Precision

Much later, *after* Turing's, Church's, Gödel's, and Post's precise formulations and *during* the age of computers and computer programming, slightly less vague, though still informal, characterizations were given by A.A. Markov, Stephen Kleene, and Donald Knuth.

Markov
According to Markov (1960, p. 1), an algorithm is a "computational process" satisfying three (informal) properties:

a) The presence of a precise prescription, leaving no possibility of arbitrary choice, and in the known sense generally understood – the algorithm is *determined*.
b) The possibility of starting from original given objects, which can vary within known limits – *applicability* of the algorithm.
c) The tendency of the algorithm to obtain a certain result, finally obtained for appropriate original given objects – the *effectiveness* of the algorithm.

These are a bit obscure: being "determined" may be akin to Rosser's "precisely predetermined." But what about "applicability"? Perhaps this simply means an algorithm must not be limited to converting one specific input to an output but must be more general. And Markov's notion of "effectiveness" seems restricted to only the second part of Rosser's notion: namely, that of "producing the answer." There is no mention of finiteness unless that is implied by being computational.

---

10 See the Online Resources for further reading on 'effective.'

Kleene

In a 1995 essay, Kleene wrote:

> [a] … a method for answering any one of a given infinite class of questions … is given by a set of rules or instructions, describing a procedure that works as follows. [b] *After* the procedure has been described, [then] if we select *any* question from the class, the procedure will then tell us how to perform successive steps, so that after a finite number of them we will have the answer to the question selected. [c] In particular, immediately after selecting the question from the class, the rules or instructions will tell us what step to perform first, unless the answer to the question selected is immediate. [d] After our performing any step to which the procedure has led us, the rules or instructions will *either* enable us to recognize that now we have the answer before us and to read it off, *or else* that we do not yet have the answer before us, in which case they will tell us what step to perform next. [e] In performing the steps, we simply follow the instructions like robots; no ingenuity or mathematical invention is required of us. (Kleene, 1995, p. 18, my enumeration)

So, for Kleene in 1995, an algorithm (informally) is:

a. A *set* of *rules or instructions* that describes *a procedure*. The procedure is one thing; its description is another: the latter is a set of imperative sentences.
b. Given a class of questions Q, a procedure P_Q for answering any member of Q, and any $q \in Q$: P_Q gives a *finite sequence* of steps (described by its rules) that answers q. So, the finiteness occurs in the *execution* of P_Q (not necessarily in P_Q itself). And P_Q does not depend on q, only on Q, which suggests, first, that the algorithm must be *general* and not restricted to a single question. (An algorithm for answering '$2 + 3 =?$' must also be able to answer all questions of the form '$x + y =?$'.) Second, it suggests that an algorithm has a goal, purpose, or teleological "function." That is, the algorithm must not just be a set of instructions that *happens* to answer the questions in Q; it must be *designed* for that purpose, because it *depends on* what Q is. (We'll investigate this important issue in Chapter 16.)
c. The algorithm takes question q as input, and either outputs q's answer ("base case") or outputs the first step to answer q ("recursive case").[11]
d. If it is the "recursive case," then, presumably, q has been reduced to a simpler question, and the process repeats until the answer is produced as a base case. Moreover, the answer is immediately recognizable. That does not necessarily require an intelligent mind to recognize it. It could be merely that the algorithm halts with a message that the output is, indeed, the answer. In other words, the output is of the form "the answer to q is q_i, where q_i is the answer and the algorithm halts, or q_i is a one-step-simpler question and then the algorithm tells us what the next step is." In a so-called "trial-and-error machine" (to be discussed in Section 11.10), the output is of the form "my current guess is that the answer to q is q_i, and then the algorithm tells us what the next step is." (We'll see such an algorithm in Section 7.7.)
e. The algorithm is complete or independent in the sense that it contains all information for executing the steps. We, the executor, do not (have to) supply anything else. In particular, we do not (have to) accept any further, unknown, or unforeseen input from any other source (i.e. no "oracle" or interaction with the external world). We'll return to these ideas in Chapters 11 and 16.

---

11 See Section 7.4.3.2 for an explanation of these scare-quoted terms.

Knuth

Donald Knuth goes into considerably more detail, albeit still informally (Knuth, 1973, "Basic Concepts: Section 1.1: Algorithms," pp. xiv–9, esp. pp. 1–9). He says that an algorithm is "a finite set of rules which gives a sequence of operations for solving a specific type of problem," with "five important features" (Knuth, 1973, p. 4):

> **1) Finiteness**. An algorithm must always terminate after a finite number of steps. (Knuth, 1973, p. 4)

Note the double finiteness: a finite number of rules in the text of the algorithm *and* a finite number of steps to be carried out. Moreover, algorithms must halt. (Halting is not guaranteed by finiteness; see point 5.)

Interestingly, Knuth also says that an algorithm is a *finite* "computational method," where a "computational method," more generally, is a "procedure," which only has the next four features (Knuth, 1973, p. 4).[12]

> **2) Definiteness**. Each step … must be precisely defined; the actions to be carried out must be rigorously and unambiguously specified …" (Knuth, 1973, p. 5).

This seems to be Knuth's analogue of the "precision" that Rosser and Markov mention. The best examples often come from the humor that they engender:

Precision: In a "Zits" comic strip from 14 March 2009, teenage Jeremy's mother hands him his laundered clothes, saying "Take these clothes. Carry them in your hands and walk in that direction until you come to the stairs. Make a 90° right turn. Walk up 22 steps. Turn right, and then enter the first room you come to. Put the clothes somewhere within those four walls." Her husband comments, "You don't leave a lot of room for interpretation, do you?" to which she replies, "After picking up several tons of laundry off that stairway, I've learned to be specific." (We'll return to this example in Section 10.5 and to this issue in Section 16.7.)

Detail: In a "Hagar the Horrible" comic strip from 1992, Helga asks Lucky Eddie to take out the garbage. He doesn't return, so Helga goes outside to find him still holding the garbage. She realizes that she has to add detail: "Now empty the garbage into the barrel and come back inside."

And Figure 7.3 is a real-life example of an ambiguous instruction.

> **3) Input.** An algorithm has zero or more inputs …" (Knuth, 1973, p. 5).

Curiously, only Knuth and Markov seem to mention this explicitly, with Markov's "applicability" property suggesting that there must be at least *one* input. Why does Knuth say *zero* or more? If algorithms are procedures for computing functions, and if functions are sets of input-output *pairs*, then wouldn't an algorithm *always* have to have input? Presumably, Knuth wants to allow for the possibility of a program that simply outputs some information. Perhaps Knuth has in mind the possibility of the input being internally stored in the computer rather than having to be obtained from the external environment. An example of this[13] would be an algorithm for computing the *n*th digit in the decimal expansion of a real number: there do not need to be any explicit inputs; the algorithm can just generate each digit in turn. Or perhaps this is how constant functions (functions whose output is constant, no matter what their input is) are handled. (We'll return to this in

---

12 See the Online Resources for further reading on algorithms vs. procedures.
13 Which Matti Tedre reminded me of (personal communication, 2018).

To connect contact wire
1. Turn head counterclockwise and remove. (A)
2. Slide battery in halfway and insert contact (on wire end) between the coils of the positive (+) spring, pushing until contact snaps onto spring. (B)
3. Remove plastic cap from negative (-) spring.
4. Slide battery fully in and replace lantern head.

Figure 7.3 A real-life example of an ambiguous instruction. Whose head should be removed? https://www .shopyourway.com/energizer-6v-led-utility-lantern/162752012.

Section 11.8.1.) It is worth noting, however, that Hartmanis and Stearns, 1965, p. 288 – the founding document of the field of computational complexity – allows their multi-tape Turing Machines to have at most one tape, which is an output-only tape; there need not be any input tapes. And, if there is only *at most* one output tape, there need not be *any* input or output at all! However, Knuth disagrees (see the next point).

4) Output. An algorithm has one or more outputs …" (Knuth, 1973, p. 5).

That there must be at least one output echoes Rosser's property [2] ("certain to produce the answer") and Markov's notion (c) of "effectiveness" ("a certain result"). But Knuth characterizes outputs as "quantities which have a specified relation to the inputs" (Knuth, 1973, p. 5): The "relation" would no doubt be the functional relation between inputs and outputs, but if there is no input, what kind of a relation would the output be in? (Cf. Copeland and Shagrir, 2011, pp. 230–231.)

Others have noted that, while neither inputs nor outputs are *necessary*, they are certainly useful:

> There remains, however, the necessity of getting the original definitory information from outside into the device, and also of getting the final information, the results, from the device to the outside. (von Neumann, 1945, Section 2.6, p. 3).

> Do computations have to have inputs and outputs? The mathematical resources of computability theory can be used to define 'computations' that lack inputs, outputs, or both. But the computations that are generally relevant for applications are computations with both inputs and outputs. (Piccinini, 2011, p. 741, note 11)

The computer has to have something to work on ("definitory information," or input), and it has to let the human user know what it has computed ("the final information, the results," or output). It shouldn't just sit there silently computing. In other words, there has to be input and output if the computer is not to be "solipsistic."

> **Philosophical Digression:** Solipsism is, roughly, the view that I am the only thing that exists, or that I (or my mind) cannot have knowledge of the external world. So, a computer with no input or output would only have "knowledge" of things "inside" itself. See Thornton, 2004, Avramides, 2020. We'll return to solipsism in Section 11.8.4.

Newell has suggested that there must be input iff there is output:

> **Read** is the companion process to **write**, each being necessary to make the other useful. **Read** only obtains what was put into expressions by **write** at an earlier time; and a **write** operation whose result is never **read** subsequently might as well not have happened. (Newell, 1980, p. 163)

However, there are circumstances where **read** would take input from the external world, not necessarily from previous output. And the last clause suggests that while output is not necessary, it is certainly useful! Of course, a partial function (Section 7.2.2) will lack outputs for those inputs for which it is undefined.

> **5) Effectiveness.** … All of the operations to be performed in the algorithm must be sufficiently basic that they can in principle be done exactly and in a finite length of time by a man [sic] using pencil and paper. (Knuth, 1973, p. 6)

Note, first, how the term 'effective' has many different meanings among all these characterizations of "algorithm," ranging from it being an unexplained term, through being synonymous with 'algorithm,' to naming very particular – and very different – properties of algorithms.

Second, it is not clear how Knuth's notion of effectiveness differs from his notion of definiteness. Both seem to have to do with the preciseness of the operations.

Third, Knuth brings in another notion of finiteness: finiteness in time. Note that an instruction to carry out an infinite sequence of steps in a finite time could be accomplished by doing each step twice as fast as the previous step; or each step might only take a finite amount of time, but the number of steps required might take longer than the expected life of the universe, as in computing a perfect, non-losing strategy in chess (Zobrist, 2000, p. 367). These may have interesting theoretical implications (explored in Section 11.5) but do not seem very practical. Knuth (1973, p. 7) observes that "we want *good* algorithms in some loosely defined aesthetic sense. One criterion of goodness is the length of time taken to perform the algorithm …."

Finally, the "gold standard" of "a [hu]man using pencil and paper" seems clearly to be an allusion to Turing's analysis (Turing, 1936), which we will examine in great detail in the next chapter.

Summary

We can summarize these informal observations as follows:

An algorithm for executor E to accomplish goal G is

1. a procedure P, i.e. a finite set (or sequence) of statements (or rules, or instructions), such that each statement S is:
   (a) composed of a finite number of symbols (or uninterpreted marks) from a finite alphabet
   (b) and unambiguous for E – i.e.
       i. E knows how to do S
       ii. E can do S
       iii. S can be done in a finite amount of time

iv. and, after doing *S*, *E* knows what to do next –
2. *P* takes a finite amount of time (i.e. *P* halts),
3. and *P* ends with *G* accomplished.

The important thing to note is that the more one tries to make precise these *informal* requirements for something to be an algorithm, the more one recapitulates Turing's motivation for the formulation of a Turing Machine. In Chapter 8, we will look at exactly what Turing did.

But first we are going to look at five great insights of CS.

7.4   Five Great Insights of CS

This section presents five great insights of CS. The first three help make precise the vague notion of algorithm that we have been looking at. The fourth links the vague notion with a precise one. Together, they define the smallest possible language in which you can write any procedure for any computer. (And by 'computer' here, I merely mean anything – machine or human – that can execute an algorithm.) The fifth brings in engineering concerns.

7.4.1   Insight 1: Representation

> … he came to us knowing *yes* and *no*, and those can go a long way once you find the right questions. —George R.R. Martin, 2011, p. 432

The first great insight is this:

All the information about any computable problem can be represented using only *two nouns*, e.g. '0' and '1.'

Here are some of the people who contributed to this insight:

- Sir Francis Bacon, around 1605, developed an encoding of the alphabet by any objects "capable of a twofold difference."[14] Bacon used 'a' and 'b,' but he also suggested that coding could be done "by Bells, by Trumpets, by Lights and Torches, by the report of Muskets, and any instruments of like natures" (http://home.hiwaay.net/~paul/bacon/advancement/book6ch1.html). And, of course, once you've represented the *alphabet* in a binary coding, then anything capable of being represented in *text* can be similarly encoded (Quine, 1987, "Universal Library," pp. 223–235, https://urbigenous.net/library/universal_library.html).
- Leibniz gave an "Explanation of Binary Arithmetic" in 1703 (http://www.leibniz-translations.com/binary.htm).
- Famously, Samuel F.B. Morse not only invented the telegraph but also (following in Bacon's footsteps) developed his eponymous binary code in the mid-1800s (http://en.wikipedia.org/wiki/Morse_code). Arguably, however, Morse code (traditionally conceived as having only two symbols, "dot" and "dash") is not strictly binary because there are "blanks," or time lapses, between those symbols (Gleick, 2011, p. 20, footnote; Bernhardt, 2016, p. 29).

---

14 Bacon, *Advancement of Learning*, http://home.hiwaay.net/~paul/bacon/advancement/book6ch1.html; for discussion, see Cerf, 2015, p. 7.

- Going beyond language, the philosopher Frank P. Ramsey, in a 1929 essay on "a language for discussing … facts" – perhaps something like Leibniz's *characteristica universalis* (Section 6.5) – suggested that "all [of the terms of the language] may be best symbolized by numbers. For instance, colours have a structure, in which any given colour may be assigned a place by three numbers …. *Even smells may be so treated* …" (Ramsey, 1929, pp. 101–102, my italics). (For more examples, see http://www.cse.buffalo.edu/~rapaport/111F04/greatidea1.html.)
- In 1936, as we will see in Chapter 8, Turing made essential use of '0' and '1' in the development of Turing Machines.
- Finally, the next year, Claude Shannon (in his development of the mathematical theory of information) used "The symbol 0 … to represent … a closed circuit, and the symbol 1 … to represent … an open circuit" (Shannon, 1937, p. 4) and then showed how propositional logic could be used to represent such circuits. Moreover,

> Up until … [the time of publication of Shannon's "Mathematical Theory of Communication" (Shannon, 1948)], everyone thought that communication was involved in trying to find ways of communicating written language, spoken language, pictures, video, and all of these different things – that all of these would require different ways of communicating. Claude said no, *you can turn all of them into binary digits*. And then you can find ways of communicating the binary digits. (Robert Gallager, quoted in Soni and Goodman, 2017)

There is nothing special about the *symbols* '0' and '1.' As Bacon emphasized, any other bistable[15] pair suffices, as long as they can flip-flop between two easily distinguishable states, such as the *numbers* 0 and 1, "on/off," "magnetized/de-magnetized," "high voltage/low voltage," etc. Strictly speaking, these can be used to represent *discrete* things; *continuous* things can be approximated to any desired degree, however.

This limitation to *two* nouns is not necessary: Turing's original theory had no restriction on how many symbols there were. There were only restrictions on the nature of the symbols (they couldn't be too "close" to each other; i.e. they had to be distinguishable) and that there be only finitely many. And some early computers used decimal notation. But if we want to have a minimal language for computation, having only two symbols *suffices*, and making them '0' and '1' (rather than, say, 'a' and 'b' – not to mention "the report of Muskets"!) is *mathematically convenient*.[16]

7.4.2 Insight 2: Processing

Turing is also responsible for providing the *verbs* of our minimal language. Our second great insight is this:

> Every algorithm can be expressed in a language for a computer (namely, a Turing Machine) consisting of
>
> - an arbitrarily long, paper tape divided into squares
>   (like a roll of toilet paper, except you never run out (Weizenbaum, 1976)),
> - with a read/write head,
> - whose only nouns are '0' and '1,'

---

15 That is, something that can be in precisely one of two states.
16 See the Online Resources for further reading on binary representation.

- **and whose only five verbs (or basic instructions) are:**
  1. **move-left-1-square**
  2. **move-right-1-square**
  3. **print-0-at-current-square**
  4. **print-1-at-current-square**
  5. **erase-current-square**

The exact verbs depend on the model of Turing Machine.[17] The two "move" instructions could be combined into a single verb that takes a direct object (i.e. a function that takes a single input): move(location). And the "print" and "erase" instructions could be combined into another single transitive verb: print(symbol), where "symbol" could be either '0,' '1,' or 'blank' (here, erasing would be modeled by printing a blank). In Section 8.12, we'll see Turing do something similar.

Digression: Erasing. Wang, 1957, p. 80, notes that

> there are many things which we can do when we permit erasing but which we cannot do otherwise. Erasing is dispensable only in the sense that all functions which are computable with erasing are also computable without erasing. For example, if we permit erasing, … only the … answer appears on the tape at the end of the operation, everything else having been erased.

Deciding how to count the number of verbs is an interesting question. In the previous formulation, do we have three nouns ('0,' '1,' 'blank') and only one transitive verb ('print(symbol)')? Or do we have only two nouns ('0,' '1') but two verbs ('print(symbol),' 'erase')? Gurevich (1999, pp. 99–100) points out that

> at one step, a Turing machine may change its control state, print a symbol at the current tape cell[,] and move its head. … One may claim that every Turing machine performs only one action at a time, but that action [can] have several parts. The number of parts is in the eye of the beholder. You counted three parts. I can count four parts by requiring that the old symbol is erased before the new symbol is printed. Also, the new symbol may be composed, e.g. '12'. Printing a composed symbol can be a composed action all by itself.

And Fortnow (2018) suggests that there are four verbs: move left, move right, read, write. In any case, we can certainly get by with only two (slightly complex) verbs or five (slightly simpler) verbs (or six, if you include "read"). But either version is pretty small.[18]

7.4.3 Insight 3: Structure

7.4.3.1 Structured Programming (I)

We have two nouns and only two or three verbs. Now we need some grammatical rules to enable us to put them together. The software-engineering concept of "structured programming" does the trick. This is a style of programming that avoids the use of the 'go to' command: In early programming languages, programmers found it useful to "go to" – or "jump" to – another location in the program, sometimes with the ability to return to where the program jumped *from* (but not

---

17 The ones cited here are taken from John Case's model described in Schagrin et al., 1985, Appendix B,http://www.cse.buffalo.edu/~rapaport/Papers/schagrinetal85-TuringMachines.pdf.
18 See the Online Resources for another minimal set of operations.

always). This resulted in what was sometimes called "spaghetti code," because if you looked at a flowchart of the program, it consisted of long, intertwined strands of code that were hard to read and harder to ensure that they were correct. Edsger W. Dijkstra (1968) wrote a letter to the editor of the *Communications of the ACM*, headlined "Go To Statement Considered Harmful," arguing against the use of such statements. This resulted in an attempt to better "structure" computer programs so that the use of 'go to' could be minimized: Corrado Böhm and Giuseppe Jacopini showed how it could be completely eliminated (Böhm and Jacopini, 1966; Harel, 1980). This gives rise to the third insight (and the third item needed to form our language):

> **Only three *rules of grammar* are needed to combine any set of basic instructions (verbs) into more complex ones:**
>
> 1. **sequence**: first do this; then do that
> 2. **selection** (or choice):
>   **if** such-&-such is the case,
>   **then** do this
>   **else** do that
> 3. **repetition** (or looping): **while** such-&-such is the case **do** this
>
>   … where "this" and "that" can be any of the basic instructions, or any complex instruction created by application of any grammatical rule.

Dijkstra, 1972, esp. Section 7, is the classic discussion of structured programming based on sequence, selection, and repetition, along with top-down design and stepwise refinement, with several examples worked out in detail.[19]

7.4.3.2 Digression – Recursive Definitions

This third insight can be thought of as a "recursive" definition of "instruction."

A recursive definition of some concept C consists of two parts. The first part, called the "base case," gives you some explicit examples of C. These are not just any old examples but are considered the simplest or most basic ("atomic") instances of C – the building blocks from which all other, more complex ("molecular") instances of C can be constructed.

The second part of a recursive definition of C consists of rules (algorithms!) that tell you how to construct those more complex instances of C. But these rules don't simply tell you how to construct the complex instances from just the base cases. Rather, they tell you how to construct the complex instances of C *from any instances of C that have already been constructed*. The first complex instances, of course, will be constructed directly from the base cases. But others, even more complex, will be constructed from the ones that were constructed directly from the base cases, and so on. What makes such a definition "recursive" is that simpler instances of C "recur" in the definitions of more complex instances.

So, the base case of a recursive definition tells you how to *begin*. And the recursive case tells you how to *continue*. Recursive definitions can be found outside of CS. Here are two examples:

1. According to some branches of Judaism, a person p is Jewish if (a) p was converted to Judaism (base case) or (b) p's mother was Jewish (recursive case).
2. "Organisms originate either through synthesis of non-living materials [base case] or through reproduction, either sexual or asexual [recursive case]"
   (Northcott and Piccinini, 2018, p. 2).

---

19 See the Online Resources for further reading on the structure insight.

Recursive definitions sometimes *seem* to be circular: After all, we seem to be defining instances of *C* in terms of instances of *C*! But really we are defining "new" (more complex) instances of *C* in terms of *other*, "older" (i.e. already constructed), simpler instances of *C*, which is not circular at all. (It would only be circular if the *base cases* were somehow defined in terms of themselves. But they are not "defined"; they are given, by fiat.)

So, the structural insight above is a recursive definition of the notion of an "instruction": the base cases of the recursion are the primitive verbs of the Turing Machine ('move(location)' and 'print(symbol)'), and the recursive cases are given by sequence, selection, and repetition.

Here are two analogies. (1) A crossword puzzle can be solved recursively: begin by filling in those words (or phrases) whose answers you know (for example, a 10-letter word for "first president of the US"). Recursive steps consist in using these "axioms" to "prove theorems": i.e. to use the letters from already-filled-in answers as additional clues (or "premises") for new words. This analogy needs to be taken with a grain of salt, however: some answers that you might know "axiomatically" might also be filled in as "provable theorems." On the other hand, even formal systems can have different axiomatizations, such that an axiom of one formalization might be a theorem of another. (What about "cheating" in the sense of looking up an answer? That's an appeal to an "oracle"; see Section 11.9.)

(2) Jigsaw puzzles can be solved recursively: the base case of the recursion consists in building the frame. A recursive step is to form a "molecular" piece that consists of two "atomic" pieces that fit together. Further recursions consist of finding two molecular pieces that fit together.

7.4.3.3 Structured Programming (II)

There are optional, additional instructions and grammatical rules:

An explicit "halt" instruction: This is not strictly necessary, because it can always be simulated by having the program execute a command that does nothing and does not have a "next" step. We will see such a program when we look at Turing Machines in Chapter 8. However, a "halt" instruction can sometimes make a program simpler or more elegant.

An "exit" instruction: This allows a program to exit from a loop under certain conditions before the body of the loop is completed. Again, this can provide simplicity or elegance.

> **Abstraction:** A structured programming language … must provide a mechanism whereby the language can be extended to contain the abstractions which the user requires. A language containing such a mechanism can be viewed as a general-purpose, indefinitely-high-level language. (Liskov and Zilles, 1974, p. 51)

Two varieties of abstraction are worth noting:

1. *Procedural abstraction (named procedures)*: Define new (typically, complex) actions by giving a single name to a (complex) action. This is very powerful in terms of human readability and comprehension, and even in terms of machine efficiency.[20]
2. *Abstract data types*: Procedural abstraction allows the programmer to define new *verbs* in terms of "old" ones. A related technique is the use of abstract data types, which allows the programmer to define new *nouns* in terms of "old" ones (Liskov and Zilles, 1974; Aho et al., 1983). Moreover, as is especially clear in object-oriented programming, new "nouns" require new "verbs":

> [A] consequence of the concept of abstract data types is that most of the abstract operations in a program will belong to the sets of operations characterizing abstract types. (Liskov and Zilles, 1974, p. 52)

---

20 "As Alfred North Whitehead wrote, 'Civilisation advances by extending the number of important operations which we can perform without thinking about them.'" (B. Hayes, 2014b, p. 22).

We'll return to abstraction in Chapter 13.[21]

Recursion: Recursion can be an elegant replacement for repetition: a recursive instruction tells you how to compute the output value of a function in terms of previously computed *output* values instead of in terms of its *input* value. Of course, the base case (i.e. the output value of the function for its initial input) has to be given to you in a kind of table-lookup. (We'll look at recursion more closely in Section 7.6.)

7.4.4   Insight 4: The Church-Turing Computability Thesis

We now have our language: any algorithm for any computable problem can be expressed in this language (for a Turing Machine) that consists of the two nouns '0' and '1,' two (or three) verbs ('move(location),' 'print(symbol),' erase), and the three grammatical rules of sequence, selection, and repetition.

But is it a *minimal* language? In other words, is that really all that is needed? Can your interaction with, say, a spreadsheet program or Facebook be expressed in this simple (if not "simple-minded"!) language? There's no doubt that a spreadsheet program written in this language would be very long, very hard to read, and perhaps inefficient. But that's not the point. The question is, could it be done? And the answer is our next great insight. In one word, 'yes':

> **Nothing besides our two nouns, two (or three) verbs, and three grammar rules is necessary.**

Such a statement, as part of a recursive definition, is sometimes called a "closure" clause (http://faculty.washington.edu/keyt/InductiveDefinitions.pdf). Here is another way to put this:

> **The informal notion of computability can be identified with (anything logically equivalent to) Turing Machine computability.**

That is, an algorithm is definable as a program expressible in (anything equivalent to) our minimal language.

This idea was almost simultaneously put forth both by Church (1936b) in terms of his lambda calculus (see Section 6.5) and by Turing (1936). Consequently, some people call it 'Church's Thesis'; others call it 'Turing's Thesis'; and, as you might expect, some call it 'the Church-Turing Thesis,' in part because Turing proved that Church's lambda calculus is logically equivalent to Turing Machines. For this reason, Robert Soare (2009) has advocated calling it, more simply and more neutrally, the 'Computability Thesis.'

But it is only a proposed definition or explication of 'computable' or 'algorithm': it proposes to identify an *informal*, intuitive notion of effective computability or algorithm with a *formal*, mathematically precise notion of (anything logically equivalent to) a Turing Machine. (To be clear, I have *not* given such a formal, mathematically precise notion *yet*; we'll get closer in Section 7.6.)[22]

How do we know that Turing Machine computability captures (all of) the intuitive notion(s) of effective computability? After all, there are other analyses of computation, such as Church's lambda calculus. There are two reasons for preferring Turing's over Church's: first, Turing's is easier to understand, because it follows from his analysis of how humans compute. Second – and this is "merely" an appeal to authority – Gödel preferred Turing's analysis, not only to Church's but also

---

21 See the Online Resources for further reading on abstraction.
22 Or see the Online Resources for Insight 4.

to his own![23] Church's lambda calculus (which John McCarthy later used as the basis of the Lisp programming language) had as its basic, or atomic, steps formal operations on function formulas that some people – Gödel in particular – did not find to be intuitively computable. The same could be said even for Gödel's own theory of recursive functions. But Turing's basic operations were, by design, simple things that any human could easily do: put a mark at specific location on a piece of paper, and shift attention to a different location.

The lambda calculus and Turing Machines are not the only theories of computation. Here is a list of some others:

Post Machines are like Turing Machines but treat the tape as a queue (https://en.wikipedia.org/wiki/Post%E2%80%93Turing_machine).

Markov algorithms were later used as the basis of the Snobol programming language (https://en.wikipedia.org/wiki/Markov_algorithm).

Post productions were later used as the basis of production systems in AI (Post, 1941, 1943; Soare, 2012, p. 3293).

Herbrand-Gödel recursion equations were later used as the basis of the Algol family of programming languages (see Section 7.6).

μ-recursive functions (see Section 7.6).

register machines (Shepherdson and Sturgis, 1963).

Any programming language including, besides those already mentioned, Pascal, C, C+, Java, etc.

However, languages like HTML don't count, because they are not "Turing-complete" – i.e. not logically equivalent to a Turing Machine – usually because they lack one or more of the three grammar rules. Such languages are *weaker* than the language for Turing Machines. (The question of whether there are models of computation that are *stronger* than Turing Machines is the topic of Chapter 11.)

There are two major reasons to believe the Computability Thesis:

1. *Logical evidence*:
   All of the formalisms that have been proposed as precise, mathematical analyses of computability are not only *logically* equivalent (i.e. any function that is computable according to one analysis is also computable according to each of the others) but also *constructively* equivalent (i.e. they are inter-compilable, in the sense that you can write a computer program that will translate (or compile) a program in any of these languages into an equivalent program in any of the others).[24] Here is how Turing expressed it in a paper published the year after his magnum opus:[25]

   > Several definitions have been given to express an exact meaning corresponding to the intuitive idea of 'effective calculability' as applied for instance to functions of positive integers. The purpose of the present paper is to show that the computable functions introduced by the author are identical with the λ-definable functions of Church and the general recursive functions due to Herbrand and Gödel and developed by Kleene. It is shown that every λ-definable function is computable and that every computable function is general recursive. … If these results are taken in conjunction with an already available proof that every general recursive function is λ-definable we shall have the required equivalence of computability with λ-definability ....

---

23 See the Online Resources for further reading on this point.

24 I am indebted to my former colleague John Case's lectures (SUNY Buffalo, ca. 1983) on the theory of computation for this phrasing.

25 In the following quote, 'λ' is the lowercase Greek letter "lambda."

The identification of 'effectively calculable' functions with computable functions is possibly *more convincing than* an identification with the λ-definable or general recursive functions. For those who take this view the formal proof of equivalence provides a justification for Church's calculus, and allows the 'machines' which generate computable functions to be replaced by the more convenient λ-definitions. (Turing, 1937, p. 153, my italics).[26]

2. *Empirical evidence*:

   All algorithms that have been devised so far can be expressed as Turing Machines; i.e. there are no known intuitively effective-computable algorithms that are not Turing Machine computable.

But this has not stopped some philosophers and computer scientists from challenging the Computability Thesis. Some have advocated forms of computation that "exceed" Turing Machine computability. We will explore these options in Chapters 10 and 11.

Can the Computability Thesis be proved? Most scholars say 'no,' because any attempt to prove it mathematically would require that the *informal* notion of computability be formalized for the purposes of the proof. Then you could prove that *that formalization* was logically equivalent to Turing Machines. But how would you prove that that formalization was "correct"? This leads to an infinite regress.[27]

7.4.5 Insight 5: Implementation

The final insight is this:[28]

The first three insights can be physically implemented ...

That is, Turing Machines can be physically implemented. And, presumably, such a physical implementation would be a computer. This was what Turing attempted when he designed the ACE (recall Section 6.4.4).

In fact, not only can the previous insights be physically implemented, but they can be physically implemented

... using only *one* kind of "logic gate":

either a NOR-gate or a NAND-gate.[29] "Nor" and "nand" are connectives of propositional logic, each of which suffices by itself in the sense that all other connectives ("not," "and," "or," "if-then," etc.) can be defined in terms of them. Typically, however (as Tedre pointed out), real computers use several different kinds of gates, for the sake of efficiency.

Moreover, as we have seen, there does not appear to be any limitation on the "medium" of implementation: most computers today are implemented electronically, but there is work on DNA, optical, etc., computers, and there have even been some built out of Tinker Toys (http://www.computerhistory.org/collections/catalog/X39.81).

This brings in the engineering aspect of CS. But it also brings limitations imposed by physical reality: limitations of space, time, memory, etc. Issues concerning what is feasibly or efficiently

---

26 See the Online Resources for further reading on Turing Machine equivalents.
27 See the Online Resources for further reading on the Computability Thesis.
28 First suggested to me by Peter Denning (personal communication, 2014).
29 As Matti Tedre (personal communication, 2018) pointed out to me.

computable in *practice* (as opposed to what is theoretically computable in *principle*) – complexity theory, worst-case analyses, etc. – and issues concerning the use of heuristics come in here.

Turing Award-winner Alan Kay divides this insight into a "triple whammy of computing":

1. Matter can be made to remember, discriminate, decide and do
2. Matter can remember descriptions and interpret and act on them
3. Matter can hold and interpret and act on descriptions that describe anything that matter can do. (Guzdial and Kay, 2010)

He later suggests that the third item is the most "powerful," followed by the first and then the second, and that issues about the limits of computability and multiple realizability are implicit in these. (We'll return to physical computation in Chapter 16.)[30]

7.5 Structured Programming[31]

Structured programming eliminates the "go to" command in favor of our three grammar rules. Let's see how this can be done.

We can begin with a (recursive) definition of 'structured program': as with all recursive definitions, we need to give a base case (consisting of two "basic programs") and a recursive case (consisting of four "program constructors"). We will use the capital and lowercase Greek letters 'pi' (Π, π) to represent programs.

7.5.1 Basic Programs

There are two kinds of basic structured programs:

1. The *empty program* $\pi = $ **begin end.** is a basic structured program.
2. Let F be a "primitive operation" that is (informally) computable.
   Then the 1-operation program $\pi = $ **begin** F **end.** is a basic structured program.

Note that this definition does not specify which operations are primitive; they will vary with the programming language. One example might be an assignment statement (which will have the form "$y \leftarrow c$," where y is a variable and c is a value that is assigned to y). Another might be the *print* and *move* operations of a Turing Machine.

Structured programming could also be applied to a language whose primitive operations are *not* informally computable. Presumably, however, we want our algorithms to be "computation preserving" in the same way that we want the rules of inference of a logic to be *truth*-preserving. Thus, just as a logic's axioms should be true, the primitive operations should be (at least informally) "computable." As we will see in Chapter 8, Turing spends a lot of time justifying his choice of primitive operations.

Compare the situation with Euclidean geometry: if the primitive operations are limited to those executable using only compass and straightedge, then an angle cannot be trisected. But of course, if the primitive operations also include measurement of an angle using a protractor, then calculating one-third of an angle's measured size will do the trick. (We'll return to this in Section 11.4.)

That means structured programming is a *style* of programming, not a particular programming *language*. It is a style that can be used with *any* programming language.

---

30 See the Online Resources for further reading on physical computation.
31 The material in this section and the next is based on lectures given by John Case at SUNY Buffalo around 1983, which in turn were based on Clark and Cowell, 1976.

7.5.2 Program Constructors

The recursive case for structured programs specifies how to construct more complex programs from simpler ones. The simplest ones, of course, are the basic programs: the empty program and the one-operation programs. So, in good recursive fashion, we begin by constructing slightly more complex programs from these. Once we have both the basic programs and the slightly more complex programs constructed from them, we can combine them – using the recursive constructs that follow – to form even more complex ones, using these techniques:

Let π, π' be (simple or complex) programs, each of which contains exactly one occurrence of **end**.

Let P be a "Boolean test." (A Boolean test, such as "$x > 0$," is sometimes called a 'propositional function' or 'open sentence.'[32] The essential feature of a Boolean test is that it is a function whose output value is "true" or else is "false." P must also be [informally] computable, and, again, Turing spends a lot of time justifying his choices of tests.)

And let y be an integer-valued variable.

Then the following are also (more complex) structured programs:

1. $\Pi = $ **begin** π; π' **end.** is a (complex) structured program.
   Such a Π is the "*linear concatenation*" of π followed by π'. It is Böhm and Jacopini's "sequence" grammar rule.
2. $\Pi = $ **begin**
       if P
         then π
         else π'
     end.
   is a (complex) structured program.

   Such a Π is a "*conditional branch*": if P is true, then π is executed; if P is false, then π' is executed. It is Böhm and Jacopini's "selection" grammar rule.
3. $\Pi = $ **begin**
       while $y > 0$ **do**
         begin
           π;
           $y \leftarrow y - 1$
         end
     end.
   is a (complex) structured program.

   Such a Π is a "*count loop*" (or "for-loop," or "bounded loop"): the simpler program π is repeatedly executed while (i.e. as long as) the Boolean test "$y > 0$" is true (i.e. until it becomes false). Eventually it will become false, because each time the loop is executed, y is decremented by 1, so eventually y must become equal to 0. Thus, an infinite loop is avoided. This is one kind of Böhm and Jacopini's "repetition" grammar rule.
4. $\Pi = $ **begin**
       while P **do** π
     end.

   is a (complex) structured program.

---

32 It is a "propositional function" because it can be thought of as a function whose input is a proposition and whose output is a truth value. It is an "open sentence" in the sense that it contains a variable instead of a constant. (In English, that would be a pronoun instead of a proper name.)

Such a Π is a *"while-loop"* (or "free" loop, or "unbounded" loop): The simpler program π is repeatedly executed while (i.e. as long as) the Boolean test P is true (i.e. until P is false). Note that unlike the case of a count loop, a while loop can be an infinite loop, because there is no built-in guarantee that P will eventually become false (because, in turn, there is no restriction on what P can be, as long as it is a Boolean test). In particular, if P is the constantly-true test "true" – or a constantly true test such as "1=1" – then the loop will be guaranteed to be infinite. This is a more powerful version of repetition.

7.5.3 Classification of Structured Programs

We can classify structured programs based on the previous recursive definition:

1. π *is a* **count-program**
   (or a "for-program," or a "bounded-loop program") $=_{def}$
   (a) π is a basic program, or
   (b) π is constructed from count-programs by:
      - linear concatenation, or
      - conditional branching, or
      - count looping
   (c) Nothing else is a count-program.
2. π *is a* **while-program**
   (or a "free-loop program," or an "unbounded-loop program") $=_{def}$
   (a) π is a basic program, or
   (b) π is constructed from while-programs by:
      - linear concatenation, or
      - conditional branching, or
      - count-looping, or
      - while-looping
   (c) Nothing else is a while-program.

The inclusion of count-loop programs in construction-clause (b) for while-programs is not strictly needed, because all count-loops are while-loops (just let the P of a while-loop be "$y > 0$" and let the π of the while-loop be the linear concatenation of some other π′ followed by "$y \leftarrow y - 1$"). So count-programs are a proper subclass of while-programs: while-programs include all count-programs *plus* programs constructed from while-loops that are not also count-loops.

7.6 Recursive Functions

Now let's look at one of the classic analyses of computable functions: a recursive definition of non-negative integer functions that are intuitively computable – i.e. functions whose inputs are non-negative integers, also known as "natural numbers." But first, what is a "natural number"?

7.6.1 A Recursive Definition of Natural Numbers

Extensionally, the set $\mathbb{N}$ of natural numbers $= \{0, 1, 2, \ldots\}$. Intensionally, they are the numbers defined (recursively!) by *Peano's axioms*.[33]

---

33 See the Online Resources for further reading on Peano's axioms.

P1 Base case: $0 \in \mathbb{N}$

That is, 0 is a natural number.

P2 Recursive case:

If $n \in \mathbb{N}$, then $S(n) \in \mathbb{N}$,

where S is a one-to-one function from $\mathbb{N}$ to $\mathbb{N}$ such that $(\forall n \in \mathbb{N})[S(n) \neq 0]$.

$S(n)$ is called "the *successor* of n." So, the recursive case says that every natural number has a successor that is also a natural number. The fact that S is a *function* means each $n \in \mathbb{N}$ has only *one* successor. The fact that S is *one-to-one* means no *two* natural numbers have the *same* successor. And the fact that 0 is not the successor of any natural number means both that S is not an "onto" function and that 0 is the "first" natural number.

P3 Closure clause: nothing else is a natural number.

We now have a set of natural numbers

$$\mathbb{N} = \{0, \ S(0), \ S(S(0)), \ S(S(S(0))), \ \ldots\}$$

and, as is usually done, we let $1 =_{def} S(0)$, $2 =_{def} S(S(0))$, etc. The closure clause guarantees that there are no other natural numbers besides 0 and its successors: suppose there *were* an $m \in \mathbb{N}$ that was neither 0 nor a successor of 0, nor a successor of any of 0's successors. Without the closure clause, such an m could be used to start a "second" natural-number sequence: $m, \ S(m), \ S(S(m)), \ \ldots$. So, the closure clause ensures that no proper *superset* of $\mathbb{N}$ is also a set of natural numbers. Thus, in a sense, $\mathbb{N}$ is "bounded from above." But we also want to "bound" it from *below*; i.e. we want to say that $\mathbb{N}$ is the *smallest* set satisfying **P1**–**P3**. We do that with one more axiom:

P4 Consider an allegedly proper (hence, smaller) subset **M** of $\mathbb{N}$.

Suppose

1) $0 \in \mathbf{M}$

and that

2) for all $n \in \mathbb{N}$, if $n \in \mathbf{M}$, then $S(n) \in \mathbf{M}$.

Then $\mathbf{M} = \mathbb{N}$.

Stated thus, **P4** is the axiom that underlies the logical rule of inference known as "mathematical induction":

> **From** the fact that 0 has a certain property **M**
> (i.e. if 0 is a member of the class of things that have property **M**), **and**
> **from** the fact that, for any natural number n, if n has the property **M**, then its successor also has property **M**,
> **then** it may be inferred that *all* natural numbers have that property.[34]

7.6.2 Recursive Definitions of Recursive Functions

There are various kinds of recursive functions. To define them, we once again begin with "basic" functions that are informally computable, and then we recursively construct more complex functions from them. In this section, we will define these basic functions and the ways they can be combined. In the next section, we will define the various kinds of recursive functions.

1. **Basic functions:**

Let $x, x_1, \ldots, x_k \in \mathbb{N}$.

---

34 For further discussion of **P4**, see http://www.cse.buffalo.edu/~rapaport/191/F10/lecturenotes-20101110.html.

(a) **Successor:** $S(x) = x + 1$

That is, $x + 1$ is the successor of x. You should check to see that S satisfies Peano's axiom **P2**.

(b) **Predecessor:** $P(x) = x \dot- 1$, where

$$a \dot- b =_{def} \begin{cases} a - b, & \text{if } a \geq b \\ 0, & \text{otherwise} \end{cases}$$

The odd-looking arithmetic operator is a "minus" sign with a dot over it, sometimes called "monus." So, the predecessor of x is $x - 1$, except for $x = 0$, which is its own predecessor.

(c) **Projection:**[35] $P_k^j(x_1, \dots, x_j, \dots, x_k) = x_j$

That is, P_k^j picks out the jth item from a sequence of k items.

The basic functions (a)–(c) intuitively correspond to the basic operations of a Turing Machine: (a) the successor function corresponds to move(right), (b) the predecessor function corresponds to move(left) (where you cannot move any further to the left than the beginning of the Turing Machine tape), and (c) the projection function corresponds to reading the current square of the tape.[36]

Digression: An alternative to predecessor as a basic function is the family of *constant* functions $C_q(x_1, \dots, x_k) = q$ for each $q \in \mathbb{N}$ (Kleene, 1952, p. 219; Soare, 2009, Section 15.2, p. 397; Soare, 2016, p. 229).

Both predecessor and monus can be defined recursively: Where $n, m \in \mathbb{N}$, let

$$P(0) = 0$$
$$P(S(n)) = n$$

and let

$$n \dot- 0 = n$$
$$n \dot- S(m) = P(n \dot- m)$$

For more details, see https://en.wikipedia.org/wiki/Monus#Natural_numbers.

And while we're at it, we can define addition recursively, too:

$$n + 0 = n$$
$$n + S(m) = S(n + m)$$

---

2. **Function constructors:**

Let $g, h, h_1, \dots, h_m$ be (basic or complex) recursive functions. Then the following are also (complex) recursive functions:

(a) f *is defined from* $g, h_1, \dots, h_m$ *by* **generalized composition** $=_{def}$

$$f(x_1, \dots, x_k) = g(h_1(x_1, \dots, x_k), \dots, h_m(x_1, \dots, x_k))$$

This can be made a bit easier to read by using the symbol $\bar{x}$ for the sequence $x_1, \dots, x_k$. If we do this, then generalized composition can be written as follows:

$$f(\bar{x}) = g(h_1(\bar{x}), \dots, h_m(\bar{x})),$$

which can be further simplified to

$$f(\bar{x}) = g(\overline{h(\bar{x})})$$

---

[35] Sometimes called 'identity' (Kleene, 1952, p. 220; Soare, 2012, p. 3280; Soare, 2016, p. 229).

[36] We'll return to this analogy in Section 8.10.2. An analogous comparison in the context of "register machines" is made in Shepherdson and Sturgis, 1963, p. 220.

Note that $g(h(x))$ – called "function composition" – is sometimes written '$g \circ h$.' So, roughly, if g and h are recursive functions, then so is their (generalized) composition $g \circ h$. This is analogous to structured programming's notion of linear concatenation (i.e. sequencing): first compute h; then compute g.

(b) f is defined from g, h, i by **conditional definition** $=_{def}$

$$f(x_1, \ldots, x_k) = \begin{cases} g(x_1, \ldots, x_k), & \text{if } x_i = 0 \\ h(x_1, \ldots, x_k), & \text{if } x_i > 0 \end{cases}$$

Using our simplified notation, we can write this as

$$f(\overline{x}) = \begin{cases} g(\overline{x}), & \text{if } x_i = 0 \\ h(\overline{x}), & \text{if } x_i > 0 \end{cases}$$

This is analogous to structured programming's notion of conditional branch (i.e. selection): if a Boolean test (in this case, "$x_i = 0$") is true, then compute g, else compute h. (Note, by the way, that "$x_i = 0$" can be written: $P^i_k(x_1, \ldots, x_k) = 0$.)

(c) f is defined from g, h by **primitive recursion** $=_{def}$

$$f(x_1, \ldots, x_k, y) = \begin{cases} g(x_1, \ldots, x_k), & \text{if } y = 0 \\ h(x_1, \ldots, x_k, f(x_1, \ldots, x_k, y - 1)), & \text{if } y > 0 \end{cases}$$

Using our simplified notation, this becomes

$$f(\overline{x}, y) = \begin{cases} g(\overline{x}), & \text{if } y = 0 \\ h(\overline{x}, f(\overline{x}, y - 1), & \text{if } y > 0 \end{cases}$$

Note, first, that the "$y = 0$" case is the base case and the "$y > 0$" case is the recursive case. Second, note that this combines conditional definition with a computation of f based on f's value for its *previous* output. This is the essence of recursive definitions of functions: instead of computing the function's output based on its *current input*, the output is computed on the basis of the function's *previous output* (Allen, 2001). This is analogous to structured programming's notion of a count-loop: while $y > 0$, decrement y and then compute f.

(d) f is defined from $g, h_1, \ldots, h_k, i$ by **while-recursion** $=_{def}$

$$f(x_1, \ldots, x_k) = \begin{cases} g(x_1, \ldots, x_k), & \text{if } x_i = 0 \\ f(h_1(x_1, \ldots, x_k), \ldots, h_k(x_1, \ldots, x_k)), & \text{if } x_i > 0 \end{cases}$$

Again, using our simplified notation, this can be written as

$$f(\overline{x}) = \begin{cases} g(\overline{x}), & \text{if } x_i = 0 \\ f(h(\overline{x})), & \text{if } x_i > 0 \end{cases}$$

This is analogous to structured programming's notion of a while-loop (i.e. repetition): while a Boolean test (in this case, "$x_i > 0$") is true, compute h, and loop back to continue computing f, but when the test becomes false, then compute g.

An Example of a Function Defined by While-Recursion:

The Fibonacci sequence is

$$0, 1, 1, 2, 3, 5, 8, 13, \ldots$$

where each term after the first two terms is computed as the sum of the previous two terms. This can be stated recursively:

- The first two terms of the sequence are 0 and 1.
- Each subsequent term in the sequence is the sum of the previous two terms.

(Continued)

(Continued)

This can be defined using while-recursion as follows:

$$f(x) = \begin{cases} 0, & \text{if } x = 0 \\ 1, & \text{if } x = 1 \\ f(x-1) + f(x-2), & \text{if } x > 1 \end{cases}$$

We can make this look a bit more like the official definition of while-recursion by taking $h_1(x) = P(x) = x \dot{-} 1$ and $h_2(x) = P(P(x)) = P(x \dot{-} 1) = (x \dot{-} 1) \dot{-} 1 = x \dot{-} 2$. In other words, the two base cases of f are projection functions, and the recursive case uses the predecessor function twice (the second time, it is the predecessor of the predecessor).

(e) *f is defined from h by the μ-**operator*** [pronounced: "mu"-operator] $=_{def}$

$$f(x_1, \ldots, x_k) = \mu z[h(x_1, \ldots, x_k, z) = 0]$$

where:

$\mu z[h(x_1, \ldots, x_k, z) = 0] =_{def}$

$$\begin{cases} min \{z : \begin{cases} h(x_1, \ldots, x_k, z) = 0 \\ \textbf{and} \\ (\forall y < z)[h(x_1, \ldots, x_k, y) \text{ has a non-0 value]} \end{cases} \}, & \textbf{if } \text{such } z \text{ exists} \\ \text{undefined}, & \textbf{if } \text{no such } z \text{ exists} \end{cases}$$

This is a complicated notion but one well worth getting an intuitive understanding of. It may help to know that it is sometimes called "unbounded search" (Soare, 2012, p. 3284).

Let me first introduce a useful notation. If $f(x)$ has a value – i.e. if it is defined (in other words, if an algorithm that computes f halts) – then we will write $f(x) \downarrow$. And, if $f(x)$ is undefined – i.e. if it is only a "partial" function (in other words, if an algorithm for computing f goes into an infinite loop) – then we will write $f(x) \uparrow$.

Now, using our simplified notation, consider the sequence

$$h(\bar{x}, 0), \ h(\bar{x}, 1), \ h(\bar{x}, 2), \ \ldots, \ h(\bar{x}, n), \ h(\bar{x}, z), \ \ldots, \ h(\bar{x}, z')$$

Suppose each of the first $n + 1$ terms of this sequence halts with a non-zero value, but thereafter each term halts with value 0; that is,

$h(\bar{x}, 0) \downarrow \neq 0$

$h(\bar{x}, 1) \downarrow \neq 0$

$h(\bar{x}, 2) \downarrow \neq 0$

$\ldots$

$h(\bar{x}, n) \downarrow \neq 0$

but:

$h(\bar{x}, z) \downarrow = 0$

$\ldots$

$h(\bar{x}, z') \downarrow = 0$

The μ-operator gives us a description of that smallest or "min"imal z (i.e. the first z in the sequence) for which h halts with value 0. So the definition of μ says, roughly,

$\mu z[h(\overline{x}, z) = 0]$ is the smallest z for which $h(\overline{x}, y)$ has a non-0 value for each $y < z$, **but** for which $h(\overline{x}, z) = 0$, **if** such a z exists;
otherwise (i.e. if no such z exists), $\mu z[h(\overline{x}, z)]$ is undefined.

So, f is defined from h by the μ-operator if you can compute $f(\overline{x})$ by computing the smallest z for which $h(\overline{x}, z) = 0$.

If h is intuitively computable, then, to compute z, we just have to compute $h(\overline{x}, y)$ for each successive natural number y until we find z. So definition by μ-operator is also intuitively computable.

7.6.3 Classification of Recursive Functions

Given these definitions, we can now classify computable functions:

1. f is a **while-recursive function** $=_{def}$
   (a) f is a basic function, or
   (b) f is defined from while-recursive functions by
      i. generalized composition, or
      ii. conditional definition, or
      iii. while-recursion
   (c) Nothing else is while-recursive.
   This is the essence of the Böhm-Jacopini Theorem: any computer program (i.e. any algorithm for any computable function) can be written using only the three rules of grammar: sequence (generalized composition), selection (conditional definition), and repetition (while-recursion).

2. f is a **primitive-recursive** function $=_{def}$
   (a) f is a basic function, or
   (b) f is defined from primitive-recursive functions by
      i. generalized composition, or
      ii. primitive recursion
   (c) Nothing else is primitive-recursive.
   The *primitive-recursive* functions and the *while-recursive* functions overlap: both include the basic functions and functions defined by generalized composition (sequencing).
   The *primitive-recursive* functions also include the functions defined by primitive recursion (a combination of selection and count-loops), but nothing else.
   The *while-recursive* functions include (along with the basic functions and generalized composition) functions defined by conditional definition (selection) and those defined by while-recursion (while-loops).

3. f is a **partial-recursive** function $=_{def}$
   (a) f is a basic function, or
   (b) f is defined from partial-recursive functions by
      i. generalized composition, or
      ii. primitive recursion, or
      iii. the μ-operator
   (c) Nothing else is partial-recursive.

4. f is a **recursive** function $=_{def}$
   (a) f is partial-recursive, and
   (b) f is a total function
      (i.e. defined for all elements of its domain)

> **Terminology:** Unfortunately, the terminology varies with the author. For example, primitive recursive functions were initially called just "recursive functions." Now it is the while-recursive functions that are usually just called "recursive functions," or sometimes "*general* recursive functions" (to distinguish them from the *primitive* recursive functions). And partial recursive functions are sometimes called "μ-recursive functions" (because they are the primitive recursive functions augmented by the μ-operator). For the history of this and some clarification, see Soare, 2009, Sections 2.3–2.4, p. 373–373; and Section 15.2, pp. 396–397; and http://mathworld.wolfram.com/RecursiveFunction.html.

How are all of these notions related? First, here are the relationships among the various kinds of recursive functions: as we saw, there is an overlap between the *primitive-recursive* functions and the *while-recursive* functions, with the basic functions and the functions defined by generalized composition in their intersection.

The *partial-recursive* functions are a superset of the *primitive-recursive* functions: the partial-recursive functions consist of the primitive-recursive functions together with the functions defined with the μ-operator.

The *recursive* functions are a subset of the *partial-recursive* functions: the recursive functions are the partial-recursive functions that are also total functions.

Second, here is how the *recursive* functions and the *computable* functions are related:

> f is *primitive-recursive* if and only if f is *count-program-computable*.
> f is *partial-recursive* iff f is *while-program-computable*.
> And both of partial-recursive and while-program-computable functions are logically equivalent to being Turing Machine computable, lambda-definable, Markov-algorithmic, etc.

7.7 Non-Computable Functions

7.7.1 The Halting Problem

> You can build an organ which can do anything that can be done, but you cannot build an organ which tells you whether it can be done.
> —John von Neumann (1966), cited in Dyson, 2012a.

Have we left anything out? That is, are there any other functions besides these? Yes! The "Halting Problem" provides an example of a *non-computable* function: i.e. a function that cannot be defined using any of the mechanisms of Section 7.6.

> **Digression on Non-Computable Functions:** On the history of the Halting Problem, see the Digression at the end of Section 8.9.6.
> The Halting Function is not the only non-computable function. Two other famous non-computable functions are Hilbert's 10th Problem (Martin Davis, 1978) and the Busy Beaver function (Radó, 1962). In fact, there are infinitely many non-computable functions. There are also infinitely many *computable* functions but "only" *countably* infinitely many, whereas there are *uncountably* infinitely many *non*-computable functions. (See the Online Resources for further reading.)

Recall that a function *is* computable if and only if there is an algorithm (i.e. a computer program) that computes it. So, the Halting Problem asks whether there is an *algorithm* (e.g. a program for a Turing Machine) – call it the "Halting Algorithm," A_H – that computes the following *function* $H(C, i)$ (call it the "Halting Function"):

$H(C, i)$ takes as input *both*

1. an algorithm (or computer program) C
   (which we can suppose takes an integer as input),
   and
2. C's input i
   (which would be an integer),

   and $H(C, i)$ outputs

- "halts," if C halts on i
- "loops," if C loops on i.

A *formula* for H is

$$H(C, i) = \begin{cases} \text{"halts"}, & \text{if } C(i) \downarrow \\ \text{"loops"}, & \text{if } C(i) \uparrow \end{cases}$$

And our question is, is there an *algorithm* A_H that computes H? Can we write such a program?

In terms of the "function machine" from Section 7.2.4, we are asking whether there is a "function machine" (i.e. a computer) whose internal "mechanism" (i.e. whose program) is A_H. When you input the pair $\langle C, i \rangle$ to *this* "function machine" and turn its "crank," it should output "halts" if *another* function machine (namely, the function machine for C) successfully outputs a value when you give it input i, and it should output "loops" if the function machine for C goes into an infinite loop and never outputs any final answer. (It may, however, output some messages, but it never halts with an answer to $C(i)$.)

Here's another way to think about this: A_H is a kind of "super"-machine that takes as input not only an integer i but also another *machine C*. When you turn A_H's "crank," A_H first feeds i to C, and then A_H turns C's "crank." If A_H detects that C has successfully output a value, then A_H outputs "halts"; otherwise, A_H outputs "loops."

This would be very useful for introductory computer-programming teachers or software engineers in general! After all, one thing you never want in a computer program is an unintentional infinite loop. Sometimes, you might want an *intentional* one, however: you *don't* want an automated teller machine to halt – you *do* want it to behave in an infinite loop so that it is always ready to accept new input from a new customer. And you don't want a program that calculates the infinite decimal expansion of π to halt. However, it would be very useful to have a *single* handy program that could quickly check to see if *any* program that someone writes has an infinite loop in it. But no such program can be written! The function that H would have to implement is not computable. It can be proved that the assumption that it *is* computable leads to a contradiction. "The reason, essentially, is that the scope of the task covers *all* computer programs, **including the program of the very computer attempting the task**" (Copeland, 2017, p. 61, original italics, my boldface).

It is important to be clear that it *can* be possible to write a program that will check if another program has an infinite loop. In other words, given a program C_1, there might be another program H_1 that will check whether C_1 – but not necessarily any *other* program – has an infinite loop. What *cannot* be done is this: to write a single program A_H that will take *any program C whatsoever* and tell you whether C will halt or not.[37]

---

37 See the Online Resources for more on the difference between these two kinds of halting programs.

Note that we can't answer the question whether C halts on i by just running C on i: if it halts, we know that it halts. But if it loops, how would we know that it loops? After all, it might just be taking a long time to halt.

There are two ways that we might try to write A_H:

1. You can imagine that $A_H(C, i)$ works as follows:
   - A_H gives C its input i and then runs C on i.
   - If C halts on i, then A_H outputs "halts";
     otherwise, A_H outputs "loops."
   So, we might write A_H as follows:

   > **algorithm** $A_H^1(C, i)$:
   > **begin**
   >   **if** $C(i) \downarrow$
   >   **then** output 'halts'
   >   **else** output 'loops'
   > **end.**

   This matches our *formula* for function H.
2. But here's another way to write A_H:

   > **algorithm** $A_H^2(C, i)$:
   > **begin**
   >   output 'loops'; {i.e. make an initial *guess* that C loops}
   >   **if** $C(i) \downarrow$
   >   **then** output 'halts'; {i.e. *revise* your guess}
   > **end.**

"Trial-and-error" programs like A_H^2 will prove useful in our later discussion of hypercomputation (Chapter 11). But this approach won't work here because we're going to need to convert our program for H to another program called A_H^*, and A_H^2 can't be converted that way, as we'll see. More importantly, A_H^2 doesn't really do the required job: it doesn't give us a *definitive* answer to the question of whether C halts, because its initial answer is not really "loops," but something like "hasn't halted yet."

The answer to our question about whether such an algorithm A_H exists or can be written is negative: there is no program for $H(C, i)$. In other words, $H(C, i)$ is a non-computable function. Note that it *is* a function: there exists a perfectly good set of input-output pairs that satisfies the extensional definition of 'function' and looks like this:

$$H = \{\langle C_1, i_1 \rangle, \text{"halts"} \rangle, \ \ldots \ , \langle C_j, i_k \rangle, \text{"loops"} \rangle, \ \ldots \}$$

The next section sketches a proof that H is not computable. The proof takes the form of a "*reductio ad absurdum*" argument. So, our proof will assume that H *is* computable and derive a contradiction. If an assumption implies a contradiction, then – because no contradiction can be true – the assumption must have been wrong. So, our assumption that H is computable will be shown to be false.

Logical Digression: A "*reductio ad absurdum*" argument is one that "reduces" a claim to "absurdity" in order to refute the claim. If you want to show that a claim P is *false*, the strategy is to *assume* – "for the sake of the argument" – that P is *true* and then derive a contradiction C (i.e. an "absurdity") from it. If you can thus show that $P \rightarrow C$, then – because you know that $\neg C$ is the case (after all, C is a contradiction, hence false; so $\neg C$ must be true) – you can conclude that $\neg P$, thus refuting P. The rule of inference that sanctions this is "Modus Tollens"; see Section 4.8.2.

7.7.2 Proof Sketch that *H* Is Not Computable

Step 1

Assume that function *H is* computable.

So, there is an *algorithm A_H* that computes *function H*.

Now consider another algorithm, A_H^*, that is just like algorithm A_H, except that

- if *C halts* on *i*, then A_H^* *loops*

> (Remember: if *C* halts on *i*, then, by *C*'s definition, A_H does *not* loop, because A_H outputs "halts" and then halts.)

   and

- if *C loops* on *i*, then A_H^* outputs "loops" and halts (just like A_H does).

Here is how we might write A_H^*, corresponding to the version of A_H that we called 'A_H^1' earlier:

> **algorithm** $A_H^{1*}(C, i)$:
> **begin**
>   **if** $C(i) \downarrow$
>   **then while** true **do begin end**
>   **else** output 'loops'
> **end.**

Here, 'true' is a Boolean test that is always true. (As we noted earlier, you could replace it by something like '1=1,' which is also always true.)

Note that we cannot write a version of A_H^2 that might look like this:

> **algorithm** $A_H^{2*}(C, i)$:
> **begin**
>   output 'loops'; {i.e. make an initial guess that *C* loops}
>   **if** $C(i) \downarrow$
>   **then while** true **do begin end**   {i.e. if *C* halts, then loop}
> **end.**

Why not? Because if *C halts*, the only output we will ever see is the message that says *C loops*! That initial, incorrect guess is never revised. So, we'll stick with A_H (i.e. with A_H^1) and with A_H^* (i.e. with A_H^{1*}).

Note that if A_H exists, so does A_H^*. That is, we can turn A_H into A_H^* as follows: if A_H were to output "halts," then let A_H^* go into an infinite loop. That is, replace A_H's "output 'halts' " by A_H^*'s infinite loop. This is important because we are going to show that, in fact, A_H^* does *not* exist; hence, neither does A_H.

Step 2

Returning to our proof sketch, the next step is to code *C* as a number so it can be treated as input to itself.

What? Why do that? Because this is the way to simulate the idea of putting the *C* "machine" into the A_H machine and then having the A_H machine "turn" *C*'s "crank."

So, how do you "code" a program as a number? This is an insight due to Kurt Gödel. To code *any* text (including a computer program) as a number in such a way that you could also *de*code it, begin by coding each symbol in the text as a unique number (e.g. using the ASCII code). Suppose these numbers, in order, are $L_1, L_2, L_3, \ldots, L_n$, where L_1 codes the first symbol in the text, L_2 codes the second, ..., and L_n codes the last symbol.

Then compute the following number:

$$2^{L_1} \times 3^{L_2} \times 5^{L_3} \times 7^{L_4} \times \ldots \times p_n^{L_n}$$

where p_n is the nth prime number and where the ith factor in this product is the ith prime number raised to the L_ith power.

By the "Fundamental Theorem of Arithmetic" (http://mathworld.wolfram.com/Fundamental TheoremofArithmetic.html), the number that is the value of this product can be uniquely factored, so those exponents can be recovered, and then they can be decoded to yield the original text. (Turing has an even simpler way to code symbols; we'll discuss his version in detail in Section 8.12.) [38]

Step 3

Now consider $A_H^*(C, C)$. This step is called "diagonalization." It looks like a form of self-reference, because it looks as if we are letting C take itself as input to itself – but actually C will take its own Gödel number as input. That is, suppose you (1) code up program C as a Gödel number, (2) use it as input to the program C itself (after all, the Gödel number of C is an integer, and thus it is in the domain of the function that C computes, so it is a legal input for C), and (3) let A_H^* do its job on *that* pair of inputs.

By the definition of A_H^*:

1. if program C *halts* on input C, then $A_H^*(C, C)$ *loops*;

   and

2. if program C *loops* on input C, then $A_H^*(C, C)$ *halts* and outputs "loops."

Step 4

Now code A_H^* by a Gödel number! And consider $A_H^*(A_H^*, A_H^*)$. This is another instance of diagonalization. Again, it may look like some kind of self-reference, but it really isn't, because the first occurrence of 'A_H^*' names an algorithm, but the second and third occurrences are just numbers that happen to be the code for that algorithm. [39]

In other words, (1) code up A_H^* by a Gödel number, (2) use it as input to the program A_H^* itself, and then (3) let A_H^* do its job on that pair of inputs.

Again, by the definition of A_H^*:

1. if program A_H^* *halts* on input A_H^*, then $A_H^*(A_H^*, A_H^*)$ *loops*;

   and

2. if program A_H^* *loops* on input A_H^*, then $A_H^*(A_H^*, A_H^*)$ *halts* and outputs "loops."

Final Result

But A_H^* outputting "loops" in clause (2) means A_H^* halts!

So, if A_H^* *halts* (outputting "loops"), then – by clause (1) – it *loops*. And if A_H^* *loops*, then – by clause (2) – it *halts*. In other words, *it loops if and only if it halts*; i.e. it *does* loop if and only if it does *not* loop!

But that's a contradiction!

So, there is no such program as A_H^*. But that means there is no such program as A_H. In other words, *the Halting Function H is not computable.*

---

38 See the Online Resources for further reading on Gödel numbering.
39 My notation here, cumbersome as it is(!), is nonetheless rather informal but – I hope – clearer than it would be if I tried to be even more formally precise.

7.8   Summary

Let's take stock of where we are. We asked whether CS is the science of *computing* (rather than the science of *computers*). To answer that, we asked what computing, or computation, is. We have now seen one answer to that question: computation is the process of executing an algorithm to determine the output value of a function, given an input value. We have seen how to make this informal notion precise, and we have also seen that it is an "interesting" notion in the sense that not all functions are computable.

But this was a temporary interruption of our study of the history of computers and computing. In the next chapter, we will return to that history by examining Alan Turing's formulation of computability. [40]

7.9   Questions for the Reader

1. To the lists of features of algorithms in Section 7.3, Gurevich, 2012, p. 4, adds "isolation":

   > Computation is self-contained. No oracle is consulted, and nobody interferes with the computation either during a computation step or in between steps. The whole computation of the algorithm is determined by the initial state.

   (a) Is this related to Markov's "being determined" feature, or Kleene's "followed … like robots" feature, or Knuth's "definiteness" feature?
   (b) Does "isolation" mean a program that asks for input from the external world (or from a user, who, of course, is in the external world!) is not doing computation? (We'll discuss this in Chapters 11 and 16, but you should start thinking about it now.)
2. Gurevich has another "constraint": "Computation is symbolic (or digital, symbol-pushing)" (p. 4). That is, computation is syntactic. (See Section 16.9 for a discussion of what that means.) Does that mean computation is not mathematical (because mathematics is about *numbers*, not numerals)? Does it mean computers cannot have real-world effects? (We'll return to these topics in Chapter 16.)
3. Vardi, 2012 argues that Turing Machines are not models of algorithms:

   > [C]onflating algorithms with Turing machines is a misreading of Turing's 1936 paper …. Turing's aim was to define *computability*, not algorithms. His paper argued that every function on natural numbers that can be computed by a human computer … can also be computed by a Turing machine. There is no claim in the paper that Turing machines offer a general model for algorithms.

   Do you agree?
4. Harry Collins described an "experimenter's regress":

   > [Y]ou can say an experiment has truly been replicated only if the replication gets the same result as the original, a conclusion which makes replication pointless. Avoiding this, and agreeing that a replication counts as "the same procedure" even when it gets a different result, requires recognising the role of tacit knowledge and judgment in experiments. (*The Economist*, 2013)

---

40 See the Online Resources for further reading on computability.

Let's consider an experiment as a mathematical binary relation whose input is, say, the experimental setup and whose output is the result of the experiment. In that case, if a replication of the experiment always gets the same result, then the relation is a function.

Can scientific experiments be considered (mathematical) functions? In that case, does it make any sense to replicate an experiment in order to confirm it?

5. Piccinini, 2020a, p. 13, says that "non-digital computation [such as takes place in the brain via spike trains] is *different* from digital computation in that its vehicles are different from strings of digits." (See also Piccinini, 2007a.) Is it the case that everything can be represented in binary? Can what the spike trains do be done (simulated?) by binary digits?

6. Should other verbs be added to the Processing Insight? Is "read" a verb on a par with the ones cited? (Is "read" even needed?) Should Boolean tests be included as verbs?

7. In Section 2.5.1, we saw that a theorem is true only relative to the truth of the premises of its proof. And in Section 7.5.1, we saw that computation was similar:

> Computability is a relative notion, not an absolute one. All computation, classical or otherwise, takes place relative to some set or other or primitive capabilities. The primitives specified by Turing in 1936 occupy no privileged position. One may ask whether a function is computable relative to these primitives or to some superset of them.
> (Copeland, 1997, p. 707; see also Copeland and Sylvan, 1999, pp. 46–47)

In Section 7.5, definition (2), I said that primitive operations had to be computable, at least in an informal sense. After all, there we were trying to define what it meant to be computable. But another way to proceed would be to say that primitive operations are computable *by definition*.

But does this allow *anything* to be a primitive operation, even something that really shouldn't be (informally) computable? What if the primitive operation is, in fact, *non*-computable? Could we have a kind of "computation" in which the recursive portions are based on a non-computable (set of) primitive operation(s)? (We'll return to these ideas in Section 11.9.) [41]

8. A research question:

> … every physical process instantiates a *computation* insofar as it progresses from state to state according to dynamics prescribed by the laws of physics, i.e. by systems of differential equations. (Fekete and Edelman, 2011, p. 808)

This suggests the following very odd and very liberal definition: something is a *computation* $=_{def}$ it is a progression from state to state that obeys a differential equation. This definition is liberal because it seems to go beyond the limitations of a Turing Machine-like algorithm. That's not necessarily bad; for one thing, it subsumes both analog and discrete computations under one rubric.

Are Turing Machine algorithms describable by differential equations?

---

41 See the Online Resources for further reading on "relative computability."

8

Turing's Analysis of Computation

[A] human calculator, provided with pencil and paper and explicit instructions, can be regarded as a kind of Turing machine.
—Alonzo Church (1937)

Turing's 'Machines.' These machines are *humans* who calculate.
—Ludwig Wittgenstein (late 1940s), in Wittgenstein, 1980, p. 191e, Section 1096

[Wittgenstein's] quotation, though insightful, is somewhat confusingly put. Better would have been: these machines are Turing's mechanical model of *humans* who calculate.
—Saul A. Kripke (2013, p. 96, footnote 12)

Why … did Wittgenstein not make the converse point that 'Turing's humans are really machines that calculate'?
—S.G. Shanker (1987, p. 619)

8.1 Introduction

What Turing did around 1936 was to give a cogent and complete logical analysis of the notion of "computation." Thus it was that although people have been computing for centuries, it has only been since 1936 that we have possessed a satisfactory answer to the question: "What is a computation?"
—Martin Davis (1978, p. 241)

In this chapter, we continue our look at the nature of computation. If there is a single document that could be called the foundational document of CS, it would be Alan Mathison Turing's 1936 article "On Computable Numbers, with an Application to the *Entscheidungsproblem*," which appeared in the journal *Proceedings of the London Mathematical Society, Series 2*. In this paper, Turing (who was only about 24 years old at the time) accomplished (at least) five major goals:

1. He gave what is considered the clearest and most convincing mathematical analysis of computation (what is now called, in his honor, a "Turing Machine").
2. He proved that there were some functions that were *not* computable, thus showing that computation was not a trivial property.
3. He proved that the Turing Machine analysis of computation was logically equivalent to Church's lambda-calculus analysis of computation.

Philosophy of Computer Science: An Introduction to the Issues and the Literature, First Edition. William J. Rapaport.
© 2023 John Wiley & Sons, Inc. Published 2023 by John Wiley & Sons, Inc.

4. He formulated a "universal" Turing Machine, which is a mathematical version of a programmable computer.
5. And (as I suggested in Section 3.16.5) he wrote the first AI program.

Thus, arguably, in this paper, he created the modern discipline of CS.

Because this paper was so important and so influential, it is well worth reading. Fortunately, although parts of it are tough going (and it contains some errors),[1] much of it is very clearly written. It is not so much that the "tough" parts are difficult or hard to understand, but they are full of nitty-gritty details that have to be slogged through. Fortunately, Turing has a subtle sense of humor, too.

In this chapter, I will provide a guide to reading *parts* of Turing's paper slowly and carefully by actively thinking about them.[2] We will concentrate on the informal expository parts and some of the technical parts. Those are, of course, of interest but are rather difficult to follow and incorrect in some parts, and most can be skimmed on a first reading. In particular, we will concentrate on Turing's Sections 1–6, studying the simple examples of Turing Machines carefully (you can skim the complex ones); and Section 9, part I, which elaborates on what it is that a *human* computer does. We will look briefly at Section 7, which describes the Universal Turing Machine, and Section 8, which describes the Halting Problem. (You can skim these sections; but please note: that's 'ski**m**,' not 'ski**p**'!)

8.2 Slow and Active Reading

One of the best ways to read is to read slowly and actively. This is especially true when you are reading a technical paper, and even more especially when you are reading mathematics. Reading slowly and actively means (1) reading each sentence slowly, (2) thinking about it actively, and (3) making sure you understand it before reading the next sentence.

One way to make sure you understand it is to ask yourself *why* the author said it or *why* it might be true. (Recall Section 2.4.2 on the importance of asking "why.") If you don't understand it (after reading it slowly and actively), then you should reread all of the previous sentences to make sure you really understood them. Whenever you come to a sentence that you really don't understand, you should ask someone to help you understand it.

Of course, it could also be the case that you don't understand a passage because it isn't true, or doesn't follow from what has been said, or is confused in some way – and *not* because it's somehow your fault that you don't understand it! When you read, imagine that what you're reading is like a computer program and you are the computer that has to understand it. Except, of course, you're an independently intelligent computer, and, if you don't understand something, you can challenge what you read. In other words, treat reading as an attempt to "debug" what the author wrote! (On the value of slow and active reading in general, see https://cse.buffalo.edu/~rapaport/howtostudy .html#readactively.)

8.3 Title: "The *Entscheidungsproblem*"

The last word of the title – '*Entscheidungsproblem*' – is a German noun that (as we saw in Section 6.5) was well known to mathematicians in the 1930s; '*Entscheidung*' means "decision,"

---

1 Some of which Turing himself corrected (Turing, 1938). For a more complete "debugging," see Davies, 1999.
2 See the Online Resources for another slow reading of Turing's paper.

'-*s*' represents the possessive,[3] and '*problem*' means "problem." So, an *Entscheidungsproblem* is a decision problem, and *the* Decision Problem was the problem of finding an algorithm that would (a) take two things as input: (1) a formal logic L and (2) a proposition ϕ_L in the language for that logic, and that would (b) output either 'yes,' if ϕ_L was a theorem of that logic, or else 'no,' if $\neg\phi_L$ was a theorem of that logic (i.e. if ϕ_L was not a theorem of L). In other words, the Decision Problem was the problem of finding a general algorithm for deciding whether any given proposition was a theorem.

Wouldn't that be nice? Mathematics could be completely automated: given any mathematical proposition, one could apply this general algorithm to it, and you would be able to know if it were a theorem or not. Turing was fascinated by this problem, and he solved it in the negative by showing that no such algorithm existed. We've already seen how: he showed that there was at least one problem (the Halting Problem) for which there was no such algorithm. Along the way, he invented CS!

8.4 Paragraph 1

8.4.1 " 'Computable' Numbers"

Let's turn to the first sentence of the first paragraph:

> The "*computable*" numbers may be described briefly as *the real numbers* whose expressions as a decimal are *calculable by finite means*.
> (Turing, 1936, paragraph 1, sentence 1, p. 230, my italics)[4]

And let's consider the italicized phrases:

"Computable": The word 'computable' occurs in quotes here because Turing is using it in an informal, intuitive sense. It is the sense that he will make mathematically precise in the rest of the paper.

Real numbers: *Real* numbers are all of the numbers on the continuous number line, consisting of

1. the *rational* numbers, which consist of:
    (a) the *integers*, which – in turn – consist of
        i. the (non-negative) natural numbers (0, 1, 2, …), and
        ii. the negative natural numbers ($-1, -2, \ldots$), and
    (b) all other numbers that can be expressed as a ratio of integers
        – i.e. fractions
    and
2. the *irrational* numbers (i.e. those numbers that *cannot* be expressed as a ratio of integers, such as π, $\sqrt{2}$, etc.).

But the real numbers do not include the "*complex*" numbers, such as $\sqrt{-1}$. Every real number can be expressed "as a decimal," i.e. in decimal notation. For instance:

$$1 = 1.0 = 1.00 = 1.000 \text{ (etc.)}$$

---

3 Just as in English, so '*Entscheidungs*' means "decision's."
4 In the rest of this chapter, all quotations are from Turing, 1936 unless otherwise noted and will be identified only by section, paragraph, and page numbers of the original version, occasionally with my interpolations in brackets. All quotations are reprinted with permission of the publisher.

$$\tfrac{1}{2} = 0.5 = 0.50 = 0.500 = 0.5000 \text{ (etc.)}$$

$$\tfrac{1}{3} = 0.33333\ldots$$

$$\tfrac{1}{7} = 0.142857142857\ldots$$

These are all rational numbers and examples of "repeating" decimals. But the reals also include the irrational numbers, which have *non*-repeating decimals, such as:

$$\pi = 3.1415926535\ldots$$
$$\sqrt{2} = 1.41421356237309\ldots$$

Finitely calculable: Given a real number, is there an algorithm for computing its decimal representation? If so, then its "decimal [is] calculable by finite means" (because algorithms must be finite, as we saw in Section 7.3).

8.4.2 "Written By a Machine"

Now, if we were really going to do a slow (and active!) reading, we would next move on to sentence 2. But in the interests of keeping this chapter shorter than a full book (and so as not to repeat everything in Petzold, 2008), we'll skip to the last sentence of the paragraph:

> According to my definition, a number is computable if its decimal can be written down *by a machine*. (para 1., last sent., p. 230, my italics.)

This is probably best understood as an alternative way of expressing the first sentence: to be "calculable by finite means" is to be capable of being "written down by a machine." Perhaps the latter way of putting it extends the notion a bit, because it suggests that if a number is calculable by finite means, then that calculation can be done automatically: i.e. by a machine – without human intervention. And that, after all, was the goal of all of those who tried to build calculators or computing machines, as we saw in Chapter 6. So, Turing's goal in this paper is to give a mathematical analysis of what can be accomplished by any such machine (and then to apply the results of this analysis to solving the Decision Problem).

Question to Think About: As we'll see later, a Turing Machine is an abstract mathematical notion, not a real, physical device. Robin Gandy – Turing's only Ph.D. student – argued "that Turing's analysis of computation by a human being does not apply directly to mechanical devices" (Gandy, 1980). (Commentaries on what has become known as "Gandy's Thesis" include Sieg and Byrnes, 1999 (which simplifies and generalizes Gandy's paper); Israel, 2002, Shagrir, 2002.)

In Turing's sentence discussed in this section, is he referring to physical machines or to Turing Machines?

8.5 Paragraph 2

8.5.1 "Naturally Regarded as Computable"

In §§9, 10 I give some arguments with the intention of showing that the *computable* numbers include all numbers which could *naturally be regarded as computable*. (para. 2, sent. 1, p. 230, my italics.)

We will look at some of those arguments later, but right now, let's focus on the phrase 'naturally be regarded as computable.' This refers to the same informal, intuitive, pre-theoretical notion of computation that his quoted use of 'computable' referred to in the first sentence. It is the sense in which Hilbert wondered about which mathematical problems were decidable, the sense in which people used the phrase "effective computation," the sense in which people used the word 'algorithm,' and so on. It is the sense in which *people* (mathematicians, in particular) can compute.

However, the *first* occurrence of 'computable' in this sentence refers to the *formal* notion that Turing will present. Thus, this sentence is an expression of Turing's computability thesis.

8.5.2 "Definable Numbers"

Once again, we'll skip to the last sentence of the paragraph:

> The *computable* numbers do not, however, include all *definable* numbers, and an example is given of a definable number which is not computable.
> (para. 2, last sent., p. 230, my italics)

It is much more interesting if not all functions – or numbers – are computable. Any property that *everything* has is not especially interesting. But if there is a property that *only some* things have (and others lack), then we can begin to categorize those things and thus learn something more about them.

So Turing is promising to show us that computability is an interesting (because not a universal) property. And he's not going to do that by giving us some abstract argument; rather, he's actually going to show us a non-computable number (and, presumably, show us *why* it's not computable). We've already seen what this is: it's the (Gödel) number for an algorithm (a Turing Machine) for the Halting Problem.

8.6 Section 1, Paragraph 1: "Computing Machines"

Let's move on to Turing's Section 1, "Computing Machines." We'll look at the first paragraph and then jump to Turing's Section 9 before returning to this section.

Here is the first paragraph of Section 1:

> We have said that the computable numbers are those whose decimals are calculable by finite means. This requires rather more explicit definition. No real attempt will be made to justify the definitions given until we reach §9. (Section 1, para. 1, p. 231.)

This is why we will jump to that section in a moment. But first let's continue with the present paragraph:

> For the present I shall only say that the justification [of the definitions] lies in the fact that the *human memory is necessarily limited*. (Section 1, para. 1, p. 231, my italics.)

Turing's point – following Hilbert – is that we humans do not have infinite means at our disposal. We all eventually die, and we cannot work infinitely fast, so the number of calculations we can make in a single lifetime is finite.

But how big is "finite"? Let's suppose, for the sake of argument, that a typical human (named 'Pat') lives as long as 100 years. And let's suppose that from the time Pat is born until the time Pat dies, Pat does nothing but compute. Obviously, this is highly unrealistic, but I want to estimate the *maximum* number of computations that a typical human could perform. The *actual* number will, of course, be far fewer. How long does a computation performed by a human take? Let's suppose that the simplest possible computation is computing the successor of a natural number, and let's suppose it takes as long as 1 second. In Pat's lifetime, approximately 3,153,600,000 successors can be computed (because that's approximately the number of seconds in 100 years). Are there any problems that would require more than that number of computations? Yes! It has been estimated that the number of possible moves in a chess game is 10^{125}, which is about 10^{116} times as large as the largest number of computations that a human could possibly perform. In other words, we humans are not only finite, we are *very* finite!

But computer scientists and mathematicians tend to ignore such human limitations and pay attention only to the mathematical notion of finiteness. Even the mathematical notion, which is quite a bit larger than the actual human notion (for more on this, see Knuth, 2001), is still smaller than infinity, so the computable numbers, as Turing defines them, include quite a bit.

8.7 Section 9: "The Extent of the Computable Numbers"

8.7.1 Turing's Computability Thesis

I want to skip now to Turing's Section 9, "The Extent of the Computable Numbers," because this section contains the most fascinating part of Turing's analysis. We'll return to his Section 1 later. He begins as follows:

> No attempt has yet been made [in Turing's article] to show that the "computable" numbers include all numbers which would *naturally* be regarded as computable. (Section 9, para. 1, p. 249, my italics.)

Again, Turing is comparing two notions of computability: the technical notion (signified by the first occurrence of the word 'computable' – in "scare quotes") and the informal or "natural" notion. He is going to argue that the first includes the second. Presumably, it is more obvious that the second (the "natural" notion) includes the first (the technical notion): i.e. that if a number is technically computable, then it is "naturally" computable. The less obvious inclusion is the one that is more in need of support, that if a number is "naturally" computable, then it is technically computable. What kind of argument would help convince us of this? Turing says,

> All arguments which can be given are bound to be, fundamentally, appeals to intuition, and for this reason rather unsatisfactory mathematically. (Section 9, para. 1, p. 249.)

Why is this so? Because one of the two notions – the "natural" one – is informal. Thus, no formal, logical argument can be based on it. This is why the Computability Thesis (i.e. Turing's thesis) is a *thesis* and not a *theorem* – it is not something formally provable. Nonetheless, Turing will give us "appeals to intuition," i.e. *in*formal arguments, in fact, three kinds, as he says in the next paragraph:

> The arguments which I shall use are of three kinds.
>
> (a) A direct appeal to intuition.

(b) A proof of the equivalence of two definitions (in case the new definition has a greater intuitive appeal).

(c) Giving examples of large classes of numbers which are computable.
(Section 9, para. 2, p. 249.)

In this chapter, we will only look at (a), his *direct* appeal to intuition.

Let's return to the last sentence of paragraph 1:

> The real question at issue is "What are the possible processes which can be carried out in computing a number?" (Section 9, para. 1, p. 249.)

If Turing can answer this question, even informally, then he may be able to come up with a formal notion that captures the informal one. That is his "direct appeal to intuition."

8.7.2 "Writing Symbols on Paper"

Turing notes about "Type (a)" – the "direct appeal to intuition" – that "this argument is only an elaboration of the ideas of §1" (p. 249). This is why we have made this digression to Turing's Section 9 from his Section 1; when we return to his Section 1, we will see that it summarizes his Section 9.

The first part of the answer to the question, "What are the possible processes which can be carried out in computing a number?" – i.e. the first intuition about "natural" computation – is this:

> Computing is normally done by writing certain symbols on paper.
> (Section 9(I), para. 1, p. 249.)

So, we need to be able to write symbols on paper. Is this true? What kind of symbols? And what kind of paper?

Is It True?

Is *computing* normally done by writing symbols on paper? We who live in the twenty-first century might think this is obviously false: computers don't *have to* write symbols on paper in order to do their job. They do have to write symbols when we ask the computer to print a document, but they don't when we are watching a YouTube video. But remember that Turing is analyzing the "natural" notion of computing: the kind of computing that *humans* do. This includes arithmetic computations, and those typically *are* done by writing symbols on paper (or, perhaps, by imagining that we are writing symbols on paper, as when we do a computation "in our head").

What Kind of Paper?

> We may suppose this paper is divided into squares like a child's arithmetic book. (Section 9(I), para. 1, p. 249.)

In other words, we can use graph paper! Presumably, we can put one symbol into each square of the graph paper. So, for example, if we're going to write down the symbols for computing the sum of 43 and 87, we could write it like this:

	1	
	4	3
+	8	7
1	3	0

We write '43' in two squares, and then we write '+87' in three squares beneath this, aligning the ones and tens columns. To perform the computation, we compute the sum of 7 and 3 and write it as follows: the ones place of the sum ('0') is written below '7' in the ones column, and the tens place of the sum ('1') is "carried" to the square above the tens place of '43.' Then the sum of 1, 4, and 8 is computed and written as follows: the ones place of that sum – namely, '3' (which is the tens place of the sum of 43 and 87) – is written below '8' in the tens column. Then the tens place of that sum – namely, '1' (which is the hundreds place of the sum of 43 and 87) – is written in the square to the left of that '3.' As we have just seen, Turing notes that …

> In elementary arithmetic the two-dimensional character of the paper is sometimes used. But such a use is always avoidable, and I think that it will be agreed that the two-dimensional character of paper is no essential of computation. (Section 9(I), para. 1, p. 249.)

In other words, we could have just as well (if not just as easily) written the computation something like this:

1		4	3	+	8	7	=	1	3	0

Here, we begin by writing the problem '43+87' in five successive squares, followed, perhaps, by an equal sign. And we can write the answer in the squares following the equal sign, writing the carried '1' in an empty square somewhere else, clearly separated (here, by a blank square) from the problem. So, the use of two-dimensional graph paper has been avoided (at the cost of some extra bookkeeping). As a consequence, Turing can say:

> I assume then that **the computation is carried out** on one-dimensional paper, *i.e.* **on a tape divided into squares**. (Section 9(I), para. 1, p. 249, my boldface.)

We now have our paper on which we can write our symbols: the famous tape of what will become a Turing Machine! Note, though, that Turing has not yet said anything about the *length* of the tape; at this point, it could be finite.[5]

It is, perhaps, worth noting that the tape doesn't have to be this simple. As Kleene observed (although Shagrir, 2022, pp. 38–39 disagrees!),

> [T]he computer is [not] restricted to taking an ant's eye view of its work, squinting at the symbol on one square at a time. … [T]he Turing-machine squares can correspond to whole sheets of paper. If we employ sheets ruled into 20 columns and 30 lines, and authorize 99 primary symbols, there are $100^{600} = 10^{1200}$ possible square conditions, and we are at the opposite extreme. The schoolboy [sic] doing arithmetic on $8\frac{1}{2}$ by 12" sheets of ruled paper would never need, and could never utilize, all this variety.
>
> Another representation of a Turing machine tape is as a stack of IBM cards, each card regarded as a single square for the machine. (Kleene, 1995, p. 26)

---

5 See the Online Resources for further reading on writing symbols on paper.

What Kind of Symbols?

> I shall also suppose that *the number of symbols which may be printed is finite.* (Section 9(I), para. 1, p. 249, my italics.)

This is the first item that Turing has put a limitation on. Actually, Turing is a bit ambiguous here: there might be infinitely many different kinds of symbols, but we're only allowed to *print* a finite number of them. Or there might only be a finite number of different kinds of symbols – with a further vagueness about how many of them we can print: if the tape is finite, then we can only print a finite number of the finite amount of symbols, but if the tape is infinite, we could print infinitely many of the finite amount of symbols. But it is clear from what he says next that he means there are only a finite number of different kinds of symbols.

Why finite? Because

> If we were to allow an infinity of symbols, then there would be symbols differing to an arbitrarily small extent. (Section 9(I), para. 1, p. 249.)

There are two things to consider here: Why would this be the case? And why does it matter? The answer to both of these questions is easy: if the human who is doing the computation has to be able to identify and distinguish among infinitely many symbols, surely some of them may get confused, especially if they look a lot alike! Would they have to look alike? A footnote at this point suggests why the answer is 'yes':

> If we regard a symbol as literally printed on a square we may suppose that the square is $0 \leq x \leq 1, 0 \leq y \leq 1$. The symbol is defined as a set of points in this square, viz. the set occupied by printer's ink. (Section 9(I), para. 1, p. 249, footnote.)

That is, we may suppose that the square is, say, 1 cm by 1 cm. Any symbol has to be printed in this space. Imagine that each symbol consists of very tiny points of ink. To be able to print infinitely many different kinds of symbols in such a square, some of them are going to differ from others by just a single point of ink, and any two such symbols are going to "differ to an arbitrarily small extent" and, thus, be impossible for a human to distinguish. So, "the number of symbols which may be printed" must be finite in order for the human to be able to easily read them.

Is this really a limitation?

> The effect of this restriction of the number of symbols is not very serious. (Section 9(I), para. 1, p. 249.)

Why not? Because

> It is always possible to use sequences of symbols in the place of single symbols. Thus an Arabic numeral such as 17 or 999999999999999 is normally treated as a single symbol. Similarly in any European language words are treated as single symbols … (Section 9(I), para. 1, pp. 249–250.)

In other words, the familiar idea of treating a sequence of symbols (a "string" of symbols, as mathematicians sometimes say) as if it were a single symbol allows us to construct as many symbols as we want from a finite number of building blocks. That is, the rules of place-value notation (for Arabic numerals) and spelling (for words in European languages) – i.e. rules that tell us how to "concatenate" our symbols (to string them together) – give us an arbitrarily large number (although still finite!) of symbols.

What about non-European languages? Turing makes a (possibly politically incorrect) comment:

> … (Chinese, however, attempts to have an enumerable infinity of symbols). (Section 9(I), para. 1, p. 250.)

Chinese writing is pictographic and thus would seem to allow for symbols that run the risk of differing by an arbitrarily small extent, or, at least, that do not have to be constructed from a finite set of elementary symbols. As Turing also notes, using a finite number of basic symbols and rules for constructing complex symbols from them does not necessarily avoid the problem of not being able to identify or differentiate them:

> The differences from our point of view between the single and compound symbols is that the compound symbols, if they are too lengthy, cannot be observed at one glance. This is in accordance with experience. We cannot tell at a glance whether 9999999999999999 and 999999999999999 are the same. (Section 9(I), para. 1, p. 250.)

And probably you can't, either! So doesn't this mean, even with a finite number of symbols, that we're no better off than with infinitely many? Although Turing doesn't say so, we can solve this problem using the techniques he's given us: don't try to write 15 or 16 occurrences of '9' inside one, tiny square – write each '9' in a separate square! And then count them to decide which sequence of them contains 15 and which contains 16, which is exactly how *you* "can tell … whether 9999999999999999 and 999999999999999 are the same."

Incidentally, Kleene (1995, p. 19) observes that Turing's emphasis on not allowing "an infinity of symbols" that "differ … to an arbitrarily small extent" marks the distinction between "*digital* computation rather than *analog* computation."

The other thing that Turing leaves unspecified here is the *minimum* number of elementary symbols we need. The answer, as we saw in Section 7.4.1, is two (they could be a blank and '1,' or '0' and '1,' or any other two symbols). Turing himself will use a few more (just as we did in our addition example, allowing for the 10 single-digit numerals together with '+' and '=').

8.7.3 States of Mind

So, let's assume that, to compute, we only need a one-dimensional tape divided into squares and a finite number of symbols (minimally, two). What else?

(*) The behaviour of *the computer* at any moment is determined by the symbols which *he* is observing, and *his* "state of mind" at that moment. (Section 9(I), para. 2, p. 250, my label and italics.)

I have always found this to be one of the most astounding and puzzling sentences! 'computer'? 'he'? 'his'? But it is only astounding or puzzling to those of us who live in the late twentieth/early twenty-first century, when computers are machines, not humans! Recall the ad from the 1892 *New York Times* for a (human) computer (Section 6.1). In 1936, when Turing was writing this article, *computers were still humans, not machines*. So, throughout this paper, whenever Turing uses the word 'computer,' he means a *human* whose job it is to compute. I strongly recommend replacing (in your mind's ear, so to speak) each occurrence of the word 'computer' in this paper with the word 'clerk.'[6]

So, "the behavior of the *clerk* at any moment is determined by the symbols which he [or she!] is observing." In other words, the clerk decides what to do next by looking at the symbols, and *which* symbols the clerk looks at *partially* determines what the clerk will do. Why do I say 'partially'? Because the clerk also needs to know what to do with them: if the clerk is looking at two numerals, should they be added? Subtracted? Compared? The other information that the clerk needs is his or her "state of mind." What is that? Let's hold off on answering that question till we see what else Turing has to say.

> We may suppose that *there is a bound B to the number of symbols or squares which the computer [the clerk!] can observe at one moment.* If he[!] wishes to observe more, he must use successive observations. (Section 9(I), para. 2, p. 250, my italics.)

This is the second kind of finiteness: we have a finite number of different kinds of symbols and a finite number of them that can be observed at any given time. This upper bound B can be quite small; in fact, $B = 1$ in most formal, mathematical presentations of Turing Machines. But Turing is allowing for B to be large enough that the clerk can read a single word without having to spell it out letter by letter, or a single numeral without having to count the number of its digits (presumably, the length of '9999999999999999' exceeds any reasonable B for humans). "Successive observations" will require the clerk to be able to move his or her eyes one square at a time to the left or right.

> We will also suppose that *the number of states of mind which need to be taken into account is finite.* (Section 9(I), para. 2, p. 250, my italics.)

Here, we have a third kind of finiteness. But we still don't know exactly what a "state of mind" is. Turing does tell us that:

> If we admitted an infinity of states of mind, some of them will be "arbitrarily close" and will be confused. (Section 9(I), para. 2, p. 250.)

– just as is the case with the number of symbols. And he also tells us that "the use of more complicated states of mind can be avoided by writing more symbols on the tape" (p. 250), but why that is the case is not at all obvious at this point. Keep in mind, however, that we have jumped ahead from Turing's Section 1, so perhaps something that he said between then and now would have clarified this. Nevertheless, let's see what we can figure out. (For more on the notion of bounds, see Sieg, 2006, p. 16.)

8.7.4 Operations

So, a clerk who is going to compute needs only a (possibly finite) tape divided into squares and a finite number of different kinds of symbols; the clerk can look at only a bounded number of them at a time; and the clerk can be in only a finite number of "states of mind" at a time. Moreover,

---

6 To be able to use a word that sounds like 'computer' without the twenty-first century implication that it is something like a Mac or a PC, some writers, such as Sieg (1994), use the nonce word 'compu*tor*' to mean a human who computes. I prefer to call them 'clerks.' And, of course, despite Turing's use of male pronouns, the clerk could be of any gender (as many of them were in subsequent decades; recall Section 6.2).

what the clerk can do (the clerk's "behavior") is determined by the observed symbols and his or her "state of mind."

What kinds of behaviors can the clerk perform?

> Let us imagine the operations performed by the computer [the clerk] to be split up into *"simple operations"* which are so elementary that it is not easy to imagine them further divided. (Section 9(I), para. 3, p. 250, my italics.)

These are going to be the basic operations, the ones that all other operations will be constructed from. What could they be? This is an important question, because this will be the heart of computation.

> Every such operation consists of some *change of the physical system* consisting of the computer [the clerk] and his[!] tape. (Section 9(I), para. 3, p. 250, my italics.)

So, what "changes of the physical system" can the clerk make? The only things that can be changed are the clerk's state of mind (i.e. the clerk can change his or her mind, so to speak) and the tape, which would mean changing a symbol on the tape or changing which symbol is being observed. What else could there be? That's all we have to manipulate: the clerk, the tape, and the symbols. And all we've been told so far is that the clerk can write a symbol on the tape or observe one that's already written. Turing makes this clear in the next sentence:

> We know the state of the system if we know the sequence of symbols on the tape, which of these are observed by the computer [by the clerk] (possibly with a special order), and the state of mind of the computer [of the clerk]. (Section 9(I), para. 3, p. 250.)

The "physical system" is the clerk, the tape, and the symbols. The only things we can know, or need to know, are

- which symbols are on the tape,
- where they are (their "sequence"),
- which are being observed and in which order (the clerk might be looking from left to right, from right to left, and so on), and
- what the clerk's (still mysterious) "state of mind" is.

Here is the first "simple operation":

> We may suppose that in a simple operation not more than *one symbol is altered*. Any other changes can be split up into simple changes of this kind.
> (Section 9(I), para. 3, p. 250, my italics.)

Altering a single symbol in a single square is a "simple" operation, i.e. a "basic" operation (or "basic program" in the sense of our discussion in Chapter 7). Alterations of sequences of symbols can be accomplished by altering the single symbols in the sequence. How do you alter a symbol? You replace it with another one; i.e. you write down a (possibly different) symbol. (And perhaps you are allowed to erase a symbol, but that can be thought of as writing a special "blank" symbol, 'b.') However, the ability to erase has a downside: it destroys information, making it difficult, if not impossible, to *reverse* a computation (B. Hayes, 2014b, p. 23; recall Section 7.4.2).

Which symbols can be altered? If the clerk is looking at the symbol in the first square, can the clerk alter the symbol in the 15th square? Yes, but only by *first* observing the 15th square and *then* changing it:

> The situation in regard to the squares whose symbols may be altered in this way is the same as in regard to the observed squares. We may, therefore, without loss of generality, assume that the squares whose symbols are changed are always "observed" squares. (Section 9(I), para. 3, p. 250.)

But wait a minute! If the clerk has to be able to find the 15th square, isn't that a kind of operation?

8.7.5 Another "Simple Operation"

Yes:

> Besides these changes of symbols, the simple operations must include changes of distribution of observed squares. The new observed squares must be immediately recognisable by the computer [by the clerk]. (Section 9(I), para. 4, p. 250.)

And how does the clerk do that? Is "finding the 15th square" a "simple" operation? Maybe. How about "finding the 9999999999999999th square"? No:

> I think it is reasonable to suppose that they can only be squares whose distance from the closest of the immediately previously observed squares does not exceed a certain fixed amount. Let us say that each of the new observed squares is within L squares of an immediately previously observed square. (Section 9(I), para. 4, p. 250.)

So here we have a fourth kind of boundedness or finiteness: the clerk can only look a certain bounded distance away. How far can the distance be? Some plausible lengths are the length of a typical word or small numeral (so L could equal B). The minimum is, of course, one square (taking $L = B = 1$). So, another "simple" operation is looking one square to the left or to the right (and, of course, the ability to repeat that operation, so that the clerk can, eventually, find the 15th or 9999999999999999th square).

8.7.6 "Immediate Recognisability"

What about a different kind of candidate for a "simple" operation – "find a square that contains the special symbol x":

> In connection with "immediate recognisability," it may be thought that there are other kinds of square which are immediately recognisable. In particular, squares marked by special symbols might be taken as immediately recognisable. Now if these squares are marked only by single symbols there can be only a finite number of them, and we should not upset our theory by adjoining these marked squares to the observed squares. (Section 9(I), para. 5, pp. 250–251.)

So, Turing allows such an operation as being "simple," because it doesn't violate the finiteness limitations. But he doesn't have to allow it. How would the clerk be able to find the only square that contains the special symbol x (assuming that there is one)? By first observing the current square. If x isn't on that square, then observe the next square to the left. If x isn't on that square, then observe the square to the right of the first one (by observing the square two squares to the right of the current one). And so on, moving back and forth, until a square with x is found. What if the clerk needs to find a sequence of squares marked with a sequence of special symbols?

> If, on the other hand, they [i.e. the squares marked by special symbols] are marked by a sequence of symbols, we cannot regard the process of recognition as a simple process. (Section 9(I), para. 5, p. 251.)

I won't follow Turing's illustration of how this can be done. Suffice it to say that it is similar to what I just sketched out as a way of avoiding having to include "finding a special square" as a "simple" operation, and Turing admits as much:

> If in spite of this it is still thought that there are other "immediately recognisable" squares, it does not upset my contention so long as these squares can be found by some process of which my type of machine is capable. (Section 9(I), para. 5, p. 251.)

In other words, other apparently "simple" operations that can be analyzed into some combination of the simplest operations of writing a symbol and observing are acceptable. It is worth noting that this can be interpreted as a claim that "subroutines" can be thought of as single operations – this is the "procedural abstraction" or "named procedure" operation discussed in Section 7.4.3.3.

8.7.7 Summary of Operations

Turing now summarizes his analysis of the minimum that a human computer (a "clerk") needs to be able to do in order to compute:

> The simple operations must therefore include:
>
> *(a)* Changes of the symbol on one of the observed squares.
> *(b)* Changes of one of the squares observed to another square within L squares of one of the previously observed squares.
>
>   It may be that some of these changes necessarily involve a change of state of mind. The most general single operation must therefore be taken to be one of the following:
>
> *(A)* A possible change *(a)* of symbol together with a possible change of state of mind.
> *(B)* A possible change *(b)* of observed squares, together with a possible change of state of mind. (Section 9(I), para. 6, p. 251.)

In other words, the two basic operations are *(A)* to write a symbol on the tape (and to change your "state of mind") and *(B)* to look somewhere else on the tape (and to change your "state of mind"). That's it: writing and looking! Well, and "changing your state of mind," which we haven't yet clarified but will, next.

8.7.8 "States of Mind"

Conditions and Actions

How does the clerk know which of these two things (writing or looking) to do? Turing's next remark tells us:

The operation actually performed is determined, as has been suggested on p. 250, by the state of mind of the computer [i.e. of the clerk] and the observed symbols. In particular, they determine the state of mind of the computer [i.e. of the clerk] *after* the operation is carried out. (Section 9(I), para. 7, p. 251, my italics.)

The passage on p. 250 that Turing is referring to is the one that I marked '(*)' and called 'astounding' earlier; it says roughly the same thing as the present passage. So, what Turing is saying here is that the clerk should

- *first* consider his or her state of mind *and* where he or she is currently looking on the paper – i.e. consider the current *condition* of the clerk and the paper,
- *then* decide what to do next
  (either write something there or look somewhere else) – i.e. perform an *action*, and
- *finally*, change his or her state of mind.

Of course, after doing that, the clerk is in a (possibly) new condition – a (possibly) new state of mind and looking at a (possibly) new location on the paper – which means the clerk is ready to do the next thing.

States of Mind, Clarified
Now, think about a typical computer program, especially an old-fashioned one, such as those written in (early versions of) Basic or Fortran, where each line of the program has a line number and a statement to be executed (a "command"). The computer (here I mean the machine, not a clerk!) starts at the first line number, executes the command, and then (typically) moves to the next line number. In "atypical" cases, the command might be a "jump" or "go to" command, which causes the computer to move to a different line number. At whatever line number the computer has moved to after executing the first command, it executes the command at that new line number. And so on.

But if you compare this description with Turing's, you will see that *what corresponds to the line number of a program is Turing's notion of a "state of mind"*! What corresponds to the currently observed symbol? It is the current input to the program! (Or, perhaps slightly more accurately, it is the current state of all "switches" or registers.)

So, let's paraphrase Turing's description of the basic operation that a clerk performs when computing. We'll write the paraphrase in terms of a computer program that the clerk is following:

> The operation performed is determined by the current line number of the program and the current input. The simple operations are: (a) print a symbol and (b) move 1 square left or right on the tape (which is tantamount to accepting new input), followed by changing to a new line number.

We can also say this in a slightly different way:

> If the current line number is N and the current input is I,
> then print or move (or both) and go to line N'.

And a program for such a computer will consist of lines of "code" that look like this:

Line N: **if** input $= I$
        then begin
            print (or move);
            go to Line N'
        end

8.7.9 Turing Machines, Turing's Thesis, and AI

The Turing Machine

I said earlier that passage (*) was "astounding"; here is its sequel:

> We may now construct a *machine* to do the work of this *computer*.
> (Section 9(I), para. 8, p. 251, my italics.)

Remember: a "computer" (for Turing) is not a machine but a human clerk who computes. And what Turing is now saying is that the *human* can be replaced by a *machine*: i.e. by what we now call a computer (a mechanical device). This sentence marks the end of Turing's analysis of what a *human* computer does and the beginning of his mathematical construction of a *mechanical* computer that can do what the human does. His description of it here is very compact; it begins as follows:

> To each state of mind of the computer [of the clerk!] corresponds an "*m*-configuration" of the machine. (Section 9(I), para. 8, p. 251.)

So, an *m*-configuration is something in the machine that corresponds to a clerk's "state of mind"; it is a line number of a program.[7] But considering modern computers and programs, programs are separate from the computers that run them, so what could Turing mean when he says that an *m*-configuration belongs to a machine? He means the machine is "hardwired" (as we would now say) to execute exactly one program (exactly one algorithm). Being hardwired, no separate program needs to be written out; it is already "compiled" into the machine. It is the "gears" of the "function machine" of Section 7.2.4. A "Turing Machine" can do one and only one thing; it can compute one and only one function, using an algorithm that is hardwired into it.

Turing continues:

> The *machine* scans B squares corresponding to the B squares observed by the *computer*.
> (Section 9(I), para. 8, p. 251, my italics.)

In more modern language, the *computer* scans B squares corresponding to the B squares observed by the *clerk*. The clerk is limited to observing a maximum of B squares on the tape, as we saw earlier (in an earlier quote from p. 250). The machine analogue of that is to read, or "scan," B squares.

Turing continues:

> In any move the machine can change a symbol on a scanned square or can change any one of the scanned squares to another square distant not more than L squares from one of the other scanned squares. Section 9(I), para. 8, pp. 251–252.)

In other words, the machine (the Turing Machine, or modern hardwired computer) can pay attention to B squares at a time, and each line of its program allows it to print a new symbol on any of those squares or move to any other square that is no more than L squares away from any of the B squares. Modern treatments simplify this: the machine is scanning a single square ($B = 1$), and each line of its program allows it to print a new symbol on that square or to move one square ($L = 1$) to its left or right (or both print and move).

---

7 Why '*m*'? It could stand for 'man,' on the grounds that this is a machine analogue of a (hu)man's state of mind; or it could stand for 'mental,' on the grounds that it is an analogue of a state of mind. But I think it most likely stands for 'machine,' because it is a configuration, or state, of a machine. Of course, Turing might have intended it to be ambiguous among all these options.

Which "move" should the machine make?

> The move which is done, and the succeeding configuration, [i.e. the next *m*-configuration; i.e. the next step in the algorithm], are determined by the scanned symbol and the [current] *m*-configuration. (Section 9(I), para. 8, p. 252.)

That is, the move that the machine should make, as well as the *next m*-configuration (i.e. the next step in the algorithm) are determined by the currently scanned symbol and the *current m*-configuration. Or, put in terms of computer programs, the instruction on the current line number together with the current input together determine what to do *now* (print, move, or both) and what to do *next* (which instruction to carry out next).

Digression: When we think of a machine that prints on a tape, we usually think of the tape as moving through a stationary machine. But in the case of the Turing Machine, it is the machine that moves, not the tape! The reason for this is simple: the machine is simulating the actions of a human computer, who writes on different parts of a piece of paper – it is the human who moves, not the paper.

Turing's (Computability) Thesis

> The machines just described do not differ very essentially from computing machines as defined in §2, and corresponding to any machine of this type a computing machine can be constructed to compute the same sequence, that is to say the sequence computed by the computer. (Section 9(I), para. 8, p. 252.)

As for the first clause, please recall that we are in the middle of a very long digression in which we have skipped ahead to Turing's Section 9 from Turing's Section 1; we have not yet read Turing's Section 2. When we do, we will see a more detailed version of the machines that Turing has just described for us here in Section 9.

The next clause is a bit ambiguous. When Turing says "any machine of *this* type," is he referring to the machines of Section 9 or the machines of Section 2? It probably doesn't matter, because he has said that the two kinds of machines "do not differ very essentially" from each other. But I think he is, in fact, referring to the machines of Section 9; "computing machines" are the ones that are "defined in Section 2."

The last phrase is of more significance: these ("computing") *machines* (of Turing's Section 2) "compute the same sequence … computed by the" *clerk*. In other words, whatever a human clerk can do, these machines can also do. What a human clerk can do (i.e. which sequences, or functions, a human clerk can compute) is captured by the *informal* notion of algorithm or computation. "These machines" are a formal counterpart of that informal notion. So this last phrase is a statement of Turing's Thesis (i.e. the Computability Thesis).

What about the other direction? Can a human clerk do everything that one of these machines can do? Or are these machines in some way more powerful than humans? I think the answer should be fairly obvious: given the way the machines are constructed on the basis of what it is that humans can do, surely a human could follow one of the programs for these machines. So humans can do everything that one of the machines can do, and – by Turing's Thesis – these machines can do everything that humans can do (well, everything that is computable in the informal sense). But these are contentious matters, and we will return to them when we consider the controversies surrounding hypercomputation (Chapter 11) and AI (Chapter 18).

Turing Machines as AI Programs

As we have seen, in order to investigate the *Entscheidungsproblem*,

> … Turing asked in the historical context in which he found himself *the* pertinent question: namely, what are the possible processes **a human being** can carry out (when computing a number or, equivalently, determining algorithmically the value of a number theoretic function)? (Sieg, 2000, p. 6; original italics, my boldface)

That is,

> Turing machines appear [in Turing's paper] as a result, a codification, of his analysis of calculations by humans. (Gandy, 1988, p. 82)

This strategy underlies much of CS, as Alan Perlis observed:

> The intent [of a first computer science course should be] to reveal, through … examples, how analysis of some *intuitively* performed *human* tasks leads to *mechanical* algorithms accomplishable by a *machine*. (Perlis, 1962, p. 189, my italics)

But not just CS in general. The branch of CS that analyzes how humans perform a task and then designs computer programs to do the same thing is AI; so, in Section 9, **Turing has developed the first AI program!** After all, he showed that human computation is mathematically computable; i.e. he showed that a certain kind of human cognitive process was computable – and that's one of the definitions of AI.

One of the founders of AI, John McCarthy, made a similar observation:

> The subject of computation is essentially that of artificial intelligence since the development of computation is in the direction of making machines carry out ever more complex and sophisticated processes, i.e. to behave as intelligently as possible. (McCarthy, 1963, Section 4.2, p. 38).

This follows from McCarthy's earlier definition (see Section 3.16.5) of computation as the science of how to get machines to carry out intellectual processes.

Turing was not unaware of this aspect of his work:

> One way of setting about our task of building a 'thinking machine' would be to take a man [sic] as a whole and to try to replace all the parts of him by machinery. (Turing, 1948, p. 420, as cited in Proudfoot and Copeland, 2012, p. 2)

But that's almost exactly what Turing's analysis of human computation in his 1936 paper does (at least in part): it takes a human's computational abilities and "replaces" them by (abstract) machinery.

8.8 "Computing Machines"

We have come to the end of our digression into Turing's Section 9. Let's return to his Section 1, "Computing Machines."

8.8.1 "Man" and "Machine"

In paragraph 2, Turing gives a more detailed presentation of his abstract computing machine, the outcome of his detailed analysis from Section 9 of *human* computing. He begins as follows:

> We may compare a man in the process of computing a real number to a machine which is only capable of a finite number of conditions $q_1, q_2, \ldots, q_R$ which will be called "*m*-configurations." (Section 1, para. 2, p. 231.)

Why "may" we do this? Turing will give his justification in his Section 9, which we have just finished studying. By a "man," Turing of course means a human, not merely a male human. And, as we have already seen, an *m*-configuration is a line of a computer program, i.e. a step in an algorithm. Here, Turing is saying that each such algorithm has a finite number (namely, R) of steps, each labeled q_i. Put otherwise, (human, or informal) computation can be "compared to" (and, by Turing's Thesis, identified with) a finite algorithm.

What else is needed?

> The machine is supplied with a "tape" (the analogue of paper) running through it, and divided into sections (called "squares") each capable of bearing a "symbol." (Section 1, para. 2, p. 231.)

There are a couple of things to note here. First, from our study of Turing's Section 9, we know why this is the case and what, exactly, the tape, squares, and symbols are supposed to be and why they are the way they are. But second, why does he put those three words in "scare quotes"? There are two possible answers. I suspect that the real answer is that Turing hasn't, at this point in his paper, explained in detail what they are; that comes later, in his Section 9.

But there is also a mathematical or logical reason: in Turing's formal notion of a computing machine, the concepts of "tape," "squares," and "symbols" are really *undefined* (or *primitive*) *terms* in exactly the same way that 'point,' 'line,' and 'plane' are undefined (or primitive) terms in Euclidean plane geometry. As Hilbert famously observed, "One must be able to say at all times – instead of points, lines, and planes – tables, chairs, and beer mugs." So, here, too, one must be able to say at all times – instead of tapes, squares, and symbols – tables, chairs, and beer mugs. (But I'll use place settings instead of chairs; it will make more sense, as you will see.) A Turing Machine, we might say, must have a table.[8] Each table must have a sequence of place settings associated with it (so we must be able to talk about the nth place setting at a table). And each place setting can have a beer mug on it; there might be different kinds of beer mugs, but they have to be able to be distinguished from each other so we don't confuse them. In other words, *it is the logical or mathematical* **structure** *of a computing machine that matters, not what it is made of*. So, a "tape" doesn't have to be made of paper (it could be a table), a "square" doesn't have to be a regular quadrilateral that is physically part of the "tape" (it could be a place setting at a table), and "symbols" only have to be such that a "square" can "bear" one (e.g. a numeral can be written on a square of the tape, or a beer mug can be placed at a place setting belonging to a table).[9]

Turing continues:

> At any moment there is just one square, say the r-th, bearing the symbol $\mathfrak{S}(r)$ which is "in the machine." We may call this square the "scanned square."
> (Section 1, para. 2, p. 231.)

---

8 The kind with four legs, not a "machine table"!
9 See the Online Resources for further reading on Hilbert's idea.

First, '𝖲' is just the capital letter '*S*' in a font called "German Fraktur" or "black letter." It's a bit hard to read, so I will replace it with '*S*' in what follows (even when quoting Turing).

Note, second, that this seems to be a slight simplification of his Section 9 analysis, with $B = 1$. Second, being "in the machine" might be another undefined (or primitive) term merely indicating a relationship between the machine and something else. But what?

Grammatical Digression: Turing's lack of punctuation allows for some ambiguity. On the one hand, if 'which' had been preceded by a comma, then "which is 'in the machine'" would have been a "non-restrictive relative clause" that refers to the *square*. On the other hand, with no comma, the "which" clause is a "restrictive" relative clause modifying '*symbol S(r)*.' (On relative clauses, see " 'Which' vs. 'that' " at http://www.cse.buffalo.edu/~rapaport/howtowrite.html# whichVthat.)

The "something else" might be the *scanned square*, or it might be the *symbol* (whatever it is, '0,' '1,' or a beer mug) that is in the machine. I think it is the former, from remarks that he makes next:

> The symbol on the scanned square may be called the "scanned symbol." The "scanned symbol" is the only one of which the machine is, so to speak, "directly aware." (Section 1, para. 2, p. 231.)

Here, Turing's scare quotes around 'directly aware,' together with the hedge 'so to speak,' clearly indicate that he is not intending to anthropomorphize his machine. His machines are not really "aware" of anything; only humans can really be "aware" of things. But the machine analogue of human awareness is: being a scanned symbol. There is nothing anthropomorphic about that: either a square is being scanned (perhaps a light is shining on a particular place setting at the table) or it isn't, and either there is a symbol on the scanned square (there is a beer mug at the lighted place setting) or there isn't.

> However, by altering its *m*-configuration the machine can effectively remember some of the symbols which it has "seen" (scanned) previously. (Section 1, para. 2, p. 231.)

What does this mean? Let's try to paraphrase it: "By altering the line number of its program, the computing machine can effectively …" – can effectively do what? It can "remember previously scanned symbols." This is to be contrasted with the *currently* scanned symbol. How does the machine "remember" by altering a line number? Well, how would it "remember" what symbol was on, say, the third square if it's now on the fourth square? It would have to move left one square and scan the symbol that's there. To do that, it would have to have an instruction to move left. And to do *that*, it would need to go to that instruction, which is just another way of saying that it would have to "alter its *m*-configuration."[10]

> The *possible* behaviour of the machine at any moment is determined by the *m*-configuration q_n and the scanned symbol $S(r)$. (Section 1, para. 2, p. 231, my italics.)

It is only a *possible* behavior, because a given line of a program is only executed when control has passed to that line. If it is not being executed at a given moment, then it is only *possible* behavior, not *actual* behavior. The machine's *m*-configuration is the analogue of a line number of a program, and

---

10 See the Online Resources for further discussion of this passage.

the scanned symbol is the analogue of the machine's input. (We'll return to this in Section 8.9.1, below, and Chapters 11 and 16.)

> This pair $q_n, S(r)$ will be called the "configuration": thus the configuration determines the possible behaviour of the machine. (Section 1, para. 2, p. 231.)

Giving a single name ('configuration') to the *combination* of the *m*-configuration *and* the currently scanned symbol reinforces the idea that the *m*-configuration alone is an analogue of a line number and that this combination is the condition (or antecedent) of a conditional statement (a condition-action pair): line q_n begins, "**if** the currently scanned symbol is $S(r)$, **then** ...," or "**if** the current instruction is the one on line q_n **and if** the currently scanned symbol is $S(r)$, **then** ...."
 What follows the '**then**'? That is, what should the machine do if the condition is satisfied?

> In some of the configurations in which the scanned square is blank (*i.e.* bears no symbol) the machine **writes** down a new symbol on the scanned square: in other configurations it **erases** the scanned symbol. The machine may also **change the square** which is being scanned, but only by shifting it one place to right or left. In addition to any of these operations **the *m*-configuration may be changed**.
> (Section 1, para. 2, p. 231, my boldface)

So, we have five operations:

1. write a new symbol
2. erase the scanned symbol
3. shift 1 square left
4. shift 1 square right
5. change *m*-configuration.

There are four things to note:

(a) The symbols are left unspecified (which is why we can feel free to add a "blank" symbol), although, as we have seen, they can be limited to just '0' and '1' (and maybe also 'b').
(b) Turing has, again, simplified his Section 9 analysis, letting $L = 1$.
(c) "Change *m*-configuration" is essentially a "jump" or "go to" instruction.
(d) There is no "halt" command. (In Section 8.9.5, we will see why this is not needed.)

 Turing next clarifies what symbols are needed. Recall that the kind of computation that Turing is interested in is the computation of the decimal of a real number.

> Some of the symbols written down will form the sequence of figures which is the decimal of the real number which is being computed. The others are just rough notes to "assist the memory." It will only be these rough notes which will be liable to erasure. (Section 1, para. 2, pp. 231–232.)

So, either we need symbols for the 10 Arabic numerals (if we write the real number in decimal notation) or we only need symbols for the 2 binary numerals (if we write the real number in binary notation). Any other symbols are merely used for bookkeeping, and they (and only they) can be erased afterward, leaving a "clean" tape with only the answer on it.

There is one more thing to keep in mind: every real number (in decimal notation)[11] has an infinite sequence of digits to the right of the decimal point, even if it is an integer or (a non-integer) rational number, which are typically written with either no digits, or a finite number of digits, in the decimal expansion (1, 1.0, 2.5, etc.). If the number is an integer, this is an infinite sequence of '0's; for example, $1 = 1.000000000000 \ldots$ (which I will abbreviate as $1.\overline{0}$). If the number is rational, this is an infinite sequence of some repeating subsequence; for example:

$$\tfrac{1}{2} = 0.500000000000 \ldots = 0.5\overline{0}$$

$$\tfrac{1}{3} = 0.333333333333 \ldots = 0.\overline{3}$$

$$\tfrac{1}{7} = 0.142857142857 \ldots = 0.\overline{142857}$$

And if the number is irrational, this is an infinite, *non*-repeating sequence; for example:

$$\sqrt{2} = 1.41421356237309 \ldots$$
$$\pi = 3.1415926535 \ldots$$

This means one of Turing's computing machines should *never halt* when computing (i.e. writing out) the decimal of a real number. It should only halt if it is writing down a *finite* sequence, and it can do this in two ways: it can write down the finite sequence and then halt. Or it can write down the finite sequence and then go into an infinite loop (either rewriting the last digit over and over in the same square, or just looping in a do-nothing operation such as the empty program).

8.8.2 Closure Clause: Turing's Thesis

Finally,

> It is my contention that these operations include all those which are used in the computation of a number. (Section 1, para. 3, p. 232.)

This is another statement of Turing's version of the Computability Thesis: to compute, all you have to do is arrange the operations of writing and shifting in a certain way. The way they are arranged – what is now called "the control structure of a computer program" – is controlled by the "configuration" and the change in m-configuration (or, in modern structured programming, by Böhm and Jacopini's three control structures (i.e. grammar rules) of sequence, selection, and while-repetition). For Turing, it goes unsaid that all computation can be reduced to the computation of a number; this is the insight we discussed in Section 7.4.1 that all the information about any computable problem can be represented using only '0' and '1'; hence, any information – including pictures and sounds – can be represented numerically. (But it is also important to realize that this kind of universal binary representation of information doesn't have to be thought of as a number, because the two symbols don't have to be '0' and '1'!)

8.9 Section 2: "Definitions"

We are now ready to look at the section in which Turing's "computing machines" are defined.

8.9.1 "Automatic Machines"

Turing begins by giving us a sequence of definitions. The first is the most famous:

> If at each stage the motion of a machine (in the sense of Section 1) is *completely* determined by the configuration, we shall call the machine an "automatic machine" (or *a*-machine). ("Automatic machines," para. 1, p. 232.)

---

11 Similar remarks can be made for binary notation.

Turing may have called such a machine an '*a*-machine.' *We* now call them – in his honor – 'Turing Machines.' (Alonzo Church (1937) seems to have been the first person to use this term, in his review of Turing's paper.)

Clearly, such a machine's "motion" (or behavior) is at least *partly* determined by its configuration (i.e. by its *m*-configuration, or line number, together with its currently scanned symbol). Might it be determined by anything else? For all that Turing has said so far, maybe such a machine's human operator could "help" it along by moving the tape for it, or by writing something on the tape. This definition rules that out by limiting our consideration to such machines whose "motion" "is *completely* determined by the configuration." So, a human operator is not allowed to "help" it in any way: no cheating allowed!

8.9.2 "Choice Machines"

What about machines that do get outside help?

> For some purposes we might use machines (choice machines or *c*-machines) whose motion is only partially determined by the configuration (hence the use of the word "possible" in Section 1). When such a machine reaches one of these ambiguous configurations, it cannot go on until some arbitrary choice has been made by an external operator. ("Automatic machines," para. 2, p. 232.)

Turing's explanation of the use of 'possible' may be slightly different from mine. But I think they are consistent. In the previous statements, Turing used 'possible' to *limit* the kind of operations that a Turing Machine could perform. Here, he is introducing a kind of machine that has another kind of possible operation: writing, moving, or changing *m*-configuration *not* as the result of an explicit instruction *but* as the result of a "choice … made by an external operator." Note that this external operator *doesn't have to be a human*; it *could* be another Turing Machine! Such *c*-machines – which allow for external (or "interactive") input – are closely related to "oracle" machines, which Turing introduced in his doctoral dissertation. We will return to this in Chapter 11.

8.9.3 "Computing Machines"

Turing gives us some more definitions:

> If an *a*-machine prints two kinds of symbols, of which the first kind (called figures) consists entirely of 0 and 1 (the others being called symbols of the second kind), then the machine will be called a computing machine.
> ("Computing machines," para. 1, p. 232.)

The principal definition here is that of 'computing machine,' a special case of an *a*- (or Turing) machine that outputs its results as a binary numeral (in accordance with the Representation insight of Section 7.4.1). Here, Turing is simplifying his Section 9 analysis of human computation, restricting the symbols to '0' and '1.' Well, not quite, because he also allows "symbols of the second kind," used for bookkeeping purposes or intermediate computations. However, any symbol of the second kind could be replaced – at the computational cost of more processing – by sequences of '0's and '1's.

Turing continues:

> If the machine is supplied with a blank tape and set in motion, starting from the correct initial *m*-configuration, the subsequence of the symbols printed by it which are of the first kind will be called the *sequence computed by the machine*.
> ("Computing machines," para. 1, p. 232.)

Here, he seems to be allowing for some of the symbols of the *second* kind to remain on the tape, so that only a *sub*sequence of the printed output constitutes the result of the computation. In other words, these secondary symbols need not be erased. One way to think of this is to compare it to the way we write decimal numerals greater than 999: namely, with the punctuation aid of non-numerical symbols ("of the second kind"). In the United States, for example, they are the 'comma' and a 'decimal point': 1,234,567.89

In the previous paragraph, I almost wrote, "to remain on the tape *after the computation halts*." But does it halt? It can't – because every real number has an *infinite* decimal part! The secondary symbols could still be erased, during the computation; that's not of great significance (obviously, it's easier to not erase them and to just ignore them). The important point to remember is that computations of decimal representations of real numbers never halt. We'll return to this in a moment.

One more small point that simplifies matters:

> The real number whose expression as a binary decimal is obtained by prefacing this sequence by a decimal point is called the *number computed by the machine*. ("Computing machines," para. 1, p. 232.)

What about the part of the expression that is to the *left* of the decimal point? It looks as if the only numbers that Turing is interested in computing are the reals between 0 and 1 (presumably including 0 but excluding 1).[12] Does this matter? Not really; first, all reals can be mapped to this interval, and, second, any other real can be computed simply by computing its "non-decimal" part in the same way. Restricting our attention to this subset of the reals simplifies the discussion without loss of generality. (We'll return to this in Section 8.9.7.)

8.9.4 "Complete Configurations"

Two more definitions:

> At any stage of the motion of the machine, the number of the scanned square, the complete sequence of all symbols on the tape, and the *m*-configuration will be said to describe the *complete configuration* at that stage. The changes of the machine and tape between successive complete configurations will be called the *moves* of the machine. ("Computing machines," para. 2, p. 232.)

Three points to note: first, "at any stage of the motion of the machine" only a *finite* number of symbols will have been printed, so it is perfectly legitimate to speak of "the complete sequence of all symbols on the tape" even though every real number has *infinitely* many numerals after the decimal point.

---

12 Or possibly including 1, if it is written as $0.\overline{9}$.

Second, the sequence of all symbols on the tape probably includes all occurrences of 'ꞵ' that do not occur after the last non-blank square (i.e. that *do* occur *before* the last non-blank square); otherwise, there would be no way to distinguish the sequence $\langle 0,0, 1, ꞵ, 0 \rangle$ from the sequence $\langle ꞵ, 0, ꞵ, 0, ꞵ, 1, 0 \rangle$.

Third, we now have three notions called 'configurations'; let's summarize them for convenience:

1. *m*-configuration = line number, q_n, of a program for a Turing Machine.
2. Configuration = the pair: $\langle q_n, S(r) \rangle$,
   where $S(r)$ is the symbol on the currently scanned square, r.
3. Complete configuration = the triple:
   $\langle r$, the sequence of all symbols on the tape,[13] $q_n \rangle$.

8.9.5 "Circular and Circle-Free Machines"

We now come to what I have found to be one of the most puzzling sections of Turing's paper. It begins with the following definitions:

> If a computing machine never writes down more than a finite number of symbols of the first kind, it will be called *circular*. Otherwise it is said to be *circle-free*. (para. 1, p. 233.)

Let's take this slowly: a computing machine is a Turing Machine that only prints a binary representation of a real number together with a few symbols of the second kind. If such a machine "*never* writes down more than a finite number of" '0's and '1's, then, trivially, it has *only* written down a *finite* number of such symbols. *That means it has halted!* And in that case, Turing wants to call it 'circular'! But to my ears, at least, 'circular' sounds like 'looping,' which, in turn, sounds like it means "*not* halting."

And if it *does* write down more than a finite number of '0's and '1's, then, trivially, it writes down *infinitely* many of them. *That means it does not halt!* In that case, Turing wants to call it 'circle-free'! But that sounds like 'loop-free,' which, in turn, sounds like it means it *does* halt. Other commentators have made the same observation:

> In Turing's terminology, circularity means that the machine never writes down more than a finite number of symbols (halting behaviour). A non-circular machine is a machine that never halts and keeps printing digits of some computable sequence of numbers. (De Mol and Primiero, 2015, pp. 197–198, footnote 11)

What's going on? Before looking ahead to see if, or how, Turing clarifies this, here's one guess: the only way a Turing Machine can print a finite number of "figures" (Turing's name for '0' and '1') and still "be circular" (which I am interpreting to mean "loop") is for it to keep repeating printing – i.e. to "overprint" – some or all of them: i.e. for it to "circle back" and print some of them over and over again. (In this case, no "halt" instruction is needed!)

And the only way a Turing Machine can print infinitely many "figures" and also be "circle-free" is for it to continually print new figures to the right of the previous one that it printed (and thus not "circle back" to a previous square, overprinting it with the same symbol that's on it).

Is that what Turing has in mind? Let's see.

---

13 As described in our previous paragraph.

8.9.6 "Circular" Machines

The next paragraph says,

> A machine will be circular if it reaches a configuration from which there is no possible move or if it goes on moving, and possibly printing symbols of the second kind, but cannot print any more symbols of the first kind. The significance of the term "circular" will be explained in Section 8. (para. 2, p. 233.)

The first sentence is rather long; let's take it phrase by phrase: "A machine will be circular" – i.e. will print out only a finite number of figures – "if [Case 1] it reaches a configuration from which there is no possible move ...." That is, it will be circular if it reaches a line number q_n and a currently scanned symbol $S(r)$ from which there is no possible move. How could that be? Easy: if there's no line of the program of the form "Line q_n: **If** currently scanned symbol $= S(r)$ **then** ...." In that case, the machine stops,[14] because there's no instruction telling it to do anything.[15]

That's even more paradoxical than my earlier interpretation; here, he is clearly saying that a machine is circular if it halts! Of course, if you are the operator of a Turing Machine and you are only looking at the tape (and not at the machinery), would you be able to tell the difference between a machine that was printing the same symbol over and over again on the same square and a machine that was doing nothing?[16] Probably not. So, from an external, behavioral point of view, these would seem to amount to the same thing.

But Turing goes on: a machine will also be circular "... if [Case 2] it goes on moving, and possibly printing [only] symbols of the second kind" but not printing any more "figures." Here, the crucial point is that the machine does not halt but goes on moving. It might or might not print *anything*, but if it does, it only prints secondary symbols. So we have the following possibilities: a machine that keeps on moving, spewing out square after square of blank tape; or a machine that keeps on moving, occasionally printing a secondary symbol. In either case, it has only printed a *finite* number of *figures*. Because it has, therefore, *not* printed an *infinite* decimal representation of a real number, it has, for all practical purposes, halted – at least in the sense that it has finished its task, though it has not succeeded in computing a real number.

Once again, a machine is circular if it halts (for all practical purposes; it's still working but just not doing anything significant). This isn't what I had in mind in my earlier interpretation. But it does seem to be quite clear, no matter how you interpret what Turing says, that he means that *a circular machine is one that does not compute a real number*, either by halting or by continuing on but doing nothing useful (not computing a real number). Machines that *do* compute real numbers are "circle-free," but they must also never halt; they must loop forever, in modern terms, but continually do useful work (computing digits of the decimal expansion of a real number):

> A machine that computes a real number in this sense was called *circle-free*; one that does not (because it never prints more than a finite number of 0s and 1s) was called *circular*. (Davis, 1995c, p. 141)

In other words, a "good" Turing Machine is a "circle-free" one that does *not* halt and that continually computes a real number. This seems to be contrary to modern terminology and the standard

---

14 At this point, I cannot resist recommending, once again, that you read E.M. Forster's wonderfully prescient, 1909(!) short story, "The Machine Stops."

15 Another possibility is that line q_n says, **If** currently scanned symbol $= S(r)$, **then** go to line q_n. In that case, the machine never stops, because it forever loops (circles?) back to the same line.

16 The machine described in the text and the machine described in the previous footnote have this property.

analysis of "algorithms" that we saw in Section 7.3. And how does this fit in with Turing's claim at the beginning of his paper that "the 'computable' numbers may be described briefly as the real numbers whose expressions as a decimal are calculable *by finite means*" (my italics)? The only way I can see to make these two claims consistent is to interpret "by finite means" to refer to the number of steps in an algorithm, or the amount of time needed to carry out one step, or the number of operations needed to carry out one step (in case any of the steps are not just basic operations). It cannot mean, as we have just seen, that the entire task can be completed in a finite amount of time (see Section 11.5) or that it would necessarily halt.

Finally, what about the allusion to Turing's Section 8 (uninformatively titled "Application of the Diagonal Process")? In that section, which we will not investigate, Turing proves that the Halting Problem is not computable (more precisely, that a Gödel-like number of a program for a problem akin to the Halting Problem is not a computable number). And, pretty obviously, his proof is a little bit different from the one that we sketched in Section 7.7 because of the difference between our modern idea that only Turing Machines that halt are "good" and Turing's idea that only Turing Machines that are circle-free are "good."

Digression: A Possible Explanation of 'Circular': It is interesting to note that, in French, 'circular' would normally be translated as '*circulaire*.' Turing wrote a summary of his 1936 paper in French. In that document, instead of calling machines that halted without computing a real number '*circulaire*,' he called them '*méchant*' – 'malicious'! Perhaps he was having second thoughts about the term 'circular' and wanted something more perspicuous. For more information on the French summary, see Corry, 2017.

For more on "circularity," see Petzold, 2008, Ch. 10, who notes, by the way, that the concept of "halting" was introduced into the modern literature by Martin Davis (Petzold, 2008, p. 179), "despite the fact that Turing's original machines never halt!" (Petzold, 2008, p. 329). Here is a slightly different observation:

> The halting theorem is often attributed to Turing in his 1936 paper. In fact, Turing did not discuss the halting problem, which was introduced by Martin Davis in about 1952. (Copeland and Proudfoot, 2010, p. 248, col. 2)

This is clarified in Bernhardt, 2016:

> The halting problem is probably the most well-known undecidable decision problem. However, this is not the problem that Turing described in his paper.
>
> As Turing described his machines, they did not have accept states [i.e. they did not halt]. They were designed to compute real numbers and so would never stop if computing an irrational number. The notion of a Turing machine was changed [from Turing's original *a*-machines] to include accept states by Stephen Kleene and Martin Davis. Once you had this new formulation of a Turing machine, you could consider the halting problem. Davis 1958 gave the halting problem its name. (Bernhardt, 2016, pp. 120–121; see also p. 142)

For an analysis of these notions in modern terms, see van Leeuwen and Wiedermann, 2013.

8.9.7 "Computable Sequences and Numbers"

Here are Turing's final definitions from this section. First:

> A *sequence* is said to be computable if it can be computed by a circle-free machine. (p. 233, my italics.)

Although this is presented as a definition of 'computable sequence,' it can, in fact, be understood as another statement of the Computability Thesis. Being "computable by a circle-free machine" is a very precise mathematical concept. In this definition, I think Turing is best understood as suggesting that this precise concept should replace the informal notion of being "computable." Alternatively, Turing is saying here that he will use the word 'computable' in this very precise way.

Next:

> A *number* is computable if it differs by an integer from the number computed by a circle-free machine. (p. 233, my italics.)

Circle-free machines compute (by printing out) a sequence of figures (a sequence of '0's and '1's). Such a sequence can be considered a decimal (actually, a binary) representation of a number between 0 and 1 (including 0 but not including 1). Here, Turing is saying that *any* real number can be said to be computable if it has the same decimal part (i.e. the same part after the decimal point) of a number representable as a computable sequence of figures. So, for instance, $\pi = 3.1415926535\ldots$ differs by the integer 3 from the number $0.1415926535\ldots$, which is computable by a circle-free Turing Machine; hence, π is also computable.

8.10   Section 3: "Examples of Computing Machines"

We are now ready to look at some "real" Turing Machines – more precisely, "computing machines," which, recall, are "automatic" a-machines that print only figures ('0,' '1') and maybe symbols of the second kind. Hence, they compute real numbers. Turing gives us two examples, which we will look at in detail.

8.10.1   Example I

> A machine can be constructed to compute the sequence $010101\ldots.$
> (Section 3(I), para. 1, p. 233.)

Actually, as we will see, it prints

$0b1b0b1b0b1b\ldots$

What real number is this? First, note that it is a *rational* number of the form $0.\overline{01}$. Treated as being written in binary notation, it $= \frac{1}{3}$; treated as being written in decimal notation, it $= \frac{1}{99}$.

> The machine is to have the four m-configurations "b," "c," "f," "e" and is capable of printing "0" and "1." (p. 233.)

The four line numbers are (in more legible italic font): b, c, f, e.

> The behaviour of the machine is described in the following table in which "R" means "the machine moves so that it scans the square immediately on the right of the one it was scanning previously." Similarly for "L." "E" means "the scanned symbol is erased" and "P" stands for "prints." (p. 233.)

This is clear enough. Note that it is the Turing Machine that moves, not the tape! As we noted in Section 8.7.9, when you do a calculation with pencil and paper, your hand moves; the paper doesn't! Of course, a pencil is really only an *output* device that prints and erases (recall the epigraph to Section 8.13).[17] To turn it into a full-fledged computer (or, at least, a physical Turing Machine), you need to add eyes (for input), hands (for moving left and right), and a mind (for "states of mind").

Before going on with this paragraph, let's look at the "table."[18] In later writings by others, such tables are sometimes called 'machine tables'; *they are computer programs for Turing Machines*, written in a "Turing Machine programming language" for which Turing is now giving us the syntax and semantics.[19]

However, it is important to keep in mind that the Turing Machine does not "consult" this table to decide what to do. We humans would consult it in order to simulate the Turing Machine's behavior. But the Turing Machine itself simply behaves *in accordance with* that table, not by *following* it. The table should be thought of as a mathematical-English description of the way that the Turing Machine is "hardwired" to behave. (We'll revisit this idea in Sections 10.4 and 12.4.4.)

Here's the table, written a bit more legibly than in Turing's paper:

Configuration		Behaviour	
m-config.	*symbol*	*operations*	*final m-config.*
b	None	P0, R	c
c	None	R	e
e	None	P1, R	f
f	None	R	b

This program consists of four lines. It is important to note that it is a *set* of lines, not a *sequence*: the order in which the lines are written down in the table (or "program") is irrelevant; there will never be any ambiguity as to which line is to be executed. Perhaps a better way of saying this is: there will never be any ambiguity as to which line is "causing" the Turing Machine to move.

Each line consists of two principal parts: a "configuration" and a "behavior." Each configuration, as you may recall, consists of two parts: an *m*-configuration (or line number) and a symbol (namely, the currently scanned symbol). Each behavior consists also of two parts: an "operation" (one or more of *E, L, R,* or *P*) and a "final *m*-configuration" (i.e. the next line number to be executed).

> This table (and all succeeding tables of the same kind) is to be understood to mean that *for a configuration described in the first two columns the operations in the third column are carried out successively, and the machine then goes over into the m-configuration described in the last column.* (p. 233, my italics.)

That is, each line of the program should be understood as follows: "Under the conditions described by the configuration, do the operation and then go to the instruction at the final *m*-configuration." Or, to use Turing's other terminology: "If your current state of mind is the one listed in the current *m*-configuration, and if the symbol on the current square being scanned is the one in the *symbol* column, then do the *operation* and change your state of mind to the one in the *final m-configuration* column."

---

17 A "Rhymes with Orange" cartoon from 11/19/2009 suggests that the pencil point is the cursor, the eraser is the delete key, and it can be rebooted with a pencil sharpener.
18 Not to be confused with our table of place settings and beer mugs from Section 8.8.1!
19 That is, the grammar and meaning; see Sections 9.4.3, 13.1.1.

A further qualification:

> When the second column [i.e. the symbol column] is left blank, it is understood that the behaviour of the third and fourth columns applies for any symbol and for no symbol. (p. 233.)

That is the situation we have in this first example, where 'None' is the entry in each row of the symbol column. So the only condition determining the behavior of this Turing Machine is its current "state of mind," i.e. its current line number.

Finally, we need to know what the initial situation is:

> The machine starts in the *m*-configuration *b* with a blank tape. (p. 233.)

Perhaps '*b*' stands for "begin," with subsequent "states of mind" (in alphabetical as well as sequential order) being *c*, *e*, and *f* ('*f*' for "final"? What happened to '*d*'?).

Let's trace this program. We start with a blank tape, which I will show as follows:

ƀƀƀƀƀƀƀƀƀƀƀ …

We are in state *b*.

Looking at the table, we see that if we are in state *b*, then (because any symbol that might be on the tape is irrelevant), we should do the sequence of operations *P*0, *R*. Turing hasn't told us what '*P*0' means, but because '*P*' means "print," it's pretty obvious that this means "print 0 on the currently scanned square."

Note, too, that he hasn't told us which square is currently being scanned! It probably doesn't matter, because all squares on the tape are blank. If the tape is infinite (or endless) in both directions, then each square is indistinguishable from any other square, at least until something is printed on one square. However, it's worth thinking about some of the options: one possibility is that we are scanning the first, or leftmost, square; this is the most likely option and the one that I will assume in what follows. But another possibility is that we are scanning some other square somewhere in the "middle" of the tape. That probably doesn't matter, because Turing only told us that it would compute the sequence '010101…'; he didn't say *where* it would be computed!

There is one further possibility, not very different from the previous one: the tape might not have a "first" square – it might be infinite in both directions! Thus, we need to consider something that Turing hasn't mentioned: how long is the tape? As far as I can tell, Turing is silent in this paper about the length of the tape. For all that he has told us, it could be infinitely long. In fact, the informal ways that Turing Machines are usually introduced often talk about an "infinite" tape. But many mathematicians and philosophers (not to mention engineers!) are not overly fond of actual infinities. The more mathematically precise way to describe it is as an "arbitrarily long" tape. That is, the tape is as long as you need it to be. For most computations (the ones that really do halt with a correct answer), the tape will be finite. Since no real machine can print out an infinitely long decimal, no real machine will require an infinite tape, either. In real life, you can only print out a finite initial segment of the decimal part of a real number; i.e. it will always be an approximation, but you can make the approximation as close as you want by just printing out a few more numbers. So instead of saying that the tape is infinitely long, we can say that, at any moment, the tape only has a finite number of squares, *but* there is no limit to the number of extra squares we are allowed to add on at one (or maybe both) ends. (As my former teacher and colleague John Case used to put it, if we run out of squares, we can always go to an office-supply store, buy some extra squares, and staple them onto our tape!) People don't have infinite memory, and neither do Turing Machines or, certainly, real computers. The major difference between Turing Machines, on the one hand,

and people and real computers, on the other hand, is that Turing Machines can have a tape (or a memory) that is as large as you need, while people and real computers are limited.

So, let's now show our initial tape as follows, where the currently scanned square is underlined:

$\underline{b}\ b\ b\ b\ b\ b\ b\ b\ b\ b$ ...

Performing the two operations on line *b* converts our initial tape to this one

$0\ \underline{b}\ b\ b\ b\ b\ b\ b\ b\ b$ ...

and puts us in state *c*. That is, we next execute the instruction on line *c*.

Looking at line *c*, we see that, no matter what symbol is on the current square (it is, in fact, blank), we should simply move right one more square and change our mind to *e*. So now our tape will look like this:

$0\ b\ \underline{b}\ b\ b\ b\ b\ b\ b\ b$ ...

Because we are now in state *e*, we look at line *e* of the program, which tells us that, no matter what, if anything, is on the current square, print '1' there, move right again, and go into state *f*. So our tape becomes

$0\ b\ 1\ \underline{b}\ b\ b\ b\ b\ b\ b$ ...

Now we are in state *f*, and looking at line *f*, we see that we merely move right once again, yielding

$0\ b\ 1\ b\ \underline{b}\ b\ b\ b\ b\ b$ ...

And we go back into state *b*. But that starts this cycle all over again; we are indeed in an infinite loop! One more cycle through this turns our tape into

$0\ b\ 1\ b\ 0\ b\ 1\ \underline{b}\ b\ b\ b$ ...

Clearly, repeated cycles through this infinitely looping program will yield a tape consisting entirely of the infinite sequence 010101… with blank squares separating each square with a symbol printed on it:

$0\ b\ 1\ b\ 0\ b\ 1\ b\ 0\ b\ 1\ b$ ...

Can this program be written differently?

> If (contrary to the description in §1) we allow the letters L, R to appear more than once in the operations column we can simplify the table considerably.
> (Section 3(I), para. 2, p. 234.)

In "the description in §1" (p. 231), Turing allowed the machine to "change the square which is being scanned, but only by shifting it *one* place to right or left" (my italics). Now, he is allowing the machine to move *more* than one place to the right or left; this is accomplished by allowing a *sequence* of moves. Here is the modified program:

m-config.	*symbol*	*operations*	*final m-config.*
	None	P0	b
b	0	R, R, P1	b
	1	R, R, P0	b

Note that there is only one *m*-configuration (i.e. only one line number); another way to think about this is that the program has only one instruction. Turing would say that this machine never changes its state of mind. But that one instruction is, of course, more complex than the previous ones. This one is what would now be called a 'case' statement: in case there is no current symbol, print 0; in case the current symbol = 0, move right two squares and print 1; and in case the current symbol = 1, move right two squares and print 0 – and, in all cases, remain in the same state of mind.

Exercises for the Reader:

1. I urge you to try to follow this version of the program, both for practice in reading such programs and to convince yourself that it has the same behavior as the first one.
2. Another interesting exercise is to write a program for a Turing Machine that will print the sequence 010101... *without* intervening blank squares.

So, our machine has "compute[d] the sequence 010101 ...." Or has it? It has certainly written down that sequence. Is that the same thing as "computing" it?

And here is another question: earlier, I said that 010101... was the binary representation of $\frac{1}{3}$ and the decimal representation of $\frac{1}{99}$. Have we just computed $\frac{1}{3}$ in base 2? Or $\frac{1}{99}$ in base 10?

Even if you are inclined to answer 'yes' to the question of whether writing is the same as computing, you might be more inclined to answer 'no' to the question of whether we have computed $\frac{1}{3}$ in base 2 or $\frac{1}{99}$ in base 10. Although Turing may have a convincing reason (in his Section 9) to say that computing consists of nothing more than writing down symbols, surely there has to be more to it than that; surely just writing down symbols is only *part* of computing. The other parts have to do with *which* symbols get written down, *in what order*, and *for what reason*. If I asked you to compute the decimal representation of $\frac{1}{99}$, how would you know that you were supposed to write down 010101...? Surely, *that* is the heart of computation. Or is it? (We'll return to this in Sections 13.3, 16.2, and 16.4.6.)

At this point, however, we should give Turing the benefit of the doubt. After all, *he* did not say that we were going to compute $\frac{1}{99}$, only that we were going to "compute" 010101..., and, after all, "computing" that sequence really just *is* writing it down; it's a trivial, or basic, or elementary, or primitive computation (choose your favorite adjective). Moreover, arguably, Turing only showed us this trivial example so that we could clearly see the *format* of his Turing Machine programs before getting a more complex example.

Before turning to such a more complex program, let's consider the syntax (specifically, the grammatical structure) of these programs a bit more. Each line of the program has the following general form

$$q_B \; S \; O \; q_E$$

where

1. q_B is an initial (or **B**eginning) *m*-configuration (a line number).
2. S is the symbol on the currently scanned square (possibly a blank).
3. O is an operation (or a sequence of operations) to be performed (where the operations are Px, E, L, R, and where x is any legally allowed symbol).[20]
4. q_E is a final (or **E**nding) *m*-configuration.

---

20 In our first program, the only symbols were '0' and '1'; we will see others in subsequent examples.

And the semantics (i.e. the meaning or interpretation) of this program line is:

> **if** the Turing Machine is in *m*-configuration q_B, **and**
> **if either** the current input = *S* **or** no input is specified,
> > **then begin**
> > > 1. do the sequence of operations *O*;
> > > { where each operation is *either*:
> > > • *Print x* on the current square,
> > > > (where printing *x* overwrites whatever is currently printed on the square), *or*
> > > • *E*rase the symbol that is on the current square,
> > > > (where erasing results in a blank square, even if the square is already blank), *or*
> > > • move *Left* one square, *or*
> > > • move *Right* one square }
> > > 2. go to *m*-configuration q_E
> > **end**

8.10.2 Example II, Paragraph 1

We now come to "a slightly more difficult example":

> As a slightly more difficult example we can construct a machine to compute the sequence 001011011101111011111 . . . . (Section 3(II), para. 1, p. 234.)

First, note that the sequence to be computed consists of the subsequences

> 0, 1, 11, 111, 1111, 11111, …

That is, it is a sequence beginning with '0,' followed by the numbers 1, 2, 3, 4, 5, … written in base 1 (i.e. as "tally strokes") – with each term separated by a '0.'

But this seems very disappointing! It seems that this "more difficult" computation is still just writing down some symbols without "computing" anything. Perhaps. But note that what is being written down (or "computed") here are the natural numbers. This program will begin counting, starting with 0, then the successor of 0, the successor of that, and so on. But as we saw in Section 7.6, the successor function is one of the basic recursive functions: i.e. one of the basic computable functions.

Being able to (merely!) write down the *successor* of any number, being able to (merely!) write down the *predecessor* of any non-0 number, and being able to *find* a given term in a sequence of numbers are the only basic recursive (or computable) functions. Turing's "slightly more difficult example" will show us how to compute the first of these. Devising a Turing Machine program for computing the predecessor of the natural number *n* should simply require us to take a numeral represented as a sequence of *n* occurrences of '1' and erase the last one. Devising a Turing Machine program for computing the *j*th term in a sequence of *k* symbols should simply require us to move a certain number of squares in some direction to find the term (or, say, the first square of a sequence of squares that represents the term, if the term is complex enough to have to be represented by a sequence of squares).

And any other recursive function can be constructed from these basic functions by generalized composition (sequencing), conditional definition (selection), and while-recursion (repetition), which are just "control structures" for how to find a path (so to speak) through a Turing Machine program – i.e. ways to organize the sequence of *m*-configurations that the Turing Machine should go through.

So, it looks as if **computation really is nothing more than writing things down, moving around (on a tape), and doing so in an order that will produce a desired result!** As historian Michael Mahoney suggested, the shortest description of Turing's accomplishment might be that Turing

> showed that any computation can be described in terms of a machine shifting among a finite number of states in response to a sequence of symbols read and written one at a time on a potentially infinite tape. (Mahoney, 2011, p. 79)

We'll return to this idea in Section 9.5.

Let's now look at this "slightly more difficult" program:

> The machine is to be capable of five *m*-configurations, viz., "*o*", "*q*", "*p*", "*f*", "*b*" and of printing "ə", "*x*", "0", "1",
> (Section 3(II), para. 1, p. 234, substituting italics for German Fraktur letters)

The first two printable symbols are going to be used only for bookkeeping purposes.[21] So, once again, Turing is really restricting himself to binary notation for the important information.

Continuing:

> The first three *symbols* on the tape will be "əə0"; the other *figures* follow on alternate squares. (Section 3(II), para. 1, p. 234, my italics.)

It may sound as if Turing is saying that the tape comes with some pre-printed information. But when we see the program, we will see that, in fact, the first instruction has us print 'əə0' on the first three squares before beginning the "real" computation. Had the tape come with pre-printed information, perhaps it could have been considered "innate" knowledge,[22] although a less cognitive description could simply have been that the manufacturer of the tape had simplified our life, knowing that the first thing that the program does to a completely blank tape is to print 'əə0' on the first three squares before beginning the 'real' computation. Because that only has to be done once, it might have been simpler to consider it pre-printed on the tape.

Note that Turing calls these 'symbols' in the first clause and then talks about 'figures' in the second clause. Figures, you may recall from Section 8.9.3, are the numerals '0' and '1.' So, Turing seems to be saying that all subsequent occurrences of '0' and '1' will occur on "alternate squares." What happens on the other squares? He tells us:

> On the intermediate squares we never print anything but "*x*." These letters serve to "keep the place" for us and are erased when we have finished with them. We also arrange that in the sequence of figures on alternate squares there shall be no blanks.
> (Section 3(II), para. 1, p. 234.)

It sounds as if the final tape will begin with 'əə0'; during the computation, subsequent squares will have '0' or '1' interspersed with '*x*'; and at the end of the computation, those subsequent squares

---

21 The inverted 'e' is called a 'ə'; it is used in phonetics to represent the sound "uh," as in 'but.' Turing uses it merely as a bookkeeping symbol with no meaning.

22 That is, knowledge that it was "born" with (or, to use another metaphor, knowledge that is "hardwired"). For more on innate knowledge, see Samet and Zaitchik, 2017.

will only have '0' or '1,' and no blanks. Of course, at the end, we could go back and erase the initial occurrences of 'ǝ', so there would only be "figures" and no other symbols.

Here is the program:

Configuration		Behaviour	
m-config.	*symbol*	*operations*	*final m-config.*
b		Pǝ, R, Pǝ, R, P0, R, R, P0, L, L	*o*
o	1	R, Px, L, L, L	*o*
	0		*q*
q	Any (0 or 1)	R, R	*q*
	None	P1, L	*p*
p	x	E, R	*q*
	ǝ	R	*f*
	None	L, L	*p*
f	Any	R, R	*f*
	None	P0, L, L	*o*

I think it will be helpful to restate this program in a more readable format:

b **begin**
> print 'ǝǝ0' on the first 3 squares;
> P0 on the 5th square;
> move left to the 3rd square {which has '0' on it};
> go to line *o*
end

o **if** current symbol = 1
then begin
> move right;
> Px;
> move left 3 squares;
> go to line *o* {i.e. stay at *o*}
> **end**
else if current symbol = 0
> **then** go to line *q*

q **if** current symbol = 0 **or** current symbol = 1
then begin
> move right 2 squares;
> go to line *q*
> **end**
else if current square is blank
> **then begin**
> > P1;
> > move left;
> > go to line *p*
> > **end**

p   **if** current symbol = *x*
    then begin
           erase the *x*;
           move right;
           go to line *q*
    end
    else if current symbol = ə
    then begin
           move right;
           go to line *f*
    end
    else if current square is blank
    then begin
           move left 2 squares;
           go to line *p*
    end

f   **if** current square is not blank
    then begin
           move right 2 squares;
           go to line *f*
    end
    else begin
           P0;
           move left 2 squares;
           go to line *o*
    end

Note that no line of the program ends with the machine changing its state of mind to *m*-configuration *b*. So that line of the program, which is the one that initializes the tape with 'əə0' on the first three squares, is only executed once. Note also that whenever an instruction ends with a command to stay in the same *m*-configuration (i.e. to go to that very same line), we are in a loop. A structured version of the program would use a **while…do** control structure, instead.

There are some odd things to consider in lines *o*, *q*, *p*: What happens if the machine is in state *o* but the current symbol is not a "figure"? What happens in state *q* if the current symbol is 'ə' or '*x*'? And what happens in state *p* if the current symbol *is* a "figure"? Turing doesn't specify what should happen in these cases. One possibility is that he has already determined that none of these cases *could* occur. Still, modern software engineering practice would recommend that an error message be printed out in those cases. In general, in a computer program, when a situation occurs for which the program does not specify what should happen, anything is legally allowed to happen, and there is no way to predict what will happen: "garbage in, garbage out."

8.10.3   Example II, Paragraph 2

Turing goes on "to illustrate the working of this machine" with "a table … of the first few complete configurations" (p. 234.) Recall that a "complete configuration" consists of information about which square is currently being scanned, the sequence of all symbols on the tape, and the line number of the instruction currently being executed. Rather than use Turing's format, I will continue to use the format that I used for Example I, adding the line number at the beginning, using underscoring to indicate the currently scanned square, and assuming that any squares not shown are blank; any blank square that is between two non-blank squares (if there are any) will be indicated by our symbol for a blank that has been made visible: ♭. You are urged to compare my trace of this program with Turing's.

So, we begin with a blank tape. What is the machine's initial state of mind, its initial *m*-configuration? Turing has forgotten to tell us! But it is fairly obvious that *b* is the initial *m*-configuration, and, presumably, we are scanning the leftmost square (or, if the tape is infinite in both directions, then we are scanning any arbitrary square), and, of course, all squares are blank:

$b : \underline{\flat}, \flat, \ldots$

The initial instruction tells us to print ə, move right, print another ə, move right again, print 0, move right two more squares, print another 0, move two squares back to the left, and go into state *o*. After doing this sequence of primitive operations, our complete configuration looks like this:

$o : ə, ə, \underline{0}, \flat, 0, \flat, \ldots$

Digression on Notation: To help you in reading Turing's paper, my notation for the initial situation should be compared with his. Here is his:

$$\vdots$$
$$b$$

He has an invisible blank, followed by a colon, with the *m*-configuration '*b*' underneath the (invisible) blank, marking the currently scanned square.

Instead, I have '*b:*' preceding a sequence of (visible) blanks, the first one of which is marked as being the scanned square.

Turing then shows the second configuration:

$$ə \quad ə \quad 0 \quad\quad 0 \quad :$$
$$o$$

Turing has two occurrences of 'ə' followed by two '0's that are separated by an (invisible) blank, with the *m*-configuration '*o*' underneath the currently scanned square (which contains the first '0'), followed by a colon to mark the end of this complete configuration.

Instead, I have '*o:*' preceding a sequence consisting of the two occurrences of 'ə,' followed by a '0' that is marked as being the scanned square, followed by a (visible) blank, followed by the second '0.'

We are now in *m*-configuration *o*, and the currently scanned square contains '0,' so the second case (i.e. the bottom row) of this second instruction tells us merely to go into state *q*. The "operations" column is left empty, so there is no operation to perform. It is worth noting that although there does not always have to be an operation to perform, *there does always have to be a final state to go into: i.e. a next instruction to perform*. So, the tape looks exactly as it did before, except that the machine is now in state *q*:

$q : ə, ə, \underline{0}, \flat, 0, \flat, \ldots$

Because the machine is now in state *q* and still scanning a '0,' the first case (i.e. the top row) of this third instruction tells us to move two squares to the right but to stay in state *q*. So the tape now looks like this:

$q : ə, ə, 0, \flat, \underline{0}, \flat, \ldots$

Because the machine is still in state q and still scanning a '0' (although the currently scanned *square* is different), we perform the same (third) instruction, moving two more squares to the right and staying in state q:

$q \,:\, \text{ə}, \text{ə}, 0, \flat, 0, \flat, \underline{\flat}, \dots$

The machine is still in state q, but now there is no scanned symbol, so the second case (bottom line) of the third instruction is executed, resulting in a '1' being printed on the current square, and the machine moves left, going into state p.

Whenever the machine is in state p and scanning a blank (as it is now), the third case (last line) of the fourth instruction is executed, so the machine moves two squares to the left and stays in state p:

$p \,:\, \text{ə}, \text{ə}, 0, \underline{\flat}, 0, \flat, 1, \dots$

Now the machine is in state p scanning a blank, so the same instruction is executed: it moves two more squares to the left and continues in state p:

$p \,:\, \text{ə}, \underline{\text{ə}}, 0, \flat, 0, \flat, 1, \dots$

But now it is the second case (middle line) of the fourth instruction that is executed, so the machine moves right and goes into state f:

$f \,:\, \text{ə}, \text{ə}, \underline{0}, \flat, 0, \flat, 1, \dots$

When in state f scanning any symbol (but not a blank), the machine moves two squares to the right, staying in f:

$f \,:\, \text{ə}, \text{ə}, 0, \flat, \underline{0}, \flat, 1, \dots$

Again, it moves two squares to the right, staying in f:

$f \,:\, \text{ə}, \text{ə}, 0, \flat, 0, \flat, \underline{1}, \dots$

And again:

$f \,:\, \text{ə}, \text{ə}, 0, \flat, 0, \flat, 1, \flat, \underline{\flat}, \dots$

But now it executes the second case of the last instruction, printing '0,' moving two squares to the left, and returning to state o:

$o \,:\, \text{ə}, \text{ə}, 0, \flat, 0, \flat, \underline{1}, \flat, 0, \dots$

Now, for the first time, the machine executes the first case of the second instruction, moving right, printing 'x,' moving three squares to the left but staying in o:

$o \,:\, \text{ə}, \text{ə}, 0, \flat, \underline{0}, \flat, 1, x, 0, \dots$

At this point, you will be forgiven if you have gotten lost in the "woods," having paid attention only to the individual "trees" and not seeing the bigger picture.[23] Recall that we are trying to count – to produce the sequence 0, 1, 11, 111, … with '0's between each term:

0 0 1 0 11 0 111 0 …

We started with a blank tape

$\flat\flat\flat \dots$

---

23 My apologies for the mixed metaphor.

and we now have a tape that looks like this:

 əə0b0b1x0b …

Clearly, we are going to have to continue tracing the program before we can see the pattern that we are expecting; Turing, however, ends his tracing at this point. But we shall continue; however, I will only show the complete configurations without spelling out the instructions (doing that is left to the reader). Here goes, continuing from where we left off:

```
o :  ə  ə  0  b  0  b  1  x  0  …
q :  ə  ə  0  b  0  b  1  x  0  …
q :  ə  ə  0  b  0  b  1  x  0  …
q :  ə  ə  0  b  0  b  1  x  0  …
q :  ə  ə  0  b  0  b  1  x  0  b  b  …
p :  ə  ə  0  b  0  b  1  x  0  b  1  …
p :  ə  ə  0  b  0  b  1  x  0  b  1  …
q :  ə  ə  0  b  0  b  1  b  0  b  1  …
q :  ə  ə  0  b  0  b  1  b  0  b  1  …
q :  ə  ə  0  b  0  b  1  b  0  b  1  b  b  b …
p :  ə  ə  0  b  0  b  1  b  0  b  1  b  1  b …
```

Hopefully, now you can see the desired pattern beginning to emerge. The occurrences of 'x' get erased, and what's left is the desired sequence but with blank squares between each term *and* with two leading occurrences of 'ə'. You can see from the program that there is no instruction that will erase those 'ə's; the only instructions that pay any attention to a 'ə' are (1) the second case of *m*-configuration *p*, which only tells the machine to move right and to go into state *f*; and (2) the first case of *m*-configuration *f*, which, when scanning any symbol, simply moves two squares to the right (but in fact, that configuration will never occur!).

In the third paragraph, Turing makes some remarks about various notation conventions that he has adopted, but we will ignore these, because we are almost finished with our slow reading. I do want to point out some other highlights, however.

8.11 Section 4: "Abbreviated Tables"

In this section, Turing introduces some concepts that are central to programming and software engineering.

> There are certain types of process used by nearly all machines, and these, in some machines, are used in many connections. These processes include copying down sequences of symbols, comparing sequences, erasing all symbols of a given form, etc. (Section 4, para. 1, p. 235.)

In other words, certain sequences of instructions occur repeatedly in different programs and can be thought of as being single "processes": copying, comparing, erasing, etc.

Turing continues:

> Where such processes are concerned we can abbreviate the tables for the *m*-configurations considerably by the use of "skeleton tables." (Section 4, para. 1, p. 235.)

The idea is that skeleton tables are descriptions of more complex sequences of instructions that are given a single name. This is the idea behind "subroutines" (or "named procedures") and "macros" in modern computer programming. (Recall Section 7.4.3.3.) If you have a sequence of instructions that accomplishes what might better be thought of as a single task (e.g. copying a sequence of symbols), and if you have to repeat this sequence many times throughout the program, it is more convenient (for the human writer or reader of the program!) to write this sequence down only once, give it a name, and then refer to it by that name whenever it is needed. There is one small complication: each time this named abbreviation is needed, it might require that parts of it refer to squares or symbols on the tape that will vary depending on the current configuration, so the one occurrence of this named sequence in the program might need to have variables in it:

> In skeleton tables there appear capital German letters and small Greek letters. These are of the nature of "variables." By replacing each capital German letter throughout by an *m*-configuration and each small Greek letter by a symbol, we obtain the table for an *m*-configuration. (Section 4, para. 1, pp. 235–236.)

Of course, whether one uses capital German letters, small Greek letters, or something more legible or easier to type is an unimportant implementation detail. The important point is this:

> The skeleton tables are to be regarded as nothing but abbreviations: they are not essential. (Section 4, para. 2, p. 236.)

8.12 Section 5: "Enumeration of Computable Sequences"

Another highlight of Turing's paper that is worth pointing out occurs in his Section 5: a way to convert every program for a Turing Machine into a number. Let me be a bit more precise about this before seeing how Turing does it.

First, it is important to note that, for Turing, there really is no difference between one of his *a*-machines (i.e. a Turing Machine) and the *program for* it. Turing Machines are "hardwired" to perform exactly one task, as specified in the program (the "table," or "machine table") for it. So, converting a program to a number is the same as converting a Turing Machine to a number. Second, "converting to a number" – i.e. assigning a number to an object – really means you are *counting*. So, in this section, Turing shows that you can count Turing Machines by assigning a number to each one. Third, if you can count Turing Machines, then you can only have a countable number of them. But there are *un*countably many real numbers, so there will be some real numbers that are not computable! (Recall our Section 7.7.1.)

Here is how Turing counts Turing Machines. First (using the lowercase Greek letter "gamma," γ):

> A computable sequence γ is determined by a description of a machine which computes γ. Thus the sequence 001011011101111… is determined by the table on p. 234, and, in fact, any computable sequence is capable of being described in terms of such a table. (Section 5, para. 1, p. 239)

"A description of a machine" is one of the tables such as those we have been looking at; i.e. it is a computer program for a Turing Machine.

But as we have also seen, it is possible to write these tables in various ways. So, before we can count them, we need to make sure we don't count any twice because we have confused two different ways of writing the same table with being two different tables. Consequently:

> It will be useful to put these tables into a kind of standard form.
> (Section 5, para. 2, p. 239.)

The first step in doing this is to be consistent about the number of separate operations that can appear in the "operations" column of one of these tables. Note that in the two programs we have looked at, we have seen examples in which there were as few as 0 operations and as many as 10 (not to mention the variations possible with skeleton tables). So:

> In the first place let us suppose that the table is given in the same form as the first table, e.g. I on p. 233. [See our Section 8.10.] That is to say, that the entry in the operations column is always of one of the forms $E : E, R : E, L : Pa : Pa, R : Pa, L : R : L :$ or no entry at all. The table can always be put into this form by introducing more m-configurations.
> (Section 5, para. 2, p. 239.)

In other words, the operation in the operations column will be exactly one of:

> erase
> erase and then move right
> erase and then move left
> print symbol a
> print a and then move right
> print a and then move left
>> (where 'a' is a variable ranging over all the possible symbols
>> in a given program)
> move right
> move left
> do nothing

"Introducing more m-configurations" merely means a single instruction such as

> b 0 $P1, R, P0, L$ f

can be replaced by two instructions:

> b 0 $P1, R$ f_1
> f_1 $P0, L$ f

where 'f_1' is a new m-configuration not appearing in the original program. Put otherwise, a *single* instruction consisting of a *sequence* of operations can be replaced by a *sequence* of instructions, each consisting of a *single* operation. (For convenience, presumably, Turing allows *pairs* of operations, where the first member of the pair is either E or P and the second is either R or L. So a single instruction consisting of a sequence of (pairs of) operations can be replaced by a sequence of instructions, each consisting of a single operation or a single such pair.)

   Numbering begins as follows:

> Now let us give numbers to the m-configurations, calling them $q_1, \ldots, q_R$ as in §1. The initial m-configuration is always to be called q_1. (Section 5, para. 2, p. 239.)

So, each *m*-configuration's number is written as a subscript on the letter '*q*.'

The numbering continues:

> We also give numbers to the symbols $S_1, \dots, S_m$ and, in particular,
> blank $= S_0, 0 = S_1, 1 = S_2$. (Section 5, para. 2, pp. 239–240.)

So, each symbol's number is written as a subscript on the letter '*S*.' Note that Turing singles out three symbols for special treatment: '0,' '1,' and what I have been writing as ♭. (Turing is finally making the blank visible.)

At this point, we have the beginnings of our "standard forms," sometimes called 'normal' forms (which Turing labels N_1, N_2, N_3):

> The lines of the table are now [one] of [the following three] form[s]

m-config.	*Symbol*	*Operations*	*Final m-config.*	
q_i	S_j	PS_k, L	q_m	(N_1)
q_i	S_j	PS_k, R	q_m	(N_2)
q_i	S_j	PS_k	q_m	(N_3)

> (Section 5, para. 2, p. 240.)

The three "normal forms" are

N_1 *m*-configuration $q_i =$ **if** currently scanned symbol is S_j,
   then begin
     print symbol S_k;
     move left;
     go to q_m
   end

N_2 *m*-configuration $q_i =$ **if** currently scanned symbol is S_j,
   then begin
     print symbol S_k;
     move right;
     go to q_m
   end

N_3 *m*-configuration $q_i =$ **if** currently scanned symbol is S_j,
   then begin
     print symbol S_k;
     go to q_m
   end

As Turing notes in the following passage (which I will not quote but merely summarize), erasing (*E*) is now going to be interpreted as printing a blank (PS_0), and a line in which the currently scanned symbol is S_j and the operation is merely to move right or left is now going to be interpreted as overprinting the very same symbol (PS_j) and then moving. So, all instructions require printing something – either a visible symbol or a blank symbol – and then either moving or not moving. As Turing notes,

> In this way we reduce each line of the table to a line of one of the forms $(N_1), (N_2), (N_3)$.
> (Section 5, para. 3, p. 240.)

Turing simplifies even further, eliminating the 'print' command and retaining only the symbol to be printed. After all, if all commands involve printing something, you don't need to write down '*P*'; you only need to write down what you're printing. So each instruction can be simplified to a 5-tuple consisting of the initial *m*-configuration, the currently scanned symbol (and there will always be one, even if the "symbol" is blank, because the blank has been replaced by 'S_0'), the symbol to be printed (again, there will always be one, even if it's the blank), and the final *m*-configuration:

> From each line of form (N_1) let us form an expression $q_i S_j S_k L q_m$; from each line of form (N_2) we form an expression $q_i S_j S_k R q_m$; and from each line of form (N_3) we form an expression $q_i S_j S_k N q_m$. (Section 5, para. 4, p. 240.)

Presumably, *N* means something like "no move." A slightly more general interpretation is that not only do we always print something (even if it's a blank), but we also always move somewhere, except that sometimes we "move" to our current location. This standardization is consistent with our earlier observation (in Section 7.4.2) that the only two verbs that are needed are 'print(symbol)' and 'move(location).'

Next:

> Let us write down all expressions so formed from the table for the machine and separate them by semi-colons. In this way we obtain a complete description of the machine. (Section 5, para. 5, p. 240.)

Turing's point here is that the set of instructions can be replaced by a single string of 5-tuples separated by semicolons. There are two observations to make. First, because the machine table is a *set* of instructions, there could (in principle) be several different strings (i.e. descriptions) for each such set, because strings are *sequences* of symbols. Second, Turing has here introduced the now-standard notion of using a semicolon to separate lines of a program; however, this is not quite the same thing as the convention of using a semicolon to signal *sequencing*, because the instructions of a Turing Machine program are not an ordered sequence of instructions (even if, whenever they are written down, they have to be written down in some order).

So, Turing has developed a standard encoding of the lines of a program: an *m*-configuration encoded as q_i (forget about *b*, *f*, etc.), a pair of symbols encoded as S_j, S_k (the first being the scanned input, the second being the printed output; again, forget about things like '0,' '1,' '*x*,' etc.), a symbol (either *L*, *R*, or *N*) encoding the location to be moved to, and another *m*-configuration encoded as q_m. Next, he gives an encoding of these standardized codes:

> In this description we shall replace q_i by the letter "*D*" followed by the letter "*A*" repeated *i* times, and S_j by "*D*" followed by "*C*" repeated *j* times. (Section 5, para. 5, p. 240.)

Before seeing *why* he does this, let's make sure we understand *what* he is doing. The only allowable *m*-configuration symbols in an instruction are: $q_1, \dots, q_l$, for some *l* that is the number of the final instruction. What really matters are, first, that each instruction can be assumed to begin and end with an *m*-configuration symbol, and, second, which one it is, which can be determined by the subscript on *q*. In this new encoding, "*D*" simply marks the beginning of an item in the 5-tuple, and the *i* occurrences of letter '*A*' encode the subscript. Similarly, the only allowable symbols are: $S_1, \dots, S_n$, for some *n* that is the number of the last symbol in the alphabet of symbols. Here, what really matters are, first, that in each instruction the second and third items in the 5-tuple can be assumed to be symbols (including a visible blank!), and, second, which ones they are, which can be determined by the subscript on S. In our new encoding, "D" again marks the beginning the next item in the 5-tuple, and the j occurrences of 'C' encode the subscript.

Turing then explains that

> This new description of the machine may be called the *standard description* (S.D). It is made up entirely from the letters "A", "C", "D", "L", "R", "N", and from "$;$". (Section 5, para. 5, p. 240.)

So, for example, this two-line program

$$q_3S_1S_4Rq_5$$

$$q_5S_4S_0Lq_5$$

will be encoded by an S.D consisting of this 38-character string:

> *DAAADCDCCCCRDAAAAA; DAAAAADCCCCDLDAAAAA*

The next step in numbering consists in replacing these symbols by numerals:

> If finally we replace "A" by "1", "C" by "2", "D" by "3", "L" by "4", "R" by "5", "N" by "6", "$;$" by "7" we shall have a description of the machine in the form of an arabic [sic] numeral. The integer represented by this numeral may be called a *description number* (D.N) of the machine. (Section 5, para. 6, p. 240)

Just as Gödel numbering is one way to create a number corresponding to a string, "Turing numbering" is another. The D.N of the machine in our previous example is this numeral:

> 31113232222531111173111113222234311111

which, written in the usual notation with commas, is

> 31,113,232,222,531,111,173,111,113,222,234,311,111

or, in words, 31 undecillion, 113 decillion, 232 nonillion, 222 octillion, 531 septillion, 111 sextillion, 173 quintillion, 111 quadrillion, 113 trillion, 222 billion, 234 million, 311 thousand, one hundred eleven. That is the "Turing number" of our 2-line program!

Turing observes that:

> The D.N determine the S.D and the structure of the machine uniquely. The machine whose D.N is n may be described as $\mathcal{M}(n)$. (Section 5, para. 6, pp. 240–242.)

Clearly, given a D.N, it is trivial to decode it back into an S.D in only one way. Equally clearly (and almost as trivially), the S.D can be decoded back into a program in only one way. Hence, "the structure of the machine" encoded by the D.N is "determine[d] ... uniquely" by the D.N. However, because of the possibility of writing a program for a machine in different ways (permuting the order of the instructions), two different D.Ns might correspond to the same machine, so there will in general be distinct numbers n, m (i.e. $n \neq m$) such that $\mathcal{M}(n) = \mathcal{M}(m)$. That is, "the" Turing Machine whose number $= n$ might be the same machine as the one whose number $= m$; a given Turing Machine might have two different numbers. Alternatively, we could consider that we have here two different machines that have exactly the same input-output behavior and that execute exactly the same algorithm. Even in that latter case, where we have more machines than in the former case, the machines are enumerable; i.e. we can count them.

Can we also count the sequences that they compute? Yes; Turing explains why (with Turing's explanation in italics and my comments interpolated in brackets):

> *To each computable sequence* [i.e. to each sequence that is printed to the tape of a Turing Machine] *there corresponds at least one description number* [we have just seen why there might be more than one], *while to no description number does there correspond more than one computable sequence* [i.e. each machine prints out exactly one sequence; there is no way a given machine could print out two different sequences, because the behavior of each machine is completely determined by its program, and no program allows for any arbitrary, free, or random "choices" that could vary what gets printed on the tape]. *The computable sequences and numbers* [remember: every sequence corresponds to a unique number.[24]] *are therefore enumerable* [i.e. countable]. (Section 5, para. 7, p. 241)

Next, on p. 241, Turing shows how to compute the D.N of program I (the one that printed the sequence $\overline{01}$). And he gives a D.N without telling the reader what program corresponds to it. (Exercise for the reader: decode it!)

Finally, he alludes to the Halting Problem:

> A number which is a description number of a circle-free machine will be called a *satisfactory* number. In §8 it is shown that there can be no general process for determining whether a given number is satisfactory or not. (Section 5, para. 10, p. 241.)

A "satisfactory" number is the number of a circle-free Turing Machine, i.e. a Turing Machine that never halts and that does compute the infinite decimal representation of a real number. That is, a "satisfactory" number is the number of a Turing Machine for a *computable* number. So, in Turing's Section 8, he is going to show that there is "no general process" – i.e. no Turing Machine that can decide (by computing) – "whether a given number is satisfactory": i.e. whether a given number is the number of a *circle-free* Turing Machine. It is easy to determine if a given number is the number of a Turing Machine: just decode it, and see if the result is a syntactically correct Turing Machine program. But even if it is a syntactically correct Turing Machine program, there will be no way to decide (i.e. to compute) whether it halts or not. (Remember: for Turing, *halting* is bad, *not* halting is *good*; in modern presentations of computing theory, halting is good, not halting is (generally considered to be)[25] *bad*.)

8.13 Section 6: "The Universal Computing Machine"

> A man provided with paper, pencil, and rubber [eraser], and subject to strict discipline, is in effect a universal machine.
> —Alan Turing (1948, p. 416)[26]

---

24 Although, because of a curiosity of decimal representation, some *numbers* correspond to more than one *sequence*. The classic example is that $1 = 1.\overline{0} = 0.\overline{9}$.
25 But see Chapter 11!
26 Compare the epigraph from Church at the beginning of this chapter.

> In fact we have been universal computers ever since the age we could follow instructions.
> —Chris Bernhardt (2016, p. 12)

Although Turing's Section 6 is at least one of, if not *the* most important section of Turing's paper, we will only look at it briefly in this chapter. You are encouraged to consult Petzold, 2008 for aid in reading it in detail.

Turing begins with this claim:

> It is possible to invent a single machine which can be used to compute any computable sequence. (Section 6, para. 1, p. 241.)

Instead of needing as many Turing Machines as there are computable numbers, we only need one. Recall that our first "great insight" was that all information can be represented using only '0' and '1' (Section 7.4.1). That means all information we would want to compute with – not only numbers but language, images, sounds, etc. – can be represented by a sequence of '0's and '1's: i.e. as a computable number (in binary notation). So, Turing's claim is that there is a *single* machine that can be used to compute *anything* that is computable.

Most of you own a physical implementation of one. Indeed, most of you own several, some of which are small enough to be carried in your pocket! They are made by Apple, Dell, et al., and they come in the form of laptop computers, smartphones, etc. They are general-purpose, programmable computers.

> If this [single] machine $\mathcal{U}$ is supplied with a tape on the beginning of which is written the S.D of some computing machine $\mathcal{M}$, then $\mathcal{U}$ will compute the same sequence as $\mathcal{M}$. (pp. 241–242)

Your laptop or smartphone is a physical implementation of one of these $\mathcal{U}$s. A program or "app" that you download to it is an S.D (written in a different programming language than Turing's) of a Turing Machine that does only what that program or "app" does. The computer or smartphone that runs that program or "app," however, can also run other programs – in fact, many of them. That's what makes it "universal":

> But to do all the things a smartphone can do without buying one, ... [a] consumer would need to buy the following: A cellphone .... A mobile e-mail reader .... A music player .... A point-and-shoot camera .... A camcorder .... A GPS unit .... A portable DVD player .... A voice recorder .... A watch .... A calculator .... *In a smartphone, all those devices are reduced to software.* (Grobart, 2011, my italics)

A Turing Machine is to a Universal Turing Machine as a music box is to a player piano: a music box (or Turing Machine) can only play (or execute) the tune (or program) that is hardwired into it. Player pianos (or Universal Turing Machines) can play (or execute) any tune (or program) that is encoded on its piano-roll (or tape).

Digression: Here's a related question: "Why is a player piano *not* a computer?" (Kanat-Alexander, 2008). Alternatively, when is a Universal Turing Machine a player piano? The "instructions" on the piano roll cause certain keys to be played; you can think of each key as a Turing Machine tape cell, with "play" or "don't play" analogous to "print-one" or "print-zero." One difference is that a player piano would be a *parallel* machine, because you can play chords. For discussion of the music-box analogy, see Sloman, 2002.

How does Turing's universal computer work? Pretty much the same way a modern computer works. Just as a program (an "app") is stored somewhere in the computer's memory, the S.D of a Turing Machine is written at the beginning of the universal machine's tape. The operating system of the computer fetches (i.e. reads) an instruction and executes it (i.e. "simulates its behavior" (Dewdney, 1989, p. 315)) and then repeats this "fetch-execute" cycle until there is no next instruction. Similarly, the single program for the universal machine fetches the first instruction on its tape, executes it, and then repeats this cycle until there is no next instruction on its tape.

The details of how it does that are fascinating but beyond our present scope.[27] However, here is one way to think about this: suppose we have *two* tapes. Tape 1 will be the one we have been discussing so far, containing input (the symbols being scanned) and output (the symbols being printed). Tape 2 will contain the computer's program, with each square representing a "state of mind." The computer can be thought of as starting in a square on Tape 2, executing the instruction in that square (by reading from, and writing to, a square on Tape 1 and then moving to a(nother) square on Tape 1), and then moving to a(nother) square on Tape 2, and repeating this "fetch-execute" loop. In reality, Turing Machines only have one tape, and the instructions are not written anywhere; rather, they are "hardwired" into the Turing Machine. Any written version of them is (merely) a description of the Turing Machine's behavior (or of its "wiring diagram"). But if we encode Tape 2 on a portion of Tape 1, then we have a "stored-program" – or universal – computer.[28]

8.14 The Rest of Turing's Paper

Sections 1–5 of Turing's paper cover the nature of computation, defining it precisely and stating what is now called "Turing's (computability) thesis." Sections 6 and 7 of Turing's paper cover the universal machine. Section 8 covers the Halting Problem.

We have already examined Section 9 in detail; that was the section in which Turing analyzed how humans compute and then designed a computer program that would do the same thing.

Section 10 shows how it can be that many numbers that one might *think* are *not* computable are, in fact, computable. Section 11 proves that Hilbert's *Entscheidungsproblem* "can have no solution" (p. 259). And the Appendix proves that Turing's notion of computation is logically equivalent to Church's.

Except for modern developments and some engineering-oriented aspects of CS, one could create an undergraduate degree program in CS based solely on this one paper that Turing wrote in 1936!

(See the Online Resources for further information on Turing, including dramatizations, his legacy, and implementations.)

---

27 But see the further readings in the Online Resources.
28 Any two-tape Turing Machine is equivalent to a one-tape Turing Machine (Dewdney, 1989, Ch. 28).

9

Computers: A Philosophical Perspective[1]

> What is computation? *By virtue of what is something a computer? Why do we say a slide rule is a computer but an egg beater is not?* These are … the philosophical questions of computer science, inasmuch as they query foundational issues that are typically glossed over as researchers get on with their projects.
> —Patricia S. Churchland & Terrence J. Sejnowski (1992, p. 61, italics added)

> … everyone who taps at a keyboard, opening a spreadsheet or a word-processing program, is working on an incarnation of a Turing Machine …
> —*Time* magazine, 29 March 1999, cited in M.D. Davis, 2006a, p. 125

9.1 What Is a Computer?

Around 1980, a professor brought a keyboard, a television-sized monitor, and a dial-up telephone modem to a programming class to demonstrate some computer programs. One student asked if that's all a computer was: an electric typewriter hooked up to a TV and a telephone![2] In this chapter, armed with the results of our previous investigations into the history of computers and the nature of computation, we return to the question of what a computer is. Of course, as we saw in Section 6.2, the earliest computers were humans! (To the extent that CS is the study of computers, does that mean it is, at least in part, a study of what humans are?) Note, however, that one of the questions we will be looking at is whether the brain is a computer, so perhaps the issue of humans as computers has only been reformulated. In any case, when the question is asked today, it is generally assumed to refer to computing *machines*, and that is primarily the way we will understand it in this chapter.

According to AI pioneer Arthur L. Samuel, in a 1953 article introducing computers to radio engineers who might not have been familiar with them,

> a computer … can be looked at from two different angles, which Professor Hartree has called the "anatomical" and the "physiological," that is, "of what is it made?" and "how does it tick?" (Samuel, 1953, p. 1223, citing Hartree, 1949, p. 56)

Samuel then goes on to describe the anatomy in terms of things like magnetic cores and vacuum tubes. Clearly, the anatomy has changed since then, so defining 'computer' "anatomically" in such terms doesn't seem to be the right way to go: it's too changeable. What's needed is a

---

1 An earlier draft of this chapter appeared as Rapaport, 2018.
2 See the "Agnes" comic at http://www.gocomics.com/agnes/2013/3/7 for a similar idea.

Philosophy of Computer Science: An Introduction to the Issues and the Literature, First Edition. William J. Rapaport.
© 2023 John Wiley & Sons, Inc. Published 2023 by John Wiley & Sons, Inc.

"physiological" – or functional – definition. At the very least, we might say that **a computer is a physical machine** (where, perhaps, it doesn't matter too much what it is made of)[3] **that is, perhaps, designed** (i.e. engineered) **to compute** (i.e. to do computations) **and that, perhaps, interacts with the world**.

But does a computer *have* to be a "machine"? Does it *have* to be "engineered"? If the brain is a computer, then it would seem that computers could be *biological* entities (which, arguably, are not machines)[4] that *evolved* (which, arguably, means they were not engineered). (At least, not engineered by *humans*. Dennett (2017) would say that they *were* engineered – by Mother Nature using the natural-selection algorithm.) So, we will also ask whether the brain is a computer.

But is it even correct to limit a computer to a *physical* device? Aren't Turing Machines computers? Should we distinguish a "real" computer from a mathematical abstraction such as a Turing Machine? But arguably, my iMac – which is surely a computer if anything is – *isn't* a Turing Machine; rather, it can be *modeled* by a (Universal) Turing Machine. And to the extent that Turing Machines don't interact with the world, so much the worse for them as a model of what a computer is. (But see Section 11.8 on interaction.)[5]

But what about a "virtual" computer? According to Denning and Martell (2015, p. 212), "A virtual machine is a simulation of one computer by another. The idea comes from the simulation principle behind Alan Turing's Universal Machine." That is, a virtual machine is a (usually single-purpose) Turing Machine that is simulated by a Universal Turing Machine. Let t be a Turing Machine. Let u be a Universal Turing Machine. Encode all Turing Machines using, say, Turing's coding scheme. Then encode t's code onto u's tape, along with t's data, and let u simulate t. As Copeland (1998, p. 153) says, "the universal machine will perform every operation that t will, in the same order as t (although interspersed with sequences of operations not performed by t)." The virtual t machine is a software version (a software implementation?) of the hardware t machine.

Of course, nothing prevents the Turing Machine that is being simulated by the Universal Turing Machine from itself being a Universal Turing Machine! For example, for an introductory course I once taught, I wrote a very simple Pascal program that added two integers. This program was compiled (i.e. implemented) using the "P88 Assembly Language Simulator" – a virtual machine whose programming language was "P88 Assembly Language," a very simple assembly language designed for instructional purposes (Biermann, 1990). That assembly language was written (i.e. implemented) in another virtual machine[6] whose programming language was a dialect of Pascal called MacPascal, which was, in turn, implemented in MacOS assembly language, which was implemented in the machine language that was implemented on a physical Mac II computer (Rapaport, 2005b). Note that, ultimately, there is a physical substrate in these cases.

> **Question for the Reader:** When two integers are input to my original Pascal program and their sum is output, "where" does the actual addition take place? Is it my Pascal program that adds the two integers? Or is it "really" the Mac II computer that adds them? Or is it one (or all?) of the intermediate implementations?

If the purpose of computers is to compute, what kind of computations do they perform? Are they restricted to mathematical computations? Is that really a restriction? The binary-representation insight (Section 7.4.1) suggests that any (computable) information can be represented as a binary

---

3 A computer probably could not be made out of, say, oatmeal.
4 Although Marvin Minsky famously said that the brain was a "meat machine" (O'Toole, 2020).
5 Thanks to my colleague Stuart C. Shapiro for many of these points.
6 See the Online Resources for further reading on virtual machines.

numeral; hence, any computation on such information could be considered a mathematical computation.

And what about the difference between a "hardwired" Turing Machine that can only compute one thing and a "programmable" Universal Turing Machine that can compute anything that is computable? And is a "programmable" computer the same as a "stored-program" computer? Or what about the difference between a real, physical computer that can only compute whatever is *practically* computable (i.e. subject to reasonable space and time constraints) and an abstract, Universal Turing Machine that is not thus constrained?

And what about Churchland and Sejnowski's egg beaters? Or rocks? Surely they are *not* computers. Or are they? In short, what is a computer?

9.2 Informal Definitions

9.2.1 Reference-Book Definitions

If you ask a random person what a computer is, they might try to describe their laptop. If you look up 'computer' in a reference book,[7] you will find things like this (from the *Encyclopedia of Computer Science*):

> A *digital computer* is a machine that will accept data and information presented to it in a discrete form, carry out arithmetic and logical operations on this data, and then supply the required results in an acceptable form. (Morris and Reilly, 2000, p. 539)

Or this (from the *OED*, http://www.oed.com/view/Entry/37975):

> **computer,** *n.*
> 1. A person who makes calculations or computations; a calculator, a reckoner; *spec[ifically,]* a person employed to make calculations in an observatory, in surveying, etc. Now chiefly *hist[orical]*. [earliest citation dated 1613]
> 2. A device or machine for performing or facilitating calculation. [earliest citation dated 1869]
> 3. An electronic device (or system of devices) which is used to store, manipulate, and communicate information, perform complex calculations, or control or regulate other devices or machines, and is capable of receiving information (data) and of processing it in accordance with variable procedural instructions (programs or software); *esp[ecially]* a small, self-contained one for individual use in the home or workplace, used esp. for handling text, images, music, and video, accessing and using the Internet, communicating with other people (e.g. by means of email), and playing games. [earliest citation dated 1945]

We'll come back to these in Section 9.2.5.

9.2.2 John von Neumann's Definition

In his "First Draft Report on the EDVAC," which – along with Turing's 1936 paper – may be taken as one of the founding documents of computer science, John von Neumann gives the following definition:

---

7 Our caution in Section 5.1 about dictionary definitions also holds for encyclopedias!

An *automatic computing system* is a (usually highly composite) device, which can carry out instructions to perform calculations of a considerable order of complexity ... ... The instructions ... must be given to the device in absolutely exhaustive detail. They include all numerical information which is required to solve the problem under consideration .... All these procedures require the use of some code to express ... the problem ..., as well as the necessary numerical material .... [T]he device ... must be able to carry them out completely and without any need for further intelligent human intervention. At the end of the required operations the device must record the results again in one of the forms referred to above. (von Neumann, 1945, Section 1.0, p. 1)

Other comments (in this section of von Neumann, 1945, as well as later, in Section 5.0 [pp. 6ff]) indicate that the code should be binary and hence that the computer is a "digital" device (Section 1.0, p. 1). This definition hews closely to being a physical implementation of a Turing Machine, with clear allusions to the required algorithmic nature of the instructions, and with a requirement that there be both input and output (recall Section 7.3).

9.2.3   Arthur Samuel's Definition

Samuel's "physiological" – or functional – definition of a computer is this:

an information or data *processing* device which *accepts* data in one form and *delivers* it in an altered form. (Samuel, 1953, p. 1223, my italics)

This seems to be a very high-level description – perhaps *too* high a level: it omits any mention of computation or algorithms. It does mention that the "delivered" data must have been "processed" from the "accepted" data by the "device"; so it's not just a *function* that relates the two forms of data – it's more of a function *machine*. But there's no specification of the *kind* of processing it does.

Partly because of this, and on purpose, it also doesn't distinguish between analog and digital computers. Samuel resolves this by adding the modifier 'digital,' commenting that "Any operation which can be reduced to arithmetic or to simple logic can be handled by such a machine. There does not seem to be any theoretical limit to the types of problems which can be handled in this way" (Samuel, 1953, p. 1224) – a nod, perhaps, to our binary-representation insight (Section 7.4.1). Still, this doesn't limit the processing to *algorithmic* processing. It does, however, allow the *brain* to be considered a computer: "when the human operator performs a reasonably complicated numerical calculation he [sic][8] is forcing his brain to act as a digital computer" (Samuel, 1953, p. 1224).[9]

A bit later (p. 1225), he does say that the processing must be governed by rules; this gets closer to the notion of an algorithm, although he (so far) puts no constraints on the rules. It is only after he discusses the control unit of the computer and its programming (pp. 1226ff) that he talks about the kinds of control structures (loops, etc.) involved with algorithms. So, perhaps we could put all of this together and say that, for Samuel, a (digital) computer is a physical device that algorithmically processes digital data.

Further on, he adds the need for input and output devices (p. 1226). Are these really needed? Are they part of the abstract, mathematical model of a computer – namely, a Turing Machine? Your first reaction might be to say that the tape serves as both input and output device. But the tape is an integral part of the Turing Machine; it is really more like the set of internal switches of

---

8 The use of the male gender here is balanced by Samuel's earlier statement that computers have "advantages in terms of the reductions in clerical manpower *and woman power*" (Samuel, 1953, p. 1223, my italics).
9 Compare the quotation from Chalmers, 1993 at the end of Section 9.7.1.

a physical computer, whereas physical computers normally have input and output devices (think of keyboards and monitors) as separate, additional components: the Mac Mini, for example, is sold without a keyboard or monitor. This is related to the necessity (or lack thereof!) of inputs and outputs that we discussed in Section 7.3.3. A computer with no input-output devices can only do batch processing of pre-stored data (if that – the Mac Mini can't do anything if there's no way to tell it to start doing something). Computers that interact with the external world require input-output devices, and that raises the question of their relationship to Turing Machines (a discussion that we will begin in Chapter 10). Briefly, interacting computers that halt or have only computable input are simulable by Turing Machines; interacting computers with *non*-computable input are equivalent to Turing's oracle machines, which we will look at in Section 11.9.

9.2.4 Martin Davis's Characterization

The computer scientist Martin Davis (2000, pp. 366–367) suggests (but does not explicitly endorse) the idea that a computer is simply any device that "carries out" an algorithm. Of course, this depends on what 'carries out' means: surely it has to include as part of its meaning that the internal mechanism of the device must operate in accordance with – must behave exactly like – one of the logically equivalent mathematical models of computation. Surely any computer does that. But is anything that does that a computer? Can a computer be defined (merely) as a set of registers with contents or switches with settings? If they are binary switches, each is either on or off; computation changes the contents (the settings). Do some of the register contents or switch settings have to be interpreted as data, some as program, and the rest as irrelevant (and some as output?). Who (or what) does the interpreting?

9.2.5 Summary

One common thread in such definitions (ignoring the ones that are only of historical interest) is that computers are

1. devices or machines …
2. … that take input (data, information),
3. process it (manipulate, operate, calculate, or compute with it) …
4. … in accordance with instructions (a program),
5. and then output a result (presumably, more data or information but also including control of another device).

There are some other features that are usually associated with "computers": the kind we are interested in must be, or typically are

Automatic: There is no human intervention (beyond, perhaps, writing the program). Of course, the Holy Grail of programming is to have self-programmed computers, possibly to include having the "desire" or "intention" to program themselves (as in science fiction). Humans might also supply the input or read the output, but that hardly qualifies as "intervention." (We will explore "intervention" – in the guise of "interactive" computing – in Section 11.8.)

General purpose: A computer must be capable of *any* processing that is "algorithmic" by means of a suitable program. This is the heart of Turing's universal machine. Recall that a Turing Machine "runs" only *one* program. The Universal Turing Machine is also a Turing Machine, so it, too, also runs only one program: namely, the fetch-execute cycle that enables the *simulation* of *another* (i.e. *any* other) single-program Turing Machine.

Physically efficient: Many lists of computer features say that computers are *electronic*. But that is a matter of "anatomy." Modern computers are, as a matter of fact, electronic, but there is work on quantum computers, optical computers (https://en.wikipedia.org/wiki/Optical_computing), DNA computers, etc.[10] So, being electronic is not essential. The crucial ("physiological") property is, rather, to be constructed in such a way as to allow for high processing speeds or other kinds of physical efficiencies. Turing (1950, Section 4, p. 439) noted this point:

> Importance is often attached to the fact that modern digital computers are electrical . . . . Since Babbage's machine was not electrical, and since all digital computers are in a sense equivalent, we see that this use of electricity cannot be of theoretical importance. Of course electricity usually comes in where fast signalling is concerned, so that it is not surprising that we find it in [digital computers] . . . . The feature of using electricity is thus seen to be only a very superficial similarity.

digital: Computers should process information expressed in a discrete, symbolic form (typically alpha-numeric form, but perhaps also including graphical form). The contrast is typically with being "analog," where information is represented by means of continuous physical quantities.

algorithmic: What about the "calculations," the "arithmetic and logical operations"? Presumably, these need to be algorithmic, although neither the *OED* nor the *Encyclopedia of Computer Science* definition says so. And it would seem that the authors of those definitions have in mind calculations or operations such as addition, subtraction, etc.; maybe solving differential equations; Boolean operations involving conjunction, disjunction, etc.; and so on. These require the data to be numeric (for math) or propositional (or truth-functional – for Boolean and logical operations), at least in some "ultimate" sense: that is, any *other* data (pictorial, etc.) must be encoded as numeric or propositional or needs to allow for other kinds of operations.

There are clear cases of things that *are* computers, both *digital* and *analog*. For example, Macs, PCs, etc. are clear cases of digital computers. And slide rules and certain machines at various universities are clear cases of analog computers. However, these may be mostly of historical interest, don't seem to be programmable – i.e. universal, in Turing's sense – and seem to be outside the historical development explored in Chapter 6.

And there seem to be clear cases of things that are *not* computers: I would guess that most people would not consider egg beaters, rocks, walls, ice cubes, or solid blocks of plastic to be computers (note that I said 'most' people!). And there are even clear cases of devices for which it might be said that it is not clear whether, or in what sense, they are computers, such as Atanasoff and Berry's ABC: recall the patent lawsuit discussed in Section 6.4.4.

So: What is a computer? What is the relation of a computer to a Turing Machine and to a Universal Turing Machine? Is the (human) brain a computer? Is your smartphone a computer? *Could* a rock or a wall be considered a computer? Might *anything* be a computer? Might *everything* – such as the universe itself – be a computer? Or are some of these just badly formed questions?[11]

---

10 See the Online Resources for Ch. 3 for further reading on quantum and DNA computing.
11 See the Online Resources for further reading on the nature of computers.

9.3 Computers, Turing Machines, and Universal Turing Machines

> All modern general-purpose digital computers are physical embodiments of the same logical abstraction[:] Turing's universal machine.
> —J. Alan Robinson (1994, pp. 4–5)

9.3.1 Computers as Turing Machines

Let's try our hand at a more formal definition of 'computer.' An obvious candidate for such a definition is this:

(DC0) **A computer is *any physical device that computes*.**

Because a Turing Machine is a mathematical model of what it means to compute, we can make this a bit more precise:

(DC1) **A computer is *an implementation of a Turing Machine*.**

A Turing Machine, as we have seen, is an abstract, mathematical structure. We will explore the meaning of 'implementation' in Chapter 13. For now, it suffices to say that an implementation of an abstract object is (usually) a physical object that satisfies the definition of the abstract one. (The hedge word 'usually' is there to allow for the possibility of non-physical – or "virtual" – software implementations of a Turing Machine.) So, a physical object that satisfies the definition of a Turing Machine would be an "implementation" of one. Of course, no physical object can satisfy that definition if part of the definition requires it to be "perfect" in the following sense:

> A Turing machine is like an actual digital computing machine, except that (1) it is error free (i.e. it always does what its table says it should do), and (2) by its access to an unlimited tape it is unhampered by any bound on the quantity of its storage of information or "memory." (Kleene, 1995, p. 27)

The type-(1) limitation of "real" Turing Machines – being error free – does not obviate the need for program verification (see Chapter 15). Even an "ideal" Turing Machine could be poorly programmed.

The type-(2) limitation of a "real" (physical) Turing Machine is not a very serious one, given (a) the option of always buying another square and (b) the fact that no computation could require an actual infinity of squares (else it would not be a finite computation). The more significant type-(2) limitation is that some computations might require more squares than there could be in the universe (as is the case with NP computations such as playing perfect chess; see Section 20.3.2).

So let's modify our definition to take care of this:

(DC2) A computer is a *"practical"* implementation of a Turing Machine

where, adapting Turing's terminology, 'practical' is intended to allow for those physical limitations:

> If we take the properties of the universal [Turing] machine in combination with the fact that the machine processes and rule of thumb processes [i.e. algorithmic processes] are synonymous we may say that the universal machine is one which, when supplied with the appropriate instructions, can be made to do any rule of thumb process. This feature is paralleled in digital computing machines such as the ACE. They are in fact *practical* versions of the universal machine. (Turing, 1947, p. 383, my italics)

Let's now consider two questions:

- Is a Turing Machine a computer?
- Is a Mac (or a PC, or any other real computer) a practical implementation of a Turing Machine?

The first question we can dismiss fairly quickly: Turing Machines are not physical objects, so they can't be computers. A Turing Machine is, of course, a mathematical model of a computer. (But a virtual, software implementation of a Turing Machine *is*, arguably, a computer.)

The second question is trickier. Strictly speaking, the answer is 'no,' because Macs (and PCs, etc.) don't behave the way Turing Machines do. They actually behave more like another mathematical model of computation: a register machine. Register machines, however, are logically equivalent to Turing Machines; they are just another mathematical model of computation (Shepherdson and Sturgis, 1963). Moreover, other logically equivalent models of computation are even further removed from Turing Machines or register machines: How might a computer based on recursive functions work? Or one based on the lambda calculus? (Think of Lisp Machines.) This suggests a further refinement to our definition:

> (DC3) **A computer is a practical implementation of *anything logically equivalent to* a Turing Machine.**

There is another problem, however: computers, in any informal sense of the term (think laptop or even mainframe computer), are programmable. Turing Machines are not!

But *Universal* Turing Machines are! The ability to store a program on a Universal Turing Machine's tape makes it programmable; i.e. the Universal Turing Machine can be changed from simulating the behavior of one Turing Machine to simulating the behavior of a different one. A computer in the modern sense of the term really means a *programmable* computer, so here is a slightly better definition:

> (DC4) **A (programmable) computer is a practical implementation of anything logically equivalent to a *Universal* Turing Machine.**

(There will be some further refinements later.)

9.3.2 Stored Program vs. Programmable

A program need not be stored physically in the computer: It could "control" the computer via a wireless connection from a different location. The ability to store a program *in* the computer *along with* the data allows for the program to change *itself*. Moreover, a hardwired, non-universal computer could be programmed by rewiring it. (This assumes that the wires are manipulable. We'll return to this point in Section 12.3.) That's how early mainframe computers (like ENIAC) were programmed. So, this raises another question: what exactly is a "stored-program" computer, and does it differ from a "programmable" computer?

In my experience, the phrase 'stored program' refers to the idea that a computer's program can be stored in the computer itself (e.g. on a Turing Machine's tape) and changed, either by storing a different program or by modifying the program itself (perhaps while it is being executed, and perhaps being (self-)modified by the program itself). However, when I asked a colleague[12] "Who first came up with the notion of 'stored program' " (fully expecting him to say either Turing or von Neumann), he replied – quite reasonably – "Jacquard" (see Section 6.4.3). On this understanding, the phrase 'stored-program computer' becomes key to understanding the difference between

---

12 Stuart C. Shapiro, personal communication, 7 November 2013.

software and hardware (or programmed vs. hardwired computer) – see Chapter 12 for more on this – and becomes a way of viewing the nature of the Universal Turing Machine.

Here is von Neumann on the concept:

> If the device [the "very high speed automatic digital computing system" (Section 1.0, p. 1)] is to be *elastic*, that is as nearly as possible *all purpose*, then a distinction must be made between the specific instructions given for and defining a particular problem, and the general control organs which see to it that these instructions – no matter what they are – are carried out. The former must be **stored** in some way… the latter are represented by definite operating parts of the device. By the *central control* we mean this latter function only ….
> (von Neumann, 1945, Section 2.3, p. 2; italics in original, my boldface)

The "specific instructions" seem clearly to refer to a specific Turing Machine's program as encoded on the tape of a Universal Turing Machine. The "central control" seems clearly to refer to the Universal Turing Machine's fetch-execute program. So, if this is what is meant by "stored program," it pretty clearly refers to the that a Universal Turing Machine works.

Vardi (2013) defines 'stored-program' in terms of "uniform handling of programs and data," which he says can be "traced back to Gödel's arithmetization of provability." (Copeland (2013) objects to this; Vardi, 2017 replies.) The commonality between both ideas is that of representing two different things in *the same notation*: both Randell's and Vardi's programs and data can be represented by '0's and '1's; both logic and arithmetic can be represented by numbers (or numerals; indeed, by '0's and '1's!). And insofar as the brain might be a computer, it is worth noting that it, too, represents everything in a single "notation": neuron firings (Piccinini, 2020b, pp. 213–214).

There is a second aspect of this commonality – storing both data and program (represented in the same notation) *in the same place*: programs and data can be stored in different sections of a single Turing Machine tape; arithmetical operations can be applied to both numbers and logical propositions; and all neuron firings are in the brain. If we reserve 'stored program' to refer to Vardi's commonality, it certainly seems to describe the principal feature of a Universal Turing Machine (even if Turing shouldn't be credited with the invention of the commonality). Clearly, a stored-program computer is programmable. Are all programmable computers stored-program computers?[13]

* * * * *

In the next three sections, we will look at three recent attempts in the philosophical literature to define 'computer.' In Section 9.7, we will briefly consider two non-standard, alleged examples of computers: brains and the universe itself.

9.4 John Searle's "Pancomputationalism": Everything Is a Computer

9.4.1 Searle's Argument

John Searle's presidential address to the American Philosophical Association, "Is the Brain a Digital Computer?" (Searle, 1990),[14] covers a lot of ground and makes a lot of points about the nature of computers, the nature of the brain, the nature of cognition, and the relationships among them. In

---

13 See the Online Resources for further reading on the meaning of 'stored program.'
14 Searle, 1990 was reprinted with a few changes as Chapter 9 of Searle, 1992.

this section, we are going to focus on what Searle says about the nature of computers, with only a few side glances at the other issues.

Here is Searle's argument relevant to our main question about what a computer is:

1. Computers are described in terms of 0s and 1s.
   (See Searle, 1990, p. 26; Searle, 1992, pp. 207–208.)
2. Therefore, being a computer is a syntactic property.
   (See Searle, 1990, p. 26; Searle, 1992, pp. 207.)
3. Therefore, being a computer is not an *"intrinsic" property* of physical objects.
   (See Searle, 1990, pp. 27–28; Searle, 1992, p. 210.)
4. Therefore, *we can ascribe* the property of being a computer *to* any object.
   (See Searle, 1990, p. 26; Searle, 1992, p. 208.)
5. Therefore, everything is a computer.
   (See Searle, 1990, p. 26; Searle, 1992, p. 208.)

Of course, this doesn't quite answer our question, "What is a computer?" Rather, the interpretation and truth value of 1–5 will depend on what Searle thinks a computer is. Let's look at exactly what Searle says about these claims.

9.4.2 Computers Are Described in Terms of 0s and 1s

Taken literally, he is saying that computers are described in terms of certain *numbers*. Instead, he might have said that computers are described in terms of the *numerals* '0' and '1.' Keep this distinction in mind as we discuss Searle's argument (recall Section 6.7.1). After briefly describing Turing Machines as devices that can perform the actions of printing '0' or '1' on a tape and of moving left or right on the tape, depending on conditions specified in their program, Searle says this:

> If you open up your home computer you are most unlikely to find any '0's and '1's or even a tape. But this does not really matter for the definition. To find out if an object is really a digital computer, it turns out that we do not actually have to look for '0's and '1's, etc.; rather we just have to look for something that **we** could *treat as* or *count as* or *could be used to* function as '0's and '1's. (Searle, 1990, p. 25, my boldface, Searle's italics; Searle, 1992, p. 206)

So, according to Searle, a computer is a physical object that can be *described* as a Turing Machine. Recall from Section 8.8.1 that anything that satisfies the definition of a Turing Machine *is* a Turing Machine, whether it has a paper tape divided into squares with the symbols '0' or '1' printed on them or whether it is a table and placemats with beer mugs on them. All we need is to be able to "treat" some part of the physical object as *playing the role of* the Turing Machine's '0's and '1's. So far, so good.

Or is it? Is your home computer *really* a Turing Machine? Or is it a device whose behavior is "merely" *logically equivalent* to that of a Turing Machine? That is, is it a device that can compute all and only the functions that a Turing Machine can compute, even if it does so differently from the way a Turing Machine does? Recall that there are lots of different mathematical models of computation: Turing Machines and recursive functions are the two we have looked at. Suppose someone builds a computer that operates in terms of recursive functions instead of in terms of a Turing Machine. That is, it can compute successors, predecessors, and projection functions, and it can combine these using generalized composition, conditional definition, and while-recursion, instead of printing '0's and '1's, moving left and right, and combining these using "go to" instructions (changing from one *m*-configuration to another). These two computers (the Turing Machine computer and the recursive-function computer), as well as your home computer (with a "von Neumann" architecture, whose method of computation uses the primitive machine-language instructions and control structures of, say, an Intel chip), are all logically equivalent to a Turing Machine in the sense of having the same input-output behavior, but their internal behaviors are radically different. To use a terminology from an earlier chapter, we can ask, are recursive-function computers, Turing Machines, Macs, and PCs not only *extensionally* equivalent but also *intensionally* equivalent? Can we really describe the recursive-function computer and your home computer in terms of a Turing Machine's '0's and '1's? Or are we limited to showing that anything that the recursive-function computer and your home computer can compute can also be computed by a Turing Machine (and vice versa) – but not necessarily *in the same way*?

Here is an analogy to help you see the issue: consider translating between French and English. To say in French that it is snowing – i.e. to convey in French the same information that 'It is snowing' conveys in English – you say, *Il neige*. To say that 'il' means "it" and 'neige' means "is snowing" is very much (perhaps exactly) like describing the recursive-function machine's behavior (analogous to the French sentence) using '0's and '1's (analogous to the English sentence).

But here is a different example: in English, if someone says, 'Thank you,' you might reply, 'You're welcome.' In French, if someone says *Merci*, you might reply: *Je vous en prie*. Does '*merci*' "mean" (the same as) 'thank you'? Does '*Je vous en prie*' "mean" (the same as) 'You're welcome'? Have we translated the English into French in the way we might "translate" a recursive-function algorithm into a Turing Machine's '0's and '1's? Not really: although '*merci*' is *used* in much the same way in French that 'thank you' is *used* in English, there is no part of '*merci*' that *means* (the same as) 'thank' or 'you'; and the literal translation of '*je vous en prie*' is something like 'I pray that of you.' There is a way of communicating the same information in both French and English, but the phrases used are not literally inter-translatable.

So, something might be a computer without being "described in terms of '0's and '1's," depending on exactly what you mean by 'described in terms of.' Perhaps Searle should have said something like this: computers are described in terms of the primitive elements of the mathematical model of computation that they implement. But let's grant him the benefit of the doubt and continue looking at his argument.

9.4.3 Being a Computer Is a Syntactic Property

Let's suppose, for the sake of argument, that computers are described in terms of '0's and '1's. Syntax is the study of the properties of, and relations among, symbols or uninterpreted marks on paper (or on some other medium); a rough synonym is 'symbol manipulation' (see Section 16.9). Note that *numerals* are symbols; *numbers* aren't. So, such a description of a computer is *syntactic*. This term (which pertains to symbols, words, grammar, etc.) is usually contrasted with 'semantic' (which pertains to meaning), and Searle emphasizes that contrast early in his essay when he says that "syntax is not the same as, nor is it by itself sufficient for, semantics" (Searle, 1990, p. 21). But now Searle uses the term 'syntactic' as a contrast to being *physical*. Just as there are many ways to be computable (Turing Machines, recursive functions, lambda-calculus, etc.) – all of which are equivalent – so there are many ways to be a carburetor. "A carburetor … is a device that blends air and fuel for internal combustion engines" (http://en.wikipedia.org/wiki/Carburetor), *but it doesn't matter what it is made of*, as long as it can perform that blending "function" (purpose). "[C]arburetors can be made of brass or steel" (Searle, 1990, p. 26); they are "multiply realizable" – that is, you can "realize" (or make) one in "multiple" (or different) ways. They "are defined in terms of the production of certain *physical* effects" (Searle, 1990, p. 26).

> But the class of computers is defined **syntactically** in terms of the *assignment* of '0's and '1's. (Searle, 1990, p. 26, Searle's italics, my boldface; Searle, 1992, p. 207)

In other words, if something is defined in terms of symbols, like '0's and '1's, then it is defined in terms of syntax, not in terms of what it is physically made of (its "anatomy").

Hence, being a computer is a syntactic property, not a physical property. It is a property that something has by virtue of … of what? There are two possibilities, given what Searle has said. First, perhaps being a computer is a property that something has by virtue of *what it **does**, its function or purpose* (its "physiology"). Second, perhaps being a computer is a property that something has by virtue of *what someone **says** that it does, how it is described*. But what something *actually* does may be different from what someone *says* it does.

Does Searle think something is a computer by virtue of its *function* or by virtue of its *syntax*? Recall our thought experiment from Section 3.11: suppose you find a black box with a keyboard and a screen in the desert, and by experimenting with it, you determine that it displays on its screen the greatest common divisor (GCD) of two numbers that you type into it. It certainly seems to *function* as a computer (as a Turing Machine for computing GCDs). And you can probably describe it in terms of '0's and '1's, so you can also *say* that it is a computer. It seems that *if something functions as a computer, you can describe it in terms of '0's and '1's.*

What about the converse? If you can *describe* something in terms of '0's and '1's, does it *function* as a computer? Suppose the black box's behavior is inscrutable: the symbols on the keys are unrecognizable, and the symbols displayed on the screen don't seem to be related in any obvious way to the input symbols. But suppose someone manages to invent an interpretation of the symbols in terms of which the box's behavior can be described as computing GCDs. Is "computing GCDs" really what it does? Might it not have been created by some extraterrestrials solely for the purpose of entertaining their young with displays of pretty pictures (meaningless symbols), and it is only by the most convoluted (and maybe not always successful) interpretation that it can be described as computing GCDs?

You might think the box's *function* is more important for determining what it is. Searle thinks our ability to *describe* it syntactically is more important! After all, whether or not the box was *intended* by its creators to compute GCDs or entertain toddlers, if it can be accurately *described* as computing GCDs, then, in fact, it computes GCDs (as well as, perhaps, entertaining toddlers with pretty pictures).

Here is an alternative view of this:

> Suppose that a student is successfully doing an exercise in a recursive function theory course which consists in implementing a certain Turing Machine program. There is then no reductionism involved in saying that he [sic] is carrying out a Turing Machine program. He intends to be carrying out a Turing Machine program. … Now suppose that, unbeknownst to the student, the Turing Machine program he is carrying out is an implementation of the Euclidean algorithm [for computing GCDs]. His instructor, looking at the pages of more or less meaningless computations handed in by the student, can tell from them that the greatest common divisor of 24 and 56 is 8. The student, not knowing the purpose of the machine instructions he is carrying out, cannot draw the same conclusion from his own work. I suggest that the instructor, but not the student, should be described as carrying out the Euclidean algorithm. (This is a version … of Searle's Chinese room argument …)[15] (Goodman, 1987, p. 484)

---

15 We will discuss the Chinese Room Argument in Section 18.6.

Again, let's grant this point to Searle. He then goes on to warn us:

But this has two consequences which might be disastrous:

1. The same principle that implies multiple realizability would seem to imply universal realizability. If computation is defined in terms of the assignment of syntax then everything would be a digital computer, because any object whatever could have syntactical ascriptions made to it. You could describe anything in terms of '0's and '1's.
2. Worse yet, syntax is not intrinsic to physics. The ascription of syntactical properties is always relative to an agent or observer who treats certain physical phenomena as syntactical.

(Searle, 1990, p. 26; Searle, 1992, pp. 207–208)

Let's take these in reverse order.

9.4.4 Being a Computer Is Not an Intrinsic Property of Physical Objects

According to Searle, being a computer is not an intrinsic property of physical objects, because being a computer is a syntactic property, and "syntax is not intrinsic to physics." What does that mean, and why does Searle think it is true?

What is an "intrinsic" property? Searle doesn't tell us, although he gives some examples:[16]

[G]reen leaves intrinsically perform photosynthesis[;] … hearts intrinsically pump blood. It is not a matter of us arbitrarily or "conventionally" assigning the word "pump" to hearts or "photosynthesis" to leaves. There is an actual fact of the matter. (Searle, 1990, p. 26; Searle, 1992, p. 208)

So, perhaps "intrinsic" properties are properties that something "really" has as opposed to merely being *said* to have, much the way our black box in the previous section may or may not "really" compute GCDs but can be *said* to compute them. But what does it mean to "really" have a property? As you might expect, the philosophical analysis of "intrinsic" properties is controversial. Perhaps the simplest characterization is this: an entity *e* has a property *P* intrinsically if *e* has *P* "independently of the nature of its environment" (Figdor, 2009, p. 3, footnote 3).

Why does Searle think syntax is not "intrinsic" to physics? Because " 'syntax' is not the name of a physical feature, like mass or gravity. … [S]yntax is essentially an observer relative notion" (Searle, 1990, p. 27; Searle, 1992, p. 209). I think that what Searle is saying here is that we can analyze physical objects in different ways, no one of which is "privileged" or "more correct"; i.e. we can carve nature into different joints in different ways. On some such carvings, we might count an object as a computer; on others, we wouldn't. By contrast, an object has mass independently of how it is described: *having* mass is *not* relative to an observer; hence, it is intrinsic. *How* its mass is measured *is* relative to an observer.

But couldn't being a computer be something like that? There may be lots of different ways to measure mass, but an object always has a certain quantity of mass, no matter whether you measure it in grams or other units. In the same way, there may be lots of different ways to measure length, but an object always has a certain length, whether you measure it in centimeters or inches. Similarly, an object (natural or artifactual) will have a certain structure, whether you describe it as a computer or something else. If that structure satisfies the definition of a Turing Machine, then it *is* a Turing Machine, no matter how anyone *describes* it.[17]

---

16 See the Online Resources for further reading on intrinsic properties.
17 See the Online Resources for further reading on mass and length.

Searle anticipates this reply:

> [S]omeone might claim that the notions of "syntax" and "symbols" are just a manner of speaking and that what we are really interested in is the existence of systems with discrete physical phenomena and state transitions between them. On this view we don't really need '0's and '1's; they are just a convenient shorthand. (Searle, 1990, p. 27; Searle, 1992, p. 210)

Compare this to my previous example: someone might claim that specific units of measurement are just a manner of speaking and that what we are really interested in is the actual length of an object; on this view, we don't really need centimeters or inches; they are just a convenient shorthand.

Searle replies:

> But I believe, this move is no help. **A physical state of a system is a computational state only relative to the assignment to that state of some computational role, function, or interpretation.** The same problem arises without '0's and '1's because notions such as computation, algorithm and program do not name intrinsic physical features of systems. Computational states are not *discovered within* the physics, they are *assigned* to the physics. (Searle, 1990, p. 27, my boldface, Searle's italics; Searle, 1992, p. 210)

But this just repeats his earlier claim; it gives no new reason to believe it. He continues to insist that being a computer is more like "inches" than like length.

So, we must ask again: why does Searle think syntax is not intrinsic to physics? Perhaps if a property is intrinsic to some object, that object can only have the property in one way. For instance, color is presumably not intrinsic to an object, because an object might have different colors depending on the conditions under which it is perceived. But the physical structure of an object that causes it to reflect a certain wavelength of light is always the same; that physical structure is intrinsic. On this view, here is a reason syntax might not be intrinsic: the syntax of an object is, roughly, its abstract structure.[18] But an object might be able to be understood in terms of several different abstract structures (and this might be the case whether or not human observers assign those structures to the object). If an object has no unique syntactic structure, then syntax is not intrinsic to it. But if an object has (or can be assigned) a syntax of a certain kind, then it does have that syntax even if it also has another one. And if, under one of those syntaxes, the object is a computer, then it *is* a computer.

But that leads to Searle's next point.

9.4.5 We Can Ascribe the Property of Being a Computer to Any Object

There is some slippage in the move from "syntax is not intrinsic to physics" to "we can ascribe the property of being a computer to any object." Even if syntax is not intrinsic to the physical structure of an object (perhaps because a given object might have several different syntactic structures), why must it be the case that *any* object *can* be ascribed the syntax of being a computer?

One reason might be this: every object has (or can be ascribed) *every* syntax. That seems to be a very strong claim. To refute it, however, all we would need to do is to find an object O and a syntax S such that O lacks (or cannot be ascribed) S. One possible place to look would be for an O whose "size" in some sense is smaller than the "size" of some S. I will leave this as an exercise for the reader: if you can find such O and S, then I think you can block Searle's argument at this point.

Here is another reason *any* object might be able to be ascribed the syntax of being a computer: there might be something special about the syntax of being a computer – i.e. about the formal

---

18 See Sections 13.2, 16.9, and 18.8.3 for further discussion of this point.

structure of Turing Machines – that does allow it to be ascribed to (or found in) any object. This may be a bit more plausible than the previous reason. After all, Turing Machines are fairly simple. Again, to refute the claim, we would need to find an object O such that O lacks (or cannot be ascribed) the syntax of a Turing Machine. Again, I will leave this as an exercise for the reader, but we will return to it later (when we look at the nature of "implementation" in Chapter 13). Searle thinks we cannot find such an object.

9.4.6 Everything Is a Computer

> Unlike computers, ordinary rocks are not sold in computer stores and are usually not taken to perform computations. Why? What do computers have that rocks lack, such that computers compute and rocks don't? (If indeed they don't?) … A good account of computing mechanisms should entail that paradigmatic examples of computing mechanisms, such as digital computers, calculators, both Universal and non-universal Turing Machines, and finite state automata, compute. … A good account of computing mechanisms should entail that all paradigmatic examples of non-computing mechanisms and systems, such as planetary systems, hurricanes, and digestive systems, don't perform computations.
> (Piccinini, 2015, pp. 7, 12)

We can ascribe the property of being a computer to any object if and only if everything is a computer:

> Thus for example the wall behind my back is right now implementing the Wordstar program, because there is some pattern of molecule movements which is isomorphic with the formal structure of Wordstar. (Searle, 1990, p. 27; Searle, 1992, pp. 208–209)

Searle does not offer a detailed argument for how this might be the case, but other philosophers have done so, and in Chapter 13, we will explore how they think it can be done. Let's assume for the moment that it can be done.

In that case, things are not good, because this trivializes the notion of being a computer. If everything has some property P, then P isn't a very interesting property: such a property doesn't help us categorize the world, so it doesn't help us understand the world:

> [A]n objection to Turing's analysis… is that although Turing's account may be necessary it is not sufficient. If it is taken to be sufficient then too many entities turn out to be computers. The objection carries an embarrassing implication for computational theories of mind: such theories are devoid of empirical content. If virtually anything meets the requirements for being a computational system then wherein lies the explanatory force of the claim that the brain is such a system?
> (Copeland, 1996, Section 1, p. 335)

"Turing's analysis" is, roughly, that x is a computer iff x is a (physical) implementation of a Turing Machine. (Recall DC0–DC4 from Section 9.3.1.) To say that this "account" is "necessary" means if x is a computer, then it is an implementation of a Turing Machine. That seems innocuous. To say it is a "sufficient" account is to say that if x is an implementation of a Turing Machine, then it is a computer. This is allegedly problematic, because, allegedly, anything can be gerrymandered to make it an implementation of a Turing Machine; hence, anything is a computer (including, for uninteresting reasons, the brain).

How might we respond to this situation? One way is to bite the bullet and accept that, under some description, any object (even the wall behind me) can be considered a computer. And not just some specific computer, such as a Turing Machine that executes the WordStar program:

> [I]f the wall is implementing Wordstar then if it is a big enough wall it is implementing any program, including any program implemented in the brain. (Searle, 1990, p. 27; Searle, 1992, p. 209)

If a big enough wall implements any program, then it implements the Universal Turing Machine!

But perhaps this is OK. After all, there is a difference between an "intended" interpretation of something and a "gerrymandered" interpretation. For instance, the intended interpretation of Peano's axioms for the natural numbers is the sequence $\langle 0,1,2,3,\ldots\rangle$. There are also many other "natural" interpretations, such as $\langle I, II, III,\ldots\rangle$, or $\langle\emptyset, \{\emptyset\}, \{\{\emptyset\}\}, \ldots\rangle$, or $\langle\emptyset, \{\emptyset\}, \{\emptyset, \{\emptyset\}\}, \ldots\rangle$, and so on. As Chris Swoyer (1991, p. 504, note 26) notes, "According to structuralism, any countably infinite (recursive) set can be arranged to form an ω-sequence that can play the role of the natural numbers. It is the structure common to all such sequences, rather than the particular objects which any happens to contain, that is important for arithmetic." But extremely contorted ones, such as a(n infinite) sequence of all numeral names in French arranged alphabetically, are hardly "good" examples. Admittedly, they *are* examples of natural numbers, but not very useful ones. (For further discussion, see Benacerraf, 1965, White, 1974.)

A better reply to Searle, however, is to say that he's wrong: some things are *not* computers. Despite what he said in the last passage quoted previously, the wall behind me is *not* a Universal Turing Machine; I really cannot use it to post to my Facebook account or write a letter, much less add $2 + 2$. It is an empirical question whether something actually behaves as a computer. And the same goes for other syntactic structures. Consider the formal definition of a mathematical group:

> A *group* =$_{def}$ a set of objects (e.g. integers) that is closed under an associative binary operation (e.g. addition), that has an identity element (e.g. 0), and is such that every element of the set has an inverse (e.g. in the case of integer n, its inverse is $-n$).

Not every set is a group. Similarly, there is no reason to believe that everything is a Turing Machine.

> In order for the system to be used to compute the addition function these causal relations have to hold *at a certain level of grain*, a level that is determined by the discriminative abilities of the user. That is why … no money is to be made trying to sell a rock as a calculator. Even if (*per mirabile*)[19] there happens to be a set of state-types at the quantum-mechanical level whose causal relations do mirror the formal structure of the addition function, microphysical changes at the quantum level are *not* discriminable by human users, hence human users could not use such a system to add. (God, in a playful mood, could use the rock to add.) (Egan, 2012, p. 46)

Chalmers (2012, pp. 215–216) makes much the same point:

> On my account, a pool table will certainly implement various a[bstract]-computations and perform various c[oncrete]-computations. It will probably not implement interesting computations such as algorithms for vector addition, but it will at least implement a few

_____
19 I.e. miraculously.

multi-state automata and the like. These computations will not be of much explanatory use in understanding the activity of playing pool, in part because so much of interest in pool are not organizationally invariant and therefore involve more than computational structure.

In other words, even if Searle's wall implements WordStar, *we* wouldn't be able to use it as such.

$$* * * * *$$

Let's take stock of where we are. Presumably, computers are things that compute. Computing is the process that Turing Machines give a precise description for. That is, computing is what Turing Machines do. And what Turing Machines do is to move around in a discrete fashion and print discrete marks on discrete sections of the space in which they move around. So, a computer is a *device* – presumably, a *physical* device – that does that. Searle agrees that computing is what Turing Machines do, and he seems to agree that computers are devices that compute. He also believes that everything is a computer; more precisely, he believes that everything can be *described as* a computer (because that's what it means to *be* a computer). And we've also seen reason to think he might be wrong about that last point.

In the next two sections, we look at two other views about what a computer is. (We will consider other responses to Searle's argument in Chapter 13.)

9.5 Patrick Hayes: Computers as Magic Paper

Let's keep straight three intertwined issues we have been looking at:

1. What is a computer?
2. Is the brain a computer?
3. Is everything a computer?

Our principal concern is with the first question. Once we have an answer to that, we can try to answer the others. As we've just seen, Searle thinks a computer is anything that is (or can be described as) a Turing Machine, that everything is (or can be described as) a computer, and, therefore, that the brain is a computer, but only trivially so, and not in any interesting sense.

The AI researcher Patrick J. Hayes (1997) gives a different definition – in fact, two of them. Here's the first:

Definition H1 By "computer" I mean *a machine which performs computations, or which computes.* (Hayes, 1997, p. 390, my italics)

A full understanding of this requires a definition of 'computation'; this will be clarified in his second definition. But there are a few points to note about this first one.

First, he prefaces it by saying

> First, I take it as simply obvious both that computers exist and that not everything is a computer, so that, contra Searle, the concept of "computer" is not vacuous. (Hayes, 1997, p. 390)

So, there are (1) machines that compute (i.e. there are things that *are* machines-that-compute), and there are (2) things that are *not* machines-that-compute. Note that (2) can be true in two ways: there might be (2a) *machines* that *don't* compute, or there might be (2b) *things* that *do* compute but *aren't machines.* Searle disputes the first possibility because he thinks everything (including, therefore, any machine) computes. But contrary to what Hayes says, Searle would probably *agree* with the second possibility, because, after all, he thinks everything (including, therefore, anything that is not a machine) computes! Searle's example of the wall that implements (or that can be interpreted as implementing) WordStar would be such a non-machine that computes. So, for Hayes's notion to contradict Searle, it must be that Hayes believes there are machines that do not compute. Perhaps a dishwasher is a machine that doesn't compute anything.[20]

Are Hayes's two "obvious" points to be understood as criteria of adequacy for any definition – criteria Hayes thinks need no argument (i.e. as something like "axioms")? Or are they intended to be more like "theorems" that follow from his first definition? If it's the former, then there is no interesting debate between Searle and Hayes; one simply denies what the other argues for. If it's the latter, then Hayes needs to provide arguments or examples to support his position.

A second thing to note about Hayes's definition is that he says a computer "*performs* computations," not "*can* perform computations." Strictly speaking, your laptop when it is turned off is not a computer by this definition, because it is not performing any computation. And, as Hayes observes,

> On this understanding, a Turing machine is not a computer, but a mathematical abstraction of a certain kind of computer. (Hayes, 1997, p. 390)

What about Searle's wall that implements WordStar? There are two ways to think about how the wall might implement WordStar. First, it might do so *statically* simply by virtue of there being a way to map every part of the WordStar program to some aspect of the molecular or subatomic structure of the wall. In that case, Hayes could well argue that the wall is *not* a WordStar computer, because it is not computing (even if it *might be able to*). But the wall might implement WordStar *dynamically*; in fact, that is why Searle thinks the wall implements WordStar …

> … because there is some pattern of molecule *movements* which is isomorphic with the formal structure of Wordstar. (Searle, 1990, p. 27, my italics; Searle, 1992, pp. 208–209)

But a pattern of *movements* suggests that Searle thinks the wall *is computing*, so it *is* a computer!

Hayes's second definition is a bit more precise, and it is, presumably, his "official" one:

Definition H2   [F]ocus on the memory. A computer's memory contains patterns … which are stable but labile [i.e. changeable], and it has the rather special property that changes to the patterns are under the control of other patterns: i.e. some of them describe changes to be made to others; and when they do, the memory changes those patterns in the way described by the first ones. … A computer is *a machine which is so constructed that patterns can be put in it, and when they are, the changes they describe will in fact occur to them.* If it were paper, it would be "magic paper" on which writing might spontaneously change, or new writing appear. (Hayes, 1997, p. 393, my italics)

There is a subtle difference between Hayes's two definitions, which highlights an ambiguity in Searle's presentation. Recall the distinction between a Turing Machine and a Universal Turing Machine: both Turing Machines and Universal Turing Machines are hardwired and compute only

---

20 A dishwasher might, however, be described by a (non-computable?) function that takes dirty dishes as input and returns clean ones as output. Aaronson, 2012 considers (semi-humorously) a "toaster-enhanced Turing machine." See Shagrir, 2022, pp. 89, 95 for discussion. We'll return to dishwashers in Section 12.4.5.

a single function. The Turing Machine computes whichever function is encoded in its machine table; it cannot compute anything else. But the one function, hardwired into its machine table, that a Universal Turing Machine computes is the fetch-execute function that takes as input a program and its data and outputs the result of executing that program on that data. In that way, a Universal Turing Machine (besides computing the fetch-execute cycle) can (in a different way) compute any computable function as long as a Turing Machine program for that function is encoded and stored on the Universal Turing Machine's tape. The Universal Turing Machine is programmable in the sense that the input program can be varied, not that its hardwired program can be.

Definition H1 seems to include physical Turing Machines (but, as Hayes noted, not abstract ones), because, after all, they compute (at least, when they are turned on and running). Definition H2 seems to *exclude* them, because the second definition requires patterns that describe changes to other patterns. That first kind of pattern is a stored program; the second kind is the data that the program operates on. So, *Definition H2 is for a Universal Turing Machine.*

Here is the ambiguity in Searle's presentation: is Searle's wall a Turing Machine or a Universal Turing Machine? On Searle's view, WordStar is a Turing Machine, so the wall must be a Turing Machine, too. So, the wall is not a computer in Definition H2. Could a wall (or a rock, or some other suitably large or complex physical object other than something like a PC or a Mac) be a *Universal Turing Machine*? My guess is that Searle would say "yes," but it is hard to see how one would actually go about programming it.

The "magic paper" aspect of Definition H2 focuses, as Hayes notes, on the memory: i.e. on the tape. It is as if you were looking at a Universal Turing Machine, but all you saw was the tape, not the read-write head or its states (*m*-configurations) or its mechanism. If you watched the Universal Turing Machine compute, you would see the patterns (the '0's and '1's) on the tape "magically" change. (This would be something like looking at an animation of the successive states of the Turing Machine tape in Section 8.10.3.)

A slightly different version of the "magic paper" idea is Alan Kay's third "computing whammy" (see Section 7.4.5):

> Matter can hold and interpret and act on descriptions that describe anything that matter can do. (Guzdial and Kay, 2010)

The idea of a computer as magic paper or magic matter may seem a bit fantastic. But there are more down-to-earth ways of thinking about this. Philosopher Richmond Thomason has said that

> ... all that a program can do between receiving an input and producing an output is to change variable assignments ... (Thomason, 2003, p. 328; cf. Lamport, 2011, p. 6)

If programs tell a computer how to change the assignments of values to variables, then a computer is a (physical) device that changes the contents of register cells (the register cells that are the physical implementations of the variables in the program). This is really just another version of Turing's machines, if you consider the tape squares to be the register cells.

Similarly, Stuart C. Shapiro points out that

> a computer is a device consisting of a vast number of connected switches. ... [T]he switch settings both determine the operation of the device and can be changed by the operation of the device. (Shapiro, 2001, p. 3)

What is a "switch"? Here is a nice description from Samuel's 1953 article:

> To bring the discussion down to earth let us consider the ordinary electric light switch in your home. This is by definition a switch. It enables one to direct electric current to a lighting fixture at will. Usually there is a detent mechanism [21] which enables the switch to remember what it is supposed to be doing so that once you turn the lights on they will remain on. It therefore has a memory. It is also a binary, or perhaps we should say a bistable device. By way of contrast, the ordinary telegrapher's key is a switch without memory since the key will remain down only as long as it is depressed by the operator's hand. But the light switch and the telegraph key are binary devices, that is, they have but two operating states. (Samuel, 1953, p. 1225)

So, a switch is a physical implementation of a Turing Machine's tape cell, which can also be "in two states" (i.e. have one of two symbols printed on it) and also has a "memory" (i.e. once a symbol is printed on a cell, it remains there until it is changed). Hayes's magic-paper patterns are just Shapiro's switch-settings or Thomason's variable assignments.[22]

Does this definition satisfy Hayes's two criteria? Surely such machines exist. I wrote this book on one of them. And surely not everything is such a machine: at least on the face of it, the stapler on my desk is not such "magic paper." Searle, I would imagine, would say that we might see it as such magic paper if we looked at it closely enough and in just the right way. And so the difference between Searle and Hayes seems to be in how one is supposed to look at candidates for being a computer: do we look at them as we normally do? In that case, not everything is a computer. Or do we squint our eyes and look at them closely in a certain way? In that case, perhaps we could see that everything could be considered a computer. Isn't that a rather odd way of thinking about things?

What about the brain? Is it a computer in the sense of "magic paper" (or magic matter)? If Hayes's "patterns" are understood as patterns of neuron firings, then, because surely some patterns of neuron firings cause changes in other such patterns, I think Hayes would consider the brain to be a computer.

9.6 Gualtiero Piccinini: Computers as Digital String Manipulators

In a series of three papers, the philosopher Gualtiero Piccinini has offered an analysis of what a computer is (Piccinini, 2007b, c, 2008) (see also Piccinini, 2015). It is more precise than Hayes's, because it talks about *how* the magic paper performs its tricks. And it is less universal than Searle's, because Piccinini doesn't think everything is a computer.

Unfortunately, there are two slightly different definitions to be found in Piccinini's papers:[23]

Definition P1 The mathematical theory of how to generate output strings from input strings in accordance with general rules that apply to all input strings and depend on the inputs (and sometimes internal states) for their application is called computability theory. Within computability theory, the activity of manipulating strings of digits in this way is called computation. *Any system*

---

21 A "detent" is "a catch in a machine that prevents motion until released" (https://www.google.com/search?q=detent).

22 See the Online Resources for further reading on computers as switch-setters.

23 We'll see a third definition in Section 9.8.

that performs this kind of activity is a computing system properly so called. (Piccinini, 2007b, p. 108, my italics)

Definition P2 *[A]ny system whose correct mechanistic explanation ascribes to it the function of generating output strings from input strings (and possibly internal states), in accordance with a general rule that applies to all strings and depends on the input strings (and possibly internal states) for its application, is a computing mechanism.* The mechanism's ability to perform computations is explained mechanistically in terms of its components, their functions, and their organization. (Piccinini, 2007c, p. 516, my italics)

These are almost the same, but there is a subtle difference between them.

9.6.1 Definition P1

Let's begin with Definition P1. It implies that a computer is any "system" (presumably, a physical device, because only something physical can actively "perform" an action) that manipulates strings of digits: i.e. that "generate[s] output strings from input strings in accordance with general rules that apply to all input strings and [that] depend on the inputs (and sometimes internal states) for their application." What kind of "general rule"? Piccinini (2008, p. 37) uses the term 'algorithm' instead of 'general rule.' This is consistent with the view that a computer is a Turing Machine and explicates Hayes's "magic trick" as being an algorithm.

The crucial point, according to Piccinini, is that the inputs and outputs must be strings of digits. This is the significant difference between (digital) computers and "analog" computers: the former manipulate strings of digits; the latter manipulate "real variables." Piccinini explicates the difference between digits and real variables as follows:

> A *digit* is a particular [i.e. a particular object or component of a device] or a discrete state of a particular, discrete in the sense that it belongs to one (and only one) of a finite number of types. ... A *string* of digits is a concatenation of digits: namely, a structure that is individuated by the types of digits that compose it, their number, and their ordering (i.e. which digit token is first, which is its successor, and so on). (Piccinini, 2007b, p. 107)[24]

Piccinini (2007c, p. 510) observes that a digit is analogous to a letter of an alphabet, so digits are like Turing's symbols that can be printed on a Turing Machine's tape. On the other hand,

> real variables are physical magnitudes that (i) *vary* over time, (ii) (are assumed to) take a *continuous range of values* within certain bounds, and (iii) (are assumed to) vary *continuously* over time. Examples of real variables include the rate of rotation of a mechanical shaft and the voltage level in an electrical wire. (Piccinini, 2008, p. 48)

So far, so good. Neither Searle nor Hayes should be upset with this characterization.

---

24 In the other two papers in his trilogy, Piccinini gives slightly different characterizations of what a digit is, but these need not concern us here; see Piccinini, 2007c, p. 510; Piccinini, 2008, p. 34.

9.6.2 Definition P2

But Piccinini's second definition adds a curious phrase. This definition implies that a computer is any system "whose correct mechanistic explanation ascribes to it the function of" manipulating digit strings according to algorithms. What is the import of that extra phrase?

It certainly seems as if this is a weaker definition. In fact, it sounds a bit Searlean, because it makes it appear as if it is not the case that a computer *is* an algorithmic, digit-string manipulator, but rather that it is anything that can be so *described* by some kind of "mechanistic explanation." And that sounds as if being a computer is something "external" and not "intrinsic."

So let's consider what Piccinini has in mind here. He says,

> Roughly, a mechanistic explanation involves a partition of a mechanism into parts, an assignment of functions and organization to those parts, and a statement that a mechanism's capacities are due to the way the parts and their functions are organized. (Piccinini, 2007c, p. 502)

As we will see in more detail in Section 18.8.3, syntax in its most general sense is the study of the properties of a collection of objects and the relations among them. If a "mechanism" is considered a collection of its parts, then Piccinini's notion of a mechanistic explanation sounds a lot like a description of the mechanism's "syntax." But syntax, you will recall, is what Searle says is not intrinsic to a system (or a mechanism).

So how is Piccinini going to avoid a Searlean "slippery slope" and deny that everything is a computer? One way he tries to do this is by suggesting that even if a system can be analyzed syntactically in different ways, only one of those ways will help us understand the system's behavior:

> Mechanistic descriptions are sometimes said to be perspectival, in the sense that the same component or activity may be seen as part of different mechanisms depending on which phenomenon is being explained .... For instance, the heart may be said to be for pumping blood as part of an explanation of blood circulation, or it may be said to be for generating rhythmic noises as part of an explanation of physicians who diagnose patients by listening to their hearts. This kind of perspectivalism does not trivialize mechanistic descriptions. Once we fix the phenomenon to be explained, the question of what explains the phenomenon has an objective answer. This applies to computations as well as other capacities of mechanisms. A heart makes the same noises regardless of whether a physician is interested in hearing it or anyone is interested in explaining medical diagnosis. (Piccinini, 2007c, p. 516)

Let's try to apply this to Searle's "WordStar wall": from one perspective, the wall is just a wall; from another, according to Searle, it can be taken as an implementation of WordStar. Compare this to Piccinini's claim that, from one perspective, a heart is a pump, and from another, it is a noisemaker. If you're a doctor interested in hearing the heart's noises, you'll consider the heart a noisemaker. If you're a doctor interested in making a medical diagnosis, you'll consider it a pump. Similarly, if you're a house painter, say, you'll consider the wall a flat surface to be colored, but if you're Searle, you'll try to consider it a computer program. (Although I don't think you'll be very successful in using it to write a term paper!)

9.7   What Else Might Be a Computer?

So, what is a computer? It would seem that almost all proposed definitions agree on at least the following:

- Computers are physical devices.
- They interact with other physical devices in the world.
- They algorithmically manipulate (physical) symbols (strings of digits), converting some into others.
- They are physical implementations of (Universal) Turing Machines in the sense that their input-output behavior is logically equivalent to that of a (Universal) Turing Machine (even though the details of their processing might not be). A slight modification of this might be necessary to avoid the possibility that a physical device might be considered a computer even if it doesn't compute: we probably want to rule out "real magic," for instance.

Does such a definition include too much? Let's assume for a moment that something like Piccinini's reply to Searle carries the day, so it makes sense to say that not everything is a computer. Still, might there be some things that intuitively aren't computers but that turn out to be computers given even our narrow characterization?

This is always a possibility. As we saw in Section 3.4.1, any time you try to make an informal concept precise, you run the risk of *including* some things under the precise concept that didn't (seem to) fall under the informal concept. You also run the risk of *excluding* some things that did. One way to react to this situation is to reject the formalization or refine it so as to minimize or eliminate the "erroneous" inclusions and exclusions. But another reaction is to bite the bullet and agree to the new inclusions and exclusions: for instance, you might even come to see that something you didn't think was a computer really was one.

In this section, we'll consider two things that may – or may not! – turn out to be computers: the brain and the universe.

9.7.1   Is a Brain a Computer?

What we [McCulloch and Pitts] thought we were doing … was treating the brain as a Turing machine ....
—Warren McCulloch, in von Neumann, 1948

[I]t is conceivable … that brain physiology would advance so far that it would be known with empirical certainty

1. that the brain suffices for the explanation of all mental phenomena and is a machine in the sense of Turing;
2. that such and such is the precise anatomical structure and physiological functioning of the part of the brain which performs mathematical thinking.

—Kurt Gödel, 1951; cited in Feferman, 2006, p. 146[25]

It is the current aim to replace, as far as possible, the human brain by an electronic digital computer.
—Grace Murray Hopper (1952, p. 243)

---

25 For commentary, see Sieg, 2007, Section 2.

> The human brain may also be accurately considered to be a symbol manipulator. That the physical mechanism for the representation of symbols are the electrochemical impulses associated with neural activity is irrelevant in this context. The important fact is that these impulses are representations of external symbols analogous to the representation of external symbols by the electronic … devices inside a digital computer.
> —Anthony Ralston (1971, p. 4)

Many people have claimed that the (human) brain *is* a computer. Searle thinks it is, but only because he thinks everything is a computer. But perhaps there is a more interesting way in which the brain is a computer. Certainly, contemporary computational cognitive science uses computers as at least a metaphor for the brain. Before computers came along, there were many other physical metaphors for the brain: the brain was considered like a telephone system or like a plumbing system.

In fact, "computationalism" is sometimes taken to be the view that the brain (or the mind) *is* a computer, or that the brain (or the mind) computes, or that brain (or mental) states and processes *are* computational states and processes:

> The basic idea of the computer model of the mind is that the mind is the program and the brain the hardware of a computational system.
> (Searle, 1990, p. 21; Searle, 1992, p. 200)

> The core idea of cognitive science is that our brains are a kind of computer …. Psychologists try to find out exactly what kinds of programs our brains use, and how our brains implement those programs. (Alison Gopnik 2009, p. 43)

> Computationalism … is the view that the functional organization of the brain (or any other functionally equivalent system) is computational, or that neural states are computational states. (Piccinini, 2010, p. 271; see also pp. 277–278)

But if one of the essential features of a computer is that it carries out computable processes by *computing* rather than (say) by some biological but non-computational technique, then it's at least logically possible that the brain is not a computer even if brain processes are comput*able*.

How can this be? A process is computable if and only if there is an algorithm (or a system of algorithms) that specifies how that process can be carried out. But it is logically possible for a process to be computable in this sense without actually being computed. Here are some examples:

1. Someone might come up with a computational theory of the behavior of the stock market, yet the actual stock market's behavior might be determined by the individual decisions made by individual investors and not by anyone or anything executing an algorithm. That is, the behavior might be comput*able* even if it is not comput*ational*.[26]
2. Calculations done by slide rules are done by analog means, yet the calculations themselves are clearly computable. Analog computations are not normally considered Turing Machine computations.

---

26 At least, this was true before the advent of algorithmic trading, as James Graham Maw reminded me (personal communication, 23 November 2021); see Fortnow, 2022, p. 83.

3. Another example might be the brain itself. Piccinini (2007a, 2020b) has argued that neuron firings (more specifically, "spike trains" – i.e. sequences of "action potential" – in groups of neurons) are not representable as digit strings. But because Piccinini believes a device is not a "digital" computer unless it manipulates digit strings, and because it is generally considered that human cognition is implemented by neuron firings, it follows that the brain's cognitive functioning – even if comput*able* – is not accomplished by digital computa*tion*. Yet if cognitive functions are computable (as contemporary cognitive science suggests – see Edelman, 2008a), then there would still be algorithms that compute cognition, even if the brain doesn't do it that way.[27]

The philosopher David Chalmers puts the point this way:

> Is the brain a [programmable] computer … ? Arguably. For a start, the brain can be "programmed" to implement various computations by the laborious means of conscious serial rule-following; but this is a fairly incidental ability. On a different level, it might be argued that learning provides a certain kind of programmability and parameter-setting, but this is a sufficiently indirect kind of parameter-setting that it might be argued that it does not qualify. In any case, the question is quite unimportant for our purposes. What counts is that the brain implements various complex computations, not that it is a computer. (Chalmers, 1993, Section 2.2, especially p. 336)

Two interesting points are made here. The first is that the brain can simulate a Turing Machine "by … conscious serial rule-following." The second is the last sentence: what really matters is that the brain can have input-output behavior that is computable, not that it "is" a computer. To say that it is a computer raises the question of what kind of computer it is: A Turing Machine? A register machine? Something *sui generis*? And these questions seem to be of less interest than the fact that its behavior is computable.

Still, if the brain computes in some way (or "implements computations"), and if a computer is, by definition, something that computes, then we might still wonder if the brain is some kind of computer. As I once read somewhere, "The best current explanation of how a brain could instantiate this kind of system of rules and representations is that it is a kind of computer." Thus, we have here the makings of an abductive argument (i.e. a scientific hypothesis) that the brain is a computer. (Recall Section 2.5.1.) Note that this is a much more reasonable argument than Searle's or than trying to model the brain as, say, a Turing Machine.[28]

9.7.2 Is the Universe a Computer?

Might the universe itself be a computer?[29] Consider Kepler's laws of planetary motion. Are they just a computable theory that describes the behavior of the solar system? If so, then a computer that calculates with them might be said to *simulate* the solar system in the same way any kind of program might be said to simulate a physical (or biological, or economic) process, or in the same

---

27 Piccinini (2020b) actually takes a slightly different view: he generalizes the notion of "computation" to include digital Turing Machine computation, analog computation, and neural computation (which he argues is neither digital nor analog), among others. On this view, the brain's cognitive functioning *is* computed – not digitally, but neurally.

28 See the Online Resources for further reading on the brain as a computer.

29 In addition to the cartoon in Figure 9.1, see also the satirical Google-like search page "Is the Universe a Computer?" http://abstrusegoose.com/115 (best viewed online!).

Figure 9.1 Source: Abstruse Goose, The Ultimate. Retrieved from https://abstrusegoose.com/219.

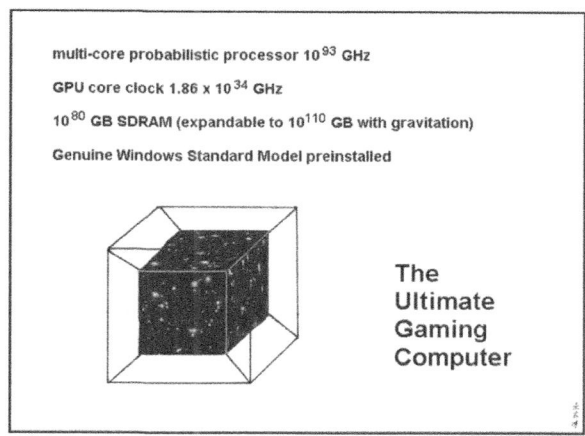

multi-core probabilistic processor 10^{93} GHz

GPU core clock 1.86×10^{34} GHz

10^{80} GB SDRAM (expandable to 10^{110} GB with gravitation)

Genuine Windows Standard Model preinstalled

The
Ultimate
Gaming
Computer

way an AI program might be said to simulate a cognitive process. (We'll return to this idea in Section 14.2 and Section 18.11, question 1.)

Or does *the solar system itself* compute Kepler's laws? If so, then the solar system would seem to be a (special-purpose) computer (i.e. a kind of Turing Machine):

> A computation is a process that establishes a mapping among some symbolic domains. … Because it involves symbols, this definition is very broad: a system instantiates a computation if its dynamics can be interpreted (by another process) as establishing the right kind of mapping.
>
> Under this definition, a stone rolling down a hillside computes its position and velocity in exactly the same sense that my notebook computes the position and the velocity of the mouse cursor on the screen (they just happen to be instantiating different symbolic mappings). Indeed, the universe in its entirety also instantiates a computation, albeit one that goes to waste for the lack of any process external to it that would make sense of what it is up to. (Edelman, 2008b, pp. 182–183)

After all, if "*biological computation* is a process that occurs in nature, not merely in computer simulations of nature" (Mitchell, 2011, p. 2), then it is at least not unreasonable that the solar system computes Kepler's Laws:

> Going further along the path of nature, suppose that we have a detailed mathematical model of some physical process such as—say—a chemical reaction; clearly we can either organise the reaction in the laboratory and observe the outcome, or we can set up the mathematical model of the reaction on a computer either as the numerical solution of a system of equations, or as a Montecarlo simulation, and we can then observe the outcome. We can all agree that when we "run the reaction" on the computer either as a numerical solution or a Montecarlo simulation, we are dealing with a computation.
>
> But why then not also consider that the laboratory experiment itself is after all only a "computational analogue" of the numerical computer experiment! In fact, the laboratory experiment will be a mixed analogue and digital phenomenon because of the actual discrete number of molecules involved, even though we may not know their number exactly. In this case, the "hardware" used for the computation are the molecules and the physical environment that they are placed in, while the software is also inscribed in the different molecules species that are involved in the reaction, via their propensities to react with each other …. (Gelenbe, 2011, pp. 3–4)

This second possibility does not necessarily follow from the first. As we just saw in the case of the brain, there might be a computational *theory* of some phenomenon – i.e. the phenomenon might be *computable* – but the phenomenon *itself* need not be *produced* computationally.

> Indeed, *computational algorithms are so powerful that they can simulate virtually any phenomena, without proving anything about the computational nature of the actual mechanisms underlying these phenomena.* Computational algorithms generate a perfect description of the rotation of the planets around the sun, although the solar system does not compute in any way. In order to be considered as providing a model of the mechanisms actually involved, and not only a simulation of the end-product of mechanisms acting at a different level, computational models have to perform better than alternative, noncomputational explanations. (Perruchet and Vinter, 2002, Section 1.3.4, p. 300, my italics)

Nevertheless, could it be the case that our solar system *is* computing Kepler's laws? Arguments along these lines have been put forth by Stephen Wolfram and Seth Lloyd.[30]

Wolfram's Argument

Wolfram, developer of the Mathematica computer program, argues as follows (Wolfram, 2002):

1. Nature is discrete.
2. Therefore, possibly it is a cellular automaton.
3. There are cellular automata that are equivalent to a Turing Machine.
4. Therefore, possibly the universe is a computer.

There are a number of problems with this argument. First, why should we believe that nature (i.e. the universe) is discrete? Presumably because quantum mechanics says it is (J.A. Wheeler, 1989, p. 314). Some distinguished physicists deny this (Weinberg, 2002), but some distinguished mathematicians give a computational reason for it (Chaitin, 2005, 2006). For those of us who are not physicists able to take a stand on this issue, Wolfram's conclusion has to be conditional: *if* the universe is discrete, *then* possibly it is a computer.

So let's suppose (for the sake of argument) that nature *is* discrete. Might it be a "cellular automaton"? The easiest way to think of a cellular automaton is as a two-dimensional Turing Machine tape for which the symbol in any cell is a function of the symbols in neighboring cells. But of course, even if a discrete universe *might* be a cellular automaton, it *need not* be. If it isn't, the argument stops here. But if it is, then – because the third premise is mathematically true – the conclusion follows validly from the premises. Premise 2 is the one most in need of justification. But even if all of the premises and (hence) the conclusion are true, it is not clear what philosophical consequences we are supposed to draw from this.[31]

Lloyd's Argument

Whereas Wolfram takes a "digital" approach, Seth Lloyd (writing with Y. Jack Ng, 2004) also argues that the universe is a computer because nature is discrete, but he takes a "quantum" approach (Piccinini and Anderson, 2020). Lloyd argues as follows:

1. Nature is discrete. (This is "the central maxim of quantum mechanics" (p. 54).)
2. In particular, elementary particles have a "spin axis" that can be in one of two directions.

---

30 See the Online Resources for further reading on the solar system as a computer.
31 See the Online Resources for further reading on cellular automata and on Wolfram.

3. ∴ They encode a bit.
4. ∴ Elementary particles store bits of information.
5. Interactions between particles can flip the spin axis;
   this transforms the stored data – i.e. these interactions are operations on the data.
6. ∴ (Because any physical system stores and processes information,)
   all physical systems are computers.
7. In particular, a rock is a computer.
8. Also, the entire universe is a computer.

Premise 1 matches Wolfram's fundamental premise and would seem to be a necessity for anything to be considered a digital computer. The next four premises also underlie quantum computing.

But the most serious problem with Lloyd's argument as presented here is premise 6. Is the processing sufficient to be considered to be Turing Machine–equivalent computation? Perhaps; after all, it seems that all that is happening is that cells change from 0s to 1s and vice versa. But that's not all that's involved in computing. (Or is it? Isn't that what Hayes's magic-paper hypothesis says?) What about the control structures – the grammar – of the computation?

And although Lloyd wants to conclude that everything in the universe (including the universe itself!) is a computer, note that this is not exactly the same as Searle's version of that claim. For Searle, everything can be interpreted as *any* computer program. For Lloyd, anything is a computer, "although they may not accept input or give output in a form that is meaningful to humans" (p. 55). So, for Lloyd, it's not a matter of interpretation. Moreover, "analyzing the universe in terms of bits and bytes does not replace analyzing it in conventional terms such as force and energy" (p. 54). It's not clear what the import of that is: does he mean the computer analysis is irrelevant? Probably not: "it does uncover new and surprising facts" (p. 54), although he is vague (in the 2004 general-audience magazine article) on what those "facts" are. Does he mean there are different ways to understand a given object? An object could be understood as a computer or as an object subject to the laws of physics. That is true, but unsurprising: animals, for instance, can be understood as physical objects satisfying the laws of quantum mechanics as well as being understood as biological objects. Does he mean force and energy can, or should, be understood in terms of the underlying computational nature of physical objects? He doesn't say.

But Lloyd does end with a speculation on what it is that the universe is computing: namely, itself! Or, as he puts it, "computation is existence" (p. 61). As mystical as this sounds, does it mean anything different from the claim that the solar system computes Kepler's Law?

And here's an interesting puzzle for Lloyd's view, relating it to issues concerning whether a computer must halt (recall our earlier discussion in Chapters 7 and 8):

> [A]ssuming the universe is computing its own evolution …, does it have a finite lifetime or not? If it is infinite, then its self-computation won't get done; it never produces an answer …. Hence, it does not qualify as a computation. (Borbely, 2005, p. 15)

Of course, Turing – as we saw in Section 8.9.5 – would not have considered this to be a problem: don't forget that his original *a*-machines only computed the decimal expansions of real numbers by *not* halting![32]

----

32 See the Online Resources for further reading on Lloyd.

9.8 Conclusion

So, finally, what is a computer?

At a bare minimum, we might say that a (programmable) computer is a practical implementation (including a virtual implementation) of anything logically equivalent to a Universal Turing Machine (DC4). Most of the definitions we have discussed might best be viewed as focusing on exactly what is meant by 'implementation' or which entities count as such implementations. This is something we will return to in Chapter 13.

Two kinds of (alleged) computers are not obviously included in this sort of definition: analog computers and "hypercomputers." Because most philosophical discussions focus on "digital" computers as opposed to analog ones, I have not considered analog computers here. By 'hypercomputer,' I have in mind any physical implementation (assuming there are any) of anything capable of "hypercomputation": i.e. anything capable of "going beyond the Turing limit"; i.e. anything that "violates" the Church-Turing Computability Thesis. The topics of hypercomputation and counterexamples to the Computability Thesis will be discussed in Chapters 10 and 11.

But one way to incorporate these other models of computation into a unified definition of 'computer' might be this:

> **(DC5) A computer is any practical implementation of anything that is logically equivalent to *at least* a Universal Turing Machine.**

In other words, if something can compute at least all Turing-computable functions[33] but might also be able to perform analog computations or hypercomputations, then it, too, is a computer.

A possible objection to this is that an adding machine, or a calculator, or a machine designed to do only *sub*-Turing computation, such as a physical implementation of a finite automaton, has at least some claim to being called a 'computer.' So another way to incorporate all such models is to go one step beyond our DC5 to

> **(DC6) A computer is a practical implementation of some model of computation.**

Indeed, Piccinini (2018, p. 2) has more recently offered a definition along these lines. He defines 'computation' as "the processing of medium independent vehicles by a functional mechanism in accordance with a rule." (See Piccinini, 2015, Ch. 7, for argumentation and more details.) This, of course, is a definition of 'computation,' not 'computer.' But we can turn it inside out to get this:

> **Definition P3** A computer is a functional mechanism that processes medium-independent vehicles in accordance with a rule.

Piccinini explicitly cites as an advantage of this very broad definition its inclusion of "not only digital but also analog and other unconventional types of computation" (p. 3) – including hypercomputation. But Piccinini (2015, Chs. 15 & 16) also distinguishes between the "mathematical" Church-Turing Computability Thesis and a "modest physical" thesis: "Any function that is physically computable is Turing-computable" (Piccinini, 2015, p. 264), and he argues that it is an "open empirical question" (p. 273) whether hypercomputers are possible (although he doubts they are). (See also Duwell, 2021.)

---

33 For convenience, from now on we will use the expression 'Turing-computable' to mean "computable by anything logically equivalent to a Turing Machine": that is, anything computable according to the classical theory of computability or recursive functions.

Recall Stuart C. Shapiro's definition, cited in Section 3.10:

> [T]he computer is a general-purpose procedure-following machine.
> (Shapiro, 2001, p. 2)

Given his broad characterization of 'procedure,' this fits with DC6. My only hesitation with these last three definitions is that they seem to be a bit too vague in their generosity, leaving all the work to the meaning of 'computation' or 'procedure' or 'rule.' But maybe that's exactly right. Despite its engineering history and despite its name, perhaps "computer science" is best viewed as the scientific study of computation, not (just) computers. Determining how computation can be done *physically* tells us what a computer is.

$$* * * * *$$

With these preliminary remarks about the nature of CS, computers, and computation as background, it is now time to look at some challenges to the Church-Turing Computability Thesis, which is the topic of the next part of the book.

9.9 Questions for the Reader

1. (This exercise was developed by Albert Goldfain.)
    (a) The following arguments are interesting to think about in relation to the question whether everything a computer. Try to evaluate them.
    Argument 1
    > **P1** A Turing Machine is a model of computation based on what a single human (i.e. a clerk) does.
    > **P2** Finite automata and push-down automata are mathematical models of computation that recognize regular languages and context-free languages, respectively.
    > **P3** Recognizing strings in a languages is also something individual humans do.
    > **C1** ∴ Turing Machines, finite automata, and push-down automata are all models of computation based on the abilities of individuals.

    Argument 2
    John Conway's "Game of Life" is a cellular-automaton model of a society (albeit a very simplistic one):[34]
    > **P1** The Game of Life can be implemented in Java.
    > **P2** Any Java program is reducible to a Turing Machine program.
    > **C1** ∴ The Game of Life is Turing-computable.

    Argument 3
    > **P1** The Game of Life can be thought of as a model of computation.
    > **P2** The Game of Life is a model of the abilities of a society.
    > **P3** The abilities of a society exceed those of an individual.
    > **C1** ∴ The abilities of a model of computation based on a society will exceed the abilities of a model based on the abilities of an individual.
    > **C2** ∴ It is not the case that every Turing Machine program could be translated to a Game-of-Life "computation."

---

34 See http://en.wikipedia.org/wiki/Conway_Game_of_Life for the rules of this game.

(b) Some of the arguments in Exercise 1 may have missing premises! To determine whether the Game of Life might be a model of computation, do a Google search using the two phrases "game of life" "turing machine."

(c) Given an integer input (remember, everything can be encoded as an integer), how could this integer be represented as live cells on an initial grid? How might "stable" structures (remember, a 2×2 grid has three neighbors each) be used as "memory"? How would an output be represented?

(d) Can Turing Machine programs be reduced to Game of Life computations?

2. Recall our discussion in Section 6.4.3 of Jacquard's looms.

> Modern programmers would say … [that Jacquard] loom programs are not computer programs: looms could not compute mathematical functions.
> (Denning and Martell, 2015, p. 83)

Looms might not have been computers, but could they have been? Even if we accept the definition of a computer (program) as one that computes mathematical functions, does it follow that Jacquard looms could not be computers? Could bits be implemented as patterns in looms?

3. Which physical processes are computing processes? Are all physical processes computations?[35] Of course, if a physical process is a computation, then presumably the physical object carrying out that process is a computer. Does this amount to saying that all physical objects that carry out processes are computers?

4. Consider this argument, adapted from Fekete and Edelman, 2011:

(a) A process is a computation iff it operates on representations.

(b) All physical processes can represent.

(c) ∴ All physical processes are computations.

Keep in mind that even if all physical processes *can* represent, it does not follow that they all *do* represent. (Or does that suggest that "computing is in the eye of the beholder. If a rock heating up in the sun is not taken as a representer, then it is not computing, but if I use how hot it is to do something else, then the hot rock is representing and so computing."[36]) Another consideration is this: computation is done over uninterpreted marks. Whether those marks represent anything is a separate matter. I might choose to interpret them as representing something; or the computational system itself might choose to (self-?)interpret them as representing something (see Schweizer, 2017).

Is this argument sound? Does this argument adequately represent Fekete and Edelman's actual argument? (See Section 7.9, #8.)

5. Is a (physical) implementation of a computation itself a computation?[37]
(See our discussion of implementation in Chapter 13.)

6.   Never mind the name change – the Apple TV and iPhone are computers to the core.
       (Gruber, 2007)

Are devices such as these computers? Choose one or more definitions of 'computer' and see if Apple TVs, iPhones, etc., are computers based on those definitions.

---

35 Thanks to Russ Abbott and Eric Dietrich for suggesting these questions.
36 Dietrich, personal communication, 28 June 2015.
37 Also due to Dietrich.

7. In Section 9.5, I considered whether a dishwasher might be a computer. What about a tree? According to César Hidalgo,

> A tree … is a computer that knows in which direction to grow its roots and leaves. Trees know when to turn genes on and off to fight parasites, when to sprout or shed their leaves and how to harvest carbon from the air via photosynthesis. As a computer, a tree begets order in the macrostructure of its branches and the microstructures of its cells. We often fail to acknowledge trees as computers, but the fact is that trees contribute to the growth of information in our planet because they compute. (Hidalgo, 2015, p. 75).

But what is his argument here? He doesn't seem to have a definition of 'computer.' Except for the last three words of this quotation, one might think his definition would be something like "a computer is an information-processing machine." Then his argument might go as follows: trees are information-processing machines (because they "contribute to the growth of information"); hence, they are computers. But those last three words suggest that his argument goes the other way: that trees are computers; hence, they contribute to the growth of information.

So, *are* dishwashers computers? Is a tree a computer? Is the human race a computer? (On the last question, see the interview with Hidalgo in O'Neill, 2015.)

8. It seems to be correct to say that a real, physical computer such as your laptop is not a Turing Machine, on the grounds that real, physical computers are finite devices (finite memory, etc.), whereas Turing Machines are infinite (infinite, or at least arbitrarily long, tape, etc.).

But could it be a finite-state machine? After all, a finite-state machine is … well … finite! At least one computer scientist has denied this:

> Another obvious distinction that is worth making explicit … is the distinction between computers (which include laptops and iPads) on the one hand and their mathematical models on the other hand. Strictly speaking, then, it is wrong to say that:

A computer is a finite state machine.

> Once again, this is like speaking about a mathematical model (the finite state machine) as if it coincides with reality (the computer). (Daylight, 2016, p. 14)

But consider this mathematical definition of a "graph" (paraphrased from https://en.wikipedia.org/wiki/Graph_(discrete_mathematics)):

> … a *graph* is an ordered pair $G = (V, E)$ comprising a set V of *vertices* … together with a set E of edges … which are … [unordered pairs of members] of V (i.e. an edge is associated with two vertices, and the association takes the form of the unordered pair of the vertices).

Now consider a real-world computer network consisting of a set V of computers and a set E of pairs of computers that are networked to each other. *Is* that computer network a graph? Or is it only *modeled* as a graph?

Similarly, could we say that a real, physical computer *is* a finite-state machine if it satisfies the definition of one? It may also have other properties that the (mathematical) definition of finite-state machine lacks. For example, the computer might be made of plastic and silicon; the definition of a finite-state machine is silent about any requirements for physical composition:

… equating a laptop with a universal Turing Machine is problematic, not primarily because the former is finite and the latter is infinite, but because the former moves when you push it and smells when you burn it while the latter can neither be displaced nor destroyed. (Daylight, 2016, p. 118)

But all properties of finite-state machines will hold for physical computers, even if there are properties of physical computers that do not hold for finite-state machines (such as ringing – or failing to ring! – a real bell if its program has a 'BEEP' command). (We'll have more to say about that kind of command in Sections 15.4.1 and 15.5.)

Part III

The Church-Turing Computability Thesis

We introduced the Church-Turing Computability Thesis as the claim that the *informal* notion of computability can be identified with any of the logically equivalent *formal* notions of Turing Machine computability, lambda-calculus computability, general recursive function computability, etc.

Here is Turing's (1939, p. 166) formulation of it (together with his footnote):

> A function is said to be "effectively calculable" if its values can be found by some purely mechanical process. Although it is fairly easy to get an intuitive grasp of this idea, it is nevertheless desirable to have some more definite, mathematically expressible definition. Such a definition was first given by Gödel at Princeton in 1934 .... These functions were described as "general recursive" by Gödel. We shall not be much concerned here with this particular definition. Another definition of effective calculability has been given by Church ..., who identifies it with λ-definability. The author has recently suggested a definition corresponding more closely to the intuitive idea .... It was stated above that "a function is effectively calculable if its values can be found by some purely mechanical process." We may take this statement literally, understanding by a purely mechanical process one which could be carried out by a machine. It is possible to give a mathematical description, in a certain normal form, of the structures of these machines. The development of these ideas leads to the author's definition of a computable function, and to an identification of computability[†] with effective calculability. It is not difficult, though somewhat laborious, to prove that these three definitions are equivalent ....
>
> † We shall use the expression "computable function" to mean a function calculable by a machine, and we let "effectively calculable" refer to the intuitive idea without particular identification with any one of these definitions. We do not restrict the values taken by a computable function to be natural numbers; we may for instance have computable propositional functions.

Recall from Chapter 4 that Popper claimed sciences must be "falsifiable" and that Kuhn claimed sciences are subject to "revolutions." Is the Computability Thesis falsifiable? In Chapter 10, we will look at two challenges to the Computability Thesis having to do with the nature of such real-life procedures as recipes. In Chapter 11, we will look at some arguments to the effect that there are forms of computation that go "beyond" Turing Machine computation. Do such forms of computation constitute a Kuhnian revolution in CS?

Philosophy of Computer Science: An Introduction to the Issues and the Literature, First Edition. William J. Rapaport.
© 2023 John Wiley & Sons, Inc. Published 2023 by John Wiley & Sons, Inc.

10

Procedures

> I believe that history will record that around the mid twentieth century many classical
> problems of philosophy and psychology were transformed by a new notion of process:
> that of a *symbolic* or *computational* process.
> —Zenon Pylyshyn (1992, p. 4)

10.1 Introduction

Algorithms – including procedures and recipes – can fail for many reasons. They can omit crucial
steps: in an "Agnes" comic, Agnes's recipe for buttered saltines is

1. Get some crackers
2. Butter them
3. Arrange on platter

After her friend comments that they are chewy, Agnes adds:

> Author's note … in recipe number one, insert the step "unwrap saltines" between step one
> and two. (http://www.gocomics.com/agnes/2011/11/7)

They can fail to be specific enough (or they can make too many assumptions). (Recall the Hagar
comic strip described in Section 7.3.3.) They can be highly context dependent or ambiguous (recall
the first instruction in Figure 7.3),[1] and so on. The general theme of the next few chapters is to
challenge various parts of the informal definition of 'algorithm':

- Does it have to be *finite*?
- Does it have to be *"effective"*? (Does it have to halt? Does it have to solve the problem? What about
  heuristics?)
- Does it have to be *unambiguous or precisely described*? (What about recipes?)

In this chapter, we will look at one kind of objection to the Computability Thesis: namely, that
there is a more general notion – the notion of a "procedure." The objection takes the form of
claiming that there are "procedures" that *are* computable in the *informal* sense but that are *not*
computable by Turing Machines.

---

1 For more humorous versions of algorithms, see the cartoons archived at http://www.cse.buffalo.edu/~rapaport/
510/alg-cartoons.html

Philosophy of Computer Science: An Introduction to the Issues and the Literature, First Edition. William J. Rapaport.
© 2023 John Wiley & Sons, Inc. Published 2023 by John Wiley & Sons, Inc.

10.2   The Church-Turing Computability Thesis

> The [Church-Turing] thesis was a great step toward understanding algorithms, but it did not solve the problem [of] what an algorithm is.
> —Andreas Blass and Yuri Gurevich (2003, p. 2)

Recall from Section 7.4.4, that "Church's Thesis" is, roughly, the claim that the informal notion of "algorithm" or "effective computation" is equivalent to (or is completely captured by, or can be completely analyzed in terms of) Church's lambda calculus. More precisely,

> **Definition 2.1. Church's Thesis (First Version, unpublished, 1934).**
> A function is effectively calculable if and only if it is λ-definable.
> (Soare, 2009, p. 372)

Later, Church reformulated it in terms of recursive functions:

> **Definition 2.2. Church's Thesis [1936].**
> A function on the positive integers is effectively calculable if and only if it is
> recursive. (Soare, 2009, p. 372; cf. Section 11.1, p. 389)

And "Turing's Thesis" is, roughly, the claim that the informal notion of "algorithm" or "computability" is equivalent to (or completely captured by, or can be completely analyzed in terms of) the notion of a Turing Machine. We saw several versions of Turing's Thesis in Chapter 8.[2] Here is Robert I. Soare's version:

> **Definition 3.1. Turing's Thesis [1936].**
> A function is intuitively computable (effectively calculable) if and only if it is computable by a Turing machine .... (Soare, 2009, p. 373)

Turing proved that Church's lambda calculus was logically equivalent to his own *a*-machines. That is, he proved that any function that was computable by the lambda calculus was also computable by a Turing Machine (more precisely, that any lambda computation could be "compiled" into a Turing machine) and vice versa – that any function that was computable by a Turing Machine was also computable by the lambda calculus (so that the lambda calculus and Turing Machines were inter-compilable). Consequently, their theses are often combined under the name the "Church-Turing Thesis."

There are other, less well-known computability theses. One is Emil Post's version:

> **Definition 5.1. [Post's Thesis, 1943, 1944].**
> A nonempty set is effectively enumerable (listable in the intuitive sense) iff it is recursively enumerable (the range of a recursive function) or equivalently iff it is generated by a (normal) production system. (Soare, 2009, p. 380)

This may look a bit different from Church's and Turing's versions, but as Soare (2009, p. 380) notes,

> Since recursively enumerable sets are equidefinable with partial computable functions ... Post's Thesis is equivalent to Turing's Thesis.

---

2 Sections 8.4.1, 8.7.1, 8.7.9, 8.8.2, and 8.9.7.

Consequently, Soare (2009, Section 12) has argued that the thesis should be called simply the "Computability Thesis," on the grounds that – given the equivalence of all mathematical models of computation (Church's, Turing's, Gödel's, Post's, etc.) – there are really many such theses and hence no reason to single out one or two names, any more than we would refer to the calculus as 'the Newton calculus' or 'the Leibniz calculus.'

On the other hand, an interesting argument to the effect that Church's Thesis should be *distinguished* from Turing's Thesis has been given by Michael Rescorla (2007): Church's Thesis asserts that intuitively computable *number-theoretic* functions are recursive. Turing's Thesis asserts that intuitively computable *string-theoretic* functions are Turing-computable. We can only combine these into a Church-Turing Computability Thesis by adding a requirement that there be a computable semantic interpretation function between strings and numbers. However, Sieg (2000) first analyzes the (informal) "calculability of number-theoretic functions" into calculability by humans "satisfying boundedness and locality conditions"; that, in turn, is analyzed into "computability by string machine"; finally, the latter is analyzed into computability by a Turing Machine. Sieg identifies "Turing's thesis" as the analysis of the first of these by the last.

The Computability Thesis, in any of its various forms, states that the *informal* notion of effective computation (or algorithm, or whatever) is equivalent to the *formal* notion of a Turing Machine program (or a lambda-definable function, or a recursive function, or whatever). The arguments in favor of the Computability Thesis are generally of two forms (Section 7.4.4). (1) All known informal algorithms are Turing-computable. (This puts it positively. To put it negatively, no one has yet found a *universally convincing* example[3] of an informally computable function that is not also Turing-computable.) (2) All of the formal, mathematical versions of computation are logically equivalent to each other.

It has also been argued that the Computability Thesis cannot be *formally proved* because one "side" of it is informal and hence not capable of being part of a formal proof. Dershowitz and Gurevich (2008) have suggested that the thesis *is* capable of being proved, by providing a set of formal "postulates" for the informal notion and then proving that Turing machines satisfy those postulates. Although this is an interesting exercise, it is not obvious that this proves the Computability Thesis. Rather, it seems to replace that Thesis with a new one: namely, that the informal notion is indeed captured by the formal postulates. But *that* thesis likewise cannot be proved for the same reason the Computability Thesis cannot: to prove it would require using an informal notion that cannot be part of a formal proof.

Others have argued that neither (1) nor (2) is even a *non*-deductively good argument for the Computability Thesis. Against (1), it can be argued that just because all *known* informal algorithms are Turing-computable, it does not follow that *all* informal algorithms are. After all, just because Aristotle's theory of physics lasted for some 2000 years until Newton came along, it did not follow that Aristotle's physics was correct; and just because Newton's theory lasted for some 200 years until Einstein came along, it did not follow that Newton's theory was correct. So, there is no inductive reason to think the Computability Thesis is correct any more than there is to think Einstein's theory is. (As to whether *any* scientific theory is "correct," on the grounds that they are all only falsifiable, see Section 4.8.2.)

But perhaps the Computability Thesis is *neither* a formally unprovable "thesis" *nor* a formally provable one, but something else altogether. In fact, Church called his statement of what we now name "Church's Thesis" "a *definition* of effective calculability" (as Turing did in 1939). It is worth quoting in full, including parts of his important footnote 3:

---

3 There are many examples, but none are universally convincing.

The purpose of the present paper is to propose a definition of effective calculability[3] which is thought to correspond satisfactorily to the somewhat vague intuitive notion in terms of which problems of this class are often stated, and to show, by means of an example, that not every problem of this class is solvable.

    [3] ... this definition of effective calculability can be stated in either of two equivalent forms, (1) that a function of positive integers shall be called effectively calculable if it is λ-definable ..., (2) that a function of positive integers shall be called effectively calculable if it is recursive .... And the proof of equivalence of the two notions is due chiefly to Kleene, ... the present author and to J.B. Rosser .... The proposal to identify these notions with the intuitive notion of effective calculability is first made in the present paper ....

    ... The fact ... that two such widely different and (in the opinion of the author) equally natural definitions of effective calculability turn out to be equivalent adds to the strength of the reasons adduced below for believing that they constitute as general a characterization of this notion as is consistent with the usual intuitive understanding of it. (Church, 1936b, p. 346)

Definitions, of course, are not susceptible to proof.

    Rather than considering it a *definition*, the philosopher and logician Richard Montague (1960, p. 430) viewed the Thesis as an *explication* of the informal notion of "effective calculability" or "algorithm." An "explication" is the *replacement* of an informal or vague term with a formal and more precise one. (The concept is due to the philosopher Rudolf Carnap (1956, pp. 7–8).) In a similar vein, the mathematician and logician Elliott Mendelson (1990, p. 229) calls the Thesis a *"rational reconstruction"* (a term also due to Carnap): "a precise, scientific concept that is offered as an equivalent of a prescientific, intuitive, imprecise notion." Mendelson goes on to argue that the Computability Thesis has the same status as the definition of a *function* as a certain *set* of ordered pairs (see Section 7.2) or as other (formal) definitions of (informal) mathematical concepts (logical validity, Tarski's definition of truth, the δ-ϵ definition of limits, etc.). Mendelson then claims that "it is completely unwarranted to say that C[hurch's] T[hesis] is unprovable just because it states an equivalence between a vague, imprecise notion ... and a precise mathematical notion" (Mendelson, 1990, p. 232). One reason he gives is that *both* sides of the equivalence are *equally* vague! He points out that "the concept of set is no clearer than that of function." Another is that the argument that all Turing Machine programs are (informally) computable is considered a proof, yet it involves a vague, informal concept. (Note that it is the converse claim that all informally computable functions are Turing-computable that is usually considered incapable of proof on these grounds.)

    It's worth comparing the formal explication of the informal notion of algorithm as a Turing Machine (or recursive functions, etc.) with other attempts to define informal concepts in scientific terms. As with any attempt at a formal explication of an informal concept (as we discussed in Sections 3.4 and 9.2), there is never any guarantee that the formal explanation will satisfactorily capture the informal notion (usually because the informal notion is informal, vague, or "fuzzy"). The formal explication might include some things that are, pre-theoretically at least, not obviously included in the informal concept, and it might exclude some things that are, pre-theoretically, included. Many of the attempts to show that there is something wrong with the Computability Thesis fall along these lines.[4]

---

4 See the Online Resources for further reading on the Computability Thesis.

> **Question for the Reader:** As we noted in Section 3.4.1, 'life' is one of these terms. One difference between the two cases is this: there are many *non*-equivalent scientific definitions of 'life.' But in the case of 'algorithm,' there are many *equivalent* formalizations: Turing Machines, recursive functions, lambda calculations, etc.
>
> What might have been the status of the informal notion if these had *not* turned out to be equivalent?

10.3 What Is a Procedure?

Herbert Simon (1962, p. 479) offers two kinds of descriptions of phenomena in the world: *state* descriptions and *process* descriptions:

> The former characterize the world as sensed; they provide the criteria for identifying objects .... The latter characterize the world as acted upon; they provide the means for producing or generating objects having the desired characteristics.

The "desired characteristics" to be produced are, presumably, given by a state description. His example of a state description is "A circle is the locus of all points equidistant from a given point"; his example of a process description is "To construct a circle, rotate a compass with one arm fixed until the other arm has returned to its starting point." (Recall our discussion in Section 3.16.3 of Euclid's *Elements*, which was originally written in terms of "process descriptions.") Process descriptions describe procedures.

State descriptions seem to be part of "science," whereas process descriptions seem to be part of "engineering" and certainly part of "computational thinking." Consider this related claim of Rescorla (2014b, Section 2, p. 1279):

> To formulate ... [Euclid's GCD algorithm], Knuth uses natural language augmented with some mathematical symbols. For most of human history, this was basically the only way to formulate mechanical instructions. The computer revolution marked the advent of rigorous *computational formalisms*, which allow one to state mechanical instructions in a precise, unambiguous, canonical way.

In other words, CS developed formal methods for making the notion and expression of procedures mathematically precise. That's what makes it a science of procedures.

Stuart C. Shapiro's (2001) more general notion of "procedure" (Section 3.10) characterizes " 'procedure' as the most general term for the way 'of going about the accomplishment of something,' " citing the Merriam-Webster *Third New International Dictionary*.[5] This includes serial algorithms as well as parallel algorithms (which are not "step by step," or serial), operating systems (which don't halt), heuristics (which "are not guaranteed to produce the correct answer"), musical scores (which are open to interpretation by individual performers), and recipes (which are also open to interpretation as well as being notoriously vague, as we will see later in this chapter). Thus, Turing Machines (or Turing Machine programs) – that is, (serial) algorithms as analyzed in Section 7.3 – are only a special case of procedures. In this chapter, we are focusing on this more general notion of 'procedure.'

------
5 http://www.merriam-webster.com/dictionary/procedure

*****

Some philosophers have challenged the Computability Thesis, arguing that there are things that *are* intuitively algorithms but that are *not* Turing Machines. In this section, we will look at two of these, due to the philosophers Carol Cleland and Beth Preston. Interestingly, both focus on recipes, although for slightly different reasons – Cleland on the fact that recipes are carried out in the real world, and Preston on the fact that they are vague and open to interpretation by chefs.

10.4 Carol Cleland: Some Effective Procedures Are Not Turing Machines

In a series of papers, Carol Cleland has argued that there are effective procedures that are not Turing Machines (Cleland, 1993, 1995, 2001, 2002, 2004). By 'effective procedure,' she means (1) a "mundane" procedure (i.e. an ordinary, everyday, or "quotidian" one) that (2) generates a causal process (i.e. a procedure that physically causes (or "effects") a change in the world).

Terminological Digression: There may be an unintentional pun here. As we have seen, the word 'effective' as used in the phrase 'effective procedure' is a semi-technical term that is roughly synonymous with 'algorithmic.' On the other hand, the *verb* 'to effect,' as used in the phrase "to effect a change (in something)," is roughly synonymous with the verbs 'to produce' and 'to cause,' and it is not directly related to 'effective' in the algorithmic sense.

And just to make things more confusing, 'effect' is also a *noun* meaning "the result of a cause." Worse, there are a verb and a noun spelled slightly differently but pronounced almost the same: '<u>a</u>ffect'! For the difference between the *verb* 'to effect' and the *noun* 'an effect,' as well as the similar-sounding verb and noun 'affect,' see Rapaport, 2013, Section 4.2.0, "affect vs. effect".

According to Cleland, there are three ways to understand the Computability Thesis:

1. It applies *only* to (mathematical) functions of integers (or, possibly, also to anything representable by – i.e. codable into – integers).
2. It applies to *all* (mathematical) functions (including real-valued functions).
3. It also applies to the production of mental and physical phenomena, such as is envisaged in AI or robotics.

She agrees that it cannot be proved but that it *can* be falsified by exhibiting an intuitively effective procedure, "*but not in Turing's sense,*" that is "more powerful" than a Turing Machine (Cleland, 1993, p. 285, my italics). Presumably, the qualification "but not in Turing's sense" simply means it must be intuitively effective yet not capable of being carried out by a Turing Machine, because, after all, that's what Turing thought his *a*-machines could do: namely, carry out any intuitively effective procedure.

But she also suggests another sense in which the Computability Thesis might be falsifiable: by exhibiting a procedure that *is* intuitively effective *in Turing's sense* yet is not Turing-computable. In other words, there might be two different kinds of counterexamples to the Computability Thesis: if Turing were alive, (1) we could show him an intuitively effective procedure that we claim is not Turing-computable, and he might agree; or (2) we could show him a procedure, and *either* he would *not* agree that it was intuitively effective (thus denying that it was a possible counterexample) *or* he could show that, indeed, it *was* Turing-computable (showing how it is not a counterexample at all). Cleland seems to be opting for (1).

Curiously, however, she goes on to say that her "mundane procedures" *are* going to be effective "in Turing's sense" (Cleland, 1993, p. 286)! In any case, they differ from Turing-computable procedures by being *causal*. ('Causal' in this context is roughly synonymous with 'physical.' When reading Cleland's article, you should continually ask yourself two questions: *Are* her "mundane procedures" causal? Are Turing Machines *not* causal?) Here is a reconstruction of her argument, with comments after some of the premises:

1. A "procedure" is a specification of something *to be followed* (Cleland, 1993, p. 287).

This includes recipes as well as computer programs. Her characterization of a procedure as something to be followed puts a focus on imperatives: you can follow an instruction that says, "Do this!" But there are other ways to characterize procedures. For example, Shapiro (2001) describes a procedure as a *way* to do something. But his focus is on the goal or end product; the way to do it – the way to accomplish that goal – might be to evaluate a function or to determine the truth value of a proposition, not necessarily to "follow" an imperative command.

You should also recall (from Section 8.10) that Turing Machines don't normally "follow" any instructions! The Turing Machine table is a description of the Turing Machine, but it is not something that the Turing Machine consults and then executes. That only happens in a *Universal Turing Machine*. But in that case, there are two different programs to consider: the program encoded on the Universal Turing Machine's tape *is* consulted and followed. But the fetch-execute procedure that constitutes the Universal Turing Machine's machine table is *not* consulted or followed.

The relationship between the stored program and the fetch-execute program gives rise to several interesting issues in epistemology and the philosophy of AI: is there any sense in which the "blindly executed" fetch-execute cycle "knows" what it is doing when it "follows" the program on the tape? (Recall the passage cited in Section 9.4.3 from Nicolas D. Goodman.) Does an AI program "understand" what it does? (Should it? Could it?) We'll examine these issues in Section 18.9 when we discuss what Daniel C. Dennett (2013b) has called "Turing's 'Strange Inversion of Reasoning.' "

2. To say that a "mundane" procedure is "effective" means, by definition, that following it always results in a certain kind of outcome (Cleland, 1993, p. 291).

The semi-technical notion of "effective" as it is used in the theory of computation is, as we have seen (Section 7.3), somewhat ambiguous. Cleland notes (1993, p. 291) that Marvin Minsky (1967) calls an algorithm 'effective' if it is "precisely described." And Church (1936b, pp. 50ff; cf. p. 357) calls an algorithm 'effective' if there is a formal system that takes a formal analogue of the algorithm's input, precisely manipulates it, and yields a formal analogue of its output. Church's notion seems to combine aspects of both Minsky's and Cleland's notions.

A non-terminating program (either one that erroneously goes into an infinite loop or one that computes all the digits in the decimal expansion of a real number) can be "effective" *at each step* even though it never halts. (We'll return to this in Section 11.8.)

3. The steps of a recipe *can* be precisely described (hence, they can be effective in Minsky's sense). (Cleland, 1993, p. 292).

This is certainly a controversial claim. Note that recipes can be notoriously vague, whereas computer programs must be excruciatingly precise:

> How do you know when a thing "just begins to boil"? How can you be sure that the milk has scorched but not burned? Or touch something too hot to touch, or tell firm peaks from stiff peaks? How do you define "chopped"? (Adam Gopnik 2009, p. 106; cf. Sheraton, 1981)

We will explore this in more detail in Section 10.5.

4. A procedure is *effective for* producing a specific output. For example, a procedure for producing fire or a procedure for producing hollandaise sauce might not be effective for producing chocolate. (Cleland, 1993, p. 293).

In other words, being effective (better: being "effective *for*") is not a *property of* a procedure but a *relation between* a procedure *and* a kind of output. This might seem to be reasonable, but a procedure for producing the truth table for conjunction might also be effective for producing the truth table for disjunction by suitably reinterpreting the symbols. (See the following "Digression on Conjunction and Disjunction.")

Digression on Conjunction and Disjunction:

Here is a truth table for the conjunction of two propositions, P and Q, using '0' to represent "false" and '1' to represent "true":

P	Q	$(P \wedge Q)$
0	0	0
0	1	0
1	0	0
1	1	1

Note that in the third column – which represents the conjunction of the first two columns – there are three '0's and one '1,' which occurs only in the line where both inputs are '1.'

And here is the analogous truth table for disjunction:

P	Q	$(P \vee Q)$
0	0	0
0	1	1
1	0	1
1	1	1

Note that the third column has three '1's and only one '0,' which occurs only in the line where both inputs are '0.'

Now suppose, instead, that we use '0' to represent "*true*" and '1' to represent "*false*." (John Case always did this in his theory of computation lectures!) Then the *first* table represents *disjunction*, and the *second* one represents *conjunction*!

Similar points are made by Peacocke, 1995, Section 1, p. 231; Shagrir, 2001; and Sprevak, 2010, Section 3.3, pp. 268–269. We'll return to this in Section 16.2.

Here are some more examples. A procedure that is effective for simulating a battle in a war might also be effective for simulating a particular game of chess (Fodor, 1978, p. 232). Or a procedure that is effective for computing with a mathematical lattice might also be effective for computing with a chemical lattice. (See Section 16.2.) And an ottoman (or a "pouf") could be (used as) either a seat or a table, yet, of course, seats and tables are usually considered mutually exclusive classes. And, most famously, certain well-known images can be used to represent either a duck or a rabbit. In cases such as these, the notion of effectiveness might not be the same as Church's, because of the possibility of interpreting the output differently. How important to the notions of (intuitively) effective computation and formal computation is the *interpretation* of the output symbols? We will explore these issues in more detail in Section 16.4.

5. The effectiveness of a recipe for producing hollandaise sauce depends on causal processes in the actual world, and these causal processes are independent of the recipe (the mundane procedure) (Cleland, 1993, p. 294).

Suppose we have an algorithm (a recipe) that takes eggs, butter, and lemon juice as input and tells us to mix them. Suppose that on Earth, the output – the result of mixing those ingredients – is an emulsion: i.e. hollandaise sauce. And suppose that on the Moon, mixing them does not result in an emulsion, so no hollandaise sauce is output (instead, the output is a messy mixture of eggs, butter, and lemon juice).

6. Therefore, mundane processes can be effective for a given output P in the *actual* world yet not be effective for P in some other *possible* world. (Cleland, 1993, pp. 293–294).

This is also plausible. Consider a "blocks-world" computer program that instructs a robot how to pick up blocks and move them onto or off of other blocks (Winston, 1977). I once saw a live demo of such a program. Unfortunately, the robot failed to pick up one of the blocks that was not correctly placed, yet the program continued to execute "perfectly" even though the output was not what was intended. (See Rapaport, 1995, p. 62. A similar situation is discussed in Dennett, 1987, Ch. 5, "Beyond Belief," p. 172. We'll return to this example in Section 16.2.)

7. Turing Machines are equally effective in all possible worlds, because they are causally inert. (Cleland, 1993, p. 294).

But here we have a potential equivocation on 'effective.' Turing Machines are effective in the sense of being step-by-step algorithms that are precisely specified, but they are not necessarily effective *for an intended output P*: it depends on the interpretation of P in the possible world!

8. Therefore, there are mundane procedures (such as recipes for hollandaise sauce) that can produce hollandaise sauce because they result in appropriate causal processes, *but* there are no Turing Machines that can produce hollandaise sauce, because Turing Machines are purely formal and therefore causally inert. QED (Cleland, 1993, p. 295).

To the objection that physical implementations of Turing Machines could be causally "ert" (so to speak),[6] Cleland replies as follows (p. 294): a Turing Machine's "actions" are not physical actions but action-*kinds*; therefore, they *are* causally *in*ert. An *embodied* (implemented?) Turing Machine *does* act: *embodied* action-*kinds are* causal actions. But that has nothing to do with their being an (abstract) Turing Machine. Alternatively, a Turing Machine's '0'-'1' outputs can be interpreted by a device that does have causal effects, such as a light switch or thermostat.

---

6 'Inert' comes from the Latin prefix 'in-,' meaning "not," and the Latin 'artem,' meaning "skill." So if 'inert' means "lacking causal power," then perhaps the non-word 'ert' could mean "having causal power"; see the *OED*'s entry on 'inert,' http://www.oed.com/view/Entry/94999.

Perhaps a procedure or algorithm that is "effective for P" is better understood as an algorithm *simpliciter*. In the actual world, it does P. In some other possible world, perhaps it does $Q (\neq P)$. In yet another possible world, perhaps it does nothing (or loops forever). And so on. For example,

> … if we represent the natural number n by a string of n consecutive 1s, and start the program with the read-write head scanning the leftmost 1 of the string, then the program,
>
> q_0 1 1 R q_0
> q_0 0 1 R q_1
>
> will scroll the head to the right across the input string, then add a single '1' to the end. It can, therefore, be taken to compute the successor function. (Aizawa, 2010, p. 229)

But if the environment (the tape) is not a string of n '1's followed by a '0,' then this does *not* compute the successor function. Compare this to Cleland's hollandaise sauce recipe being executed on the Moon. Hence, mundane procedures are *interpreted* Turing Machine programs, so they *are* computable.

Aaron Sloman (2002, Section 3.2) makes a useful distinction between "internal" and "external" processes: the former "include manipulation of cogs, levers, pulleys, strings, etc." The latter "include movements or rearrangements of various kinds of physical objects." So, a computer on Earth that is programmed to make hollandaise sauce and one on the Moon that is identically programmed will have the same internal processes but different external ones (because of differences in the external environment). A related distinction was made by linguist Noam Chomsky between *competence* ("an ideal" language user's "knowledge of his [sic] language") and *performance* ("the actual use of language in concrete situations") (Chomsky, 1965, pp. 3–4). A computer might be *competent* to make hollandaise because of its *internal* processes yet fail to do so because of *performance* limitations due to *external* environmental limitations.

Does the ability of a machine to *do* something that is not Turing-computable mean it can *compute* something that is not Turing-computable? What does physical performance have to do with computation? Surely we want to say that whatever a robot can do is computable, even if that includes cooking. But surely that's because of the "internal" processes, not the "external" ones.

> … Turing machines are not so relevant [to AI] intrinsically as machines that are designed from the start to have interfaces to external sensors and motors with which they can interact online, unlike Turing machines which at least in their main form are totally self contained, and are designed primarily to run in ballistic mode once set up with an initial machine table and tape configuration. (Sloman, 2002, Section 4.2)

This seems to be a distinction between abstract Turing Machines and robots. And Cleland's arguments seem more relevant to robots than to Turing Machines and hence have nothing really to say about the Computability Thesis (which only concerns Turing Machines and their equivalents). Indeed, Copeland and Sylvan (1999, p. 46) (see also Copeland, 1997) distinguish between two interpretations of the Computability Thesis. The one that they claim was actually stated by Church and by Turing "concerned the functions that are in principle computable by an *idealised human being unaided by machinery*." This one, they claim, is correct. The other interpretation is "that the class of well-defined computations is exhausted by the computations that can be carried out by Turing machines."

So, one possible objection to Cleland is that cooking (for example) is not something that can be carried out by "an idealized human being unaided by machinery," and hence the failure of a hollandaise sauce recipe on the Moon is irrelevant to the correct interpretation of the Computability Thesis.

Compare Cleland's hollandaise sauce example with the following: suppose we have an algorithm (a recipe) that tells us to mix eggs, butter, and lemon juice *until an emulsion is produced* and that outputs hollandaise sauce. In the actual world, an emulsion is indeed produced, and hollandaise sauce is output. But on the Moon, this algorithm goes into an infinite loop; nothing (and, in particular, no hollandaise sauce) is ever output.

One problem with this is that the "until" clause ("until an emulsion is produced") is not clearly algorithmic. How would the *computer* tell if an emulsion has been produced? This is not a clearly algorithmic, Boolean condition whose truth value can be determined by the computer simply by checking one of its switch settings (i.e. a value stored in some variable). It would need sensors to determine what the external world is like. But that is a form of *interactive* computing, which we'll discuss in Section 11.8.[7]

10.5 Beth Preston: Recipes, Algorithms, and Specifications

Introductory computer science courses often use the analogy of recipes to explain what algorithms are. Recipes are clearly procedures of some kind. But are recipes really good models of *algorithms*? Cleland has assumed that they are.

Beth Preston (2013) has a different take. She is interested in the nature of artifacts and how they are produced (or implemented) from plans (such as blueprints). Compare this to how a program (which is like a plan) is actually executed.

According to Preston, the classical view of production is that of "centralized control." The etymology of the word 'control' is of interest here: 'to control' originally meant "to check or verify (originally by comparison with a duplicate register) . . . ." (*OED*, http://www.oed.com/view/Entry/40563). The "duplicate register" was a "counter-roll"; to control something originally meant to faithfully copy it for the sake of verification or to regulate it. So, to implement a plan is to copy an abstract design into reality: i.e. to control it. (For more on this idea of verification by comparison, see the discussion of the relation between syntax and semantics in Section 18.8.3.)

A "mental design" of an artifact to be produced first exists in someone's mind. This mental design "specifies all the relevant features of the" artifact to be produced (the "copy") (Preston, 2013, p. 30) "along with a set of instructions for construction" (Preston, 2013, p. 39). Then the "actual construction" of the artifact (i.e. the copying of the mental design) …

> … is a process that faithfully follows the instructions of the construction plan, and by so doing reproduces in a material medium the features of the product specified in the design. This faithful copying relationship between the design and construction phases of production is the *control* aspect of the model. (Preston, 2013, pp. 30–31)

Compare the way in which a program controls the operations and output of a computer.

But there is a problem: a "faithful copy" requires that …

---

7 See the Online Resources for further reading on Cleland's arguments.

… *all* relevant features of the product [the artifact] be specified in the design …. In other words, the design is ideally supposed to be an algorithm (effective procedure) for realizing both the construction process and the product. (Preston, 2013, p. 39, my italics)

According to Preston, however, recipes show that this ideal model isn't realistic.

Preston (2013, p. 40) argues that recipes differ from algorithms in being open to interpretation (recall the "Zits" comic described in Section 7.3.3). First, recipes leave details open (e.g. details about ingredients, which play the same role in recipes that data structures do in programs). And they do this in several ways: for one thing, they provide for *alternatives* – e.g. use "either sour cream or yogurt." But couldn't a recipe, or a program for that matter, simply call for a data-analogue of a typed variable or subroutine here? The recipe might call not for sour cream or yogurt but instead for a "fermented milk product." In any case, non-deterministic algorithms also provide for alternatives: in an ordinary "if" statement, when more than one Boolean condition is satisfied, the first one is executed. In a "guarded if" statement, it doesn't matter which one is executed (Gries, 1985, Ch. 10). Here is an example of a non-deterministic procedure for computing the absolute value of x using a "guarded if" control structure:

> **if** $x \geq 0$ **then** return x;
> **if** $x \leq 0$ **then** return $-x$;

In this case, if $x = 0$, it does not matter which line of the program is executed. In such procedures, a detail is left open, yet we still have an *algorithm*. Or consider a program that simply says to input an integer, without specifying anything else about the integer. (As a magician might say, "Think of a number, any number …"). This could still be an algorithm. (Or maybe not! See Section 11.8.)

Next, recipes specify some ingredients *generically*: e.g. "use frosting," without specifying what kind of frosting. But compare typed variables or named subroutines. It does not matter *how* the subroutine works; all that matters is its input-output behavior (*what* it does). And it doesn't matter *what* value the variable takes, as long as it is of the correct type. Typed variables and named subroutines are "generic," yet they appear in algorithms.

And recipes provide for *optional* ingredients: e.g. "use chopped nuts if desired." But compare any conditional statement in a program that is based on user input. (On the other hand – as noted earlier – user input *may* raise issues for interactive computing; again, see Section 11.8.)

Second, recipes leave construction steps (= control structure?) open. For instance, the order of steps is not necessarily specified (at best, a *partial order* is given): e.g. "add the rest" of the ingredients, where no order for adding is given for "the rest of the ingredients." However, compare non-deterministic statements, such as the guarded-if command in the earlier example, or programs written in languages like Lisp, where the order of the functions in the program is not related in any way to the order in which they are evaluated when the program is executed. (A Lisp program is an (unordered) *set* of functions, not a(n ordered) *sequence* of instructions.)

Third, in recipes, some necessary steps (e.g. "put these cookies on a baking sheet before baking them") can be *omitted* (i.e. go unmentioned). Should the baking sheet be greased? A knowledgeable chef would know whether it has to be, so a recipe written for such a chef need not mention the obvious. But the same kind of thing can occur in a program, with preconditions that are assumed but not spelled out, or details hidden in a subroutine. (Perhaps "put cookies on baking sheet" is a call to a subroutine of the form "grease baking sheet; put cookies on it.")

Finally, recipes can provide *alternative ways* to do something: e.g. "roll in powdered sugar …" or "shake in bag with powdered sugar …·' Again, non-determinism is similar, as are subroutines:

to say "multiply x and y" is not to specify how; to say "coat in powdered sugar" is not to specify whether this should be done by rolling or shaking.

Preston claims that the cook (i.e. the CPU) is expected to do some of the design work, to supply the missing parts. So, not everything is in the design. She claims that cooks don't *faithfully follow* recipes; instead, they *improvise*, as jazz or rock musicians do. They can even change other ("fixed") parts of the recipe because of their background knowledge, based on experience. For example, they can substitute walnuts for pecans in a recipe for pecan pie. Therefore, the constructor or executor is typically *intelligent*, in contrast to an *un*intelligent CPU (or the "unintelligent" fetch-execute cycle of a Universal Turing Machine).

But here is a different interpretation of Preston's analysis: she offers the centralized control model as a description of an *algorithm* together with a CPU that produces a *process* (i.e. an algorithm being executed). But her theory of collaborative improvisation might better describe an earlier stage in the production of a process: namely, the production of an *algorithm* by a programmer from a *specification*. That is, although the execution of an algorithm might well be modeled as centralized control, nevertheless the development of an algorithm by a programmer from a specification might well be improvisatory and collaborative, precisely because specifications – like recipes – can be vague and open to interpretation. So, recipes are more like *design specifications* for computer programs than they are like *algorithms*. In fact, my counterexamples to differentiate between algorithms and recipes just show that *either* recipes are high-level programs requiring implementation in a lower level *or* recipes are specifications.[8]

10.6 Summary

So, are there good reasons for seriously doubting the Computability Thesis? We have just seen two candidates: Cleland argues that certain "mundane procedures" are effectively computable but not Turing-computable, and Preston suggests that certain recipe-like procedures of the sort typically cited as examples of effective procedures are not really algorithmic.

But against Cleland's example, we have seen that there may be a concern in how one determines what the proper output of an algorithm is, or, to put it another way, in determining the problem an algorithm is supposed to solve. Consider the recipe for hollandaise sauce that, when correctly executed on Earth, produces hollandaise sauce but does not do so when correctly executed on the Moon. Does it follow that that recipe is therefore not Turing-computable? It would seem to be Turing-computable on Earth but not on the Moon. Or is it Turing-computable *simpliciter* (e.g. no matter where it is executed), but conditions having nothing to do with the algorithm or recipe *itself* conspire to make it unsuccessful *as a recipe for hollandaise sauce* on the Moon? Is the algorithm or recipe *itself* any different? (We'll come back to this issue beginning in Section 16.3.)

And against Preston's example, we have seen that recipes are, perhaps, more like *specifications* for algorithms than they are like *algorithms*.

It can be argued that even though the common theme underlying the equivalence of Turing Machines – lambda definability, recursive functions, etc. – is "robust" and of great mathematical interest, that is not reason enough to think there might not be any *other* theory of effective computation. In the next chapter, we will look at such potential counterexamples to the Computability Thesis.

---

8 See the Online Resources for further reading on Preston's ideas.

*****

I will close this chapter with one last version of the Thesis (not to be taken too seriously!):

> The Church-Turing Thesis: A problem is computable just in case it wants to be solved. (Anonymous undergraduate student in the author's course, CSE 111, "Great Ideas in Computer Science," 2000)[9]

10.7 Questions for the Reader

1. It is often said that Turing's analysis of computation provides a characterization of what an algorithm is. However, our analysis of algorithms in Section 7.3.3 did not mention anything about computation or mathematical functions. The same is true about our presentation in Section 7.5 of structured programming, which defined a basic program as either the empty program or one with a single "primitive operation" that was left unspecified.

    What is the relationship between algorithms thus understood and computation? Are the two concepts identical? That is, are all algorithms computable, and are all computations algorithms?

    To help you think about this, consider a recipe that is fully and explicitly spelled out in complete detail, leaving nothing to the imagination or interpretation of the chef (i.e. the recipe's executor). Another example is Landesman, 2021, which suggests some primitive paper-folding operations for origami. Suppose a fully specified algorithm is provided for folding a crane out of a square sheet of paper using something like Landesman's primitive operations. (Perhaps it is a computer program for an origami robot.)

    Are these algorithms? Are they computations? Is the fact that their primitive operations are neither Turing's "move" and "print" nor recursive-function theory's successor, predecessor, and projection mean they are *not* computations? Or does the fact that, according to the binary Representation Insight of Section 7.4.1, anything – hence any primitive operation of cooking or paper folding – can be represented by binary numerals mean they *are* computations or *are* computable?

2. This follow-up question is aimed at having you think about issues that will be discussed in more detail in Chapters 14 and 16:

    Suppose that recipes and origami *are* computable procedures (as suggested in the previous question) on the grounds that computations using binary numerals can represent (or be interpreted as) their primitive operations.

    Does that mean representation or interpretation is essential to computation? Does it mean (all?) computer programs *model* what they deal with?

---

9 http://www.cse.buffalo.edu/~rapaport/111F04.html

11

Hypercomputation

Speculation that there may be physical processes – and so, potentially, machine-operations – whose behaviour cannot be simulated by the universal Turing machine of 1936 stretches back over a number of decades. Could a machine, or natural system, deserving the name 'hypercomputer' really exist? Indeed, is the mind – or the brain – some form of hypercomputer?
—B. Jack Copeland (2002, p. 462)

We now know both that hypercomputation (or super-recursive computation) is mathematically well-understood, and that it provides a theory that according to some accounts for some real-life computation … [is] better than the standard theory of computation at and below the "Turing Limit." … [S]ince it's mathematically *possible* that human minds are hypercomputers, such minds *are* in fact hypercomputers.
—Selmer Bringsjord & Konstantine Arkoudas (2004, p. 167)

The editors have kindly invited me to write an introduction to this special issue devoted to "hypercomputation" despite their perfect awareness of my belief that there is no such subject.
—Martin D. Davis (2006c, p. 4)

Church, Gödel, and Turing defined … [computation] in terms of mathematical functions … Today, I believe we are breaking out of the era where only algorithmic processes are included in the term *computation*.
—Dennis J. Frailey (2010, p. 2)

Nobody would be fired from a computer science department for axiomatizing analog computation or hypercomputation. Both are still in [the] purview of computer science.
—Marcin Miłkowski (2018, Section 3.2)

11.1 Introduction

We have seen that it can be argued that (1) CS includes (among other things) the systematic study of *computing*, (2) computing is the mathematical study of *computational algorithms*, and (3) such algorithms are best understood mathematically in terms of *Turing Machines* (or anything logically equivalent to them).

But *are* computations best understood that way?

Philosophy of Computer Science: An Introduction to the Issues and the Literature, First Edition. William J. Rapaport.
© 2023 John Wiley & Sons, Inc. Published 2023 by John Wiley & Sons, Inc.

Just as Euclidean geometry is a mathematical theory about what geometrical constructions are possible by a human using only compass and straightedge, so Turing-computation is a mathematical theory about what functions are computable by a human using only paper and pencil:[1]

> In simplest terms an effective procedure would appear to be one that does not transcend *our* computational abilities. (Shanker, 1987, p. 628, my italics)

But just as there are geometrical constructions that are possible using other means (e.g. angle trisection by use of a protractor), hypercomputation concerns whether there are functions that are "computable" by other means.

Let's first consider different kinds of "computation." We have already distinguished between analog and discrete (or digital) computation (Sections 6.4.2 and 9.2). Within the latter, we can distinguish several "levels" of computation. Several models of computation are *weaker* than Turing Machine computation; let's call them "sub-Turing computation":[2] In Sections 7.5 and 7.6, we looked at primitive recursion and count-programs. But there are models that are even weaker than those: informally, a *finite automaton* is a machine that only

> moves from left to right on a finite input tape .... [It] will have only one opportunity to scan each square in its motion from left to right, [and] nothing will be gained by permitting the device to "print" new symbols on its tape .... Thus, a finite automaton can be thought of as a very limited computing device which, after reading a string of symbols on the input tape, either accepts the input or rejects it, depending upon the state the machine is in when it has finished reading the tape. (Davis and Weyuker, 1983, pp. 149–150)

Turing Machine computation (or any logically equivalent model of computation) is at the "top" of these levels. A reasonable question to ask is whether there are levels "above" it: Is there such a thing as "super"-Turing computation? If so, how would it affect the Computability Thesis? After all, that thesis says that any (informal) notion of "computation" is equivalent to Turing computation. Sub-Turing computation can be performed by Turing Machines simply by not using all the "power" of Turing Machines. But if super-Turing computation can do "more" than classical Turing computation – perhaps even just using Turing Machines – might that be a counterexample to the Computability Thesis?

Recall our discussion of Kuhn's philosophy of science (Section 4.9): to the extent that the Church-Turing Computability Thesis is the standard "paradigm" in CS, rejection of it could be considered a Kuhnian revolutionary challenge to "normal" CS (Stepney et al., 2005; Cockshott and Michaelson, 2007, Section 2.5, p. 235). We saw in the previous chapter that it can be argued that there might be "procedures" that are not computable by a Turing Machine. But of course, this depends on what is meant by 'procedure.' Recall that several computer scientists distinguish between "algorithms," which must halt, and "procedures," which need not (Hopcroft and Ullman, 1969, p. 2; Knuth, 1973, p. 4; S.C. Shapiro, 2001). Hopcroft and Ullman (1969, p. 80, my italics) also characterize "Church's hypothesis" as the claim "that any process which could naturally be called a *procedure* can be realized by a Turing machine." Since procedures in their sense need not halt, Turing Machines need not either.

Thus, two ways to generalize the notion of computation are changing the primitive operations and relaxing one or more of the constraints on the notion of "algorithm." In this chapter, we continue our look at procedures that are allegedly not computable by a Turing Machine.

---

1 See the Online Resources for an application of this idea to origami.
2 Sub-Turing systems are sometimes referred to as "*hypo*computation" (from the Greek root 'hypo,' meaning "under"; 'hyper' means "over"). See the Online Resources for further reading.

11.2 Generic Computation

Are there other kinds of computation besides Turing computation? Piccinini says "yes": there are analog, neural, and generic computation (Piccinini, 2020a, b). He defines 'generic computation' as "the processing of medium independent vehicles by a functional mechanism in accordance with a rule" (Piccinini, 2018, p. 2; see Piccinini, 2015, Ch. 7 and Piccinini, 2020b, Ch. 6 for argumentation and details). And he explicitly cites as an advantage of this very broad definition its inclusion of "not only digital but also analog and other unconventional types of computation" (p. 3) – including hypercomputation. But Piccinini (2015, Chs. 15 & 16) also distinguishes between the "mathematical" Church-Turing Computability Thesis and a "modest physical" thesis: "Any function that is physically computable is Turing-computable" (Piccinini, 2015, p. 264), and he argues that it is an "open empirical question" (p. 273) whether hypercomputers are possible (although he doubts that they are). For an overview, see Piccinini and Maley, 2021.

11.3 Non-Euclidean Geometries and "Non-Turing Computations"

Hypercomputation might be compared to non-Euclidean geometries. There are two relevant features of Euclidean geometry. One concerns the axioms. As we saw in Section 2.4.3, replacing Euclid's "parallel postulate" with a different one yielded several varieties of "non-Euclidean" geometries. But all such geometries – Euclidean and non-Euclidean – maintained another feature:

> [W]hether a procedure literally 'can in the most general sense be carried out' … depend[s] *only on the execution of its atomic tasks.* (Webb, 1980, p. 224, original italics)

Many interesting questions in "procedural" (as opposed to axiomatic) geometry concern which geometrical figures can be constructed solely with operations enabled by certain basic devices. (Recall our earlier discussions in Sections 3.16.3 and 7.5.) The standard devices, of course, are compass and straightedge – more precisely, *collapsible* compass and *unruled* straightedge. (A collapsible compass is the familiar one that allows you to draw circles of different radii. A straightedge is a ruler without markings of inches or centimeters.) But different systems of geometry can be studied that allow for *measuring* devices. Famously, an angle cannot be trisected using only compass and straightedge. This is an impossibility proof on a par with the Halting Problem.

However, if you allow a measuring device (such as a protractor), angle trisection is trivial (Sloman, 2020). And as we will see, *if* you allow a hypercomputer, the Halting Problem *can* be solved! (But Martin Davis, 1978, p. 255 disagrees!) Moreover, just as there are alternative primitive operations for Turing-like machines, there are alternative primitive operations for geometry: a collapsible compass can be replaced with a fixed compass. (On this, see http://en.wikipedia.org/wiki/Compass_equivalence:theorem.)

The idea behind hypercomputation is similar: what do you get if you relax or change some of Turing's restrictions on what is (humanly) computable? Turing computation is a mathematical model of *human* computation. Are there "non-Turing computations"? Many say "yes," as we will see. Instead of calling these 'computation,' Martin Davis (2006c, p. 4) calls them "computation-like process[es]." I will try to reserve the term 'compute' and its cognates for Turing computation and will use "scare quotes" to signal any kind of processing that is potentially "non-Turing computation." I will also use them to refer to the informal notion that is the subject of the Computability Thesis.

11.4 Hypercomputation

In a series of papers, the logician and philosopher B. Jack Copeland (along with several co-authors) has suggested that CS "is outgrowing its traditional foundations" (such as Turing's analysis of computation) and has called for "a 21st-century overhaul of that classical analysis" (Copeland et al., 2016, pp. 34, 36), contrasting Turing computation with "hypercomputation" (Copeland, 2002, p. 461): the former is a computation of "functions or numbers … with paper and pencil in a finite number of steps by a human clerk working effectively." 'Hypercomputation' is "the computation of functions or numbers that cannot be computed in the sense of Turing (1936)." If hypercomputable functions or numbers cannot be computed by a Turing Machine, can they be "computed" at all, and, if so, how? Copeland (2002) and Copeland and Sylvan (2000, Section 8.1, esp. pp. 190–191) cite the following possibilities, among others.

First, the constraint of data as *symbols* on paper could be relaxed. For example, Cleland's "mundane" hollandaise-sauce recipe that we looked at in the previous chapter does not take symbols as either input or output. Instead, the inputs are certain food ingredients, and the output is a certain food preparation. Indeed, any computer-controlled physical process – including robotics – seems to relax this symbolic constraint.

Second, the primitive operations of the "computation" might not be executable by a human *working alone*, in the way that Turing's 1936 paper described. Here, there seem to be at least two options: one is that the human might need help that can only be given by a machine capable of doing something that a human could not do even *in principle*. This might include a relaxation of the constraints about a finite number of steps or a finite amount of time, or working with what Copeland and Sylvan (2000, p. 190) call "newer physics." (See Section 11.5.)

The other is that the human might need help in the form of information that is *not pre-stored* on the tape: this might include allowing data to be supplied *during* the computation, rather than requiring it all to be pre-stored on the Turing Machine tape. This is what happens in "interactive" computing and in Turing's "oracle" machines. (See Sections 11.8 and 11.9.)

Copeland and Sylvan also identify two kinds of relativity: *"logical" relativity* concerns the use of non-classical logics, such as relevance logics (see Section 2.5.1). Copeland and Sylvan (2000) suggest that these might lead to hypercomputation. Perhaps; but it is certainly the case that *classical* computers can compute using relevance logics (Shapiro and Rapaport, 1987; Martins and Shapiro, 1988).

"Resource" relativity includes "relativity in procedures, methods or in the devices available for computing." This includes the "newer physics" and oracle machines just mentioned. It also includes analog computing. "Relativity in procedures" might include different basic operations or instructions (in much the same way that different geometric systems might go "beyond" straightedge and compass). Does such procedural relativity necessarily go beyond (or below) Turing computability? We'll look at this in more detail in Section 11.9.

Several basic questions need to be considered: Hilbert's original constraints (finiteness, etc.) seem to require "computation" to be *humanly possible* computation. So, *are* hypercomputers really alternative models of *humanly* effective procedures? (And does 'effective' need to mean "*humanly* effective"?) Are hypercomputers *counterexamples* to the Computability Thesis? Or are they just *other models* of Turing computation? Or are they models of a more general notion of "computation": i.e. nevertheless consistent with the Computability Thesis? How realistic are hypercomputers? Can they physically exist? Finally, is the mind or brain a hypercomputer (rather than a Turing Machine computer)?

In the rest of this chapter, we'll survey a few of these systems.[3]

11.5 "Newer Physics" Hypercomputers

> According to a 1992 paper, a computer operating in a Malament-Hogarth spacetime or in orbit around a rotating black hole could theoretically perform non-Turing computations. (http://en.wikipedia.org/wiki/Hypercomputation)

As we have seen, Turing's model of computation is based on what *humans* can do. Yet it is an *idealized* human whom Turing modeled, e.g. one that has no limits on *space* (recall that the tape is infinite, or at least arbitrarily large). Cleland (2004, p. 212) points out that, in that case, one could allow other idealizations, such as no limits on *speed* of computation. Copeland (1998, p. 150) agrees: "Neither Turing nor Post, in their descriptions of the devices we now call Turing machines, made much mention of time .... They listed the primitive operations that their devices perform ... but they made no mention of the *duration* of each primitive operation."

If we relax temporal restrictions that would limit humans, then we could devise a machine that could calculate each digit of a real number's decimal expansion in half the time of the previous digit's calculation. A "Zeus machine" is a Turing Machine that "accelerates" this way: each step is executed in half the time of the previous step (Boolos and Jeffrey, 1974). Thus, an *infinite* calculation, including the Halting Problem, could be computed in a *finite* amount of time. However, as Bertrand Russell (1936, p. 143) observed of a very similar example, although this is not *logically* impossible, it *is* "medically" impossible! And Scott Aaronson (2018, Slide 19) has observed that it is physically impossible for another reason:

> [O]nce you get down to the Planck time of 10^{-43} seconds, you'd need so much energy to run your computer that fast that, according to our best current theories, you'd exceed what's called the Schwarzschild radius, and your computer would collapse to a black hole. You don't want that to happen.

So, we might choose to ignore or reject Zeus machines on the grounds that they are "medically" and physically impossible. After all, no physical, and certainly no biological, device can really accelerate that way. But then, by parity of reasoning, should we reject ordinary Turing Machines on the grounds that they, too, are physically impossible, because, after all no (physical) device can really have an infinite tape or even an arbitrarily extendable tape? If so, and if an abstract Turing Machine is mathematically possible, then surely so is an (equally abstract) accelerating Turing Machine. That would make a Zeus machine at least as plausible as a Turing Machine.

But what about the physics of the actual world – relativity theory and quantum mechanics? The relativistic hypercomputer described in the epigraph seems far-fetched and certainly not practical. Here is what Aaronson (2018, Slides 18, 20) has to say about these:

> We can also base computers on that other great theory of the twentieth century, relativity! The idea here is simple: you start your computer working on some really hard problem, and

---

3 See the Online Resources for further reading on hypercomputation.

leave it on earth. Then you get on a spaceship and accelerate to close to the speed of light. When you get back to earth, billions of years have passed on Earth and all your friends are long dead, but at least you've got the answer to your computational problem. I don't know why more people don't try it!

So OK, how about the TIME TRAVEL COMPUTER! The idea here is that, by creating a loop in time – a so-called "closed timelike curve" – you could force the universe to solve some incredibly hard computational problem, just because that's the only way to avoid a Grandfather Paradox and keep the laws of physics consistent. It would be like if you went back in time, and you told Shakespeare what plays he was going to write, and then he wrote them, and then you knew what the plays were because he wrote them … like, DUDE.

As for quantum computation, the issue is whether it allows for the "computation" of *non*–Turing-computable functions or merely makes the computation of Turing-computable functions more *efficient*, perhaps by efficiently computing *NP* problems (Folger, 2016; Aaronson, 2018).[4]

11.6 Analog Recurrent Neural Networks

A slightly different model of hypercomputation is that of Hava T. Siegelmann (1995). She proposed a "*possibly* realizable" "analog shift map" or "analog recurrent neural network" – a "highly chaotic dynamical system … which has computational power beyond the Turing limit (super-Turing); it computes exactly like neural networks and analog machines" (p. 545, my italics).

Two questions to think about in trying to understand her proposal are (1) what, if anything, neural networks and analog computers might have in common, and (2) how neural networks are different from Turing Machines. As to (1), recall that Piccinini (2020b) thinks neural "computation" is distinct from both Turing- and analog computation. And as to (2), if neural-network computations are implemented on ordinary computers, whose behavior is completely analyzable in terms of Turing Machines, how would something that "computes exactly like neural networks" be a *hyper*computer? More importantly, Martin Davis (2004, pp. 8–9) shows how "the non-computability that Siegelmann gets from her neural nets is nothing more than the non-computability she has built into them."

11.7 Objections to Hypercomputation

Indeed, Davis (2004, 2006c) thinks that most of these hypercomputers are either wrong-headed or just plain silly, essentially likening them to the "garbage in/garbage out" principle, which says that if you allow for incorrect input, you should expect incorrect output. Similarly, according to Davis, if you allow for *non-computable* input to a hypercomputer, you should expect to be able to get *non-computable* output. Davis (as we will see) argues that all examples of hypercomputation involve non-computable input.

Along the same lines, Scott Aaronson (2012) argues against hypercomputation via a parallel argument that because Turing Machines can't toast bread, a toaster-enhanced Turing Machine that "allows bread as a possible input and includes toasting it as a primitive operation" would

---

4 See the Online Resources for further reading on Zeus machines, quantum hypercomputation, and relativistic computation.

be more powerful than a classic Turing Machine. (Recall Cleland's argument (Section 10.4): is a Turing Machine that can produce hollandaise sauce more powerful than a classic Turing Machine?)

11.8 Interactive Computation

11.8.1 "Internal" vs. "External" Inputs

Let's turn from physically impossible or unrealistic machines to ones that we actually deal with on a daily basis. Recall our discussion in Section 7.3.3 about whether an algorithm can have zero inputs. I suggested that a program to generate the decimal expansion of a real number might not require any explicit inputs. In Chapter 8, we saw Turing discuss just such algorithms. But do such algorithms really have no inputs? It might be better to say that there is an ambiguity in what counts as an input. After all, a program that generates the decimal expansion of a real number doesn't need any input *from the external world*, but – because any function considered a set of ordered *pairs* must have an input *in the sense of being the first member of each such pair* – there is always an "internal" input in *that* sense. A program that has no *external* inputs would still have "internal" inputs in the functional sense. "Interactive" computation concerns programs that *do* have external inputs.

11.8.2 Batch vs. Online Processing

Computing with *no* external inputs is sometimes called "batch" processing or "computational" programs. And computing *with* external inputs is sometimes called "online" processing or "reactive" programs. Soare, 2009, Section 1.3, p. 370, discusses the batch-online distinction; Amir Pnueli (2002, lecture1.pdf), defines a "computational program" as one that is "Run in order to produce a final result on termination," and he defines

> Reactive systems … [as those] whose role is to maintain an ongoing interaction with their environment rather than produce some final value upon termination. Typical examples of reactive systems are air traffic control system[s], programs controlling mechanical devices such as a train, a plane, or ongoing processes such as a nuclear reactor.

"Batch" or "computational" processing can be understood as the behavior of a Turing Machine (including a Universal Turing Machine):

> The classic models of computation are analogous to minds without bodies. For Turing's machine, a calculation begins with a problem on its tape, and ends with an answer there. … How the initial tape … is input, and how the final one is output, are questions neither asked nor answered. These theories conform to the practice of batch computing. (Wadler, 1997, pp. 240–241)

"Online" or "reactive" processing has several varieties, all of which involve interaction with the external world – the world outside of the computer: A computer might have access to a (changeable) "offline" database: it might interact with the external world via sensors or effectors (or both, of course – recall Shapiro's observations in Section 3.10); it might interact with another computer; it might interact with a human – or any combination of these.[5]

---

5 See the Online Resources for further reading on such systems.

Arguably, even "batch-processing" Turing Machines have perceptors and effectors in the sense of having a read-write head. But these are really internal to the Turing Machine and don't necessarily "reach out" to the external world. However,

> a computer linked with a human mind is a more powerful tool than an unassisted human mind. One hope of many computer scientists is to demonstrate … that the computer/human team can actually accomplish far more than a human alone. (Forsythe, 1967a, p. 3, col. 2).

One might also ask whether such a "computer/human team" could accomplish far more than a *computer* alone, say by *interacting* with the computer while it is computing (Lohr, 2013; Steed, 2013):

> [H]umans are fundamentally social animals. This insures our survival: organisms working together can do so much more than organisms working apart or in parallel. The greatest challenge for A.I. is … the lack of attention to teaming intelligence that would allow the pairing of humans' remarkable predictive powers with A.I.'s superior bottom-up analysis of data. (Vera, 2018)

Here is the rest of what Wadler has to say:

> Today, computing scientists face their own version of the mind-body problem: how can virtual software interact with the real world? In the beginning, we merely wanted computers to extend our minds: to calculate trajectories, to sum finances, and to recall addresses. But as time passed, we also wanted computers to extend our bodies: to guide missiles, to link telephones, and to proffer menus. … Eventually, interactive models of computation emerged, analogous to minds in bodies. … A single input at initiation and a single output at termination are now superseded by multiple inputs and outputs distributed in time and space. These theories conform to the practice of interactive computing. Interaction is the mind-body problem of computing.[6] (Wadler, 1997, pp. 240–241)

Weizenbaum (1976, Ch. 5, p. 135) interestingly distinguishes between "computers" and "robots," where the latter (but not the former) "have perceptors … and effectors." So, are *interactive* computers – "robots," in Weizenbaum's sense – *hyper*computers?

11.8.3 Peter Wegner: Interaction Is Not Turing-Computable

Peter Wegner (1997) argues that "interaction machines" are strictly more powerful than Turing Machines. Wegner (1995, p. 45) identifies interaction machines with oracle machines (which we'll look at in Section 11.9) and with "modeling reactive processes" (citing Pneuli).

Interaction relaxes one of the "constraints" on Turing's analysis of computation: that of being **Isolated[:]** Computation is self-contained. No oracle is consulted, and nobody interferes with the computation either during a computation step or in between steps. The whole computation of the algorithm is determined by the initial state" (Gurevich, 2012, p. 4). This certainly suggests that interactive computation is not Turing computation. On the other hand, it could also be interpreted to mean merely that computation must be "mechanical" or "automatic," and surely this

---

6 On the mind-body problem of *philosophy*, see Sections 2.7 and 12.6.

could include the "mechanical" or "automatic" use of input from an external source (including an oracle).[7]

For example, Prasse and Rittgen (1998, p. 359) consider a program such as the following:

```
let b be a Boolean variable;
let x be an integer variable;
begin
    b ← true;
    while b = true do
        input(x);
        output(x²);
        print("Should I continue? Enter true or false:");
        input(b)
end
```

They say of a program such as this,

> Neglecting input/output, each iteration can be interpreted as a computation performed by a Turing machine. However, the influence of the (unknown) user input on the control flow makes it hard to determine what overall function is computed by this procedure (or if the procedure can be seen as a computation at all). ... The input will be determined only at run-time. The overall function is derived by integrating the user into the computation, thus closing the system. It is evident that Turing machines cannot model this behavior directly due to the missing input/output operations. Therefore, we need models that take into account inputs and outputs at run-time to describe computer systems adequately. (Prasse and Rittgen, 1998, p. 359)

Question for the Reader: We could easily write a Turing Machine program that would be a version of this while-loop. Consider such a Turing Machine with a tape that is initialized with all of the input (a sequence of *b*s and *x*s, encoded appropriately). This Turing Machine clearly is (or executes) an algorithm in the classical sense. Now consider a Turing Machine with the same program (the same machine table), but with an initially blank tape and a user who inscribes appropriate *b*s and *x*s on the tape *just before each step of the program is executed* (so the Turing Machine is not "waiting" for user input, but the input is inscribed on the tape *just in time*).

Is there a mathematical difference between these two Turing Machines? Is there anything in Turing, 1936 that rules this out?

Interestingly, Prasse and Rittgen's point is that this does *not* violate the Computability Thesis, despite Wegner's interpretation:

> Interaction machines are defined as Turing machines with input and output. Therefore, their internal behavior and expressiveness do not differ from that of equivalent Turing machines. Though Wegner leaves open the question of how the input/output mechanism works, it can be assumed that input and output involve only data transport, without any computational capabilities. Therefore, the interaction machine itself does not possess

---
7 See the Online Resources for further reading on "isolation."

greater computational power than a Turing machine. However, through communication, the computational capabilities of other machines can be utilized. Interaction can then be interpreted as a (subroutine) call.
(Prasse and Rittgen, 1998, p. 361)

Wegner and Goldin disagree and suggest that Turing disagreed, too: they discuss "Turing's assertion [in Turing, 1936] that TMs have limited power and that choice machines, which extend TMs to interactive computation, represent a distinct form of computing not modeled by TMs" (Wegner and Goldin, 2006, p. 28, col. 1). So, what is a "choice" machine, and how does it differ from a Turing Machine?

Along with his *a*(utomatic)-machines (now called 'Turing Machines'), Turing (1936) introduced *c*(hoice) machines. As we saw in Section 8.9.1, *c*-machines are Turing Machines that allow for "ambiguous configurations." Recall from Section 8.8.1 that a "configuration" is a line number together with the currently read symbol; in other words, it is the "condition" part of the condition-action expression of a Turing Machine instruction. So, an "ambiguous configuration" is a "condition" with more than one possible "action." In a *c*-machine, "an external operator" makes an "arbitrary choice" for the next action (Turing, 1936, p. 232; see our Section 8.9.1).

However,

> The 'Choice Machines' from Turing's paper are just what we now call nondeterministic Turing machines. In … [Turing, 1936, p. 252, footnote ‡], Turing showed that the choice machines can be simulated by traditional Turing machines, contradicting Wegner and Goldin's claim that Turing asserted his machines have limited power. (Fortnow, 2006).

Thus, *c*-machines (or *non*-deterministic Turing Machines) are no more powerful than *deterministic* Turing Machines, so they don't provide counterexamples to the Computability Thesis.

Fortnow (2006) notes that there is a difference between *modeling* and *simulating*. Neither of these terms have universally accepted definitions, but we can say that one way for system S_1 to *simulate* system S_2 is simply for their input-output behaviors to match (in other words, for S_1 and S_2 to compute *the same function*, although possibly in different ways). And one way for S_1 to *model* (or *emulate*) S_2 is for their *internal* behaviors to match as well: i.e. for S_1 to simulate S_2 *and* for their *algorithms* to match (in other words, for S_1 and S_2 to compute the same function *in the same way*). (We'll discuss simulation in more detail in Section 14.2.) If the only way for a Turing Machine to simulate a *c*-machine is by pre-storing the possible inputs, it is arguably not *modeling* it. The non-interactive Turing Machine with pre-stored input (what Soare (2009, Sections 1.3, 9) notes is essentially a "batch" processor) can *simulate* the interactive system even if it does not *model* it. Perhaps this is Wegner and Goldin's point. Yet another pair of terms can illuminate the relationship: an interactive Turing Machine may be *extensionally* equivalent to one with all input pre-stored, but it is not *intensionally* equivalent (Section 3.4.2).[8]

Fortnow (2006) goes on to point out that Turing Machines also only simulate but don't model many other kinds of computation, such as "random-access memory, machines that alter their own programs, multiple processors, nondeterministic, probabilistic or quantum computation." However, "Everything computable by these and other seemingly more powerful models can also be computed by the lowly one-tape Turing machine. That is the beauty of the Church-Turing thesis." The Church-Turing Computability Thesis "doesn't try to explain *how* computation can happen, just that *when* computation happens it must happen in a way computable by a Turing machine" (Fortnow, 2006, my italics).

---

8 Stuart C. Shapiro, personal communication.

It is important to keep in mind that when there are two input-output–equivalent ways to do something, it still might be the case that one of those ways has an advantage over the other for certain purposes. For example, no one would want to program an airline reservation system using the programming language of a Turing Machine! Rather, a high-level language (Java?, C++?, etc.) would be much more efficient. Similarly, it is easier to prove theorems *about* an axiomatic system of logic that has only *one* rule of inference (usually modus ponens), but it is easier to prove theorems *in* a natural-deduction system of logic, which has *many* rules of inference (usually at least two for each logical connective), even if both systems are logically equivalent. (See Section 15.4 for more on the difference between axiomatic and natural-deduction systems of logic.)

11.8.4 Can Interaction Be Simulated by a Non-Interactive Turing Machine?

11.8.4.1 The Power of Interaction

Nevertheless, interaction is indeed ubiquitous and powerful. Consider, for example, the following observation by Donald Knuth:

> I can design a program that never crashes if I don't give the user any options. And if I allow the user to choose from only a small number of options, limited to things that appear on a menu, I can be sure that nothing anomalous will happen, because each option can be foreseen in advance and its effects can be checked. But if I give the user the ability to write *programs* that will combine with my own program, all hell might break loose. (Knuth, 2001, pp. 189–190)

That is, a program does not have to pre-store all possible inputs. Here is how Herbert Simon put it, commenting on the objection to AI that …

> … "computers can only do what you program them to do." That is correct. The behavior of a computer at any specific moment is completely determined by the contents of its memory and the symbols that are input to it at that moment. This does not mean that the programmer must anticipate and prescribe in the program the precise course of its behavior. … [W]hat actions actually transpire depends on the successive states of the machine *and its inputs at each stage of the process* – neither of which need be envisioned in advance either by the programmer or by the machine. (Simon, 1977, p. 1187, my italics)

And those inputs are a function of the computer's interactions with the external world!

(The objection to AI that Simon quoted is a version of the "Lovelace objection," which we'll examine in more detail in Section 18.4.)

11.8.4.2 Simulating a *Halting* Interaction Machine

Let's consider Fortnow's position first: if an interaction machine halts, then it can be simulated by a Universal Turing Machine by pre-storing all of its inputs. Here's why.

In the theory of Turing computation, there is a theorem called the *S-m-n* Theorem. Before stating it, let me introduce some notation: first, recall from Section 8.12 that Turing Machines are enumerable – they can be counted. (In fact, they are "recursively" – i.e. computably – enumerable.) So, let 'ϕ_n' represent the nth Turing Machine (in some numbering scheme for Turing Machines), and let i represent its input. Here is the *S-m-n* Theorem:

$$(\exists \text{ Turing Machine } s)(\forall x, y, z \in \mathbb{N})[\phi_x(y, z) = \phi_{s(x,y)}(z)]$$

This says that there exists a Turing Machine s (i.e. a function s that is computable by a Turing Machine) that has the following property: for any three natural numbers x, y, z, the following is true: the xth Turing Machine, when given *both y and z* as inputs, produces the same output that the $s(x, y)$th Turing Machine does when given *only z* as input. But what is $s(x, y)$? It is a Turing Machine that already has x and y "pre-stored" on its tape!

Here is another way to say this: first, enumerate all of the Turing Machines, and let ϕ_x be the xth Turing Machine. Suppose it takes two inputs: y and z (another way to say this is that its *single* input is the ordered pair $\langle y, z \rangle$). Then there exists *another* Turing Machine $\phi_{s(x,y)}$ – i.e. we can find another Turing Machine that depends on ϕ_x's (*two*) inputs (and the dependence is itself a Turing-computable function, s) – such that $\phi_{s(x,y)}$ is input-output–equivalent to ϕ_x when y is fixed, and which is such that $\phi_{s(x,y)}$ is a Turing Machine with y (i.e. with part of ϕ_x's input) *stored internally as data*.

Here is an example of these two kinds of Turing Machines, with a program for $\phi_x(y, z)$ on the left and a program for $\phi_{s(x,y)}(z)$ on the right:

Algorithm x:
begin
    input(y, z);
        {y is input from the external world}
    output ← process(y, z);
    print(*output*)
end.

Algorithm s(x,y):
begin
    constant ← y;
        {y is pre-stored in the program}
    input(z);
    output ← process$(constant, z)$;
    print(*output*)
end.

In other words, any Turing Machine that takes input y from the external world (or as user input) can be *simulated* by a *different* Turing Machine that has y pre-stored on its tape. That is, data can be stored algorithmically in programs; the data need not be input from the external world.[9] The Turing Machine that interacts with the external world can be simulated by a different Turing Machine that doesn't. So, an interaction machine that halts is no more powerful than an ordinary, non-interacting Turing Machine.

But keep in mind the comment at the end of Section 11.8.3 about relative advantages: the interaction machine might be more useful in practice; the non-interacting machine might be easier to prove theorems about.[10]

So, any *interactive* program that halts could, in principle, be shown to be logically equivalent to a *non*-interactive program. That is, any interactive program that halts can be simulated by an "ordinary" Turing Machine by pre-storing the external input:

An interactive system is a system that interacts with the environment via its input/output behavior. The environment of the system is generally identified with the components which do not belong to it. Therefore, an interactive system can be referred to as an open system because it depends on external information to fulfil its task. *If the external resources are integrated into the system, the system no longer interacts with the environment and we get a new, closed system.* So, the difference between open and closed systems 'lies in the eye of

---

9 In our statement of the *S-m-n* Theorem, the variable z is also being input from the external world, but it is only there for technical reasons required for the proof of the theorem in the most general case. In practice, z can also be pre-stored on the tape or even omitted.
10 See the Online Resources for further reading on the *S-m-n* Theorem.

the beholder.' (Prasse and Rittgen, 1998, p. 359, col. 1, my italics; Teuscher and Sipper, 2002, p. 24, make a similar observation)

The catch is that you need to know "in advance" what the external input is going to be. Halting is important here, because once the interactive machine halts, all of its inputs are known and can then be pre-stored on the simulating machine's tape. But the *S-m-n* Theorem does say that once you know what that input is, you need only an ordinary Turing Machine, not an interactive hyper-computer.

Philosophical Digression: "Solipsism," as defined by Bertrand Russell (1927, p. 398), is "the view that from the events which I experience there is no valid method of inferring the character, or even the existence, of events which I do not experience." It is occasionally parodied as the view that *I* am the only thing that exists; *you* are all figments of my imagination. Note that *you* cannot make the same claim, because, after all, if solipsism is true, then you don't exist! There's a story that at a lecture Bertrand Russell once gave on solipsism, someone in the audience commented that it was such a good theory, why didn't more people believe it? Actually, solipsism is not really the claim that only I exist. Rather, it is the claim that I live in a world of my own, completely cut off from the external world, and so do you. This is reminiscent of the philosopher Gottfried Leibniz's "monads" (Leibniz, 1714, https://en.wikipedia.org/wiki/Monadology), but that's beyond our present scope.

"*Methodological* solipsism" is a view in the philosophy of mind and of cognitive science that says that to understand the "psychology" of a cognitive agent, it is not necessary to specify the details of the external world in which the agent is situated and that impinge on the agent's sense organs. This is not to deny that there *is* such a world or such sensory input – hence the qualifier 'methodological.' Rather, it is to acknowledge (or assume) that all that is of interest psychologically or cognitively can be studied from the surface inward, so to speak (Putnam, 1975; Fodor, 1980). That is, cognition can be studied by acting as if the brain (or the mind) only does "batch processing." (We'll come back to this in Sections 16.11 and 18.8.2.)

Consider an AI system that can understand and generate natural-language and that gets its input from the external world (i.e. from a user). The point of methodological solipsism is that we could *simulate* this by building in the input (assuming a finite input). Indeed, this can be done for *any* partial recursive function, according to the *S-m-n* Theorem. If we understood methodological solipsism as the *S-m-n* Theorem, we would have an argument for methodological solipsism from the theory of computation!

11.8.4.3 Simulating a *Non-Halting* Interaction Machine

But suppose our interaction machine does *not* halt – not because of a pernicious infinite loop, but (say) because it is running an operating system or an automated teller machine; such machines only halt when they are broken or being repaired:

> **Interactive computing.** Many systems, such as operating systems, Web servers, and the Internet itself, are designed to run indefinitely and not halt. Halting is an abnormal event for these systems. The traditional definition of computation is tied to algorithms, which halt. Execution sequences of machines running indefinitely *seem to* violate the definition. (Denning, 2010, p. 5, my italics)

But even Turing's original Turing Machines didn't halt: they computed infinite decimals, after all! The central idea behind the Halting Problem is to find an algorithm that distinguishes between programs that halt and those that don't. Whether halting is a Good Thing is independent of that. Of course, any stage in the process is a finite (i.e. halting) computation. (Recall Prasse and Rittgen's first sentence, quoted on p. 326.) Even Turing's computation of reals is a (non-halting) sequence of halting computations of successive terms of the decimal expansion.

There are two non-halting cases to consider. In the first case, the unending input stream is a number computable by a Universal Turing Machine. In this case, the interaction machine can *also* be simulated by a Universal Turing Machine. Hence, interaction in this case also does not go beyond the Computability Thesis, because – being computable – the inputs are "knowable" – i.e. computable – "in advance." So, instead of pre-storing the individual *inputs*, we can simply pre-store a copy of the *program* that generates those inputs.

In the second case, suppose not only that the Turing Machine does not halt but also that the unending input stream is *not* computable by a Turing Machine. Then the interaction machine would seem to be a hypercomputer. It is only this situation – where the input is *non*-computable (hence, *not* knowable in advance, even in principle) – that we have hypercomputation (recall Martin Davis's objection, Section 11.7). But is it? Or is this just an oracle machine? We will see in Section 11.9 why it is not obvious that oracle-machine computation is "hyper" in any interesting sense, either.

Why might such a non-halting, non-computable interaction machine be a hypercomputer? Its input stream might be random. Truly random numbers are not computable (Church, 1940; Chaitin, 2006). But de Leeuw et al., 1956 showed that "the computing power of Turing machines provided with a random number generator … could compute only functions that are already computable by ordinary Turing machines" (Davis, 2004, p. 14). Even if not random, the input stream of such an interaction machine might be non-computable. According to Copeland and Sylvan (1999, p. 51), "A coupled Turing machine is the result of coupling a Turing machine to its environment via one or more input channels. Each channel supplies a stream of symbols to the tape as the machine operates." They give a simple proof (p. 52) that there is a coupled Turing Machine "that cannot be simulated by the universal Turing machine." However, the proof involves an oracle that supplies a non-Turing computable real number, so their example falls prey to Davis's objection.[11]

11.9 Oracle Computation

> Let us suppose that we are supplied with some unspecified means of solving number-theoretic problems; a kind of oracle as it were. We shall not go any further into the nature of this oracle apart from saying that it cannot be a machine. With the help of the oracle we could form a new kind of machine (call them *o*-machines), having as one of its fundamental processes that of solving a given number-theoretic problem. More definitely these machines are to behave in this way. The moves of the machine are determined as usual by a table except in the case of moves from a certain internal configuration *o*. If the machine is in the internal configuration *o* and if the sequence of symbols marked with *l* is then the well-formed formula **A**, then the machine goes into the internal configuration **p** or **t** according as it is or is not true that **A** is dual. The decision as to which is the case is referred to the oracle.
>
> —Alan Turing (1939, pp. 172–173)[12]

---

11 See the Online Resources for further reading on interactive computing.
12 Here is Turing's explanation of some of the technical terms in this passage: "Every number-theoretic theorem is equivalent to a statement of the form '**A(n)** is convertible to 2 for every W.F.F. **n** representing a positive integer,'

An *o*(racle)-machine is a Turing Machine that can "interrogate an '*oracle*' (*external database*) during the computation" (Soare, 2009, Section 1.3, p. 370) to determine its action (including its next configuration). Moreover, the database "cannot be a machine" (Turing, 1939, p. 173). If it were a "machine" – presumably an *a*-machine – then its behavior would be computable, and vice versa.

Historical Digression: Oracle machines were first described in Turing's Ph.D. dissertation at Princeton, which was completed in 1938 and which he began *after* his classic 1936 paper was published; Church was his dissertation advisor. His dissertation can be read online at http://www.dcc.fc.up.pt/~acm/turing-phd.pdf; it was published as Turing, 1939, from which this section's epigraph was taken.

However, if the choice made by the oracle *were* computable, then *c*-machines could be considered a special case of *o*-machines. If interaction is best modeled by an *oracle* machine, then Wegner and Goldin are incorrect about *choice* machines being the ones that "extend" Turing Machines "to interactive computing" (see Section 11.8.3). In fact, according to Martin Davis (1958, pp. 20–24), Turing Machines "deal … only with closed computations. However, it is easy to imagine a machine that halts a computation at various times and requests additional information." He then discusses relative computation and *o*-machines in the form of Turing Machines that can ask whether a given integer is an element of a given set, observing that "This provides a Turing machine with a means of communication with 'the external world.' "

The external database is a "black box" that could contain the answers to questions that are not computable by an ordinary Turing *a*-machine. If a function g is computable by an *o*-machine whose oracle outputs the value of a (non–Turing-computable) function f, then it is said that g is computable *relative to f*.

The computer scientist Solomon Feferman (1992, p. 340, footnote 8) said this: "Several people have suggested to me that *interactive computation* exemplifies Turing's 'oracle' in practice. While I agree that the comparison is apt, I don't see how to state the relationship more precisely." However, Bertil Ekdahl (1999) has a nice example that illustrates how interactive computing is modeled by *o*-machines and relative computability. The essence of the example considers a simplified version of an airline-reservation program. Such a program is a standard example of the kind of interactive program that Wegner claims is not Turing computable, yet it is not obviously an *o*-machine, because it does not obviously ask an oracle for the solution to a non-computable problem. Suppose our simplified reservation program is this:

```
while true do
  begin
    input(passenger, destination);
    output(ticket(passenger, destination))
  end
```

Ekdahl observes that although writing the passenger and destination information on the input tape is computable "and can equally well be done by another Turing machine," when our reservation program then " 'asks' for two new" inputs, "**which** [inputs are] going to [be written] on the tape *is not a recursive process.* … So, the input of [passenger and destination] can be regarded as a question

---

A being a W.F.F. determined by the theorem; the property of **A** here asserted will be described briefly as '**A** is dual' " (p. 170). "Convertibility" is an equivalence relation in Church's lambda calculus.

to an *oracle*. An oracle answers questions known in advance but the answers are not possible to reckon in advance" (Ekdahl, 1999, Section 3, pp. 262–263, italics in original; my boldface). Here, the "oracle" is the reservations agent!

Conceivably, the "computation-like process" performed by the physics-challenging machines described in Section 11.5 can also be simulated (if not modeled) by oracle machines. So, the hyper-computation question seems to come down to whether *o*-machines violate the Computability Thesis. Let's look at them a bit more closely.

Feferman (1992, p. 321) notes that *o*-machines can be "generalized to that of a *B*-machine for any set *B*." Instead of Turing Machines, Feferman discusses the logically equivalent register machines of Shepherdson and Sturgis, 1963 (Section 9.3.1). Briefly, a register machine consists of "registers" (storage units), each of which can contain a natural number. In Feferman's version (1992, p. 316), for each register r_i, the machine has four basic operations:

1. $r_i \leftarrow 0$ (i.e. set r_i to 0)
2. $r_i \leftarrow r_i + 1$ (i.e. increment r_i)
3. **if** $r_i \neq 0$, **then** $r_i \leftarrow r_i - 1$ (i.e. decrement non-0 r_i)
4. **if** $r_i = 0$, **then** go to instruction j **else** go to instruction k

To turn this into a *B*-machine, we add one more kind of operation (p. 321):

5. **if** $r_k \in B$, **then** $r_i \leftarrow 1$ **else** $r_i \leftarrow 0$

In other words, a *B*-machine is an *o*-machine: a Turing Machine together with a set *B* that plays the role of the oracle. The machine's program can consult oracle *B* to see if it contains some value r_k. The fifth operation puts a 1 or a 0 into register r_i if the oracle tells it whether the value $r_k \in B$.

Essentially, this *adds* primitive operations to those of a Turing Machine (or a register machine) (Dean, 2020, Def 3.8). If these operations can be simulated by the standard primitive operations of the Turing Machine, then we haven't increased its power, only its expressivity, essentially by the use of named subroutines. (Recall Prasse and Rittgen's observation that "interaction can … be interpreted as a (subroutine) call.") Turing's *o*-machines are of this type; the call to a (possibly non-computable) oracle is simply a call to a (possibly non-computable) subroutine. So, as Prasse and Rittgen say, "the machine *itself*" is just a Turing Machine. And as Davis would say, if a non-computable input is encoded in *B*, then a non-computable output can be encoded on its tape. If *B* contains the answers to problems not solvable by the Turing Machine, then of course we have increased the machine's power.

But does that provide a counterexample to the Computability Thesis?

In fact, Feferman (1992, pp. 339–340) observes that the "built-in functions" of "actual computers" (e.g. the primitive recursive functions or the primitive operations of a Turing Machine) are "given by a 'black box' – which is just another name for an 'oracle' – and a program to compute a function *f* from one or more of these" built-in functions "is really an algorithm for computation of *f* relative to" those built-in functions.

To say that a set *A* is Turing computable from (or "Turing reducible to") a set *B* (written: $A \leq_t B$) is to say that $x \in A$ iff the *B*-machine outputs 1 when its input is *x* (where output 1 means "yes, $x \in A$"). Davis (2006b, p. 1218) notes that where *A* and *B* are sets of natural numbers, if $A \leq_t B$, and "if *B* is itself a computable set, then nothing new happens; in such a case $A \leq_t B$ just means *A* is computable. But if *B* is non-computable, then interesting things happen."[13] According to Davis (2006c), one of the *un*interesting things, of course, is that *A* will then turn out to be non-computable. The *interesting* things have to do with "degrees" of *non*-computability: "can one non-computable set be more non-computable than another?" (Davis, 2006b, p. 1218).

---

13 Knuth's expression for a similar situation is "all hell might break loose" (Knuth, 2001, pp. 189–190).

What does that mean? Recall that Gödel's Incompleteness Theorem shows that there is a true statement of arithmetic that cannot be proved from Peano's axioms. What if we add that statement as a new axiom? Then we can construct a different true statement of arithmetic that cannot be proved from this new set of axioms. And we can continue in this matter, constructing ever more powerful theories of arithmetic with no end. Turing's dissertation and invention of oracles essentially applied the same kind of logic to computability. Consequently, Feferman (1992, p. 321) observes that

> the arguments for the Church-Turing Thesis lead one strongly to accept a relativized version: $(C\text{-}T)^r$ [a set] A is effectively computable from [a set] B if (and only if) $A \leq_T B$.

Feferman then says that "Turing reducibility gives the most general concept of relative effective computability" (p. 321). And here is Feferman on the crucial matter:

> Uniform global recursion provides a much more realistic picture of computing over finite data structures than the absolute computability picture, for finite data bases are constantly being updated. As examples, we may consider … airline reservation systems. (Feferman, 1992, p. 342)

He does, however, go on to say that "while notions of relativized (as compared to absolute) computability theory are essentially involved in actual hardware and software design, the bulk of methods and results of recursion theory have so far proved to be irrelevant to practice" (Feferman, 1992, p. 343). That certainly is congenial to Wegner's complaints.

On the other hand, Feferman (1992, p. 315) also claims that "notions of relative (rather than absolute) computability" (i.e. notions based on Turing's o-machines rather than on his a-machines) have "primary significance for practice" and that these relative notions are to be understood as "generalization[s] … of computability [and "of the Church-Turing Thesis"] to arbitrary structures." So this seems to fly in the face of Wegner's claims that interaction is something *new* while agreeing with the substance of his claims that interaction is more central to modern computing than Turing Machines are.

Soare agrees:

> Almost all the results in theoretical computability use relative reducibility and o-machines rather than a-machines and most computing processes in the real world are potentially online or interactive. Therefore, we argue that Turing o-machines, relative computability, and online computing are the most important concepts in the subject, more so than Turing a-machines and standard computable functions since they are special cases of the former and are presented first only for pedagogical clarity to beginning students. (Soare, 2009, Abstract, p. 368)

This is an interesting passage, because it could be interpreted by hypercomputation advocates as supporting their position and by anti-hypercomputationalists as supporting theirs! In fact, a later comment in the same paper suggests the pro-hypercomputational reading:

> The original implementations of computing devices were generally offline devices such as calculators or batch processing devices. However, in recent years the implementations have been increasingly online computing devices which can access or interact with some external database or other device. The Turing o-machine is a better model to study them because the Turing a-machine lacks this online capacity. (Soare, 2009, Section 9, p. 387)

He also says (referring to Turing, 1939 and Post, 1943),

> The theory of relative computability developed by Turing and Post and the *o*-machines provide a precise mathematical framework for database [or interactive] or online computing just as Turing *a*-machines provide one for offline computing processes such as batch processing. (Soare, 2009, Section 1.3, pp. 370–371).

And he notes that oracles can model both client-server interaction as well as communication with the Web. However, the interesting point is that all of these are *extensions* of Turing Machines, not entirely new notions.

More importantly, Soare does not disparage, object to, or try to "refute" the Computability Thesis; rather, he celebrates it (Soare, 2009, Section 12). This certainly suggests that some of the things that Copeland and Wegner say about hypercomputation are a bit hyperbolic; it suggests that both the kind of hypercomputation that takes non-computable input (supplied by an oracle) to produce non-computable output as well as the kind that is interactive are both well-studied and simple extensions of classical computation theory.

Soare's basic point on this topic seems to be this:

> **Conclusion 14.3** The subject is primarily about incomputable objects not computable ones, and has been since the 1930's. The single most important concept is that of relative computability to relate incomputable objects. (Soare, 2009, Section 14, p. 395)

> Turing's oracle machine was developed by Post into Turing reducibility .... It is the most important concept in computability theory. Today, the notion of a local machine interacting with a remote database or remote machine is central to practical computing. (Soare, 2012, p. 3290)

This is certainly in the spirit of hypercomputation without denigrating the Computability Thesis.[14]

11.10 Trial-and-Error Computation

11.10.1 Introduction

There is one more candidate for hypercomputation that is worth looking at for its intrinsic interest. It goes under many names: "trial-and-error computation," "inductive inference," "Putnam-Gold machines," and "limit computation." Here, the "constraint" that is relaxed is that we change our interpretation of what counts as the output of the "computation."

Here is how Putnam introduced "trial and error predicates." First, a "predicate" can be thought of as a Boolean-valued function. Next, as in Section 7.6.2, we'll let the notation $\bar{x}$ represent an *n*-tuple of variables $x_1, \ldots, x_n$, for some *n*. Then (paraphrasing Putnam, 1965, p. 49) a predicate P is a *trial and error predicate* $=_{def}$ there is a computable function f such that for every $\bar{x}$,

$$P(\bar{x}) = 1 \quad \text{iff} \quad \lim_{y \to \infty} f(\bar{x}, y) = 1,$$

and

$$P(\bar{x}) = 0 \quad \text{iff} \quad \lim_{y \to \infty} f(\bar{x}, y) = 0,$$

---

14 See the Online Resources for further reading on oracle machines.

where

$$\lim_{y \to \infty} f(\overline{x}, y) = k \ =_{def} \ \exists w \forall z [z \geq w \supset f(\overline{x}, z) = k]$$

Function f takes as input $n + 1$ natural numbers (n xs plus one y), and it outputs a natural number. Each value of y ($y = 0,1,2,\ldots$) will, in general, yield a different value for f, but at some point (at w, in fact), no matter how large y gets, f will remain constant with value k. In other words, no matter what initial value (or values) the function f takes, the predicate P is true (or false) iff, in the "limit" (i.e. at w or "beyond"), the function $f = 1$ (or $f = 0$). That is, "the eventual value of" f is 1 or 0 (Welch, 2007, p. 770).

Putnam "modifies" the notion of Turing computability …

> … by (1) allowing the procedure to "change its mind" any finite number of times (in terms of Turing Machines: we visualize the machine as being given an integer (or an n-tuple of integers) as input. The machine then "prints out" a finite sequence of "yesses" and "nos." The *last* "yes" or "no" is always to be the correct answer.); and (2) we give up the requirement that it be possible to tell (effectively) if the computation has terminated[.] I.e. if the machine has most recently printed "yes," then we know that the integer put in as input must be in the set *unless the machine is going to change its mind*; but we have no procedure for telling whether the machine will change its mind or not.
>
> The sets for which there exist decision procedures in this widened sense are decidable by "empirical" means – for, if we always "posit" that the most recently generated answer is correct, we will make a finite number of mistakes, but we will eventually get the correct answer. (Note, however, that even if we have gotten to the correct answer (the end of the finite sequence) we are never *sure* that we have the correct answer.) (Putnam, 1965, p. 49)

In general, a trial-and-error machine is a Turing Machine with input i that outputs a sequence of responses such that it is the *last* output that is "the" desired output of the machine (rather than the first, or only, output). But you don't allow any way to tell effectively if you've actually achieved the desired output: i.e. if the machine has really halted. The philosopher and psychologist William James once said, in a very different context, that …

> … the faith that truth exists, and that our minds can find it, may be held in two ways. We may talk of the *empiricist* way and of the *absolutist* way of believing in truth. The absolutists in this matter say that we not only can attain to knowing truth, but we can *know when* we have attained to knowing it; whilst the empiricists think that although we may attain it, we cannot infallibly know when. To *know* is one thing, and to know for certain *that* we know is another. (James, 1897, Section V, p. 465)

To paraphrase James,

> The faith that a problem has a computable (or algorithmic) solution exists, and that our computers can find it, may be held in two ways. We may talk of the trial-and-error way and of the Turing-algorithmic way of solving a problem. The Turing algorithmists in this matter say that we (or Turing Machines) not only can solve computable problems, but we can know when we (or they) have solved them; while the trial-and-error hypercomputationalists think that although we (or our computers) may solve them, we cannot infallibly know when. For a computer to produce a solution is one thing, and for us to know for certain that it has done so is another.

Recall from Section 8.9.3 that Turing called the marks printed by a Turing Machine that were *not* to be taken as output "symbols of the second kind," used only for bookkeeping. Peter Kugel (1986) takes up this distinction:

> We distinguish an output from a result. An output is anything M ["an idealized general-purpose computing machine"] prints, whereas a result is a selection, from among the things it prints, that we agree to pay attention to. … The difference between a computing procedure and a trial and error procedure is this[:] When we run M_p [M running under program p] as a computing procedure, we count its *first* output as its result. When we run it as a trial and error procedure, we count its *last* output as its result. (Kugel, 1986, pp. 139–140).

In a similar vein, Kugel (2002) notes that a distinction can be made between a Turing *Machine* and Turing *machinery*. Sub-Turing computation, although not *requiring* all the power of a Turing *Machine*, can be accomplished using Turing *machinery*. As Hintikka and Mutanen (1997, p. 175) put it, "there is more than one sense in which the same idealized hardware [i.e. Turing *machinery*] can be used to compute a function." (Here, 'compute' does not refer to Turing computation, because trial-and-error computability "is wider than recursivity.")

In a Turing *Machine*, the *first* output is the result of its computation. But there is nothing preventing the use of Turing *machinery* and taking the *last* output of its operation as its result. You can't say that the operation of such Turing *machinery* is computation if you accept the Computability Thesis, which identifies computation with the operation of a Turing Machine. But if a trial-and-error machine could do super-Turing "computation," then it would be a hypercomputer that uses Turing *machinery* (and would not require "newer physics").

Recall our discussion of the Halting Problem. In Section 7.7.1, we contrasted two alleged algorithms for determining whether a program C halts on input i:

Algorithm $A_H^1(C, i)$:
begin
  if $C(i)$ halts
    then output 'halts'
    else output 'loops'
end.

Algorithm $A_H^2(C, i)$:
begin
  output 'loops'; {i.e. make an initial *guess* that C loops}
  if $C(i)$ halts
    then output 'halts'; {i.e. *revise* your guess}
end.

Algorithm A_H^1 *can* be converted to the self-referential A_H^{1*} and thereby used to show that the Halting Problem is not Turing computable (Step 1 of our proof sketch). But A_H^2 could *not* be so converted. It is an example of a trial-and-error procedure: it makes an initial guess about the desired output and then keeps running program C on a number of "trials." If the trials produce "errors" or don't come up with a desired response, then continue to run more trials.

As Hintikka and Mutanen (1997, p. 181) note, the Halting Problem algorithm in its trial-and-error form is not computable, "even though it is obviously mechanically determined in a perfectly natural sense." They also note that this "perfectly natural sense" is part of the informal notion of computation that the Computability Thesis asserts is identical to Turing computation, and hence they conclude that the Computability Thesis "is not valid" (p. 180). Actually, they're a bit more cautious, claiming that the informal notion is "ambiguous": "We doubt that our pretheoretical ideas of mechanical calculability are so sharp as to allow for only one explication" (p. 180).

So, a trial-and-error machine uses Turing *machinery* to perform hypercomputations. However, trial-and-error computation is equivalent to computations by o-machines that solve the halting problem!

If the computation is to determine whether or not a natural number n as input belongs to some set S, then it turns out that sets for which such "trial and error" computation is available are exactly those … that are computable relative to … an oracle that provides correct answers to queries concerning whether a given Turing machine … will eventually halt. (Martin Davis, 2006a, p. 128)

So, trial-and-error computation falls prey to the same objections as other forms of hypercomputation. However, because trial-and-error computation only requires an ordinary, physically plausible Turing Machine and no special oracle, it does have some other uses, which are worth looking at. Whether these are legitimate kinds of hypercomputation is something left for you to decide! [15]

11.10.2 Does "Intelligence" Require Trial-and-Error Machines?

A trial-and-error machine can "compute" the uncomputable, but we can't reliably use the result. But what if we *have* to? When we learn to speak, we don't wait (we *can't* wait) until we fully understand our language before we start (before we *have* to start) to use it. Similarly, when we reason or make plans, we must also draw conclusions or act on the basis of incomplete information. Herbert Simon (1996a) called this "satisficing" or "bounded rationality" (Section 5.6).

One of the claims of hypercomputationalists is that some phenomena that are not Turing computable are (or might be) "computable" in some extended sense. And one of these phenomena is "intelligence," or cognition. Siegelman's version that we looked at in Section 11.5, based on neural networks, is one of these. Another, based on trial-and-error computation, is what we will look at now.

Terminological Digression: 'Intelligence' is the term that many people use – including, famously, Turing (1950) – and it is enshrined in the phrase 'artificial intelligence.' However, I prefer the more general term 'cognition,' because the concept that both terms attempt to capture has little or nothing to do with "intelligence" in the sense of IQ tests. So, when you see the words 'intelligence' or 'intelligent' in the following, try substituting 'cognition' or 'cognitive' to see whether the meaning differs. In Chapter 18 (especially Section 18.2.2), we'll go into much more detail on what I prefer to call "computational cognition."

Kugel (2002) argues that AI will be possible using digital computers – and not require fancy quantum computers or other kinds of non-digital computers – by using those digital computers in only a non–Turing-computational way. He begins his argument by observing that intelligence in general, and artificial intelligence in particular, requires "initiative," which he roughly identifies with the absence of "discipline," defined, in turn, as the ability to follow orders. (This is reminiscent of Beth Preston's views on improvisation, which we discussed in Section 10.5.) Thus, perhaps, intelligence and AI require the ability to break rules! Computation, on the other hand, requires such "discipline" (after all, as we have seen, computation certainly includes the ability to follow orders or, at least, to behave in accordance with orders).

Moreover, Kugel argues that Turing made the same point. But did he? Kugel quotes the following sentence:

Intelligent behaviour presumably consists in a *departure* from the completely disciplined behaviour involved in computation, but a rather *slight* one, which does not give rise to random behaviour, or to pointless repetitive loops. (Turing, 1950, p. 459, my italics)

---

15 See the Online Resources for further reading on trial-and-error computation.

However, the larger context of this passage makes it clear that Turing is thinking of a learning machine. So the "slight departure" he refers to is not so much a lack of discipline as it is the Universal Turing Machine's ability to change its behavior: i.e. to change the software that it is running. It can't change its hardware (i.e. its fetch-execute cycle). But because the program a Universal Turing Machine is executing is inscribed on the same tape it can print on, the Universal Turing Machine can change that program! There is no difference between a program stored on the tape and the data also stored on the tape. (There is a difference, of course, between a *hardwired* program and data.)

This is *not* to say that computing is *not* enough for intelligence. Turing (1947) claimed that infallible entities could not be intelligent but that fallibility allows for intelligence:

> … fair play must be given to the machine. Instead of it sometimes giving no answer we could arrange that it gives occasional wrong answers. But the human mathematician would likewise make blunders when trying out new techniques. It is easy for us to regard these blunders as not counting and give him another chance, but the machine would probably be allowed no mercy. *In other words then, if a machine is expected to be infallible, it cannot also be intelligent.* There are several mathematical theorems which say almost exactly that. (Turing, 1947, p. 394, my italics)

A few years later, Turing said something similar:

> [O]ne can show that however the machine [i.e. a computer] is constructed there are bound to be cases where the machine fails to give an answer [to a mathematical question], but a mathematician would be able to. On the other hand, the machine has certain advantages over the mathematician. Whatever it does can be relied upon, assuming no mechanical 'breakdown,' whereas the mathematician makes a certain proportion of mistakes. I believe that this danger of the mathematician making mistakes is an unavoidable corollary of his [sic] power of sometimes hitting upon an entirely new method. (Turing, 1951b, p. 256)

Digression: It's not obvious what Turing was alluding to when he said, "there are bound to be cases where the machine fails to give an answer, but a mathematician would be able to." One possibility is that he's referring to Gödel's Incompleteness Theorem (see Section 6.5, footnote 14). If a Turing Machine is programmed to prove theorems in Peano arithmetic, then, by Gödel's theorem, there will be a *true* statement of arithmetic that it cannot *prove* to be a *theorem* – i.e. to which it "fails to give an answer" in one sense. A human mathematician, however, could show *by other means* (but not prove as a theorem!) that the undecidable statement was *true* – i.e. the human "would be able to" give an answer to the mathematical question, in a different sense. That is, there are two ways to "give an answer": an answer can be given by "syntactically proving a theorem" or by "semantically showing a statement to be true." For more on syntax vs. semantics, see Section 18.8.3. For more on the mathematical abilities of humans vs. machines, see the Online Resources.

This gives support to Kugel's claims about fallibility. Such trade-offs are common: for example, as Gödel showed, certain formal arithmetic systems can be either consistent (infallible?) or complete (truthful?), but not both. An analogy is this: in the early days of cable TV (the late 1970s), there were typically two sources of information about what shows were on: *TV Guide* magazine and the local newspaper. The former was "consistent" or "infallible" in the sense that everything it *said was* on

TV was, indeed, on TV; but it was incomplete, because it did not list any cable TV shows. The local newspaper, on the other hand, was "complete" in the sense that it included all broadcast as well as all cable TV shows, but it was "inconsistent" or "fallible" because it also erroneously included shows that were *not* on TV *or* cable (but there was no way of knowing which was which except by being disappointed when an advertised show was not actually on).

But the context of Turing's essays strongly suggests that what Turing had in mind was the ability of *both* human mathematicians *and* computers to learn from their mistakes, so to speak, and to develop new methods for solving problems – i.e. to change their "software." Turing (1947, p. 394) observes that this might come about by "allow[ing the computer] to have contact with human beings in order that it may adapt itself to their standards," perhaps achieving such interaction through playing chess with humans.

In a later passage, Turing suggests "one feature that … should be incorporated in the machines, and that is a 'random element' " (p. 259). This turns the computer into a kind of interactive *o*-machine that "would result in the behaviour of the machine not being by any means completely determined by the experiences to which it was subjected" (p. 259), suggesting that Turing realized it would make it a kind of hypercomputer but, presumably, one that would be only (small) extension of a Turing Machine.

Question for the Reader: Wouldn't the "random element" be one of "the experiences to which it was subjected"? If so, wouldn't the machine's behavior be completely determined by its experiences, even though the experiences would not be *predictable* and hence not simulatable by an ordinary Turing Machine?

Kugel next argues that Turing computation does not suffice for intelligence, on the grounds that if it did, such an AI agent would not be able to survive! Suppose (by way of *reductio*) that Turing computation did suffice for intelligence. And suppose a mind is a Universal Turing Machine with "instincts" (i.e. with some built-in programs) and is capable of learning (i.e. capable of computing new programs). To learn (i.e. to compute a new program), it could *either* compute a *total* computable program (i.e. one defined on all inputs) *or* compute a *partial* computable program (i.e. one that is undefined on some inputs).

Next, Kugel defines a *total machine* to be one that computes *only* total computable functions and a *universal machine* to be one that computes *all* total computable functions and, presumably, all *partial* computable functions. Is a Universal Turing Machine "total" or "universal" in Kugel's sense? According to Kugel, it can't be both: the set of total computable functions is enumerable $(f_1, f_2, …)$. Let P_i be a program that computes f_i, and let P be a program (machine?) that runs each P_i. Next, let P' compute $f_n + 1$. Then P' is a total computable function, but it is not among the P_i, and hence it is not computed by P. That is, if P computes *only* total functions, then it can't compute *all* of them (Kugel, 2002, p. 577, note 6).

According to Kugel, a Universal Turing Machine is a "universal" machine (so it also computes *partial* functions). If the mind is a Universal Turing Machine, then there are partial functions whose values it can't compute for some inputs. And this, says Kugel would be detrimental to survival. If the mind were total, then there would be functions that it couldn't compute *at all* (namely, partial ones). This would be equally detrimental.

But, says Kugel, there is a third option: let the mind be a Universal Turing Machine with "pre-computed" or "default" values for those undefined inputs. Such a machine is not a Turing Machine; it is a trial-and-error machine, because it relies on intermediate outputs when it can't wait for a final result. That is, it "satisfices," because its "rationality" is "bounded," as Simon might have put it. In other words, hypercomputation in the form of trial-and-error computation, according to Kugel, is necessary for cognition.[16]

11.10.3 Inductive Inference

Is there a specific aspect of cognition that is not Turing computable but that *is* trial-and-error computable? Arguably, yes: language learning.

Language learning is an example of learning a function from its values. Such learning is called "computational learning theory" or "inductive inference." Given the initial outputs of a function $f – f(1), f(2), \dots$ – try to infer (or guess, or compute, or "compute") what function f is. This is an abstract way of describing the problem that a child faces when learning its native language: $f(t)$ is the parts of the language that the child has heard up to time t, and f is the grammar of the language.

Is learning a language computable (or hypercomputable)? Trial-and-error machines are appropriate to model this. E. Mark Gold investigated the conditions under which a *class* of languages could be said to be "learnable." Gold, 1965 presents the mathematics behind trial-and-error machines:

> A class of problems is called decidable if there is an algorithm which will give the answer to any problem of the class after a *finite* length of time. The purpose of this paper is to discuss the classes of problems that can be solved by *infinitely* long decision procedures in the following sense: An algorithm is given which, for any problem of the class, generates an infinitely long sequence of guesses. The problem will be said to be *solved in the limit* if, after some finite point in the sequence, all the guesses are correct and the same …(From the abstract, my italics.)[17]

11.11 Summary

There are many kinds of sub-Turing, or "*hypo-,*" computation. So, if there is any serious super-Turing, or "*hyper-,*" computation, that would put classical, Turing computation somewhere in the middle. And no one disagrees that it holds a central place, given the equivalence of Turing Machines to recursive functions to lambda calculation to Post-production systems, etc., and also given its modeling of human computing and its relation to Hilbert's *Entscheidungsproblem*.

*Hyper*computation seems to come in two "flavors": what I'll call "weird" hypercomputation and what I'll call "plausible" hypercomputation (to use "neutral" terms!). In the former category, I'll put "medically impossible" Zeus machines, relativistic machines that can only exist near black holes, etc. In the latter category, I'll put trial-and-error machines, interactive machines, and o-machines; o-machines are clearly a plausible extension of Turing Machines, as even Turing knew.

Only the "plausible" kinds of hypercomputation seem useful. But both interaction machines and trial-and-error machines seem to be only minor extensions of the Turing analysis of computation, and their behavior is well understood and modelable by Turing's o-machines together with the notion of relative computability. Indeed, when you think of it (and as Feferman (1992, pp. 339–340) pointed out), all notions of computability are relative to (1) what counts as a primitive operation or basic function and (2) what count as the ways to combine them to create other operations and functions.

---

16 See the Online Resources for further reading on Kugel's views.
17 See the Online Resources for further reading on inductive inference.

Two things make Turing Machines (and their logical equivalents) central. The first is their power – they are provably more powerful than "*hypo*computational" models. The second is the fact that the different models of (classical) computation are logically equivalent to each other. Except for the physically "weird" hypercomputers, all other "plausible" models of hypercomputation not only can be seen as minimal (and natural) generalizations of the Turing Machine model but also are all logically equivalent to Turing's *o*-machines. And the main "problem" with those is Davis's "non-computable in"–"non-computable out" principle.

We might even suggest a generalized Computability Thesis:

> A function is "computable" iff it is computable by an *o*-machine.

Recall that Turing explicitly required that the oracle "cannot" be a Turing Machine. But if we relax this constraint, then when the oracle *is* Turing computable, this generalized thesis is just the classical one. When the oracle is *not* Turing computable, we can have non–Turing-computable – i.e. "hypercomputable" – output, but only at the cost of non-computable input. However, we *can* analyze different degrees of uncomputability, as Davis, Feferman, Soare, and many others have noted.

> O-machines show us that not all that is studied in computation theory is Turing-equivalent. (Aizawa, 2010, p. 230)

But note the subtle difference between saying this and saying something like "all computation is equivalent to Turing computation" (which is a version of the Computability Thesis).

Fortnow (2010) nicely refutes three of the major arguments in favor of hypercomputation (including analog computation). Of most interest to us is this passage, inspired by Turing's comment that "The real question at issue is 'What are the possible processes which can be carried out in computing a number?' " (Turing, 1936, Section 9, p. 249; see Section 8.7.2):

> Computation is about process, about the transitions made from one state of the machine to another. Computation is not about the input and the output, point A and point B, but the journey. Turing uses the computable numbers as a way to analyze the power and limitations of computation but they do not reflect computation itself. You can feed a Turing machine an infinite digits [sic] of a real number …, have computers interact with each other …, or have a computer that perform an infinite series of tasks … but *in all these cases the process remains the same, each step following Turing's model*. … So yes Virginia, the Earth is round, man has walked on the moon, Elvis is dead and *everything computable is computable by a Turing machine*. (Fortnow, 2010, pp. 3, 5, my italics)

Robert Soare makes a similar observation:

> Indeed, we claim that the common conception of mechanical procedure and algorithm envisioned over this period is exactly what Turing's computor [i.e. what we called the "clerk" in Section 8.7.3, footnote 6] captures. This may be viewed as roughly analogous to Euclidean geometry or Newtonian physics capturing a large part of everyday geometry or physics, but not necessarily all conceivable parts. Here, Turing has captured the notion of a function computable by a mechanical procedure, and *as yet there is no evidence for any kind of computability which is not included under this concept. If it existed, such evidence would not affect Turing's thesis about mechanical computability any more than hyperbolic geometry or Einsteinian physics refutes the laws of Euclidean geometry or Newtonian physics. Each simply describes a different part of the universe.* (Soare, 1999, pp. 9–10, my italics)

Perhaps the issue is not so much whether it is *possible* to compute the uncomputable (by extending or weakening the notion of Turing computation) but whether it is *practical* to do so. Davis (2006a, p. 126) finds this to be ironic:

> … computer scientists have had to struggle with the all-too-evident fact that from a practical point of view, Turing computability does not suffice. … With these [NP-complete] problems Turing computability doesn't help because, in each case, the number of steps required by the best algorithms available grows exponentially with the length of the input, making their use in practice problematical. How strange that despite this clear evidence that computability alone does not suffice for practical purposes, a movement has developed under the banner of "hypercomputation" proposing the practicality of computing the non-computable.

11.12   Questions for the Reader

1. "There are things … bees can do that humans cannot and vice versa" (Sloman, 2002, Section 3.3). Does that mean bees can do non-computable tasks? Or does 'do' mean something different from 'compute,' such as physical performance? If "doing" is different from "computing," how does that affect Cleland's arguments (see Section 10.4) against the Computability Thesis?

2. If you don't allow physically impossible computations, black-hole computations, etc., can interactive computation make the Halting Problem "computable"? Put another way, the Halting Problem is not classically computable; is it interactively "computable"?

3. The n-body problem is the problem of how to compute the behavior of n objects in space. For example, the 2-body problem concerns the relation of the Earth to the Sun (or to the Moon). The 3-body problem concerns the relation of Earth, Sun, *and* Moon. And so on. Brian Hayes (2015, esp. pp. 92–93) has suggested that one technique for simulating solutions to the n-body problem is to use an ordinary computer linked to a graphics processing unit that is far more powerful than the ordinary computer. Is such a combination like a Turing Machine with an oracle?

4. As we will see in Section 18.3, the Turing Test is interactive. If interaction is not modeled by Turing Machines, how does that affect Turing's arguments about "computing machinery and intelligence"? (If you are not yet familiar with the Turing Test, you might want to come back to this question after reading Section 18.3.)

5. There is a large philosophical literature on "extended cognition" – the view that the mind can extend beyond the boundaries of the skin to include aspects of the external world (Clark and Chalmers, 1998; Rowlands et al., 2020). And in Chapter 16, we will discuss a similar topic: "wide" computing. How might these issues relate to interactive computing?

6. You will probably need to study the mathematics of o-machines, Turing reducibility, etc., in order to give a proper answer to this question and the next, but they are worth thinking about. As I have presented it, oracles seem to play several possibly distinct roles. They can be considered subroutine calls. They can be considered input sources. And they can be considered "miraculous sources of unknowable facts" (at least, unknowable *in advance*).

   Do oracles really play all these roles? Are these roles really all distinct? And what does this conflation of roles say about my proposed "generalized Computability Thesis" in Section 11.11?[18]

7. As presented in Soare, 2016, p. 52, an oracle machine consists, in part, of a Turing Machine together with

---

18 Thanks to Robin K. Hill (personal correspondence) for raising this issue and for the quoted phrase.

an extra 'read only' tape, called the *oracle tape*, upon which is written the characteristic function of some set *A*, called *oracle*, **whose symbols … cannot be printed over** . . . .

Evaluate the following apparent paradox:
(a) Interactive computing involves inputting information from, and outputting information to, the external world.
(b) An oracle machine models interactive computing.
(c) It is the oracle that models the external world.
(d) Therefore, the oracle machine must be able to modify the oracle.
(e) But by definition, the oracle is not modifiable by the Turing machine (because it is read-only).

Part IV

Computer Programs

In Part II, we looked at the nature of computer science, computers, and algorithms, and in Part III, we looked a bit further at algorithms, focusing on challenges to the Computability Thesis.

In Part IV, we will look at computer programs – linguistic implementations of algorithms.

- Chapter 12 will look at the relations between algorithms and programs and between software and hardware.
- Chapter 13 investigates the nature of the implementation relation.
- In line with the possibility that CS is a science, Chapter 14 will ask whether computer programs can be considered scientific theories.
- And in line with the possibility that CS is a *mathematical* science, Chapter 15 will look at whether computer programs are mathematical objects that can be logically proved to be "correct."
- Finally, in Chapter 16, we will consider the important topic of the relation between computer programs and the real world that they operate and act in, along with some discussion of the nature of syntax (symbol manipulation) and semantics (meaning).

Philosophy of Computer Science: An Introduction to the Issues and the Literature, First Edition. William J. Rapaport.
© 2023 John Wiley & Sons, Inc. Published 2023 by John Wiley & Sons, Inc.

12

Software and Hardware

> **program**: /n./ 1. A magic spell cast over a computer allowing it to turn one's input into error messages. 2. An exercise in experimental epistemology. 3. A form of art, ostensibly intended for the instruction of computers, which is nevertheless almost inevitably a failure if other programmers can't understand it.
> —*The Jargon Lexicon*, http://www.jargon.net/jargonfile/p/program.html
>
> A program is fundamentally a transformation of one computer into another . . . .
> —Joseph Weizenbaum (1972, p. 610)

12.1 The Nature of Computer Programs

We have explored what an *algorithm* is; we are now going to look at computer *programs*. In the course of the next few chapters, we will consider these questions:

- What *is* a computer program?
- Do computer programs "implement" algorithms?
- What is the nature of implementation?
- What are "software" and "hardware," and how are they related?
- Can (some) computer programs be considered to be scientific theories?
- Are programs mathematical entities susceptible to mathematical proofs?

Typically, one tends to consider a computer program as an expression, in some language, of an algorithm. The language is typically a programming language such as Java, Lisp, or Fortran. And a programming language is typically required to be "Turing complete," i.e. to be able to express the primitive operations of a Turing Machine, together with all three of the Böhm-Jacopini "grammar" rules: sequence, selection, and while-loops, as discussed in §7.4.3. So, "*computer* languages," such as HTML, that lack one or more of these "control structures" are not "*programming* languages" in this sense.

An algorithm is something more "abstract," whereas a program that expresses it (or "implements" it in language) is something more "concrete." A program is more concrete than an algorithm in two ways: first, a program is a physical object, either written on paper or "hardwired" in a computer. Perhaps the relationship between an algorithm and a program is something like the relationship between a number and a numeral: just as the number "two" can be expressed with many different numerals (such as '2' or 'II') and many different words (such as 'two,' 'deux,' or 'zwei'), so a single algorithm, such as the algorithm for binary search, can be expressed in many different programming languages.

In fact, we can't really talk about algorithms (or numbers) without using some kind of language, so maybe there really aren't any of these abstract things called 'algorithms' (or numbers!), just words for them. This is an ontological view in philosophy called 'nominalism.' "Platonists" believe that mathematics deals with *numbers* – abstract entities that exist in a "Platonic" realm that is more perfect than, and independent of, the real world (Linnebo, 2018). "Nominalists," on the other hand, deny the existence of abstract numbers, and hold that mathematics deals only with *numerals* – real marks on paper, for instance (Bueno, 2020). Maybe only programs exist, some of which might be written in programming languages that can be directly used to cause a computer to execute the program (or execute the algorithm?), and some of which might be written in a natural language, such as English. (The nominalist can still talk about "algorithms," understanding them as computer programs. However, ask yourself whether a nominalist can still talk about numbers, understood as numerals: after all, there are infinitely many numbers, but only finitely many numerals.)

The second way a program is more concrete than an algorithm is that a program is more detailed. Where an algorithm might simply specify how to perform a binary search, a binary-search program for a particular computer would have to spell out the details of how that search would be physically implemented in that computer. (We'll have more to say about implementation in Chapter 13.)

In the early days of computers, programs were not typically expressed in programming languages; rather they were "hardwired" into the computer or certain physical switches were set in certain ways. These programs were physical parts of the computer's hardware, not texts. The program could be changed by re-wiring the computer (perhaps by re-setting the switches). Yet computer programs are typically considered "software," not "hardware," so was such wiring (or switch-setting) a computer program?

And what about a program written on a piece of paper? Does it differ from the very same program written on a computer file? The former just sits there doing nothing. So does the latter, but the latter can be used as input to other programs on the computer that will use the information in the program to "set the switches" so that the computer can execute the program. But is the medium on which the program is written the only difference between these two programs?[1]

12.2 Programs and Algorithms

In §7.2, we saw that a *function* defined *extensionally* as a *set* of input-output pairs satisfying the same-input/same-output constraint could be "implemented" – made more precise or more explicit – by many different functions defined *intensionally* by a *formula*, each of which is a *description of the relationship* between the input and the output. Thus, for example, the function $f = \{(0,0), (1,3), (2,6), (3,9), \ldots\}$ can be implemented by the formula $f_1(x) = 3x$ or by the formula $f_2(x) = x + x + x$, etc.

We also saw that a *formula* could be implemented by many different *algorithms*, each of which spells out the intermediate steps that *compute* the output according to the formula. Thus, for example, the formula f_1 could be computed by either of the following algorithms:

---

1 See the Online Resources for further reading on the nature of programs and programming.

Algorithm $A^1_{f_1}(x)$	**Algorithm** $A^2_{f_1}(x)$
begin	**begin**
$\quad f_1 \leftarrow 3;$	$\quad f_1 \leftarrow x;$
$\quad f_1 \leftarrow f_1 * x$	$\quad f_1 \leftarrow 3 * f_1$
end.	**end.**

And the formula f_2 could be computed by either of these algorithms:

Algorithm $A^1_{f_2}(x)$	**Algorithm** $A^2_{f_2}(x)$
begin	**begin**
$\quad f_2 \leftarrow x + x;$	$\quad f_2 \leftarrow x;$
$\quad f_2 \leftarrow f_2 + x$	$\quad f_2 \leftarrow f_2 + x;$
end.	$\quad f_2 \leftarrow f_2 + x$
	end.

And so on. Thus, our original function f could be computed by any one of those four algorithms, among infinitely many others.

One way to consider the relationship between algorithms and programs is to continue this chain of implementations: an algorithm can be implemented by a computer program written in a high-level computer programming language. That program can then be implemented in assembly language (which is computer-specific and provides more detail). The assembly-language program, in turn, can be implemented in machine language. And finally, the machine-language program can be implemented in hardware by "hardwiring" a computer – or, in more modern terminology, by using a chip designed to perform that function. Arguably, the static, hardwired program is implemented by the dynamic *process* that is created when the computer executes "the" program.[2]

Both algorithms and programs are normally considered "software," and physical implementations of them in a computer are normally considered to be "hardware." But what exactly is software, and how can it be distinguished from hardware? Many authors use 'program' and 'software' as synonyms. But if we view a program as an implementation of an algorithm (in some medium such as language or the switch settings of a computer), and if we view software as contrasted with hardware, it's not obvious that programs and software are exactly the same thing. Programs can be expressed on paper in a programming language, which seems like software. But they can also be hardwired in a physical computer, which seems like hardware. And 'software' is not usually defined in terms of algorithms.

12.3 Software, Programs, and Hardware

12.3.1 Etymology of 'Software'

The earliest use of the word 'software' in its modern sense has been traced back to the mathematician John W. Tukey (1958, p. 2):

> Today the "software" comprising the carefully planned interpretive routines, compilers, and other aspects of automative [sic] programming are at least as important to the modern electronic calculator as its "hardware" of tubes, transistors, wires, tapes and the like.

---

2 See the Online Resources for further reading on programs and algorithms.

But the word is older than Tukey's use of it: The earliest cited use (in 1782, according to the *OED*)[3] is for textiles and fabrics – literally "soft wares." A later use, dating to 1850, equated it with "vegetable and animal matters – everything that will decompose" in the realm of "rubbish-tip pickers" (F.R. Shapiro, 2000, p. 69).[4] And two years before Tukey's paper, Richard B. Carhart (1956, p. 149) equated software with the *people* who operate a computer system, and the computer system was identified as the hardware (programs or other modern notions of software were not mentioned). (On the history of software, see Mahoney, 2011, Ch. 13.)

12.3.2 Software and Music

> Is Bach's written score to the *Art of the Fugue*, perhaps with a human interpreter thrown in,
> the software of an organ?
> —Peter Suber (1988, p. 90)

Tukey's use of the term strongly suggests that the things that count as software are more abstract than the things that count as hardware. Amnon H. Eden (2005, Slide 36) considers software "as a cognitive artefact: software is conceived and designed at a level of abstraction higher than the programming language." Using a concept very similar to Eden's, Nurbay Irmak (2012) argues that software is an "abstract artifact," likening it to another abstract or cognitive artifact: musical works (§2, pp. 65ff). There are close similarities. For instance, a Turing Machine (or any hardwired computer that can perform only one task) is like a music box that can play only one tune, whereas a player piano is like a Universal Turing Machine, capable of playing any tune encoded on its "piano roll."

One difference between software and music that Irmak points out concerns "a change or a revision on a musical work once composed" (p. 67). This raises some interesting questions: recall our brief discussion of the relationships between software and improvisational music in §10.5. How should musical adaptations or jazzy versions of a piece of music be characterized? What about different players' interpretations? One pianist's version of, say, Bach's *Goldberg Variations* will sound very different from another's, yet, presumably, they are using the same "software." Are there analogies to these with respect to computer software?[5]

12.3.3 The Dual Nature of Programs

Our first main issue concerns the dual nature of programs: they can be considered both text and machine, both software and hardware. To clarify this dual nature, consider this problem:

> … Bruce Schneier authored a book entitled *Applied Cryptography*, which discusses many commonly used ciphers and included source code for a number of algorithms. The State Department decided that the book was freely exportable because it had been openly published but refused permission for export of a floppy disk containing the same source code printed in the book. The book's appendices on disk are apparently munitions legally indistinguishable from a cluster bomb or laser-guided missile. … [The] disk cannot legally leave the country, even though the original book has long since passed overseas and all the code in it is available on the Internet. (Wallich, 1997, p. 42)[6]

---

3 http://www.oed.com/view/Entry/183938; see also http://www.historyofinformation.com/expanded.php?id=936.
4 Insofar as decomposition is a form of changeability, this is consistent with Moor's definition of 'software,' as we will see in §12.4.5!
5 See the Online Resources for further reading on software and art.
6 The case is discussed at length in Colburn, 1999, 2000.

How can a program written on paper be considered a different thing from the very same program "written" on a floppy disk? What if the paper that the program was written on was Hayes's "magic paper" (§9.5)? But isn't that similar to what a floppy disk is, at least, when it is being "read" by a computer?

Is the machine-table "program" of a Turing Machine software, or is it hardware? It certainly seems to be "hardwired"; it *is* the Turing Machine, not a separable part of one. If you think that it is a kind of category mistake to talk about whether an abstract, mathematical entity such as a Turing Machine can have software or hardware, then consider this: suppose you have a physical implementation of a Turing Machine: a hardwired, single-purpose, physical computer that (let's say) does nothing but accept two integers as input and produces their sum as output. Is the program that runs this adder software or hardware? Because the machine table of such a (physical implementation of a) Turing Machine is not written down anywhere but is part of the (physical) mechanism of the machine, it certainly seems to be more like hardware than software.

Is the machine table of a *Universal* Turing Machine (i.e. its fetch-execute cycle) software or hardware? What about the program that is stored on its tape? By the logic of the previous paragraph, its fetch-execute machine table would be hardware, and its stored program would be software. Again, if you prefer to limit the discussion to physical computers, then consider a smartphone, one of whose apps is a calculator that can add two integers. Not only can the calculator do other mathematical operations, the smartphone itself can do many other things (play music, take pictures, make phone calls, etc.), *and* it can download new apps that will allow it to do many other things. So it can be considered a physical implementation of a Universal Turing Machine. By our previous reasoning, the program that is the smartphone's adder (calculator) is software, and the program that allows the smartphone to do all of these things is hardware.

12.3.4 Copyright vs. Patent

> Computers don't work the way some legal documents and court precedents say they do.
> —"PolR", 2009

Another aspect of the dual nature of programs is this: considered software, a program is (arguably) copyrightable. Considered as hardware, a program is (arguably) patentable. Yet nothing can be both copyrighted and patented!

According to a brochure published by the US Copyright Office (my italics):

> Copyright is a form of protection provided by the laws of the United States[7] to the authors of "original works of authorship" that are fixed in a tangible form of expression. An original work of authorship is a work that is independently created by a human author and possesses at least some minimal degree of creativity. A work is "fixed" when it is captured … in a sufficiently permanent medium such that the work can be perceived, reproduced, or communicated for more than a short time. … Examples of copyrightable works include [l]iterary works … *computer programs … can be registered as "literary works"*; … Copyright does not protect [i]deas, procedures, methods, systems, processes, concepts, principles, or discoveries …
> ("Copyright Basics," September 2017, https://www.copyright.gov/circs/circ01.pdf)

---

7 Title 17, *US Code*, https://www.copyright.gov/title17/.

If a computer program is an implementation of an algorithm in the medium of text, then, as a text, it is a "literary work," hence copyrightable. Why "literary"? After all, they don't seem to read like novels! But all that 'literary' means in this context is that they can be written and read. Presumably, however, the abstract algorithm is not copyrightable because it is an "idea," "procedure," or "method."

And here is a definition of 'patent' from the US Patent and Trademark Office's website (https://www.uspto.gov/):

> A patent for an invention is the grant of a property right to the inventor ... "... to exclude others from making, using, offering for sale, or selling" the invention in the United States or "importing" the invention into the United States. ... **Utility patents** may be granted to anyone who invents or discovers any new and useful process, machine, article of manufacture, or composition of matter, or any new and useful improvement thereof; ... (https://www.uspto.gov/patents/basics/patent-process-overview#step3)[8]

On the website for utility patents, we find this:

> **Specification**
> The specification is a written description of the invention .... For inventions involving computer programming, computer program listings may be submitted as part of the specification ....
> ("Nonprovisional (Utility) Patent Application Filing Guide," https://www.uspto.gov/patents-getting-started/patent-basics/types-patent-applications/nonprovisional-utility-patent)

This suggests that what is patentable is the "process" or "machine" specified by a computer program. But copyrightable programs can be "performed," i.e. executed, just like lectures, plays, movies, or music. The relation of a program to a process (i.e. a program being executed; see §3.11) is similar to the relation of a script to a performance of a play or a showing of a movie, of a musical score to a musical performance, or (perhaps) of a set of slides to a delivery of a lecture. Yet *processes* are *not* copyrightable; they are patentable.

If software, generally speaking, is copyrightable (but not patentable), and if hardware, generally speaking, is patentable (but not copyrightable), what about a virtual machine, which is a software implementation of a piece of hardware? Pamela Samuelson, Randall Davis, Mitchell D. Kapor, and J.J. Reichman (1994, p. 2324) argue "that programs should be viewed as virtual machines."[9]

They argue against the appropriateness of copyright law (Samuelson et al., 1994, p. 2350):

1. "computer programs are machines whose medium of construction is text"
2. "Copyright law does not protect the behavior of physical machines (nor their internal construction)"
3. ∴ "program behavior ... is unprotectable by copyright law on account of its functionality" (p. 2351).

---

8 The relevant laws are the US Constitution, Article I, §8; and various laws cited at https://www.uspto.gov/web/offices/pac/mpep/consolidated_laws.pdf. See also "Computer Systems Based on Specific Computational Models," https://www.uspto.gov/web/patents/classification/cpc/pdf/cpc-definition-G06N.pdf.
9 This essay is from a special issue of the *Columbia Law Review* on the legal protection of computer programs. A summary version appears as Davis et al., 1996. Other articles in that special issue elaborate on, or reply to, Samuelson et al. (1994).

But far from arguing that programs, because not copyrightable, should be patentable, they also think that patentability is inappropriate:

> The predominantly functional nature of program behavior and other industrial design aspects of programs precludes copyright protection, while the incremental nature of innovation in software largely precludes patent protection.
> (Samuelson et al., 1994, p. 2333)

They offer two arguments against patentability. Here is the first (p. 2345):

1. Patents are given for "methods of achieving results,"
2. Patents are not given "for results themselves."
3. "It is … possible to produce functionally indistinguishable program behaviors through use of more than one method."
4. ∴ A patent could be given for one method of producing a result, but that would not "prevent the use of another method."
5. If it is the result, not the method, that is the "principal source of value" of a program, then the patent on the one method would not protect the result produced by the other method

And here is their second argument against patentability (p. 2346):

1. "Patent law requires an inventive advance over the prior art"
2. But the innovations in "functional program behavior, user interfaces, and the industrial design of programs … are typically of an incremental sort."
3. ∴ Programs do not fall under patent law.

A similar analysis to the effect that neither copyright nor patent seems appropriate for computer programs was offered by Allen Newell. In a paper written for a law journal, Newell (1986) argues that the "conceptual models" of algorithms and their uses are "broken," i.e. that they are "inadequate" for discussions involving patents (and copyrights). Newell's bottom line is that there are two intellectual tasks: a *computer-scientific and philosophical task* is to devise good models ("ontologies") of algorithms and other computer-scientific items. A *legal task* is to devise good legal structures to protect these computational items:

> I think fixing the models is an important intellectual task. It will be difficult. The concepts that are being jumbled together – methods, processes, mental steps, abstraction, algorithms, procedures, determinism – ramify throughout the social and economic fabric .... The task is to get … new models. There is a fertile field to be plowed here, to understand what models might work for the law. It is a job for lawyers and, importantly, theoretical computer scientists. **It could also use some philosophers of computation, if we could ever grow some.** (Newell, 1986, p. 1035, my boldface)

Readers of this book, take note![10]

✳✳✳✳✳

In the next three sections, we will look at what three philosophers have had to say about software: James H. Moor (1978) argues that software is *changeable*. Peter Suber (1988) argues that it is *pure syntax*. And Timothy Colburn (1999, 2000, Ch. 12) argues that it is a *concrete abstraction*. Keep in mind that they may be assuming that software and computer programs are the same things.

---

10 See the Online Resources for further reading on legal protection of programs.

12.4 Moor: Software Is Changeable

12.4.1 Levels of Understanding

For very many phenomena, a single entity can be viewed from multiple perspectives (sometimes called "levels" or "stances"). Dennett (1971) suggested that a chess-playing computer or its computer program can be understood in *three* different ways:

First, from the *physical stance*, its behavior can be predicted or explained on the basis of its physical construction together with physical laws. Thus, we might say that it made (or failed to make) a certain move because logic gates #5, #7, and #8 were open or because transistor #41 was defective.

Second, from the *design stance*, its behavior can be predicted or explained based on information or assumptions about how it was designed or how it is expected to behave, assuming that it *was* designed to behave that way and isn't malfunctioning. Thus, we might say that it made (or failed to make) a certain move because line #73 of its program has an if-then-else statement with an infinite loop.

Third, from the *intentional stance*, its behavior can be predicted or explained based on the language of "folk psychology": ordinary people's informal (and not necessarily scientific) theories of why people behave the way they do, expressed in the language of beliefs, desires, and intentions. For instance, I might explain your behavior by saying that (a) you desired a piece of chocolate, (b) you believed that someone would give you chocolate if you asked them for it, so (c) you formed the intention of asking me for some chocolate. Similarly, we might say that the chess-playing computer made a certain move because (a) it desired to put my king in check, (b) it believed that moving its knight to a certain square would put my king in check, and so (c) it formed the intention of moving its knight to that position.

Each of these "stances" has different advantages for dealing with the chess-playing computer: if the computer is physically broken, the physical stance can help us repair it. If the computer is playing poorly, then the design stance can help us debug its program. If I am playing chess against the computer, then the intentional stance can help me figure out a way to beat it.

According to Moor (1978, p. 213), both computers and computer programs "can be understood on two levels": they can be understood as physical objects, subject to the laws of physics, electronics, and so on. A computer disk containing a program would be a clear example of this level. But they can also be understood on a symbolic level: a computer can be considered a calculating device, and a computer program can be considered a set of instructions. The text of the computer program that is engraved on the disk would be a clear example of this level. Moor's two levels – the physical and the symbolic – are close to Dennett's physical and design "stances."[11]

12.4.2 Programs Are Relative to Computers

Moor offers a definition of 'computer program' that is intended to be neutral with respect to the different stances of the software-hardware duality:

> a computer program is a set of instructions which a computer can follow (or at least there is an acknowledged effective procedure for putting them into a form which the computer can follow) to perform an activity. (Moor, 1978, p. 214)

---

11 See the Online Resources for further reading on Dennett's intentional stance.

Let's make this a bit more explicit to highlight its principal features:

Definition M1:
> Let C be a computer.
> Let S be a set of instructions.
> Let A be an activity.
> Then *S is a computer program for C to do A* $=_{def}$
>> 1. there is an effective procedure for putting S in a form …
>> 2. … that C can "follow" …
>> 3. … in order to perform A.

In this definition, *being a computer program* is not simply a *property* of some set of instructions. Rather, it is a *ternary* relation among a set of instructions, a computer, and an activity. In §7.3.3, we briefly looked at the role of an algorithm's *purpose*, and we will examine it in more detail beginning in §16.3. But here, I just want to focus on the role of the computer, so we'll (temporarily) ignore clause 3 and take a computer program as a *binary* relation between a set of instructions and a computer.

As a binary relation, a set of instructions that is a computer program for one computer might not be a computer program for a different computer, perhaps because the second one lacks an effective procedure for knowing how to "follow" it: one computer's program might be another's noise. For instance, the Microsoft Word program that is written for an iMac computer running MacOS X differs from the Microsoft Word program that is written for a PC running Windows, because the underlying computers use different operating systems and different machine languages. This would be so even if the "look and feel" of the two programs (i.e. what the user sees on the screen and how the user interacts with the program) were identical.

But what are the "instructions," and what does it mean to "follow" them?

12.4.3 Instructions

Presumably, the instructions must be algorithmic, though Moor does not explicitly say so. Is the set of instructions physical, i.e. hardwired? Or are the instructions written in some language? Could they be drawn, instead – perhaps as a flowchart? Could they be spoken? Here, Moor's answer seems to be that it doesn't matter, as long as there is a way for the computer to "interpret" or "understand" the instructions and thus carry them out. (In §12.5, we will see that Suber makes a similar point.) Importantly, the "way" that the computer "interprets" the instructions must itself be a computable function ("an effective procedure for putting them into a form which the computer can follow"). Otherwise, it might require some kind of "built-in," *non*-computable method of "understanding" what it is doing.

Terminological Digression: When I say that the computer has to "interpret" the instructions, I simply mean the computer has, somehow, to be able to convert the symbols that are part of the program into actions that it performs on its "switches." This is different from the distinction in CS between "interpreted" and "compiled" programs. A "compiled" program is translated into the computer's machine language all at once, and then the computer executes the machine-language version of the program, in much the same way that an entire book might be translated from one language to another. By contrast, an "interpreted" program is

---

(Continued)

(Continued)

translated step by step into the computer's machine language, and the computer executes each step before translating the next one, in much the same way that a UN "simultaneous translator" translates a speech sentence by sentence while the speaker is giving it. In both cases, the computer is "interpreting" – understanding – the instructions in the sense in which I used that word in the previous paragraph.

12.4.4 "Following Instructions."

Moor says that computers "follow instructions." And we saw in §9.8 that Stuart C. Shapiro defined a computer as "a general-purpose procedure-*following* machine" (Shapiro, 2001, p. 2, my italics). But *does* a computer "follow" instructions? Or does it merely *behave in accordance with* them?

> It is common to differentiate between satisfying a rule and following a rule (cf. Searle (1980); Wittgenstein (1958, §§185–242)). To satisfy a rule is simply to behave in such a way that fits the description of the rule – merely to conform behaviour to the rule. It is in this sense that the motion of the planets satisfy [sic] the rules embodied by classical physics. On the other hand, following a rule implies a causal link between the rule and some behaviour, and moreover that the rule is an intentional object. … [M]erely satisfying a rule is not sufficient for following the rule. (Chow, 2015, p. 1000)

Compare the human use of natural language: when we speak or write, do we "follow" the rules of grammar in the sense of consulting them (even if unconsciously) before generating our speech or writing? Or does it make more sense to say that the rules of grammar merely *describe* our linguistic behavior? We probably do both, though the latter predominates:

> We learn from psycholinguistics that … [understanding language] involves subconscious, subpersonal, automatic, extraordinarily fast processing [i.e. what we referred to as "System 1" in §3.6.1], *and that is mostly all that it involves* …. Where understanding is difficult – e.g. with multiple center embedding ["A mouse that a cat that a dog chased caught ate cheese"] – it *may* be helped by "central processor," relatively slow reasoning, leading to a conscious judgment about … [an] utterance. But such high-level processes are a very small part of language understanding. (Devitt and Porot, 2018, p. 9, italics in original)[12]

Note, however, that a computer programmed to understand and generate natural language might, in fact, speak or write by *explicitly following* rules of grammar that are encoded in its suite of natural-language-processing programs (Stuart C. Shapiro 1989; Shapiro and Rapaport 1991, 1995; Jurafsky and Martin 2000).[13] We have also seen a similar question when we considered whether the solar system "follows" Kepler's laws of planetary motion or whether the planets' movements are merely best *described* by Kepler's laws (§9.7.2).

Turing Machines – as models of hardwired, single-purpose computers – merely behave in accordance with their machine table. They don't "consult" those "instructions" and then "follow" them. On the other hand, *Universal* Turing Machines – as models of programmable, general-purpose

---

12 See the Online Resources for further information on center embedding.
13 And see our discussion in §18.6 of Searle's Chinese Room Argument.

computers – *can* be said to "follow" instructions. Behaving in accordance with their fetch-execute machine table, they *do* "consult" the instructions stored on their tape, and *follow* them.

> **Question for the Reader:** Suppose that a Universal Turing Machine (or your Mac or PC) is running a program that adds two integers. What is it doing? Is it adding two integers? Or is it carrying out a fetch-execute cycle? Or is it doing both? Or is it doing one *by* doing the other? And what exactly does it mean to do one thing "by" doing another?

Searle (1969, §2.5) identifies a related distinction concerning the instructions or rules themselves: roughly, *constitutive* rules determine or define the behavior of some system, whereas *regulative* rules "regulate antecedently or independently existing forms of behavior" (p. 33). For example, the rules of grammar that linguists discover about the natural languages that we speak are *constitutive* rules; they are descriptive of the "innate" rules that we "execute" or use unconsciously, such as "Declarative sentences of English consist of a noun phrase followed by a verb phrase." The explicit rules of grammar that we have to learn in school (or that "grammar Nazis" insist that we should "follow") are *regulative* rules; they recommend (or insist upon) a way to do things – e.g. "Prepositions should not be used to end sentences with." In the theory of computation, the program for a Turing Machine is a constitutive rule. Because the program for a *Universal* Turing Machine is its fetch-execute cycle, that program is a constitutive rule; but the program (the software) inscribed on its tape that a Universal Turing Machine is "following" is a regulative rule.[14]

12.4.5 Moor's Definitions of 'Software' and 'Hardware.'

Next, Moor distinguishes between software and hardware. The informal and traditional distinction is that a computer *program* is "software" and a *computer* is "hardware." But this raises the problem of whether the "wiring" in a hardwired computer is hardware (because it involves physical wires) or software (because it is the computer's program). And, of course, it gives rise to the problem mentioned by Wallich, cited in §12.3.3. So, Moor suggests a better set of definitions:

> For a given person and computer system the software will be those programs which can be run on the computer system and which contain instructions the person can change, and the hardware will be that part of the computer system which is not software. (Moor, 1978, p. 215)

Again, let's make this a bit more explicit to highlight its features:

Definition M2:
Let C be a computer.
Let P be a person (perhaps C's programmer).
Let S be some entity (possibly a part of C).
Then S *is software for C and P* $=_{def}$

    1. S is a computer program for C (to perform some activity A) and
    2. S is changeable by P.

and H *is hardware for C and P* $=_{def}$

    1. H is (a physical) part of C, and
    2. H is *not* software for C and P.

---

14 See the Online Resources for further reading on following instructions.

That is, a physical part of a computer will be hardware for a person and that computer if *either* it is *not* a computer program for that computer *or* it is not *changeable* by that person.

Note that *being software* is a *ternary* relation among a computer program, a person, and a computer.[15] It is not a simple property such that something either is, or else it isn't, software. In other words, software is in the eye of the beholder: one person's or one computer's software might be another's hardware!

These definitions seem to allow for the following possibilities: first, consider a computer program *J* written in Java that runs on my computer. Even if *J* is changeable by the programmer who wrote it or the lab technician who operates the computer – and therefore *software* for that person – it will be *hardware* for me if I don't know Java or don't have access to the program so that I could change it.

Second, if a programmer can "rewire" a computer (or directly set its "switches"), then that computer's program is software, even if it is a physical part of the computer: Hardware can be software (and vice versa)!

Later writers have made similar observations: Frank Vahid (2003, p. 27, original italics, my boldface) notes that, in the early days of computing, "the frequently **changing** programs, or *software*, became distinguished from the **unchanging** hardware on which they ran." Vahid suggests that "the processors, memories, and buses – what we previously considered a system's unchangeable hardware – can actually be quite soft" (p. 32). What he seems to mean by this is that *embedded systems* – "hidden computing systems [that] drive the electronic products around us" (p. 27) – can be swapped for others in the larger systems that they are components of, thus becoming "changeable" in much the way that software is normally considered. But this seems to just be the same as the old rewiring of the early days of programming (except that, instead of changing the wires or switches, it is entire, but miniature, computers that are changed).

On Moor's definition, the machine table of a Turing Machine – even though we sometimes think of it as the Turing Machine's "program" – is part of its hardware, because it is not changeable: were it to be changed (somehow), we would have a *different* Turing Machine. It is probably best to think of a Turing Machine's machine table, not as a program written in a Turing Machine programming language such as we used in Chapter 8, but as the way that the "gears" of the Turing Machine are arranged so that it behaves the way that it does. This is what Samuel (1953, pp. 1226–1227) called "fixed programming":

> By fixed programming we mean the kind of programming which controls your automatic dishwasher for example. Here the sequence of operations is fixed and built into the wiring of the control or sequencing unit. Once started, the dishwasher will proceed through a regular series of operations, washing, rinsing and drying. Of course, if one wished, one could change the wiring to alter the program.

Note, however, that modern dishwashers allow for some "programming" by pushbuttons that can alter its operations. But perhaps this is more like interactive computing (§11.8)! (And recall §9.5.)

12.5 Suber: Software Is Pattern

Peter Suber (1988, p. 94) says that he will "use 'program' and 'software' interchangeably." This is unfortunate because it seems to beg the question about whether all programs are software. In what

---

15 It is a *quaternary* relation, if we include activity *A*.

follows, we will ignore this (up to a point, as you will see), and simply try to understand what he means by 'software.' His definition is straightforward and rather different from Moor's:

> [S]oftware is pattern *per se*, or syntactical form .... (Suber, 1988, online abstract)

What does he mean by this, and why does he think that it is true? Here is his argument:

1. "Software patterns … are essentially expressed as arrays of symbols – or texts."
   (Suber, 1988, §2 ("Digital and Analog Patterns"), p. 91)

That is, all software is a text, which he calls a "digital pattern." If we think of software as a computer program written in a programming language or even expressed as arrays of '0's and '1's, this is plausible. But what about hardwired programs? He might say that they are not software. But he might also say that because there is no significant difference between an array of '0's and '1's and an array of switches in one of two positions, even such a hardwired program is a text. (But see premise 7.)

2. "The important feature of digital patterns here is … their formal articulation of parts …." (Suber, 1988, §2, p. 91)

By 'formal articulation,' I will assume that he means "syntax." So, all texts have a syntax. But do they? An array of '0's and '1's that corresponds to the binary expression of the decimal part of a real number arguably has a syntax. But what about a random array? Of course, if the syntax of an array is just the properties and relations of its elements (recall §9.4.3), then even a random array has a syntax.

3. "Each joint of articulation carries information for any machine designed to read it." (Suber, 1988, §2, p. 91)

So, the syntax of digital patterns can convey information for appropriate readers. This is close to clause 2 of Moor's Definition M1: Computers have to be able to "follow" their programs, and, to do that, a computer must be able to *read* its program.

4. The Noiseless Principle: "some order may be made of any set of data points; every formal expression has at least one interpretation. … [N]o pattern is noise to all possible machines and languages." (Suber, 1988, §3 ("First Formulation"), p. 94)

This seems to be equivalent to premise 3: if "every joint of articulation carries information," then "no pattern is noise," and vice versa.

5. ∴ The executability of software is a function of its syntax, the language that it is written in, and of the machine that runs it. (Suber, 1988, §4 ("Executability"), especially p. 97)

This follows from the previous premises: All software is a text; each text has a(t least one) syntax; each syntax has a(t least one) interpretation. A machine designed to "understand" that syntax can execute that software. Suber hypothesizes that all software must be readable and executable. Here, he is arguing that any text is executable given an appropriate syntax for it and the right language and machine to interpret that syntax.

6. Software is readable if and only if (1) it has a "physical representation … that suits the machine that is to read it" and (2) it is "in 'machine language.'" (Suber, 1988, §5 ("Readability"), p. 98)

This seems to come down to the same thing as saying that there must be a machine that is capable of reading it. Just as I can't read something written in invisible ink and in Mandarin (because I can neither see it nor parse it even if I *could* see it), so the machine has to be able to "see" the text, and it has to be able to understand it. Perhaps this is best taken as a definition of 'readable.' In any case, it does not seem to add anything over and above the previous conclusion.

7. The Sensible Principle: "any pattern can be physically embodied." (Suber, 1988, §5, p. 100)

So, even if hardwired programs are not "texts" (as we wondered in premise 1), they are "physical embodiments" of texts.

8. The Digital Principle: any "pattern" – i.e. any text, including an analog pattern – "can be reproduced by a digital pattern to an arbitrary degree of accuracy." (Suber, 1988, §6 ("Pattern *Per Se* Again"), p. 91)

9. ∴ Any text is readable. (Follows from the Sensible and the Digital Principles.)

10. ∴ Any text is executable. (Follows from line 5 and premise 6, clause (2).)

But you should ask yourself how a text that does not contain any instructions might be executable. 'Instructions' and 'executable' might not be the best terms here. Some programming languages speak, instead, of "functions" (Lisp) or "clauses" (Prolog) that are "evaluated." The question to be asked is how a text that is not an *algorithm* (such as this chapter, perhaps) might be "executable."

11. ∴ Any text is software.

There are two things to note here. First, recall premise 1. All software is text. Suber seems to have argued from that premise to its converse. So "software" and "text" are the same thing. Second, you might ask yourself how this relates to Searle's claim that everything is a computer!

Suber also notes that "software is *portable*": "one can run the same piece on this machine and then on that machine" (Suber, 1988, §7 ("Liftability"), p. 103). Moreover, because it is essentially unembodied text, software "can be ported from one substratum to another. It is *liftable*" (Suber, 1988, §7, pp. 103–104). And it is "alterable" (Suber, 1988, §8 ("The Softness of Software"), p. 105). These are what distinguish it from hardware (Suber, 1988, §8). Alterability, of course, is what Moor cites as the essence of software. For Suber, that seems to follow from its being "pattern *per se.*"

Does Suber really mean *every* text is software – even random bits or "noise"? He claims that "software patterns do not carry their own meanings" (Suber, 1988, §6, p. 103). In other words, they are purely formal syntax, meaningless marks, symbols with no intrinsic meanings. If a computer can give meaning to a text, then it can read and execute it, according to Suber. But can any text be given a meaning by some computer? Yes, according to the Noiseless Principle.

We can summarize Suber's argument as follows: software is text. As such, it has syntax but no intrinsic semantics. For to be "meaningful" – readable and executable – it has to be interpreted by something else (e.g. a computer) that can ascribe meaning to it (and that can execute its instructions). What about texts that are not programs (or not intended to be programs)? Consider a text such as this book or random noise. If there is a device that can ascribe some meaning to such a text, then it, too, is readable and has the potential to be executable. (But what would it mean to "execute" the chapter you are now reading?)

But texts need to be interpreted by a suitable computer: "They need only make a fruitful match with another pattern (embodied in a machine) [which] *we* create" (Suber, 1988, §6, p. 103, my italics). So, software is pure syntax and needs *another* piece of syntax to interpret it.

How can one piece of syntax "interpret" another? Recall from §9.4.3 that syntax is the study of the properties of, and relations among, symbols or uninterpreted marks. Roughly, semantics is the study of meaning, and – again, roughly – to say that a piece of syntax *has* a meaning is to say that it is *related* to that meaning.[16] In this view, semantics is the study of the relations between two sets of entities: the syntactic objects and their meanings. But the meanings have their own properties and relations; i.e. the meanings also have a syntax. So the syntax of the meanings can "interpret" the syntax of the software.

This is not far from Moor's definition: Both Moor and Suber require someone (or something) to interpret the syntax. Could a hardwired program and a written program both be software, perhaps because they have the same syntactic form? I think the answer is 'yes.' Here is a possible refinement: software is a pattern that is read*able* and execut*able* by a machine. This is roughly Moor's definition of a computer program. But for Suber, *all* patterns are read*able* and execut*able*. The bottom line is that Suber's notion of *software* is closer to Moor's notion of a computer *program*. The idea that software is pure syntax is consistent with the claim of Tenenbaum and Augenstein, 1981, p. 6, that information has no meaning; recall their statement cited in §3.8. We'll come back to this idea in §13.2.3.

12.6 Colburn: Software Is a Concrete Abstraction

Finally, Timothy R. Colburn (1999, 2000, Ch. 12) argues that software is not "a machine made out of text." Thus, he would probably disagree with Hayes's definition of a computer as "magic paper." Colburn says this because he believes there is a difference between software's "medium of description" and its "medium of execution." The former is the text in a formal language (something relatively abstract). The latter consists of circuits and semi-conductors (which are concrete). Consequently, Colburn says that software is a "concrete abstraction."

But is this a single thing (a "concrete abstraction") or two things: (1) something that is abstract: a "medium of description" and (2) something *else* that is concrete: a "medium of execution"? Colburn borrows the phrase from the title of an introductory CS textbook (Hailperin et al., 1999), which doesn't define it. All that Hailperin et al. say is that abstractions can be thought of "as actual concrete objects," and they give as an example a word processor, which they describe as an *abstraction* that is "merely [a] convenient way of describing patterns of electrical activity" *and* a "*thing that we can buy, sell, copy, and use*" (p. ix).

Part of Colburn's goal is to explicate this notion of a thing that can be both abstract and concrete. To do so, he offers several analogies to positions that philosophers have taken on the mind-body problem (see §2.7), so we might call this the "abstract-program/concrete-program problem." (See https://people.umass.edu/ffeldman/ChisholmMemorial.htm for an illustration of these positions.)

Consider various theories of *monism*, views that there is only one kind of thing: either minds or else brains, but not both. The view that there are only minds is called 'idealism,' associated primarily with the philosopher George Berkeley (1685–1783). The view that there are only brains is called 'materialism' or 'physicalism.' Similarly, a monist with respect to software might hold that either software is abstract or else it is concrete, but it cannot be both. According to Colburn, no matter how strong the arguments for, say, materialism might be as the best answer so far to the mind-body problem, monism as a solution to the abstract-concrete problem fails to account for its dual nature.

So let's consider various theories of *dualism*, views that there are both minds and brains. In the mind-body problem, there are several versions of dualism, differing in how they explain

---

16 That's a controversial claim among philosophers. Some philosophers deny the existence of things that are meanings. Others would say that the meaning of a piece of syntax is the role that it plays in the language that it is part of. We'll discuss this further in §13.1.2.

the relationship between minds and brains. The most famous version is called '*interactionism,*' due to Descartes. This says that (1) there are minds (non-physical substances that think and obey only psychological laws); (2) there are brains (substances that are physically extended in space and that obey only physical and biological laws); and (3) minds and brains interact. The problem for interactionism as a solution to the mind-body problem is that there is no good explanation of *how* they interact. After all, one is physical and the other isn't. So you can't give a physical explanation of how they would interact, because the laws of physics don't apply to the mind. And you can't give a psychological explanation of how they interact, because the laws of psychology don't apply to the brain (according to Cartesian interactionism). Similarly, according to Colburn, when applied to the abstract-concrete problem, an interactionist perspective fails to account for the *relation between* abstractions and concrete things, presumably because the relations themselves are either abstract, in which case they don't apply to concrete things, or they are concrete and, so, don't apply to abstractions. (There is, however, a possible way out, which we will explore in depth in Chapter 13: namely, perhaps the relationship between them is one of implementation, or semantic interpretation, not unlike Suber's theory.)

A theory intermediate between monism and dualism is called the '*dual-aspect*' theory, due to Baruch Spinoza (1632–1677). Here, instead of saying that there are two different kinds of "substance," mental substance and physical substance, it is said that there is a single, more fundamental kind of substance of which minds and brains are two different "aspects." (For Spinoza, this more fundamental substance – which he believed had more than just the two aspects that we humans are cognizant of – was "nature," which he identified with God.) As a solution to the abstract-concrete problem, Colburn points out that we would need some way to characterize that more fundamental underlying "substance," and he doesn't think any is forthcoming. Again, however, one alternative possibility is to think of how a single abstraction can have multiple implementations.[17] Yet another alternative is a dual *property* view: certain physical objects (in particular, brains) can have *both* physical *and* psychological *properties* (Chalmers, 1996).

Finally, another family of dualisms is known as '*parallelism*': there are minds, there are brains, and they are not identical (hence this is dualistic). But they do *not* interact. Rather, they operate in parallel, and so there is no puzzle about interaction. One version of parallelism ('*occasionalism*') says that God makes sure that, on every "occasion" when there appears to be interaction, every mental event corresponds to a parallel brain event (this keeps God awfully busy on very small matters!).

Another version ("*pre-established harmony*") seems to be Colburn's favored version. This says that God initially set things up so that minds and their brains work in parallel, much in the way that two clocks can keep the same time, even though neither causally influences the other. That way, God does not have to keep track of things once they have been set in motion. For Colburn, this seems to mean implementation of an algorithm as a textual program parallels its implementation in the hardware of a physical computer:

> For the abstract/concrete problem we can replace God by the programmer who, on the one hand, by his [sic] casting of an algorithm in program text, describes a world of multiplying matrices, or resizing windows, or even processor registers; but on the other hand, by his act of typing, compiling, assembling, and link-loading, he causes a sequence of physical state changes that *electronically mirrors* his abstract world. (Colburn, 1999, p. 17, my italics)

He puts this slightly differently in Colburn, 2000, p. 208 (my italics), where he says that the "sequence of physical state changes … *structurally matches* his abstract world," and he adds

---

17 See the Online Resources for some examples.

that "the abstract world of the computer programmer can be thought of as ticking along in preestablished synchrony with the microscopic physical events within the machine." The idea that the textual program and the physical state changes share a common structure is consistent with a view that a single abstraction can have two "parallel" implementations. But it is hard to imagine that the textual program (or even the abstract algorithm) can "tick along," because text – unlike the physical events – is static, not dynamic: it doesn't "tick."

A more modern take on the mind-body problem (not considered by Colburn) is "*functionalism.*" Roughly, this is the view that certain abilities or purposes (teleological "functions") of the brain are mental and *are* describable by the laws of psychology *in addition to* the laws of physics and biology (Putnam, 1960, Fodor, 1968b; Block, 1995, §11.5; Levin, 2021). Rather than taking a position on the existence (or "ontological status") of something called "the mind," functionalism holds that what makes certain brain activity *mental* in addition to being *physical* is the role it plays – its "function" – in the overall activity of the brain or the person.

12.7 Summary

Algorithms – which are abstract – can be implemented as programs (i.e. as texts written in a computer-programming language). Programs, in turn, can be implemented as part of the hardware of a computer. A given algorithm can be implemented differently in different programs, and a program can be implemented differently in different computers. Both Suber's and Colburn's theories of software focus on this *implementational* aspect. Moor's theory focuses on the *changeability* of software. Presumably, the more abstract an entity is, the easier it is to change it. So the software-hardware distinction may be more of a continuum than something with a sharp boundary. Moreover, you can't really talk about a program or software *by itself*, but you have to bring in the computer or other entity that interprets it: programs and software are *relational* notions.[18]

We need to explore the notion of implementation in more detail, which we will do in the next chapter.

12.8 Questions for the Reader

1. Turing's work clearly showed the extensive interchangeability of hardware and software in computing.
    —Juris Hartmanis (1993, p. 11)

    Tanenbaum, 2006, p. 8, points out that hardware and software are "logically equivalent" in the sense that anything doable in hardware is doable in software, and vice versa. Similar or analogous cases of such logical equivalence of distinct things are Turing Machines, the lambda-calculus, and recursive functions. Also, such an equivalent-but-different situation corresponds to the intensional-extensional distinction: two intensionally distinct things can be extensionally identical.

    How does this equivalence or "interchangeability" relate to Moor's or Colburn's definitions of software and hardware?

---

18 See the Online Resources for further reading on the software-hardware distinction.

2.　　　Academically and professionally, computer engineering took charge of the hardware, while computer science concerned itself with the software . . . .
　　　　—Michael Sean Mahoney (2011, p. 108)

If software and hardware can*not* easily be distinguished, does that mean neither can computer engineering and computer science?

3. Find a (short) article on the mind-body problem (e.g. the *Wikipedia* article at http://en .wikipedia.org/wiki/Mind-body_problem). Replace all words like 'mind,' 'mental,' etc., with words relating to 'software'; and replace all words like 'body,' 'brain,' etc., with words like 'hardware,' 'computer,' etc. Discuss whether your new paraphrased article makes sense and what this says about the similarities (or differences) between the mind-body problem and the software-hardware problem.[19]

4. Recall from §11.8.2 that Wadler (1997, pp. 240–241) said that "Interaction is the mind-body problem of computing." He was not referring to the kind of interaction that Descartes's dualism requires; rather, he was referring to computers that interact with the real world or with an oracle, such as we discussed in Chapter 11.

　　Nevertheless, how do the various positions on the mind-body problem relate to Wadler's observation?

5. Recall the discussion in §12.3.2 on the relationship of software to music, art, and literature.

　　What do you think Moor or Suber might say about it? Would Moor disagree? After all, art is not usually changeable. Would Suber be more sympathetic? And what about Colburn? Are any art forms "concrete abstractions"?

6. Do you think that "pre-established harmony" explicates "concrete abstraction"? Is the mind a "concrete abstraction"?

7. What would a functional solution to the abstract-concrete problem look like? Might we say that some hardware *functions* as a computer program?

---

19 Thanks to James Geller for this idea.

13

Implementation[1]

"I wish to God these calculations had been executed by steam!"
—Charles Babbage (1821), quoted in Swade 1993, p. 86.

[W]hy wasn't Mark I an electronic device? Again, the answer is money. It was going to take a lot of money. Thousands and thousands of parts! It was very clear that this thing could be done with electronic parts, too, using the techniques of the digital counters that had been made with vacuum tubes, just a few years before I started, for counting cosmic rays. But what it comes down to is this: if Monroe [Calculating Machine Co.] had decided to pay the bill, this thing would have been made out of mechanical parts. If RCA had been interested, it might have been electronic. And it was made out of tabulating machine parts because IBM was willing to pay the bill.
—Howard H. Aiken (quoted by I. Bernard Cohen in Chase 1980, p. 200)

… Darwin discovered the fundamental *algorithm* of evolution by natural selection, an abstract structure that can be implemented or "realized" in different materials or media.
—Daniel C. Dennett (2017, p. 138)

13.1 Introduction

On the one hand, we have a very elegant set of mathematical results ranging from Turing's theorem to Church's thesis to recursive function theory. On the other hand, we have an impressive set of electronic devices which we use every day. Since we have such advanced mathematics and such good electronics, we assume that somehow *somebody must have done the basic philosophical work of connecting the mathematics to the electronics. But as far as I can tell that is not the case. On the contrary, we are in a peculiar situation where there is little theoretical agreement among the practitioners on such absolutely fundamental questions as,* What exactly is a digital computer? What exactly is a symbol? What exactly is a computational process? *Under what physical conditions exactly are two systems implementing the same program?* —John Searle (1990, p. 25, my italics)

The concept of "implementation" has appeared in nearly every chapter so far. So, what is an implementation?

1 Portions of this chapter are adapted from Rapaport 1999, 2005b.

Philosophy of Computer Science: An Introduction to the Issues and the Literature, First Edition. William J. Rapaport.
© 2023 John Wiley & Sons, Inc. Published 2023 by John Wiley & Sons, Inc.

13.1.1 Implementation vs. Abstraction

Let's begin by contrasting "implementation" with "abstraction." Abstractions are usually thought of as being non-physical; the opposite is usually said to be something that is "concrete." But more generally, something is abstract if it *omits some details*:

> What we desire from an abstraction is a mechanism which permits the expression of relevant details and the suppression of irrelevant details. (Liskov and Zilles, 1974, p. 51)

And precisely because abstractions omit details, they are also *more general* than something that has those details filled in. The more details that are omitted, the more abstract and the more general something is. For example, "geology" is (etymologically) the study of the *Earth* and its physical structure, and "selenology" is (etymologically) the study of the *Moon* and *its* physical structure (Clarke, 1951, p. 42). If you *abstract* away from the particular heavenly body that is being studied (Earth or Moon), the result would be a more general science that studies the physical structure of a(n unspecified) heavenly body (even if it might still be *called* 'geology'; we'll return to this idea in Section 18.3.3).

When you fill in some of the details that were omitted in an abstraction, you get an implementation of the abstraction. Indeed, the word 'implement' comes from a Latin word meaning "to fill up, to complete."[2] Importantly, an implementation does not have to be "concrete"; it can itself be abstract if it doesn't fill in *all* the details. As Rosenblueth and Wiener (1945, p. 320) put it, implementation (what they call 'embodiment') is the "converse" of abstraction. But I think it is better to say that an implementation and the abstraction that it implements lie along a spectrum.

Rosenblueth and Wiener observe that in order to understand some part of the complex universe, scientists replace it "by a model of similar but simpler structure" (p. 316) – this is the technique of *abstraction*. Thus, another mark of being an abstraction is to be *simpler* than what it's an abstraction of; what it's an abstraction of (in this case, a part of the universe) will have "extra" features. These extra features might be quite important ones that are being ignored merely temporarily or for the sake of expediency, or they might be "noise" – irrelevant details, or details that arise from the *medium* of implementation. In such cases, the extra features that are not in the abstraction are often referred to as "implementation-dependent details." For example, Rescorla (2014b, Section 2, p. 1280) says, "A physical system usually has many notable properties besides those encoded by our computational model: colors, shapes, sizes, and so on."

There are, according to Rosenblueth and Wiener, two kinds of models: formal and material, both of which are abstractions (p. 316). Formal models are like mathematical models: they are formal symbol systems expressed in formal languages and understood in terms of their *syntax*. Recall from Section 9.4.1 that syntax, in its most general form, can be understood as the study of the properties of, and the relations among, the symbols of a language; e.g. the syntax of a language is its grammar. Let's call a formal system that is understood in terms of its syntax a "syntactic domain."

Material – i.e. *physical* – models, however, are more like scale models (p. 317) than like symbol systems. Because such models omit some details (e.g. scale models are smaller and usually made of different materials than what they are models of), they are "abstract" even though they are "concrete," or physical. But a material model can also be "more elaborate" than that which it models (p. 318). This suggests that "implementation-dependent details" – i.e. parts of the model that are *not* (or are not *intended* to be) representations of the complex system – are ignored. For instance, the physical matter that the model is made of, or imperfections in it, would be ignored: one does

---

2 For more on the etymology of 'implement,' see Rapaport, 1999, Section 2, pp. 110–111; and Section 4, pp. 115–116.

not infer from a plastic scale model of the solar system that the solar system is made of plastic, nor from a globe that the Earth has writing on it.[3]

Typically, implementations are *physical* "realizations" or "embodiments" of *non*-physical "abstractions." That is, implementation is typically understood as a relation between an abstract specification and a concrete, physical entity or process. But a real, *physical* airplane could be considered an implementation of a *physical* scale model "of" it. The former is complete in all details – it really flies and carries passengers – but the scale model, even though physical, lacks these abilities.

13.1.2 Implementations as Role Players

There can be multiple different implementations of a single abstraction: some merely fill in more or different details, but others might do so using different (usually physical) "stuff" – different *media*. For example, in the fairy tale, the three little pigs can be thought of as having used a single abstract blueprint to build three different versions of the "same" house out of three different materials (in three different media): straw, sticks, and bricks.

So, abstractions can be "multiply realized" – implemented in more than one way, just as many different actors can play the same role in different productions (implementations!) of the same play. Hamlet is a role; Richard Burton occupied that role in the 1964 Broadway production of *Hamlet*, and Laurence Olivier occupied it in the 1948 film. Alternatively, we could say that Burton and Olivier "implemented" Hamlet in the "medium" of human being (and a drawing of Hamlet implements Hamlet in the medium of an animated cartoon version of the play).

The implementation-abstraction distinction is also mirrored in the "occupant"-"role" distinction made in functionalist theories of the mind (Lycan, 1990, p. 77). According to those theories, mental states and processes are "functional roles" played – or implemented – by neuron firings in brains (or perhaps computational states and processes in AI computer programs; we'll come back to this idea in Section 19.2).

And mathematical "structuralists" have argued that numbers are not "things" (existing in some Platonic heaven somewhere) but are more like "roles" in a mathematical structure (defined, say, by Peano's axioms) that can be "played" by many different things, such as different sets, Arabic numerals, etc. (Benacerraf, 1965). Recall our brief mention in Section 9.4.6 of structuralism: the natural numbers can be considered a "role" that can be "played" by any "countably infinite (recursive) set ... arranged to form an ω-sequence" (Swoyer, 1991, p. 504, note 26). In the rest of that passage, Swoyer goes on to say that "a concrete realization [i.e. an implementation of the natural numbers] would be obtained by adding a domain of individuals and assigning them as extensions [i.e. as semantic interpretations] to the properties and relations in the structure." In other words, an implementation of the natural numbers is the same as a semantic interpretation of them.

13.1.3 Abstract Data Types

Computer scientists also use the term 'implementation' to refer to the relation between an *abstract* data *type* and its "implementation" or "representation" by an *abstract* data *structure* in a programming language:

> An *abstract data type* defines a class of abstract objects which is completely characterized by the operations available on those objects. This means that an abstract data type can be defined by defining the characterizing operations for that type.
> (Liskov and Zilles, 1974, p. 51)

---

3 A "Family Circus" comic from 26 July 1991 shows little Billy looking at a globe and asking, "Does the real world have writing all over it?"

So, an abstract data *type* is a kind of abstract noun – an indefinite description of the form "an entity that can perform actions $A_1, \ldots, A_k$, and that actions $A_{k+1}, \ldots, A_n$ can be performed on," where the A_i are new, abstract "verbs." The data *structures* that satisfy such a description are implementations of the abstract data type.

Although it is not entirely abstract, a data structure is also not entirely physical: it is part of the software, not the hardware. Moreover, one abstract data type could even be implemented in a different *abstract* data type. Some programming languages, such as Lisp, do not have stacks as a built-in data structure. So, a programmer who wants to write a Lisp program that requires the use of stacks must find a substitute. In Lisp, whose only built-in data structure is a linked list, the stack would have to be built out of a linked list: stacks in Lisp can be implemented by linked lists. Here's how:

First, a stack is a particular kind of abstract data type, often thought of as consisting of a set of items structured like an ordinary, physical stack of cafeteria trays: new items are added to the stack by "pushing" them on "top," and items can be removed from the stack only by "popping" them from the top. Thus, to define a stack, one needs (i) a way of referring to its top and (ii) operations for pushing new items onto the top and popping items off the top. That, more or less (mostly less, since this is informal), is a stack defined as an abstract data type.

Second, a linked list ('list,' for short) is itself an abstract data type. It is a sequence of items whose three basic operations are

1. *first(l)*, which returns the first element on the list *l*,
2. *rest(l)*, which returns a list consisting of all the original items except the first, and
3. *make-list(i, l)*, which recursively constructs a list by putting item *i* at the beginning of list *l*.

Finally, a stack *s* can be implemented as a list *l*, where $top(s) \leftarrow first(l)$, $push(i, s) \leftarrow make\text{-}list(i, l)$, and *pop(s)* returns *top(s)* and redefines the list to be *rest(l)*.

As another example of an "abstract implementation," consider a top-down-design, stepwise-refinement (i.e. a recursive development) of a computer program (Section 6.4.3): each level (each refinement) is an abstract implementation of the previous, higher-level one. Each of the more detailed implementation levels is less abstract than the previous one. A "concrete implementation" would be an implementation in a physical medium.

Question for the Reader: Could this be related to what Colburn might have had in mind when he talked about a "concrete abstraction"? (Recall our discussion in Section 12.6.)

13.1.4 The Structure of Implementation

So, abstractions omit details and can be thought of as roles. Implementations fill in some of those details and can be thought of as things that play the role specified by the abstraction. Some implementations may add "implementation-dependent" details that do not belong to the abstraction. For example, Hamlet's age is not specified in Shakespeare's play: he was supposed to be a college student, but Burton was about 39 when he played the role, and Olivier was about 41; their real ages are implementation-dependent details. Furthermore, there can be multiple implementations of a given abstraction, which differ in the "stuff" that the implementation is made of. (That "stuff" is also an implementation-dependent detail.)

To sum up, implementation is best thought of as a three-place relation:

I is an implementation *in* medium *M of* abstraction *A*.

And there are two fundamental principles concerning this relation:

Implementation Principle I: For every implementation I, there is an abstraction A such that I implements A.

Implementation Principle II: For every abstraction A, there can be more than one implementation of A.

Principle I actually follows from the nature of the three-place relation; Principle II is a generalization of the principle of "multiple realizability" (Bickle, 2020). (We will return to these two principles in Section 18.7.2.)

In the next two sections, we will look at two theories that spell out more of the details about the nature of implementation. One will use the relation between syntax and semantics to illuminate implementation. The other, due to David Chalmers, was designed to reply to Searle's (1990) pancomputationalism (Section 9.4).

13.2 Implementation as Semantic Interpretation

A theory of implementation tells us which conditions the physical system needs to satisfy for it to implement the computation. Such a theory gives us the *truth conditions* of claims about computational implementation. **This serves** not only **as a semantic theory** but also to explicate the concept (or concepts) of computational implementation as they appear in the computational sciences.
—Mark Sprevak (2018, Section 2, original italics, my boldface)

13.2.1 What Kind of Relation Is Implementation?

One main point of the previous section is that not all examples of implementation concern the implementation of something *abstract* by something *concrete*: One abstract thing can implement another abstract thing, and one concrete thing can implement another concrete thing. What we need is a more general notion. There are several candidates:

Individuation: This is the relation between the lowest level of a genus-species tree (such as "cat" or "human") and individual cats or humans: for example, my son's cats Billy and Phoebe "individuate" the species *Felis catus*; both my son and Socrates "individuate" the species "human." Individuation seems to be a kind of implementation: we could say that Billy and Phoebe are *implementations* of *Felix catus*. But not all cases of implementation are individuations.

Instantiation: This is the relation between a specific instance of something and the kind of thing it is: for example, the specific redness of my notebook cover is an instance of the color "red." Instantiation seems to be a kind of implementation: we could say that the specific instance of red that is my notebook's color is an *implementation* of the (abstract) color "red." But not all implementations are instantiations.

Exemplification: This is the relation between a (physical) object and a property that it has: for example, Bertrand Russell exemplifies the property of being a philosopher. Exemplification seems to be a kind of implementation: we could say that Bertrand Russell is an *implementation* of a philosopher. But not all implementations are exemplifications.

Reduction: This is the relation between a higher-level object and the lower-level objects that it is made of: for example, water is reducible to a molecule consisting of two atoms of hydrogen and one atom of oxygen, or, perhaps, the emotion of anger is reducible to a certain combination of neuron firings. This kind of reduction[4] seems to be a kind of implementation: we could say that water is *implemented* by H_2O or that my anger is *implemented* by certain neuron firings in my brain and nervous system. But not all implementations are reductions.

Each of these may be implementations, but not vice versa. In other words, implementation is a more general notion than any of these. But all of them can be viewed as semantic interpretations. (For more details and argumentation, see Rapaport, 1999, 2005b.)[5]

Let *A* be an "abstraction": i.e. something that omits certain details, spells out a role to be played, or is a generalization of the notion of an abstract data type. Abstractions would be characterized by their properties, including their constituents and properties (and the relations *among* the constituents). Thus, abstractions would be characterized by their "syntax." And let *M* be any abstract or concrete "medium." Then we can say that

> *I* is an implementation in medium *M* of abstraction *A*

iff

> *I* is a semantic interpretation in semantic domain *M* of syntactic domain *A*.

Implementation *I* could be either an abstraction itself or something concrete, depending on *M*.

And as we have seen, there could be a sequence of implementations (or what Brian Cantwell Smith (1987) calls a "correspondence continuum"; for discussion, see Rapaport, 1995, Section 2): a stack can be implemented by a linked list, which in turn could be implemented in the programming language Pascal, which in turn could be implemented (i.e. compiled into) some machine language *L*, which in turn could be implemented on my Mac computer. (We saw this same phenomenon in a related situation in Section 9.1. Sloman, 1998, Section 2, p. 2, makes the same point about what he calls "implementation layers.") But it could also be more than a mere sequence of implementations, because there could be a *tree* of implementations. The very same linked list could be implemented in Java as well as in Pascal, and the Java and Pascal programs could be implemented in other machine languages on other kinds of computers.

The ideas that abstractions can implement other abstractions and that there can be "continua" of implementations are a consequence of what the philosopher William G. Lycan (1990, p. 78), refers to as the "relativity" of implementation:

> … "software"/"hardware" talk encourages the idea of a bipartite Nature, divided into two levels, roughly the physiochemical and the (supervenient) "functional" or higher-organizational – as against reality, which is a multiple *hierarchy* of levels of nature …. See Nature as hierarchically organized in this way, and the "function"/"structure" distinction *goes relative*: something is a role as opposed to an occupant, a functional state as opposed to a realizer, or vice versa, only *modulo* a designated level of nature.

13.2.2 Semantic Interpretation

If implementations are semantic interpretations, we need to understand the nature of a semantic interpretation. We'll begin with the notion of a "formal system."

---

4 There are others; see Goodman, 1987, p. 480, and Dennett, 1995, Ch. 3, Section 5, for useful discussions.
5 See the Online Resources for further reading on implementation as semantic interpretation.

13.2.2.1 Formal Systems

> [T]he formal character of [a] system … makes it possible to abstract from the meaning of the symbols and to regard the proving of theorems (of formal logic) as *a game played with marks on paper according to a certain arbitrary set of rules.*
> —Alonzo Church (1933, p. 842, my italics)

Formal systems are sometimes called "symbol systems," "theories" (understood as a set of sentences), or "formal languages." A formal system consists of

1. Primitive (or atomic) "symbols" (sometimes called "tokens" or "markers," which can be thought of as being like the playing pieces in a board game).

Axioms are assumed to have no interpretation or meaning; hence my use of the term 'marker,' rather than 'symbol,' which many writers use to mean a marker *plus* its meaning (as when we say, "a wedding ring is a symbol *for* marriage" (Levesque, 2017, p. 108, my italics). The race car token in Monopoly isn't interpreted as a race car in the game; it's just a token that happens to be race-car shaped to distinguish it from the token that is top-hat shaped. (And the top-hat token isn't interpreted as a top hat in the game: even if you think it makes sense for a race car to travel around the Monopoly board, it makes no sense for a top hat to do so!) Examples of such atomic markers include the letters of an alphabet, (some of) the vocabulary of a language, (possibly) neuron firings, or even states of a computation.

2. (Recursive) rules for forming new (complex, or molecular) markers, sometimes called 'well-formed formulas' (wffs) – i.e. grammatically correct formulas – from "old" markers (that is, from previously formed markers), beginning with the atomic markers as the basic "building blocks."

These rules might be spelling rules (if the atomic markers are alphabet letters), or grammar rules (if the atomic markers are words), or bundles of synchronous neuron firings (if the atomic markers are single neuron firings). The molecular markers can be thought of as "strings" (i.e. sequences of atomic markers), or words (if the atomic markers are letters), or sentences (if the atomic markers are words).

3. Axioms: a "distinguished" subset of wffs (i.e. one that is singled out for a special purpose).

These are optional. Geometry considered as a formal system has axioms. But English considered as a formal system (Montague, 1970) doesn't need them, nor do natural-deduction systems of logic (see Section 15.2.1).

4. Recursive rules (called 'rules of inference' or 'transformation rules') for forming ("proving") new wffs (called 'theorems') from old ones (usually, but not always, beginning with the axioms).

13.2.2.2 Syntax

The "syntax" of such a system is the study of the *properties of* the markers of the system and the *relations among* them (but *not* any relations *between* the markers and anything external to the system). Among these (internal) relations are the "grammatical" relations, which specify which strings of markers are well formed (according to the rules of grammar), and the "proof-theoretic" (or "logical") relations, which specify which sequences of wffs are proofs of theorems (i.e. which wffs are derivable by means of the rules of inference).

Here is an analogy: consider a new kind of toy system consisting of Lego-like blocks that can be used to construct Transformer monsters.[6] The basic Lego blocks are the primitive markers of this

---

6 Transformers are toys that convert from trucks to robots, and back again. A Lego version wouldn't be a very practical real toy, because things made out of Lego blocks tend to fall apart. This is a thought experiment, not a real one!

system. Transformer monsters that are built from Lego blocks are the wffs of the system. And the sequences of moves that transform the monsters into trucks (and back again) are the proofs of theorems.

Turing Machines can be viewed as (implementations of) formal systems: Roughly, (1) the atomic markers correspond to the '0's and '1's of a Turing Machine, (2) the wffs correspond to the sequences of '0's and '1's on the tape during a computation, (3) the axioms correspond to the initial string of '0's and '1's on the tape, and (4) the recursive rules of inference correspond to the instructions for the Turing Machine.[7]

13.2.2.3 Semantic Interpretation

An important fact about a formal system and its syntax is that there is no mention of truth, meaning, reference, or any other "semantic" notion. These are all relations *between* the markers of a formal system and things that are *external* to the system. Semantic relations are not part of the formal system. They are also not part of the system of things that are outside of the formal system![8] We came across this idea in Section 8.8.1, when we discussed Hilbert's claim that geometry could be as much about tables, chairs, and beer mugs as about points, lines, and planes. Tenenbaum and Augenstein (1981, p. 1), note that "the concept of information in computer science is similar to the concepts of point, line, and plane in geometry – they are all undefined terms about which statements can be made but which cannot be explained in terms of more elementary concepts."

But sometimes we want to "understand" a formal system. As we saw in Section 4.5, there are two ways to do that (Rapaport, 1986, 1995). One way is to understand the system in terms of *itself* – to become familiar with the system's syntax. This can be called "syntactic understanding." Another way is to understand the system in terms of *another* system that we *already* understand. This can be called "semantic understanding":

> Material models [i.e. semantic interpretations] … may assist the scientist in replacing a phenomenon in an *unfamiliar* field by one in a field in which he [sic] is *more at home*. (Rosenblueth and Wiener, 1945, p. 317, my italics)

The "semantics" of a formal system is the study of the relations *between* the markers of the system (on the one hand) and something else (on the other hand). The "something else" might be what the markers "represent," or what they "mean," or what they "refer to," or what they "name," or what they "describe." Or it might be "the world." If the formal system is a language, then semantics studies the relations between, on the one hand, the words and sentences of the language, and on the other hand, their meanings. If the formal system is a (scientific) theory, then semantics studies the relations between the markers of the theory and the world – the world that the theory is a theory *of*. (We'll come back to this theme in the next chapter, when we consider whether (some) computer programs are scientific theories.)

Semantics, in general, requires three things:

1. A syntactic domain; call it 'SYN' – typically, but not necessarily, a formal system.
2. A semantic domain; call it 'SEM' – characterized by an "ontology."

An ontology is, roughly, a theory or description of the things in the semantic domain. It can be understood as a (syntactic!) theory of the semantic domain, in the sense that it specifies (a) the parts of the semantic domain (its members, categories, etc.) and (b) their properties and relationships (structural, as well as inferential or logical).

3. A semantic *interpretation* mapping from SYN to SEM.

---

7 See the Online Resources for further reading on syntax, semantics, and formal systems.
8 They play a role not unlike that of the "interaction" between a mind and a body, as discussed in Section 12.6.

SEM is a "model" or "interpretation" of SYN; SYN is a "theory" or a "description" of SEM:

> Physical system *P* realizes/implements computational model *M* just in case [c]omputational model *M* accurately *describes* physical system *P*. (Rescorla, 2014b, p. 1278, my italics)

Here are several examples of semantic domains that are implementations of syntactic domains:

SYNTAX		SEMANTICS
algorithms	are implemented by	computer programs (in language *L*)
computer programs (in language *L*)	are implemented by	computational processes (on machine *m*)
abstract data types	are implemented by	data structures (in language *L*)
musical scores	are implemented by	performances (by musicians)
play scripts	are implemented by	performances (by actors)
blueprints	are implemented by	buildings (made of wood, bricks, etc.)
formal theories	are implemented by	(set-theoretic) models

Digression: Syntax, Semantics, and Puzzles: We can illustrate the difference between syntax and semantics by means of jigsaw puzzles. The typical jigsaw puzzle that one can buy in a store consists of a (usually rectangular) piece of heavy cardboard or wood that has a picture printed on it and has been "jigsawed" into pieces. Each piece has a distinct shape and a fragment of the original picture on it. Furthermore, the shapes of the pieces are such that they can be put together in (usually) only one way. In other words, any two pieces are completely unrelated in terms of their shape, or they are such that they fit together to form part of the completed puzzle. A map of the United States can be used as an example of this "fitting together": the boundaries (i.e. the shape) of New York State and Pennsylvania fit together across New York's southern boundary and Pennsylvania's northern boundary, but New York and California are unrelated in this way. These properties (shapes and picture fragments) and relations (fitting together) constitute the syntax of the puzzle.

The pieces are usually stored in a box that has a copy of the picture on it (put together, but without the boundaries of the pieces indicated). The picture can be thought of as a semantic interpretation of the pieces.

The object of the puzzle is to put the pieces back together again to (re-)form the picture. There are at least two distinct ways to do this:

Syntactically: One way is to pay attention only to the syntax of the pieces. In a rectangular puzzle, one strategy is to first find the outer boundary pieces, each of which has at least

---

(Continued)

> **(Continued)**
>
> one straight edge, and then to fit them together to form the "frame" of the puzzle. Next, one would try to find pieces that fit either in terms of their shape or in terms of the pattern (picture fragment) on it. This method makes no use of the completed picture as shown on the box. If that picture is understood as a semantic interpretation of the pieces, then this syntactic method of solving the puzzle makes no use of semantics.
>
> **Semantically:** But by using the semantic information of the picture on the box, one can solve the puzzle solely by matching the patterns on the pieces with the picture on the box and then placing the pieces together according to that external semantic information.
>
> Of course, typically, one uses both techniques. But the point I want to make is that this is a nice example of the difference between syntax and semantics.

13.2.3 Two Modes of Understanding

Semantic understanding is a two-way street. Typically, we already understand SEM; thus, we can use SEM to help us understand SYN. For example, knowing something about the history and culture of an ancient civilization can help us understand its written language. But we can also use SYN to understand SEM. For example, language and scientific theories expressed in language enable us to describe and understand the world (as we discussed in Chapter 4). Rosenblueth and Wiener (1945, p. 317) observe that in the eighteenth and nineteenth centuries, mechanical models were used to understand electrical problems, but in the twentieth century, electrical models were used to understand mechanical problems! Swoyer (1991, p. 482) notes that a semantic interpretation of a language is a mapping "from the syntax to the semantics" (from SYN to SEM). But in note 27 (p. 504), he observes that in other structural representations, the mapping "runs in the opposite direction": from SEM to SYN, to use our terminology. And Horsman et al., 2014, 2017 use notions like these to analyze the relation between physical and abstract computation.

Data types are another example. In Section 13.1.3, we said that an abstract data type can be implemented by a data structure:

> A data type is an abstract concept defined by a set of logical properties. Once such an abstract data type is defined and the legal operations involving that type are specified, we may implement that data type .... An implementation may be a *hardware implementation*, in which the circuitry necessary to perform the required operations is designed and constructed as part of a computer. Or it may be a *software implementation*, in which a program consisting of already existing hardware instructions is written to interpret bit strings in the desired fashion and to perform the required operations. (Tenenbaum and Augenstein, 1981, p. 8)

But an abstract data type can itself be viewed as an implementation:

> A method of interpreting a bit pattern is often called a *data type*. ...
> ... a type is a method for interpreting a portion of memory. When a variable identifier is declared as being of a certain type, we are saying that the identifier refers to a certain portion of memory and that the contents of that memory are to be interpreted according to the pattern defined by the type. (Tenenbaum and Augenstein, 1981, pp. 6, 45)

What matters is the existence of a mapping; its *direction* is a matter of which system is being used to understand the other. The crucial issue is which system (SYN or SEM) is antecedently understood. One person's antecedently understood domain is another's in need of understanding. The antecedently understood domain can be viewed as an implementation of the domain that needs to be understood. So, (typically) SEM is an *implementation* of SYN. But sometimes SYN is best understood as an implementation of SEM.

Digression: A Recursive Definition of Understanding: We can combine the two kinds of under- standing into a recursive definition of 'understand.' After all, if one understands a domain semantically in terms of an antecedently understood domain, we might wonder how that antecedently understood domain is understood. If it is understood in terms of yet another antecedently understood domain, we run the risk of an infinite regress, unless there is one domain that is understood in terms of *itself*, rather than in terms of another domain. But if a domain is going to be understood in terms of itself, it would have to be understood in terms only of *its* properties and internal relations, and that means it would have to be understood *syntactically*. So the base case of understanding is to understand something syntactically – in terms of itself. The recursive case of understanding is to understand something semantically, in terms of something else that is already understood. (See Rapaport, 1995 for further dis- cussion. Linnebo and Pettigrew, 2011 introduce a notion of "conceptual autonomy": a theory "T_1 has *conceptual autonomy* with respect to T_2 if it is possible to understand T_1 without first understanding notions that belong to T_2" (Assadian and Buijsman, 2019, p. 566). Using this ter- minology, we could say that syntactic understanding is conceptually autonomous with respect to semantic understanding, but not vice versa.)

Consider a program written in a high-level programming language. Suppose the program has a data structure called a "person record" containing information about (i.e. a representation of) a person, something like the record in two *Bloom County* comic strips (https://www.gocomics.com/ bloomcounty/2010/08/06 and https://www.gocomics.com/bloomcounty/2010/08/09). In the first strip, a young computer hacker deletes "all traces of" his father "from the files of the I.R.S.," at which point his *father* disappears. (And the hacker says, "Even the breathtaking political, philo- sophical and religious implications of this are dwarfed by the breathtaking implications of explain- ing this to Mom.") In the second strip, the hacker enters the following into his computer:

> Enter new record: "Howard L. Jones, Age 36. Height 6ft. Race: Black. Soc. Sec.# 003-15-9003. Serial# 66-77-1140. License# 3476140. Duck-hunting permit# 78103.

– at which point his *father* pops back into existence.
So, Howard L. Jones's record looks something like this:

```
(person-record:
 (name "Howard L. Jones")
 (age 36)
 (height (feet 6))
 (race Black)
 (ssn 003-15-9003)
 (serial-number 66-77-1140)
```

```
    (license-number 3476140)
    (duck-hunting-permit 78103)
  )
```

This is merely a piece of syntax: a sequence of markers. You and I reading it might think it represents a person named 'Howard L. Jones' whose age is 36, whose height is 6 feet, and so on. But as far as the computer (program) is concerned, this record might just as well look like this (McDermott, 1980):

```
(PR:
  (g100 n456)
  (g101 36)
  (g102 (u7 6)))
  (g103 r7)
  (g104 003159003)
  (g105 66771140)
  (g106 3476140)
  (g107 78103)
)
```

And in fact, the machine-language version of this record does look much like this (Colburn, 1999, p. 8). As long as the program 'knows' how to input new data, how to compute with these data, and how to output the results in a humanly readable form, it really doesn't matter what the data look like *to us*. That is, as long as the relationships among the symbols are well specified, it doesn't matter – as far as computing is concerned – how those symbols are related to other symbols that might be meaningful *to us*. That is why it is syntax, not semantics.

Now, there are at least two ways in which this piece of syntax could be implemented. One implementation, of course, might be Jones himself in the real world:[9] a person named 'Howard L. Jones' who is 36 years old, etc. Jones – the real person – implements that data structure; he is also a semantic interpretation of it.

Another implementation is the way in which that data structure is actually represented in the computer's machine language. That is, when the program containing that data structure is compiled into a particular computer's machine language, that data structure will be represented in some other data structure expressed in that machine language. That will actually be another piece of syntax. And that machine-language syntax will be an implementation of our original data structure.

But when that machine-language program is being executed by the computer, some region of the computer's memory will be allocated to that data structure (to the computer's representation of Jones, if you will), which will probably be an array of '0's and '1's – more precisely, of bits in memory. These bits will be yet another implementation of the original data structure as well as an implementation of the machine-language data structure.

Question for Discussion: What is the relation between the human (Jones himself) and this region of the computer's memory? Does the memory location "simulate" Jones? (Do bits *simulate* people?) Does the memory location *implement* Jones? (Do bits *implement* people?) Also: the

---

9 Yes, I'm aware that this Jones is a cartoon character and hence not a real person!

> '0's and '1's in memory can be thought of as the '0's and '1's on a Turing Machine tape, and Jones can be thought of as an interpretation of that Turing Machine tape. Now, recall from Section 10.4 what Cleland said about the difference between Turing Machine programs and mundane procedures: the former can be understood independently of any real-world interpretation (i.e. they can be understood purely syntactically, to use the current terminology) – understood in the sense that we can describe the computation in purely '0'/'1' terms. (Can we? Don't we at least have to interpret the '0's and '1's in terms of a machine-language data structure, interpretable in turn in terms of a high-level programming language data structure, which is interpretable, in turn, in terms of the real-world Jones?) Mundane procedures, on the other hand, must be understood in terms of the real world (i.e. the causal world) that they are manipulating.

13.3 Chalmers's Theory of Implementation

13.3.1 Introduction

It is one thing to spell out the general structure of implementation as we did in Section 13.1 and another to suggest that the notion of semantic interpretation is a good way to understand what implementation is, as we did in Section 13.2.2. But we still need a more detailed *theory* of implementation: what is the nature of the relation between an abstraction and one of its implementations?

One reason we need such a theory is to refute Searle's (1990) claim that "*any* given [physical] system can be seen to implement *any* computation *if interpreted appropriately*" (Chalmers, 1993, p. 325, my italics). David Chalmers's (1993) essay, "A Computational Foundation for the Study of Cognition" concerns both implementation and cognition; here, we will focus only on what he has to say about implementation.

One of his claims is that we need a "bridge" between the *abstract theory* of computation and *physical* systems that "implement" them. Besides 'bridge,' other phrases that he mentions as synonyms for 'implement' are: 'realize' (i.e. make real)[10] and 'described by,' as in this passage:

> Certainly, I think that when a physical system implements an a-computation [i.e. a computation abstractly conceived], the a-computation can be seen as a description of it. (Chalmers, 2012, p. 215)

That is, the physical system that implements the abstract computation *is described by* that computation. (Here, Chalmers seems to agree with Rescorla, Section 13.2.2.)

13.3.2 An Analysis of Chalmers's Theory

According to the simplest version of Chalmers's theory of implementation,

> A physical system implements a given computation when the causal structure of the physical system mirrors the formal structure of the computation. (Chalmers, 1993, p. 326)

---

10 Note, however, that all realizations are implementations, but not vice versa. A realization makes an abstraction "real" (e.g. physical). An implementation fills in all details, but the implementation might still be abstract. The notions of implementation and abstraction are relative terms (Section 13.2.1).

Almost every word of this needs clarification! For convenience, let *P* be a "physical system," and let *C* be a "computation":

- What kind of physical system is *P*? It need not be a computer, according to Chalmers.
- What kind of computation is *C*? Is it merely an abstract *algorithm*? Is it more specifically a Turing Machine *program*? In any case, it would seem to be something that is more or less abstract (because it has a "formal structure"; recall our discussion in Section 3.9 of the meaning of 'formal').
- What *is* a "formal structure"?
- What is a "causal structure"?
- What does 'when' mean? Is this intended to be *just* a sufficient condition ("when"), or is it supposed to be a stronger *biconditional* ("when *and only when*")?
- And the most important question: what does 'mirror' mean?

So, here is Chalmers's "more detail[ed]" version (1993, p. 326, my interpolated numerals), which begins to answer some of these questions:

> A physical system implements a given computation when there exists [1] a grouping of physical states of the system into state-types and [5a] a one-to-one mapping from formal states of the computation to physical state-types, such that [3] formal states related by an abstract state-transition relation are mapped [5b] onto physical state-types [2] related by a [4] corresponding causal state-transition relation.

Let me try to clarify this. (The numerals I inserted into the previous passage correspond to the following list.) According to Chalmers, *P* implements *C* when (and maybe only when)

1. the *physical states* of *P* can be grouped into (physical-)state *types*,[11]
2. the physical-state *types* of *P* are related by a *causal* state-transition relation,
3. the *formal states* of *C* are related by an *abstract* state-transition relation,
4. the abstract state-transition relation of *C corresponds* to the causal state-transition relation of *P*, and
5. there is (a) *a 1–1 and* (b) *onto* map *from* the formal states of *C to* the physical-state types of *P*.

We still need some clarification. We have already defined "1–1" and "onto" maps in Section 7.2.2. A state is "formal" if it's part of the *abstract* – i.e. non-physical – computation *C*, and a state is "causal" if it's part of the *physical* system *P*: here, 'formal' just means "abstract," and 'causal' just means "physical." But what are abstract and causal "state-transition relations"? And what does 'correspond' mean?

Figure 13.1 may help make some of this clear. In this figure, the physical system *P* is represented as a set of dots, each of which represents a physical state of *P*. These dots are partitioned into subsets of dots: i.e. subsets containing the physical states. Each subset represents a state-type: i.e. a set of states that are all of the same type. (That takes care of part 1 of Chalmers's account.)

---

11 A word on punctuation. Chalmers calls them 'physical state-types.' This *looks* as if he is talking about "state-types" that are "physical." But he is really talking about "types" of *P*'s "physical states." In other words, '-types' has wide scope over 'physical state,' even though it doesn't look like that. So, I prefer to call them 'physical-state types' – note the position of the hyphen, which I think clarifies that no matter how you hyphenate it, what's being discussed are "types of physical states." The types themselves are *not* physical; "types" are collections of things, so they are abstract, even though the things they are collections of *are* physical.

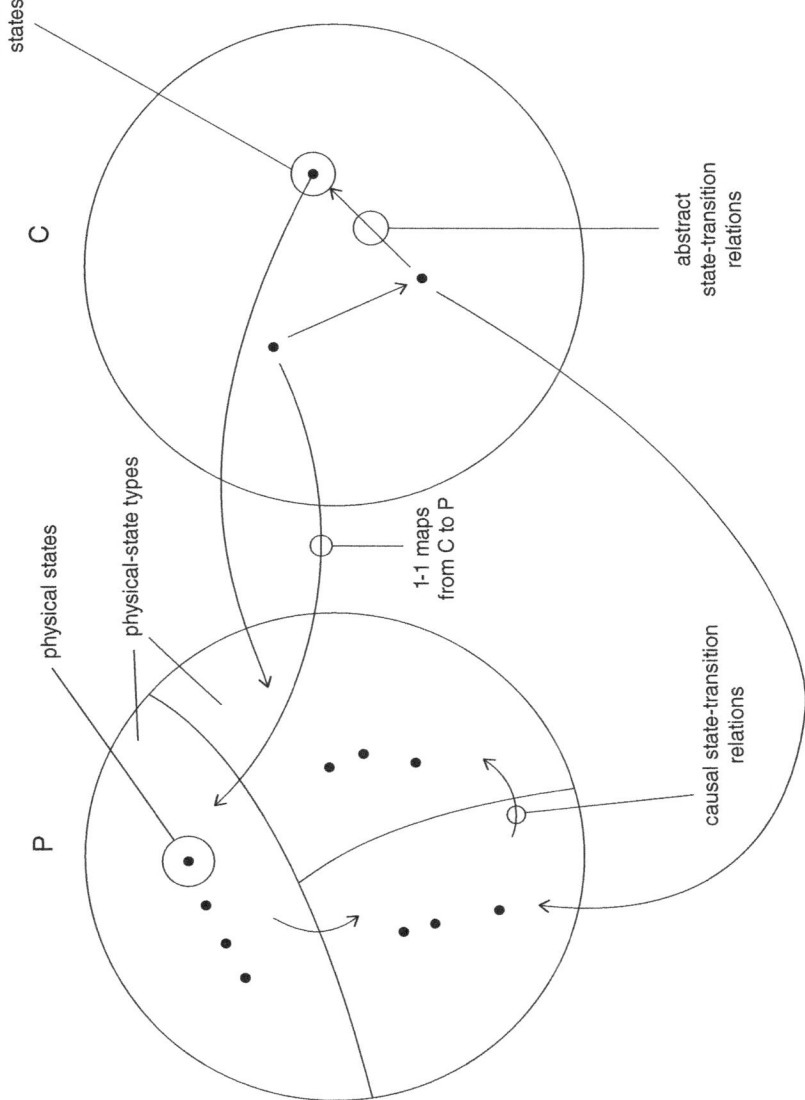

Figure 13.1 A pictorial representation of Chalmers's analysis of implementation; see the text for explanation. Source: Author's drawing.

The computation C is also represented as a set of dots. Here, each dot represents one of C's formal states. The arrows that point *from* the dots in C (i.e. from C's formal states) *to* the subsets of dots in P (i.e. to the state-types in P) represent the 1–1 map from C to P. To make it a 1–1 map, each formal state of C must map to a *distinct* physical state-type of P. (That takes care of part 5a of Chalmers's account.)

The arrows *in* set C represent the abstract state-transition relation among C's formal states (that's part 3). And the arrows *in* set P among P's subsets represent the causal state-transition relation among P's state-types (that's part 2). Finally, because C's formal states are mapped *onto* P's physical-state types, the 1–1 map is a 1–1 correspondence (this is part 5b).

Chalmers (1993, p. 326) seems to be aware of the two-sided nature of understanding. As we have seen, he says that P implements C when there is an isomorphism *from C to P*; yet on the very next page, he says that P implements a finite-state automaton M if there is a mapping *from P to M*! (Of course, the mapping is a 1–1 correspondence, and therefore it has an inverse!)

Chalmers also says that C's *abstract* state-transition relations "correspond" to P's *causal* state-transition relations. I take it that this means the 1–1 correspondence is a "homomorphism": i.e. a structure-preserving map (that's part 4). Because the map is also "onto," it is an "isomorphism." (An isomorphism is a structure- or "shape"-preserving 1–1 correspondence.) So, P and C have the same *structure*.

Digression: Homomorphism: Suppose $c_1, \ldots, c_n$ are entities that stand in relation R. Then a function f is a *homomorphism* $=_{def} f(R(c_1, \ldots, c_n)) = f(R)(f(c_1), \ldots, f(c_n))$. That is, if the c_i are related by some relation R, and if that relationship is mapped by f, then the image of $R(c_1, \ldots, c_n)$ will be the image of R applied to the images of the c_i. That's what it means to preserve structure.

We can then say that a physical system (e.g. a process) P *implements* an abstract computation C (e.g. a Turing Machine or a less-powerful finite automaton [recall Section 11.1]) or a "combinatorial-state automaton" (see the next Digression) if and only if there is a "reliably causal" isomorphism $f : C \rightarrow P$. ('Reliably' perhaps means something like "no physical breakdowns"; see Chapter 15.) Such an f is a relation between an abstract, computational model and something in the real, physical, causal world. This f is 1–1 and onto – a structure-preserving isomorphism such that the abstract, input-output and processing relations in C correspond to reliably causal processes in P. Michael Rescorla (2013, Section 1, p. 682) dubs Chalmers's view of implementation "structuralism about computational implementation." It is the fact that the *structure* of the physical system matches ("mirrors," in Chalmers's terms; more precisely, is isomorphic to) the structure of the computational system that matters. Note that P can be viewed as a semantic interpretation of C, and C can be viewed as a description of P.

Digression: Combinatorial-State Automata: According to Chalmers (1993, p. 328),

> Simple finite-state automata are unsatisfactory for many purposes, due to the monadic nature of their states. The states in most computational formalisms have a combinatorial structure: a cell pattern in a cellular automaton, a combination of tape-state and head-state in a Turing machine, variables and registers in a Pascal program, and so on. All this can be accommodated within the framework of combinatorial-state automata …, which differ from … [finite automata] only in that an internal state is specified not

> by a monadic label S, but by a vector $[S^1, S^2, S^3, \ldots]$. The elements of this vector can be thought of as the components of the overall state, such as the cells in a cellular automaton or the tape-squares in a Turing machine.
>
> See the rest of Chalmers, 1993, Section 2.1, for more details.

Three things follow from this analysis. First, *every* physical system implements *some* computation. That is, for every physical system P, there is some computation C such that P implements C. Does this make the notion of computation vacuous? No, because the fact that some P implements some C is not necessarily *the reason why* P is the *kind* of physical process that it is. (But in the case of cognition, it *might* be the reason! We'll come back to this in Chapter 18.)

But second, *not every* physical system implements *any* given computation. That is, it is not the case that for every P and for every C, P implements C. That is, there is some P and there is some C such that P does *not* implement C (because there are computations that cannot be mapped isomorphically to P). For example, it is highly unlikely that the wall behind me implements WordStar, because the computation is too complex.

Finally, a *single* physical system *can* implement *more than one* computation. That is, for any P, there might be two different computations $C_1 \neq C_2$ such that P implements *both* C_1 *and* C_2. For example, my computer, right this minute as I type this, is implementing the "vi" text-processing program as well as a clock, PowerPoint, and several other computer programs ("apps"), because each of these computations map to *different* parts of P. (We'll see other examples in Section 16.2; recall Section 10.4.)

13.3.3 Rescorla's Analysis of Chalmers's Theory

There is one aspect of Chalmers's analysis that we have not yet considered: "when" vs. "only when." Taken literally, Chalmers has offered only a *sufficient* condition for P being an implementation of C: *when* (i.e. "if") there is a 1–1 correspondence from C to P as described previously, *then* P implements C. But is this also a *necessary* condition – is P an implementation of C *only when* (i.e. only if) there is such a 1–1 correspondence? (That is, *when* (or *if*) P is an implementation of C, *then* there is such a 1–1 correspondence.)

Interestingly, Rescorla (2013) agrees that such structural identity is *necessary* for a physical system to implement a computation, but he *denies* that it is *sufficient*! That is, although any physical system that implements a computation must have the same structure as the computation, there are (according to Rescorla) physical systems that *have* the same structure as certain computations but are *not* implementations of them (Section 1, p. 683). This is because *semantic* "relations to the social environment sometimes help determine whether a physical system realizes a computation" (Abstract, p. 681). The key word here is 'sometimes': "On my position, the implementation conditions for some but not all computational models essentially involve semantic properties" (Section 2, p. 684).

Roughly, the issue concerns the "intentionality" of implementation: must P somehow be "intended" (by whom?) to implement C? Or could P be, so to speak, an "accidental" implementation of C?[12] (Recall our discussion in Section 3.4.1 of "chauvinism" vs. "liberalism" when trying to formalize informal notions.) In that case, Rescorla might say that P wasn't *really* an implementation of C. But Rescorla's position is rather more subtle. A year earlier, he had said:

---

12 Perhaps in the same way that a fictional character might only "coincidentally resemble" a real person (Kripke, 2011, pp. 56, 72).

Mathematical models of computation, such as the Turing machine, are abstract entities. They do not exist in space or time, and they do not participate in causal relations. Under suitable circumstances, a physical system *implements* or *physically realizes* an abstract computational model. Some philosophers hold that a physical system implements a computational model only if the system has semantic or representational properties [... Ladyman 2009]. Call this *the semantic view of computational implementation*. In contrast, [Chalmers 1994; Piccinini, 2006], and others deny any essential tie between semantics and physical computation. *I agree with Chalmers and Piccinini*. (Rescorla, 2012a, Section 2.1, p. 705, my italics)

But in Rescorla, 2013, p. 684, after reciting the semantic and non-semantic passage just quoted in almost the same words, he says that he "*reject*[s] *both* the semantic and the non-semantic views of computational implementation" (my italics). We will investigate this issue in more detail in Chapter 16. But in Rescorla, 2013, he provides a "counterexample to the non-semantic view" (Section 1, p. 684): an example of a physical implementation that – he claims – *requires* a representational (i.e. a semantic) feature. (It is not enough to find an implementation that merely *has* a semantics; there are plenty of those, because a semantic interpretation can always be given to one.) One example that he gives is a Scheme program for Euclid's algorithm for computing GCDs (Section 4, p. 686):

```
(define (gcd a b)
  (if (= b 0)
    a
    (gcd b (remainder a b)))))
```

This is a recursive algorithm that we can paraphrase in English as follows:

> **To** compute the GCD of integers a and b, **do** the following:
> **If** $b = 0$,
> > **then** output a
> > **else** compute the GCD of b and the remainder of dividing a by b.

Rescorla points out that "To do that, the machine *must* represent *numbers*. Thus, the Scheme program contains *content*-involving instructions ..." (Section 4, p. 687, my italics). A "content-involving instruction is specified, at least partly, in semantic or representational terms" (Section 3, p. 685; he borrows this notion from Peacocke, 1995). So, the Scheme program is specified in semantic terms (specifically, it is specified in terms of integers). Therefore, if a physical system is going to implement the program, that physical system must represent integers; i.e. it requires semantics. Hence, "The Scheme program is a counter-example to the non-semantic view of computational implementation" (Section 4, p. 687).

I can see that the machine *does* represent numbers (or can be interpreted as representing them). But why does he say that it *must* represent them? I can even see that for an agent to use such a physical computer *to compute GCDs*, the *agent must* interpret the computer as representing numbers. But surely an agent could use this computer, implementing this program, to print out interesting patterns of uninterpreted markers. (Recall the computer-in-the-desert of Section 3.11.)

To respond to this kind of worry, Rescorla asks us to consider two copies of this machine, calling them M_{10} and M_{13}. The former uses base-10 notation; the latter uses base-13. When each is given the input pair of *numerals* ('115,' '20'), each outputs the *numeral* '5.' But only the former computes the GCD of the *numbers* 115 and 20. (The latter was given the integers 187 and 26 as inputs; but their GCD is 1.) So M_{10} implements the program, but M_{13} does not; yet they are identical physical computers.

One possible response to this is that the semantics lies in the user's interpretation of the inputs and outputs, not in the physical machine. Thus, one could say that both machines *do* implement the program but that it is the user's interpretation of the inputs, the outputs, and the program's symbols that makes all the difference. After all, consider the following Scheme program:

```
(define (MYSTERY a b)
  (if (= b 0)
    a
    (MYSTERY b (remainder a b)))))
```

If we are using base-10 notation, we can interpret 'MYSTERY' as GCD; if we are using base-13 notation, we might either be able to interpret 'MYSTERY' as some other numerical function or not be able to interpret it at all. In either case, our two computers both implement the MYSTERY program.

One possible conclusion to draw from this is that any role semantics has to play is not at the level of the abstract computation, but rather at the level of the physical implementation. Rescorla's response to this might be incorporated in these remarks:

> The program's formal structure does not even begin to fix a unique semantic interpretation. Implementing the program requires more than instantiating a causal structure that mirrors relevant formal structure. (Rescorla, 2013, Section 4, p. 688)

I agree with the first sentence: we can interpret the MYSTERY program in many ways. I disagree with the term 'requires' in the second sentence: I would say that implementing the program *only* requires "instantiating the mirroring causal structure." But I would go on to say that if one wanted to use the physical implementation to compute GCDs, then one would, indeed, be required to do something extra: namely, to provide a base-10 interpretation of the inputs and outputs (and an interpretation of 'MYSTERY' as GCD).

In fact, Rescorla agrees that the semantic interpretation of 'MYSTERY' as GCD is required: "there is more to a program than meaningless signs. The signs have an intended interpretation …" (Section 4, p. 689). But it is notoriously hard (some would say logically impossible)[13] to pin down what "the intended interpretation" of any formal system is. We will return to this debate and Rescorla's example in Chapter 16.

Questions to Consider:

1. Are the inputs to the Euclidean GCD algorithm *numerals* (like '10') or *numbers* (like 10 or 13)? (Note that the base-10 numeral '10' represents the number 10, but the base-13 numeral '10' represents the number 13.)
2. What about the inputs to a computer program written in Scheme that implements the Euclidean algorithm: are *its* inputs numerals or numbers? It may help to consider this analogous question: is the input to a word-processing program the letter 'a' or an electronic signal or ASCII-code representing 'a'?

---

13 In part because of something called the Löwenheim-Skolem Theorem. For discussion, see Suber, 1997b.

Rescorla (2012b, p. 12; italics in original) gives another example of semantic computation, in the sense of a computation that requires *numbers*, not (merely) *numerals*:

> A register machine contains a set of memory locations, called *registers*. A program governs the evolution of register states. The program may individuate register states syntactically. For instance, it may describe the machine as storing *numerals* in registers, and it may dictate how to manipulate those syntactic items. Alternatively, the program may individuate register states representationally. Indeed, the first register machine in the published literature models computation *over natural numbers* [Shepherdson and Sturgis 1963, p. 219]. A program for this numerical register machine contains instructions to execute elementary arithmetical operations, such as *add 1* or *subtract 1*. A physical system implements the program only if [it] can execute the relevant arithmetical operations. A physical system executes arithmetical operations only if it bears appropriate representational relations to numbers. Thus, a physical system implements a numerical register machine program only if it bears appropriate representational relations to numbers. Notably, a numerical register machine program ignores *how* the physical system represent[s] numbers. It applies whether the system's numerical notation is unary, binary, decimal, etc. The program characterizes internal states representationally (e.g. *a numeral that represents the number 20 is stored in a certain memory location*) rather than syntactically (e.g. *decimal numeral "20" is stored in a certain memory location*). It individuates computational states through denotational relations to natural numbers. It contains mechanical rules (e.g. *add 1*) that characterize computational states through their numerical denotations.

I agree that this is a *semantic* computation. Note that it is *not* a Turing Machine (which *would* be a purely syntactic computation). And note that there cannot be a *physical* "numerical register machine" – i.e. a register machine that manipulates *numbers*, not numerals – only a *syntactic* one. This is not because there are no numbers, but because (if numbers do exist) they are not physical!

These are important questions, and we will return to them in Chapter 16. But first we need to look at two issues concerning the nature of computer programs: in the next chapter, we'll consider whether any computer programs can be considered scientific theories. Then, in Chapter 15, we'll discuss whether computer programs can be "verified" (or proved "correct").[14]

---

14 See the Online Resources for further reading on Chalmers and implementation.

14

Computer Programs as Scientific Theories

[W]ithin ten years most theories in psychology will take the form of computer programs, or of qualitative statements about the characteristics of computer programs.
—Herbert A. Simon and Allen Newell (1958, pp. 7–8)

… what's a [scientific] theory? It's a computer program for predicting [calculating] observations. And the statement that the simplest theory is best translates into saying that a concise computer program constitutes the best theory.
—Gregory Chaitin (2005, pp. 170, 175, 188; Chaitin's words in brackets)

14.1   Introduction

I haven't formalized my theory of belief revision, but I have an algorithm that does it.
—Frances L. Johnson (personal communication, February 2004)

The issue raised in this epigraph (from a former graduate student in my department) is whether an *algorithm* or a computer *program* – both of which are pretty formal, precise things – is different from a formal *theory*. Some might say that her algorithm *is* her theory. Others might say that they are distinct things and that her algorithm (merely) *expresses* – or *implements* – her theory. Roger Schank – an AI researcher famous for taking a "scruffy" (i.e. non-formal) approach to AI – used formal algorithms to express his non-formal theories. That sounds paradoxical. Does it really make sense to say that you *don't* have a formal theory of something if you *do* have a formal algorithm that implements your (perhaps informal) theory?

We have seen that algorithms are implemented in computer programs. If implementation is semantic interpretation (as I suggested in Section 13.2), then computer programs are semantic interpretations of algorithms in the medium of some programming language. However, some philosophers have argued that computer programs are theories; yet theories are more like abstractions than they are like implementations. After all, if an algorithm (merely) *expresses* a theory, then a theory is akin to an abstract idea, as in our discussion of copyrights in Section 12.3.4. And others have argued that computer programs are *simulations* or *models* of real-world situations, which sounds more like an implementation than an abstraction.

In Section 4.6, we briefly discussed the nature of scientific theories. In this chapter, we will look further into the nature of theories, models, and simulations and whether programs are scientific theories. And we will begin an investigation into the relation of a program to that which it models or simulates. (We'll continue it in Chapter 16.)

Philosophy of Computer Science: An Introduction to the Issues and the Literature, First Edition. William J. Rapaport.
© 2023 John Wiley & Sons, Inc. Published 2023 by John Wiley & Sons, Inc.

14.2 Simulations

> Computer simulations have introduced some strange problems into reality.
> —Sherry Turkle, quoted in Shieh and Turkle, 2009

Simulations are sometimes contrasted with "emulations." And sometimes a simulation is taken to be an "imitation." Let's look at these distinctions.

14.2.1 Simulation vs. Emulation

There is no standard, agreed-upon definition of either 'simulation' or 'emulation.' This sort of situation occurs unfortunately all too frequently. Therefore, it is always important for you to try to find out how a person is using such terms before deciding whether to agree with what they say about them.

Here is a paraphrase of one definition of 'simulate,' from the *Encyclopedia of Computer Science* (R.D. Smith, 2000):

> x **simulates** y means (roughly): y is a real or imagined system, and x is a model of y, and we experiment with x in order to understand y.

This is only a rough definition, because it does not say what is meant by 'system,' 'model,' or 'understand,' not to mention 'real,' 'imagined,' or 'experiment'! Typically, a computer program (x) is said to simulate some real-world situation y when program x *stands in for y*: that is, when x is a model of situation y. If we want to understand *situation y*, we can do so by experimenting with *program x*. In the terminology of Section 13.2.3, presumably the program is antecedently understood – at least it is more understandable than the situation it simulates, because it is *designed* by someone. (We'll discuss an exception to this in Chapter 17.) Perhaps the program is easier to deal with or manipulate than the real-world situation. In an extreme case, x simulates y if and only if x and y have the same input-output behavior, but they might differ greatly in some of the details of how they work.

And, paraphrasing another *Encyclopedia of Computer Science* definition (Habib, 2000), let's say that

> x **emulates** y means (roughly)
>
> *either*:
> x and y are computer systems, and x interprets and executes y's instruction set by implementing y's operation codes in x's hardware – i.e. hardware y is implemented as a virtual machine on x,[1]
>
> *or*:
> x is some software feature, and y is some hardware feature, and x simulates y, doing what y does "exactly" as y does it.

In general, x emulates y if and only if x and y have the same input-output behavior (x simulates y) and x also uses the same algorithms and data structures as y.

––––––––
1 On emulation as *simulation by a virtual machine*, see Denning and Martell, 2015, p. 212.

It is unlikely that being a simulation and being an emulation are completely distinct notions. More likely they are the ends of a spectrum, in the middle of which are xs and ys that differ in the level of detail of the algorithms and data structures that x uses to do y's job. At the "pure" simulation end of the spectrum, only x's and y's external, input-output behaviors agree; at the "pure" emulation end, all of their internal behaviors also agree. Perhaps, then, the only pure example of an emulation of y would be y itself! Perhaps, even, there is no real distinction between simulation and emulation except the degree of faithfulness to what is being simulated or emulated.

Questions for the Reader:

1. Does a Universal Turing Machine that is executing a program for some algorithm A *simulate* or *emulate* a (dedicated) Turing Machine for A?
2. Is that Universal Turing Machine "really" executing A, or is it "merely" simulating or emulating it?

14.2.2 Simulation vs. Imitation

In many cases, y is only an imagined situation, whereas x is always something real. On the other hand, it is often said that x is "merely" a simulation, which suggests that y *is* real but x is not. That is, the word 'simulation' has a connotation of "imitation" or "unreal." For example, it is often argued that a simulation of a hurricane is not a real hurricane, or that a simulation of digestion is not real digestion.

But there are cases where a simulation *is* the real thing. (Or should such simulations be called 'emulations'?) For example, although a scale model of the Statue of Liberty is not the real Statue of Liberty, a scale model of a scale model (of the Statue of Liberty) *is* itself a scale model (of the Statue of Liberty). A Xerox copy or PDF or faxed copy of a document *is* that document, even for legal purposes (although perhaps not for historical purposes;[2] see Korsmeyer, 2012). Some philosophers and computational cognitive scientists have argued that a computational simulation of cognition really is cognition (Edelman, 2008a; Rapaport, 2012b). In general, it seems, a simulation of *information* **is** that information.

There are also cases where it is difficult or impossible to tell if something is a simulation or not, such as Nick Bostrom's (2003) argument that "we are almost certainly living in a [*Matrix*-like] computer simulation" (see Sections 9.7.2 and 19.7). After all, if a program could be a scientific *theory*, then the process that comes into being when the program is executed could be a *model* of what the program is a theory of. And if some models *are* the kind of thing that they model, then a simulation of the real world could be a real world.[3]

14.2.3 Models

> … computational models are better able to describe many aspects of the universe better than any other models we know. All scientific theories can, e.g. be modeled by programs.
> —Donald E. Knuth (2001, p. 168)

Simulations and semantic theories are sometimes said to be "models." So, what is a model?

---

2 Unless a PDF *is* the original document!
3 See the Online Resources for further reading on simulation.

The notion of *model* is associated with what I have called "The Muddle of the Model in the Middle" (Wartofsky, 1966, 1979; Rapaport, 1995). There are two different uses of the term 'model': it can be used to refer to a *syntactic* domain, as in the phrase 'mathematical model' of a real-world situation. And it can be used to refer to a *semantic* domain, as in the phrase 'set-theoretic model' of a mathematical theory. And, of course, there is the real-world situation that both of them refer to in some way. The "muddle" concerns the relationships among these.

We saw the dual nature of models in Section 13.2.3, when we briefly considered what Brian Cantwell Smith (1987) called a "correspondence continuum" (Section 13.2.1): scientists typically begin with data that they then interpret or model using a formal theory; so, the data are the syntactic domain in need of understanding, and the formal theory is a semantic domain in terms of which it can be understood. The formal theory can then be modeled set-theoretically or mathematically; the formal theory now becomes the syntactic domain, and the set-theoretic or mathematical model is *its* semantic domain. But that set-theoretic or mathematical model can be interpreted by some real-world phenomenon; so, the model is now the syntactic domain, and the real world is the semantic domain. To close the circle, that real-world phenomenon consists of the same kind of data that we started with! (Recall the example in Section 13.2.3 of person records and persons.) Hence my phrase "the muddle of the model *in the middle*."

(For more on models, and an argument that computing is modeling, see Shagrir, 2022, Ch. 9.)

14.2.4 Theories

Recall our discussion in Section 4.6 of the term 'theory.' When people say in ordinary language that something is a "theory," they often mean it is mere speculation, that it isn't necessarily true. But scientists and philosophers use the word 'theory' in a more technical sense.

This is one reason people who believe in the "theory" of evolution and those who don't are often talking at cross purposes, with the former saying that evolution is a true, scientific theory:

> Referring to biological evolution as a theory for the purpose of contesting it would be counterproductive, since scientists only grant the status of theory to well-tested ideas. (Terry Holliday, Kentucky education commissioner, 2011; cited in *Science* 337 (24 August 2012): 897)

and the latter saying that if it is only a theory – if, that is, it is mere speculation – then it might not be true:

> The theory of evolution is a theory, and essentially the theory of evolution is not science – Darwin made it up. (Ben Waide, Kentucky state representative, 2011; cited in *Science* 337 (24 August 2012): 897)

They are using the word in very different senses.

Further complicating the issue, there are at least two views within the philosophy of science about what scientific theories are:

On the *syntactic* approach to theories (due to a group of philosophers known as the "Logical Positivists"; see Uebel, 2021), a theory is an abstract description of some situation (which usually is, but need not be, a real-world situation) expressed in a formal language with an axiomatic structure. That is, a theory is a formal system (see Section 13.2.2). Such a "theory" is typically considered a set of sentences (linguistic expressions, well-formed formulas) that describe a situation or that codify (scientific) laws about a situation. (This is the main sense in which the theory of evolution is a "theory.") Such a description, of course, must be expressed in some language. Typically, the theory is expressed in a formal, axiomatic language that is semantically interpreted by rules linking the sentences to "observable" phenomena. These phenomena either are directly observable – either by unaided vision or with the help of devices such as microscopes and telescopes – or are theoretical terms (such as 'electron') that are definable in terms of directly observable phenomena (such as a vapor trail in a cloud chamber).

On the *semantic* approach to theories (due largely to the philosopher Patrick Suppes; see Frigg and Hartmann, 2020), theories are the *set-theoretic models of* an axiomatic formal system. Such models are isomorphic to the real-world situation being modeled. (Weaker semantic views of theories see them as "state spaces" (http://en.wikipedia.org/wiki/State_space) or "prototypes" (http://en.wikipedia.org/wiki/Prototype), which are merely "similar" to the real-world situation.) A theory viewed semantically can clearly resemble a simulation (or an emulation).

No one seems to deny that computer programs can be simulations or models. But can they be *theories*?

14.3 Computer Programs *Are* Theories

14.3.1 Introduction

In Section 9.7, we asked whether there can be a "computational theory" of some phenomenon, such as the stock market or the brain. Presumably, a computational theory of X is expressed as a computer program.

Several computational cognitive scientists have claimed that (some) computer programs *are* theories, in the sense that the programming languages in which they are written are languages for theories and the programs are ways to express theories. The clearest statement of this comes from Herbert Simon and Allen Newell:

> Computer programs can be written that use nonnumerical symbol manipulating processes to perform tasks which, in humans, require thinking and learning. ... *These programs can be regarded as theories, in a completely literal sense, of the corresponding human processes.* These theories are testable in a number of ways: among them, by comparing the symbolic behavior of a computer so programmed with the symbolic behavior of a human subject when both are performing the same problem-solving or thinking tasks. (Simon and Newell, 1962, p. 97, my italics)

Others have said similar things:

> There is a well established list of advantages that [computer] programs bring to a theorist: they concentrate the mind marvelously; they transform mysticism into information processing, forcing the theorist to make intuitions explicit and to translate vague terminology into concrete proposals; they provide a secure test of the consistency of a theory and thereby allow complicated interactive components to be safely assembled; they are "working models" whose behavior can be directly compared with human performance. Yet, many research

workers look on the idea of *developing their theories in the form of computer programs* with considerable suspicion. The reason … [i]n part … derives from the fact that any large-scale program intended to model cognition inevitably incorporates components that lack psychological plausibility …. The remedy … is not to abandon computer programs, but to *make a clear distinction between a program and the theory that it is intended to model.* (Johnson-Laird, 1981, pp. 185–186, my italics)

[T]he … requirement – that we be able to implement [a cognitive] process in terms of an actual, running program that exhibits tokens of the behaviors in question, under the appropriate circumstances – has far-reaching consequences. One of the clearest advantages of *expressing a cognitive-process model in the form of a computer program* is, it provides a remarkable intellectual prosthetic for dealing with complexity and for exploring both the entailments of a large set of proposed principles and their interactions. (Pylyshyn, 1984, p. 76, my italics):

[T]heories of mind should be expressed in a form that can be modelled in a computer program. A theory may fail to satisfy this criterion for several reasons: it may be radically incomplete; it may rely on a process that is not computable; it may be inconsistent, incoherent, or, like a mystical doctrine, take so much for granted that it is understood only by its adherents. These flaws are not always so obvious. Students of the mind do not always know that they do not know what they are talking about. The surest way to find out is to try to devise a computer program that models the theory. (Johnson-Laird, 1988, p. 52, my italics):

The basic idea is that *a theory must be expressed in some language.* As an old saying has it, "How can I know what I think till I see what I say?" (Wallas, 1926, p. 54). If you don't express a theory in a language, how do you know what it is? And if you don't write down your theory in some language, no one can evaluate it. Scientific theories, in this view, are sets of sentences. And the sentences have to be in some language: some theories are expressed in a natural language such as English, some in the language of mathematics, some in the language of formal logic, some in the language of statistics and probability. The claim here is that *some theories can be expressed in a **programming** language.*

One advantage of expressing a theory as a computer program is that *all* details must be filled in. That is, a computer program must be a full "implementation" of the theory. Of course, there will be implementation-dependent details. There is certainly a difference between a theory and the part of the world that it is a theory of. One such difference is this:

Why should theories of all kinds make irrelevant statements – possess properties not shared by the situations they model? The reason is clearest in the case of electromechanical analogues. To operate at all, they have to obey electromechanical laws – they have to be made of something – and at a sufficiently microscopic level these laws will not mirror anything in the reality being pictured. If such analogies serve at all as theories of the phenomena, it is only at a sufficiently high level of aggregation. (Simon and Newell, 1956, p. 74)

For another example, if the theory is expressed in Java, there will be details of Java that are irrelevant to the theory itself. This is an unavoidable problem arising whenever an abstraction is implemented. It is only at a more abstract level ("a sufficiently high level of aggregation") that we can say an implementation and a corresponding abstract theory are "the same." So, one must try to ensure that such details are indeed irrelevant. One way to do so is to make sure two computer programs expressing the same theory but are written in two different programming languages – with different implementation-dependent details – have the same input-output, algorithmic, and data-structure behavior: i.e. that they fully emulate each other.[4]

In some of the passages quoted earlier, programs were said to be *models* rather than *theories*. Arguably, however, this is another advantage of expressing a theory as a computer program: you can run the program to see how it behaves and what predictions it makes. So, in a sense, *the theory becomes its own model and can be used to test itself*. As Joseph Weizenbaum (1976, pp. 144–145) says,

> … theories are texts. Texts are written in a language. Computer languages are languages too, and theories may be written in them. … Theories written in the form of computer programs are ordinary theories as seen from one point of view. … But the computer program has the advantage [over "a set of mathematical equations" or even a theory written in English] not only that it may be understood by anyone suitable trained in its language, … but that it may also be run on a computer. … A theory written in the form of a computer program is thus both a theory and, when placed on a computer and run, a model to which the theory applies.

*****

Let's look at three explicit arguments for the conclusion that computer programs can be scientific theories – two due to Herbert Simon, and one from the Supreme Court.

14.3.2 Simon and Newell's Argument from Analogies

An *analogue A* of a thing *B* is a thing different from *B* but similar, parallel, or in some way equivalent (but not equal or identical) to *B*. Analogues, in this sense – and this spelling – are related to *analogies*.

> **Spelling Digression:** 'Analog' – spelled without the 'ue' – is the typical (American) spelling for a mathematically continuous concept, usually contrasted with 'discrete.' The distinction between the two kinds of "analog(ue)s" may not be a sharp one. Simon and Newell, 1956, p. 71, suggest that analog computers (my spelling, not theirs!) work by creating analogues (again, my spelling) of the phenomenon they "represent." For more on analogy, see Hofstadter and Sander, 2013.

Simon and Newell, 1956 argued as follows: first, "All theories are analogies, and all analogies are theories" (p. 82). That is,

1. *x* is a *theory* of *y* iff *x* is an *analogue* of *y*.

More precisely, the *content* of a *theory* of *y* is identical to the *content* of an *analogue* of *y*. The only difference between them is the way in which they are *expressed*. We could equally well say that a syntactic theory of *y* (e.g. a verbal or mathematical theory of *y*) is an analogue of *y* (p. 75). In fact, their argument only needs the right-to-left direction of this premise: all analogies are theories.

---

4 In alternative terminology, by implementing the theory in two different programming languages, you "divide out" the irrelevant implementation-dependent details (Rapaport, 1999, Section 3.2; Rapaport, 2005b, p. 395).

Next, a digital "computer is programed [sic] to carry out the arithmetic computations called for in ... [a] mathematical theory. Thus, the computer is an analogue for the arithmetic process" (p. 71; see also pp. 79–82). That is,

2. (Some) computer programs are analogies.
3. ∴ (Some) computer programs are theories.

But what *kind* of theory is a computer program? According to Simon and Newell, a theory is a set of statements, but those statements could be

verbal: "Consumption increases linearly with income, but less than proportionately" (p. 69),
mathematical: "$C = a + bY$; $a > 0$; $0 < b < 1$" (p. 70, col. 1),
   or
analog:

> The idea that the flows of goods and money in an economy are somehow analogical to liquid flows is an old one. There now exists a hydraulic mechanism ... one part of which is so arranged that, when the level of the colored water in one tube is made to rise, the level in a second tube rises ..., but less than proportionately. I cannot "state" this theory here, since its statement is not in words but in water. (p. 70)

Presumably they would classify programming languages as being of the mathematical kind, from which it would follow that computer programs are theories expressed in that language. Alternatively, there seems to be no reason not to admit a fourth kind of theory: namely, one expressed computationally, e.g. in procedural language.

One reason theories expressed as computer programs may be better than theories expressed in mathematics or English ("verbally") has to do with the idea that such computational theories are analogies:

> ... what is the particular value of the computer analogy? Why not work directly toward a mathematical (or verbal) theory of human problem-solving processes without troubling about electronic computers? ... it is at least possible, and perhaps even plausible, that we are dealing here with systems of such complexity that we have a greater chance of building a theory by way of the computer program than by a direct attempt at mathematical formulation. (p. 81)

Note first that they seem to consider (some) computer programs as analogy theories, not mathematical theories! Second, computation is perhaps the best way of managing complexity (as we saw in Section 3.16.2).

There are two advantages of expressing a theory in a programming language. "First, we would experiment with various modifications of the ... program to see how closely we could simulate in detail the observable phenomena" (pp. 81–82). In other words, we can run the program to see how it behaves – to see how good a theory it is – and we can then modify the program (and then run the modified version) to make it a better theory.

Second, the program can (or, at least, should) be written in such a way that it *explains* what it is doing: "The computer, however complex its over-all program, could be programed [sic] to report, in accurate detail, a description of any part of its own computing processes in which we might be interested" (p. 82). This, of course, can make it easier not only to debug and improve the *program* but also to correct and improve the *theory*. The ability – and the desirability – of a program to explain its own behavior is also important for the *ethical* use of computer programs; we'll return to this in Section 17.6.2.

14.3.3 Simon's Argument from Prediction

In a later essay, Simon said

> These programs, which predict each successive step in behavior as a function of the current state of the memories together with the current inputs, are theories, quite analogous to the differential equation systems of the physical sciences. (Simon, 1996a, pp. 161–162)

An argument for the claim that (some) computer programs are scientific theories can, perhaps, be constructed from this:

1. Differential equation systems of the physical sciences predict successive steps in physical processes as a function of the current state together with the current inputs.
2. Anything that allows prediction (of successive steps in some process as a function of the current state together with the current inputs) is a theory.
3. ∴ Differential equation systems are theories (in physics).
4. Cognitive computer programs predict successive steps in human cognitive behavior as a function of the current state of the memories together with the current inputs.
5. ∴ (Cognitive) computer programs are theories (in psychology).

The point is that the reason we consider differential equation systems to be theories – namely, their predictive power (Section 4.4.3) – is the same reason we should consider computer programs (cognitive ones in particular, but other kinds of programs as well) to be theories.

Well, maybe not *all* computer programs. Arguably, a computer program for adding two numbers or computing income tax is not a theory. But maybe they should be considered theories expressed computationally: a theory of addition in the first case, a theory of taxation in the second!

Simon believes that computer programs are simultaneously both theories and simulations:

> Thus the digital computer provided both a means (program) for stating precise theories of cognition and a means (simulation, using these programs) for testing the degree of correspondence between the predictions of theory and actual human behavior. (Simon, 1996a, p. 160)

Thus, for Simon, computer programs are a very special kind of theory. Not only are they statements, but they are simultaneously models – instances of the very thing they describe. Well, perhaps not quite: they only become such instances when they are being executed. This duality gives them the ability to be self-testing theories. And their precision gives them the ability to pay attention to details in a way that theories expressed in English (and perhaps even theories expressed in mathematics) lack.

Simon hedges a bit, however:

> … a program was analogous to a system of differential (or difference) equations, hence could *express* a dynamic theory. (Simon, 1996a, p. 161, my italics)

So, is it the case that a program *is* a theory? Or is merely the case that a program *expresses* a theory? Perhaps this distinction is unimportant. After all, it hardly seems to matter whether a system of equations *is* a theory or merely *expresses* a theory. (The distinction is roughly akin to that between a sentence and the proposition it expresses.)[5]

---

5 See the Online Resources for further reading on Simon's views.

14.3.4 Daubert vs. Merrell-Dow

As we have seen, there are several questions to consider: What is a computational theory (of X)? Is it a theory? A *scientific* theory?

Daubert v. Merrell-Dow Pharmaceuticals, Inc. (http://openjurist.org/509/us/579) was a 1993 Supreme Court case "determining the standard for admitting expert testimony in federal courts." (See https://en.wikipedia.org/wiki/Daubert_v._Merrell_Dow_Pharmaceuticals,_Inc. for an overview.)[6] My colleague Sargur N. Srihari recommended Daubert to me after his experience being called as an expert witness on handwriting analysis, on the grounds that his computer programs that could recognize handwriting were scientific theories of handwriting analysis.

Presumably, a computer scientist is an expert on CS. But is a computer scientist who writes a computer program about (or who develops a computational theory of) X (where $X \neq$ CS) thereby an expert on X? Or must that computer scientist *become*, or work with, an expert on X? (Recall question 5 in Section 3.18 about who counts as being a computer scientist.)

Two points that we have considered about the nature of science and engineering were (1) Popper's view that a statement was scientific to the extent that it was falsifiable (Section 4.8.1) and (2) Simon's views about bounded rationality (Section 5.6). These are nicely summarized in three comments in Daubert:

> … scientists do not assert that they know what is immutably 'true' – they are committed to searching for new, *temporary* theories to explain, *as best they can*, phenomena. (Brief for Nicolaas Bloembergen et al. as Amici Curiae 9, cited in Daubert at II.B.24 in the online version, my italics)

> Science is not an encyclopedic body of knowledge about the universe. Instead, it represents a process for proposing and refining theoretical explanations about the world that are subject to further testing and refinement. (Brief for American Association for the Advancement of Science and the National Academy of Sciences as Amici Curiae 7–8, cited in Daubert at II.B.24)

> … there are important differences between the quest for truth in the courtroom and the quest for truth in the laboratory. Scientific conclusions are subject to perpetual revision. Law, on the other hand, must resolve disputes finally and quickly. (Daubert, at III.35)

Supreme Court Justice Harry Blackmun, writing in Daubert at II.B.24 and citing the first two of these quotes, stated that "in order to qualify as 'scientific knowledge,' an inference or assertion must be derived by the scientific method." So, if a computer program that can, say, identify handwriting is a good *scientific* theory of handwriting, is its creator a scientific expert on handwriting?

There are two concerns with this: first, a computer program that can identify handwriting need not be a good *scientific theory* of handwriting. It might be a "lucky guess" not based on any scientific theory, or it might not even work very well outside carefully selected samples. Second, even if it is *based on* a scientific theory of handwriting and works well on arbitrary samples, the programmer need only be a good *interpreter* of the theory, not necessarily a good handwriting scientist. However, if a computer scientist studies the nature of handwriting and develops a scientific theory of it that is then expressed in a computer program capable of, say, identifying handwriting, then it would seem to be the case that that computer scientist is (also) a scientific expert in handwriting.

---

6 The period at the end of 'Inc.' is part of the URL.

Blackmun, writing in Daubert at II.C.28, suggests four tests of "whether a theory or technique is scientific knowledge." Note that this could include a computer program as a "technique," whether or not such programs are (scientific) theories:

Testability (and falsifiability) (II.C.28): Computer programs would seem to be scientific on these grounds because they can be tested and possibly falsified by simply running the program on a wide variety of data to see if it behaves as expected.

Peer review (II.C.29): Surely a computer program can (and should!) be peer reviewed.

Error rate (II.C.30): It's not immediately clear what Blackmun might have in mind here, but perhaps it's something like this: a scientific theory's predictions should be within a reasonable margin of error. To take a perhaps overly simplistic example, a polling error of 5 ± 4 points is not a very accurate ("scientific") measurement, nor is a measurement error of 5.00000 ± 0.00001 inches if made with an ordinary wooden ruler. In any case, surely a computer program's errors should be "reasonable."

General acceptance (II.C.31): A computer program that is not based on a "generally accepted" scientific theory or on "generally accepted" scientific principles would not be considered scientific.

Whether or not Blackmun's four criteria are complete or adequate is not the point here. The more general point is that whatever criteria are held to be essential to a *theory* being considered scientific should also apply to computer *programs* that are under consideration.[7]

14.4 Computer Programs *Aren't* Theories

However, philosophers James Moor (1978, Section 4) and Paul Thagard (1984) argue that computer programs are *not* theories, on the grounds that they are neither sets of (declarative) sentences nor set-theoretic models of axiom systems.

14.4.1 Moor's Objections

Moor (1978, pp. 219–220) says that computer models simulate phenomena in the real world and that models "help [us] understand and test theories." He also warns that

> computer scientists often speak as if there is no distinction among programs, models, and theories; and discussions slide easily from programs to models and from models to theories. (p. 220)

There is no question that this is the case, as might be clear from our earlier discussion, but what picture does Moor himself provide? Presumably a theory is a kind of description of part of the real world. A model helps us understand the *theory*; hence it only *indirectly* helps us understand the *world*. Yet a (computer) model is said to simulate the *world*. Here is one way to make sense of this: a theory is a syntactic domain that has *two* semantic interpretations – one semantic domain is the real world; the other is a (computer) model of the real world. Presumably the computer model is easier to understand (and to manipulate) than the real world, which is why it can help us test the *theory*.

---

7 See the Online Resources for further reading on computer programs as scientific theories.

This is consistent with what Moor says next:

> One can have a theory, *i.e.*, a set of laws used to explain and predict a set of events, without having a model except for the subject matter itself. Also, one can have a model of a given subject matter, *i.e.*, a set of objects or processes which have an isomorphism with some portion of the subject matter, without having a theory about the subject matter. (p. 220)

So, for Moor, there are three independent things: a portion of *the real world* (the "subject matter"); a *theory* about the real world, which offers explanations, predictions, and (presumably) descriptions of (a portion of) the real world; and a *model* of the real world, which could be a set-theoretic object that is isomorphic to (a portion of) the real world. But if the theory describes the real world, then it *also* describes a model that is isomorphic to the real world. That's why I said the syntactic domain that is the theory has *both* the real world *and* the model of the real world as two semantic interpretations of it.

Now, what about computer programs as *models*? Moor says that a computer *model* "is … more than just a computer *program*" (p. 220, my italics): to turn a computer program into a computer model, he says, you need a semantic interpretation function between the program and the portion of the real world it is modeling. I think that makes sense: the program is merely a syntactic object; the parts of the program need not "wear their meanings on their sleeve," so to speak. We saw this in Section 13.2.3 when we noted that a "person record" that had slots for things like "name" and "age" (hence "obviously" modeling a real person – at least, "obviously" to a user of the program) would work just as well as a "PR" record that had slots *unobviously* labeled 'g100' or 'g101.' To know that such a computer program modeled a *person*, one would have to know that 'g100' was to be semantically interpreted as a *name* and that its value (n456, in our example) was to be semantically interpreted as the name 'Howard L. Jones.'

So far, so good. But Moor goes on to say that it is a "myth" that a computer program that is a model of a real-world phenomenon is therefore a *theory* of that phenomenon:

> The model/theory myth occurs in computer science when the model/theory distinction is blurred so that programming a computer to generate a model of a given subject matter is taken as tantamount to producing a theory about the subject matter or at the very least an embodiment of a theory. (pp. 220–221)

One of his reasons for this conclusion that programs are not theories is that "The theory must be stable independently of the computer model" (p. 221). He has already marked the distinction between a theory, a program, and a model. And he allows that programs can be *models* (of the world) if they have a semantic interpretation in terms of the world.

But why couldn't a program also be a *theory*? Theories, for Moor, must explain and predict. Wouldn't a computer program be able to do that? Suppose we want to have a scientific understanding of some portion of the real world in which we observe that certain causes always have certain effects. Suppose we have a computer program in which computer analogues of those causes always computationally yield computer analogues of those effects. Would not the program itself explain how this occurs? And would we not be able to use the program to make predictions about future effects from future causes? More to the point, why would we need a *separate* ("independent") "set of laws" (presumably expressed in declarative sentences)?

Another reason that Moor offers for why a computer program that successfully models the real world is not thereby a theory of the world is that the program might be "*ad hoc*" (p. 221). He gives as an example Joseph Weizenbaum's Eliza program that simulates a Rogerian psychotherapist not by embodying a theory of Rogerian psychotherapy, but by "superficial analysis of semantic and syntactic cues" (p. 221). But at most, that shows that not all computer programs are theories. The question of whether a computer program *can* be a theory remains open.

A third reason he provides is this:

> The program will be a collection of *instructions* which are not true or false, but the theory will be a collection of *statements* which are true or false.
> (pp. 221–222, my italics)

And this, I think, is his real reason for arguing that programs are not theories: they are procedural, not declarative. They tell you how to *do* things, not how things *are*. But why must theories be declarative? Recall our earlier discussion in Section 3.16.3 of procedural vs. declarative language. There, we saw not only that those two kinds of language can (often) be intertranslatable but also – more to the point – that the statements of programming languages such as Prolog can be interpreted both procedurally *and* declaratively. Arguably, a declarative theory expressed in Prolog would also be a computer program.

14.4.2 Thagard's Objections

Thagard (1984) argues that on both the "syntactic" and the "semantic" conceptions of what a theory is, computer programs are not theories. He also argues that programs are not "models." Rather, "a program *is* a simulation *of* a model which approximates *to* a theory" (p. 77, my italics). So, in Thagard's view, we have

R	=	some aspect of the real word
$T(R)$	=	a theory about R
$M(T(R))$	=	a model of T
$P(M(T(R)))$	=	a program that simulates M

On the syntactic theory of theories, a theory is a set of sentences (perhaps expressed as a formal system with axioms). On the semantic theory of theories, "theories are definitions of kinds of systems" (p. 77). Presumably he will argue that a program is neither a set of sentences nor a definition (of a kind of system). He has not yet said what a "model" is.

Along with Moor, Thagard argues that because programs are sets of *instructions*, which do not have truth values and hence are not sentences, programs cannot be theories in the syntactic sense (pp. 77–78). We have just seen a problem with this. Let's consider it further.

Suppose we are looking for a buried treasure. I might say, "I have a theory about where the treasure is buried: walk three paces north, turn left, walk five paces, and then dig. I'll bet you find the treasure." Is this not a theory? I suppose Thagard might say it isn't. But isn't there a sentential theory that is associated with it – perhaps something like "The treasure is buried at a location that is three paces north and five paces west of our current location"? Doesn't my original algorithm for finding the treasure carry the same information as this theory, merely expressing it differently? The argument that Thagard makes that programs can't be theories because they are not sets of declarative sentences just seems parochial. They are surely sets of (imperative) statements that have the additional benefit that they can become an instance of what they describe (alternatively: that they can control a device that becomes an instance of what they describe).

Thagard (1984, p. 78) considers a *model* to be "a set-theoretic interpretation of the sentences in a [syntactic] theory …… … a system of things … which provide an interpretation of the sentences."

But "a program is not … a system of things, nor does it … provide an interpretation for anything." Hence, a program is not a model. But a program being executed – a process – can be considered a system of (virtual) things that are interpretations of data structures in the program. If a *process* might be a *model*, then why couldn't the *program* be a *theory*?

On the semantic or "structuralist" (p. 78) view of theories, a theory "is a definition of a kind of natural system" (p. 79). Given some scientific laws (which, presumably, are declarative, truth-functional sentences, perhaps expressed in the language of mathematics), we would say that something is a certain kind of natural system "if and only if it is a system of objects satisfying" those laws. The system is *defined* as being something that satisfies those laws. But this seems very close to what a model is. In fact, Thagard says that a "real system *R* is a system of the kind defined by the theory *T*" (p. 79). But how is that different from saying that *R* is an implementation of (i.e. a model of) *T*?

Thagard's response is that, first, "a program *simulates* a system: it does not define a system" and that, second, programs contain "a host of characteristics which we know to be extraneous" to the real system that they are supposed to be like (p. 79). He says this because simulations aren't definitions: a *simulation* of the solar system, to use his example, doesn't *define* the solar system. (Recall Section 9.7.2.) This seems reasonable, but it also seems to support the idea that a *process* (not necessarily a program) is a *model* and hence that a *program* would be a *theory*.

As for the problem of implementation-dependent details, Thagard says that

> if our program [for some aspect of human cognition] is written in LISP, it consists of a series of definitions of functions. The purpose of writing those functions is not to suggest that the brain actually uses them, but to simulate at a higher level the operation of more complex processes such as image … processing. (p. 80)

In other words, the real system that is being simulated (modeled?) by the program (process?) need not have Lisp functions. But as he notes, it will have "complex processes" that do the same thing as the Lisp functions. But isn't this also true of any theory compared to the real system it is a theory of? A theory of cognitive behavior expressed in declarative sentences will have, say, English words in it, but the brain doesn't. A similar point is made by Paul Humphreys (1990, p. 501):

> Inasmuch as the simulation has abstracted from the material content of the system being simulated, has employed various simplifications in the model, and uses only the mathematical form, it obviously and trivially differs from the 'real thing,' but in the respect, there is no difference between simulations and any other kind of mathematical model ….

Thagard does admit that he "shall take models to be like theories (on the semantic conception) as being definitions of kinds of systems" (p. 80). And he notes that "a model contains specifications which are known to be false of the target real system" – i.e. implementation-dependent details! According to Thagard, the problem with this is that if you try to make a prediction about the real system based on the model, you might erroneously make it based on one of these implementation-dependent details (pp. 80–81). But that seems to be a problem endemic to any model (or any theory, for that matter).

If you make a prediction that turns out to be false, you may have to change your theory or your model. Perhaps you have to eliminate that implementation-dependent detail. But others will always crop up; otherwise, your theory or model will not merely describe or simulate the real system; it will *be* the real system. But there are well-known reasons why a life-sized map of a country is not a very good map! (We'll return to this in Section 15.6.2.)

However, Thagard's final summary is not really the wholesale rejection of programs as theories that it might first appear to be. It is more subtle:

> a program P, when executed on a computer, provides a simulation of a system of a kind defined by a model M, where M defines systems which are crude versions of the systems defined by a theory T, and the set of systems defined by T is intended to include the real system R. (p. 82)

I can live with this: both P and R are implementations of T.

15

Computer Programs as Mathematical Objects

> Mechanical computers should, Babbage thought, offer a means to eliminate at a stroke all the sources of mistakes in mathematical tables. … A printed record could … be generated …, thereby eliminating every opportunity for the genesis of errors. … Babbage boasted that his machines would produce the correct result or would jam but that they would never deceive.
> —Doron D. Swade (1993, pp. 86–87)

> We talk as if these parts [of a machine] could only move in this way, as if they could not do anything else. How is this – do we forget the possibility of their bending, breaking off, melting, and so on?
> —Ludwig Wittgenstein (1958, Section 193)

> Present-day computers are amazing pieces of equipment, but most amazing of all are the uncertain grounds on account of which we attach any validity to their output. It starts already with our belief that the hardware functions properly.
> —Edsger W. Dijkstra (1972, p. 3)

> The history of program verification … has now expanded to be about nothing less than the nature of the relationship between abstract logical systems and the physical world.
> —Selmer Bringsjord (2015, p. 265)

15.1 Introduction

Whether or not CS in its entirety is (nothing but) a mathematical or formal science (as we considered in Section 3.9), aspects of it certainly are mathematical: the theory of computation, computational complexity, and, arguably, programming languages and computer programs. In this chapter, we will examine one of the ways in which programs can be treated as mathematical objects about which theorems can be proved.

15.1.1 Bugs and Intended Behavior

The Halting Problem (Section 7.7) tells us that it is not possible to have a *single* computer program that can tell us in advance whether any given computer program will halt. However, given a *specific* computer program, there might be ways of determining whether that particular program will halt. What about other problems that computer programs might have? It would be useful to be able to know in advance whether a given computer program will work. But what does it mean to say that a program "works"? It could mean the program successfully transforms its input into output in the sense that when you start it up, it finishes. It could mean the program not only finishes but also has no logical "bugs" (such as dividing by 0 or having an unintended infinite loop) that would cause it to "crash." It could mean the program it not only finishes without bugs but yields the *correct* output. It could mean the program not only finishes without bugs or incorrect output but also does what it was intended to do.[1]

On the TV cooking competition "Chopped," chefs are sometimes eliminated from competition because they tell the judges that the dish they just prepared is, say, a puttanesca, but the judge says it isn't, on the grounds that it includes some ingredient that it shouldn't (or vice versa). Yet the dish might be delicious. Had the chef not said that it was a puttanesca, the chef might not have been chopped. Does it matter what the chef calls the dish? Is a delicious dish unsuccessful because it has a misleading name? A similar problem can occur with computer programs. Consider this:

> … we cannot, by observing its output behavior, acquire the knowledge that a physical computer is operating normally, that it is correctly computing the values of a function F, that it is executing program P, and that it is using data structure D. … [P]hysical computers can break down in various ways, and when they do, they might not physically realize the … function which the computer would correctly compute if it did not break down. … When a physical computer is functioning normally in the computation of the values of some function F, it will output the correct range value for F when it is given as input a domain value for F. However, there might be another function G which the same physical computer might be computing. When the physical computer is operating normally *in the computation of F* it is suffering a breakdown *in the computation of G*. By examining only its output behavior, one cannot determine whether it is operating normally (in the computation of F) or suffering a breakdown (in the computation of G). Similarly, by examining only its output behavior, one cannot determine whether it is operating normally (in the computation of G) or suffering a breakdown (in the computation of F). The problem is that what is breakdown behavior in the computation of F is normal behavior in the computation of G. Whether this physical computer is operating normally or suffering a breakdown is relative to which it is actually computing. (Buechner, 2018, pp. 496–497)

Does this mean we might not ever be able to decide if a computer is doing what it is "supposed" to be doing? Is this an even more serious problem than merely determining whether a program "works" (in any of the senses of 'works' that we just mentioned)? What if we are not limited to examining only the output? On the other hand, what if that is the only thing that *can* be examined (as might be the case with some "black box" machine-learning algorithms; see Sections 3.11 and 17.6.2)? (On the nature of "miscomputation," see Dewhurst, 2020. We'll return to this in Section 16.5.)

15.1.2 Proofs and Programs

Many of these issues concern the relationship of mathematics to the real world[2] and also touch on ethical problems:

1. Many, if not most, errors in software engineering occur when bridging the gap between the informal, real world and the formal world of mathematical specifications. …

---

1 See the Online Resources for further reading on the history of computer "bugs."
2 See the Online Resources for Section 4.11.

2. Even if software engineers have a clear-cut specification of how they intend their software to behave, they will at best be able to prove that a mathematical model of their software satisfies this specification, not that the software will have the desired effects in the real world. (Daylight, 2016, p. viii)

Is there a way to logically *prove* that a computer program "works"?

Recall our discussions of the dual mathematical and engineering natures of CS: Is computer programming a kind of mathematics? Or is it a kind of engineering? Many people identify computer programming with "software *engineering*." Yet many others think of a program as being like a *mathematical* proof: a formal structure, expressed in a formal language. For example, Peter Suber (1997a) compares programs to proofs this way: a program's *input* is analogous to the *axioms* used in a proof; the program's *output* is analogous to the *theorem* being proved; and the *program* itself is like the *rules of inference* that transform axioms into theorems, with the program transforming the input into the output. Or perhaps a *program* is more like the endpoint of a proof: namely, a mathematical *theorem*. In that case, just as theorems can be proved (and, indeed, *must* be proved before they are *accepted*), perhaps programs can be proved (and, perhaps, *should* be proved before they are *used*). (We'll explore these analogies further in Section 15.2.2.)

Are programs mathematical objects about which we can prove things? What *kinds* of things might be provable about them? Two answers have been given to the first of these questions: yes and no. (Did you expect anything else?)

One of the most influential proponents of the view that programs *can* be the subjects of mathematical proofs is Turing Award winner C.A.R. (Tony) Hoare (the developer of the Quicksort sorting algorithm):

> Computer programming is an exact science in that all the properties of a program and all the consequences of executing it *in any given environment* can, in principle, be found out from the text of the program itself by means of purely deductive reasoning. … When the correctness of a program, its compiler, *and the hardware of the computer* have all been established with mathematical certainty, it will be possible to place great reliance on the results of the program, and predict their properties with *a confidence limited only by the reliability of the electronics*. (Hoare 1969, pp. 576, 579; my italics)

> I hold the opinion that the construction of computer programs is a mathematical activity like the solution of differential equations, that programs can be derived *from their specifications* through mathematical insight, calculation, and proof, using algebraic laws as simple and elegant as those of elementary arithmetic. (Hoare 1986, p. 115, my italics)

Among those arguing the opposite point of view are the computer scientists Richard DeMillo, Richard Lipton, and Alan Perlis:

> … formal verifications of programs, no matter how obtained, will not play the same key role in the development of computer science and software engineering as proofs do in mathematics. (De Millo et al., 1979, p. 271)

One reason that they give for this is their view that formal proofs are long and tedious and don't always yield acceptance or belief. (One of my college friends always advised that you had to *believe* a mathematical proposition *before* you could try to *prove* it.) Hence, they argue, it is not worthwhile trying to formally verify programs.

The epigraph to their essay (slightly incorrectly quoted) is the following quotation from J. Barkley Rosser's logic textbook:

> "I should like to ask the same question that Descartes asked. You are proposing to give a precise definition of logical correctness which is to be the same as my vague intuitive feeling for logical correctness. How do you intend to show that they are the same?" …
> [T]he average mathematician … should not forget that intuition is the final authority …. (Rosser, 1978, pp. 4, 11)

The similarity to the Church-Turing Computability Thesis – which also states an equivalence between a "precise" notion and a "vague intuitive" one – is not accidental: recall from Section 7.3.2 that Rosser – a pioneer in computability theory – was one of Church's Ph.D. students.

The "Descartes" mentioned in the passage is not the real Descartes but a fictional version visited by a time-traveling mathematician who tries to convince him that the modern and formally "precise" ϵ-δ definition of a continuous curve is equivalent to the fictional Descartes's "vague intuitive" definition as something able to be drawn without lifting pencil from paper. Rosser observes that "the value of the ϵ-δ definition lies mainly in *proving* things about continuity and only slightly in *deciding* things about continuity" (p. 2, my italics). "Descartes" then says this to the time-traveling mathematician,

> "I have here an important concept which I call continuity. At present my notion of it is rather vague, not sufficiently vague that I cannot decide which curves are continuous, but too vague to permit of careful proofs. You are proposing a precise definition of this same notion. However, since my definition is too vague to be the basis for a careful proof, how are we going to verify that my vague definition and your precise definition are definitions of the same thing?" (Rosser, 1978, p. 2)

The time traveler and "Descartes" then agree that despite the informality of one definition and the formality of the other, the two definitions can be "verified" – but not "proved" – to be equivalent by seeing that they agree on a wide variety of cases. When the mathematician returns to the present, a logician points out that the mathematician's intuitive notion of proof bears the same relation to the logician's formal notion of proof as the fictional Descartes's intuitive notion of continuity bears to the mathematician's formal definition of continuity. The passage that De Millo et al. (1979, p. 271) quote is the mathematician's response to the logician in the story. (It is the real logician Rosser, in his own voice, who comments that "intuition is the final authority"!)

So, Hoare says that programs *can and should* be formally verified. DeMillo et al. (and Rosser, perhaps) suggest that they can but *need not* be. Along comes the philosopher James Fetzer, who argues that they *cannot*. More precisely, he argues that the things we *can* prove about programs are not what we think they are:

> … there are reasons for doubting whether program verification can succeed as a generally applicable and completely reliable method for guaranteeing the performance of a program. (Fetzer, 1988, p. 1049)

We'll come back to DeMillo et al. in Section 15.3.3. But first, what does it mean to formally verify a program? Before we can answer that, we need to be clear about what it means to verify – i.e. to prove – a *theorem*.[3]

---

3 See the Online Resources for further reading on programs and proofs.

15.2 Theorem Verification

15.2.1 Theorems and Proofs

A formal proof in logic or mathematics is a certain sequence of propositions, beginning with axioms and ending with a theorem. A "proposition" is what computer scientists call a 'Boolean statement': i.e. a statement that is either true or false. The proofs themselves (the *sequences* of propositions) are not Boolean-valued: they are neither true nor false; rather, they are either "correct" (the technical terms are 'valid' and 'sound') or "incorrect" (technically, 'invalid' or 'unsound'). (Recall Section 2.9.)

Syntax However, the actual truth values of the propositions in a proof are irrelevant to the *structure* of the proof. From a purely syntactic point of view, a **proof** of a theorem T has the general form of a sequence

$$\langle A_1, \dots, A_n, P_1, \dots, P_m, C_1, \dots, C_l, T \rangle$$

where

- T – the last item in the sequence of propositions that constitutes the proof – is the **theorem** to be proved.
- The A_i are **axioms**: i.e. propositions that are "given" or "assumed without argument." They are the starting points – the "basic" or "primitive" propositions – of any proof, no matter what the subject matter is. From a strictly syntactic point of view, the axioms of a formal system need not be (semantically) "true."[4]
- The P_j are **premises**: i.e. propositions about some particular subject matter that is being formalized. They are also "starting points," accepted – at least for the sake of the proof – without further argument.
- The C_k are propositions that logically follow from previous propositions in the sequence by a "rule of inference." A rule of inference can be thought of as a "primitive proof" in the same sense as the "primitive" operations of a Turing Machine or the "basic" functions in the definition of recursive functions. A **rule of inference** has the form:

> From propositions $Q_1, \dots, Q_r$
> you may infer proposition R

(See Section 2.5.1 for an example.) Just as with axioms, the rules of inference are given by fiat. The rules of inference are *syntactically valid* by definition. Note that if $n = 0$ (i.e. if there are *no* axioms, and, especially, if $m = 0$ also – i.e. if there are no premises), then there will typically have to be *lots* of rules of inference, and the demonstration is then said to be done by "natural deduction" (because it is the way that logicians "naturally" prove things) (Pelletier, 1999; Pelletier and Hazen, 2021).

A more complex proof is then recursively defined in terms of successive applications of rules of inference. Then, to say that the sequence

$$\langle A_1, \dots, A_n, P_1, \dots, P_m, C_1, \dots, C_l, T \rangle$$

---

4 For examples of "axioms" that are not "true," see some of the formal systems mentioned in the Online Resources for Section 13.2.2.

is a proof of T from the axioms and premises means (by definition) that each C_k and T follow from previous propositions in the sequence by a (syntactically valid) rule of inference. A proof is **syntactically valid** iff the final conclusion T and every intermediate conclusion C_k results from a correct application of a rule of inference to preceding propositions in the proof.

Semantics What about truth? Surely, we want our theorems to be true!

First, from a semantic point of view, the axioms are typically considered *necessarily* true by virtue of their meanings (or assumed to be true for the time being); they are usually logical tautologies. Premises, on this account, are *contingently* or *empirically assumed* to be true (but they would normally require some justification). For example, often you need to justify a premise P_j by providing a *proof* of P_j or at least some empirical evidence in its favor.

Second, just as, normally, we want our axioms and premises to be (semantically) *true*, so, normally, we want our rules of inference to be (semantically) *truth-preserving*. A rule of inference (of the previous form) is **truth-preserving** $=_{def}$ *if* each of $Q_1, \ldots, Q_r$ is true, then R is true. This does not mean R *is* true; all it means is that R is true *relative to* $Q_1, \ldots, Q_r$. As we saw with axioms, from a purely syntactic point of view, rules of inference do not *have* to be truth-preserving.[5] A truth-preserving rule of inference is said to be **semantically valid**.

Finally, in order for theorem T to be *true*, each rule of inference (and therefore the entire proof) must be semantically valid: i.e. it must be truth-preserving – its conclusion *must* be true *if* its axioms and premises are true. But of course, the axioms and premises of an argument might *not* be true. If they *are* true, *and if* the argument is semantically valid, then its conclusion *must* be true. Such a truth-preserving argument with true premises is said to be **sound**; hence, it is *unsound* iff *either* one or more of its axioms or premises is false *or* it is *syntactically* invalid. Roughly, a syntactically valid argument that is unsound because of a false axiom or premise is like a correct program whose input is "garbage"; the output of such a program is also "garbage" (as in the famous saying: "garbage in, garbage out"). And the final conclusion of a syntactically valid but semantically invalid proof need not be true.

But for a syntactically valid proof to also be *semantically valid*, the rules of inference must be truth-preserving. And for it to be *sound*, the axioms must be true. (Recall Sections 2.5.1 and 2.9.)

It is, strictly speaking, incorrect to say that a theorem is "*proved to be true.*" 'Proof' and 'theorem' are *syntactic* notions, while 'truth' is a *semantic* notion. Theorems do not have to be true: a syntactically valid proof that began with a false axiom or a false premise might end with a theorem that is also false (or it might end with a theorem that *is* true!). And a truth need not be a theorem: Gödel's Incompleteness Theorem shows that there are true propositions of arithmetic that are not provable. The conclusion of any formal proof – i.e. any theorem – is only true *relative to* the *axioms* (and premises) of the formal theory (Hempel, 1945, p. 9; Rapaport, 1984, p. 613; recall Section 2.5.1). Of course, *if* all the axioms (and premises) *are* true, *and if* all the rules of inference *are* truth-*preserving*, then the theorem that has been (syntactically) proved will be (semantically) true. (We'll come back to this point in Section 15.4.1.)

15.2.2 Programs and Proofs

How is all this applicable to *programs*? That depends on how the *logical* paraphernalia (axioms, rules of inference, proofs, theorems, etc.) line up with the *computational* paraphernalia (specifications, input, programs, output, etc.). Several different analogies can be made, but all such analogies must "be taken with a big grain of salt, since all these words can mean many things" (Scherlis and Scott, 1983, p. 207):

---

5 Again, see some of the formal systems cited in the Online Resources for Section 13.2.2.

A1: Fetzer's and Suber's Analogy (Fetzer, 1988, p. 1056, col. 1; Suber, 1997a)

Logic		Computation
axioms, premises	≈	input
rules of inference	≈	program
theorem	≈	output

On this analogy, verifying a *program* would be like proving the *inference rules*! But inference rules are not propositions, so they can't be proved. So, a slight modification of this is analogy A2:

A2:

Logic		Computation
axioms, premises	≈	input
intermediate conclusions	≈	program
theorem	≈	output

Both intermediate conclusions and program are a sequence (or at least a set) of expressions that begin with something given and end with a desired result. But you are trying to *prove* the theorem; you are not trying to prove the intermediate conclusions. And you *verify* an entire program, not just its output. So, let's consider analogy A3:

A3: Scherlis and Scott's Analogy (Scherlis and Scott, 1983, p. 207)

Logic		Computation
problem	≈	specification
proof	≈	program derivation (or verification)
theorem	≈	program

On this analogy, proving a theorem is like verifying or deriving a program. But the role of axioms (and premises) and rules of inference are not made clear. So, let's try a slightly different modification of A2:

A4:

Logic		Computation
axioms, premises	≈	input
proof	≈	program
theorem	≈	output

Just as when you prove (or derive) a theorem, you transform the axioms into the theorem, so a program transforms input into output. To verify a program is to prove that it will, indeed, transform the input into the expected output – i.e. that it will satisfy its specification. This analogy seems closer to what program verification is all about. (We'll refine it in Section 15.2.3.)

A5: There is one more analogy: the idea behind this one is that a theorem usually has the form "if antecedent *A*, then consequent *C*." And most programs can be put into the form "if you execute program *P*, then you will accomplish goal *G*." So, proving a theorem is analogous to verifying that *P* accomplishes *G*. One issue that this analogy highlights is whether the *goal* of a program is an essential part of it. We'll return to this issue beginning in Section 16.3.

15.2.3   Programs, Proofs, and Formal Systems

There is another way to think about rules of inference that clarifies the relationship between programs, proofs, and formal systems.

First, consider Gödel's observation about the importance of Turing's analysis of computation. Although here I only want to focus on Gödel's first sentence, it is worth quoting the rest because of his observations on other topics that we have discussed:

> *[D]ue to A.M. Turing's work, a precise and unquestionably adequate definition of the general concept of formal system can now be given* …. Turing's work gives an analysis of the concept of "mechanical procedure" (alias "algorithm" or "computation procedure" …). This concept is shown to be equivalent with that of a "Turing machine." A formal system can simply be defined to be any mechanical procedure for producing formulas, called provable formulas. For any formal system in this sense there exists one in the sense … [of "a system of symbols with rules for employing them" – p. 41] that has the same provable formulas (and likewise vice versa) …. [The "essence" of] the concept of formal system … is that reasoning is completely replaced by mechanical operations on formulas. (Note that the question of whether there exist finite <u>non-mechanical</u> procedures … not equivalent with any algorithm, has nothing whatsoever to do with the adequacy of the definition of "formal system" and of "mechanical procedure.") (Gödel, 1964, pp. 71–72, my italics, original underlining)

Recall our discussion in Section 13.2.2 of Turing Machines as formal systems. In what sense is a Turing Machine a kind of formal system?

As we have seen, a rule of inference tells you what kind of proposition can be inferred from other kinds. So, for instance, the rule *Modus Ponens* (also sometimes called "$\rightarrow$ elimination") tells us that a proposition of the form Q (that is, any proposition whatsoever) can be inferred from the two propositions of the forms $(P \rightarrow Q)$ and P. And the rule of *Addition* (sometimes also called "and-introduction" or "$\wedge$ introduction") tells us that a proposition of the form $(P \wedge Q)$ can be inferred from the two propositions P and Q.

Each of these rules can also be thought of as functions: Modus Ponens is the function $MP((P \rightarrow Q), P) = Q$; addition is the function $ADD(P, Q) = (P \wedge Q)$. A proof of a theorem T from axioms (and premises) $A_1, \ldots, A_n$ can then be thought of as successive applications of such inference-rule functions to the axioms and to the previous outputs of such applications. Are these functions (these rules of inference) computable? If so, then a proof can be thought of as a kind of program.

Exercise for the Reader: Show that a typical rule of inference is, indeed, computable. Hint: Can you write a Turing Machine program that has the inputs to a rule of inference encoded on its tape and that, when executed, has the output of the rule encoded on the tape?

But this raises an interesting question: Do any of our analogies capture this relationship? Is verifying a program really like proving a theorem? The relationship I have just outlined suggests that if a program is like a proof, then verifying a program is actually more like checking a proof to show that it is syntactically valid. To check a proof for validity is to check whether each proposition in the sequence of propositions that constitutes the proof is either a basic proposition (an axiom) or follows from previous propositions in the sequence by a rule of inference (by the application of an inference-rule function).

Let's consider a program that is a *sequence* of instructions:

begin S_1; ... ; S_n **end**.

Annotate each S_i with two Boolean-valued comments, P_i, Q_i, so that the program looks like this:

begin $\{P_1\}$ S_1 $\{Q_1\}$; ... ; $\{P_n\}$ S_n $\{Q_n\}$ **end**.

P_1 will play the role of an axiom, and each proposition of the form

If P_i is the case, and if S_i is executed, then Q_i is the case

will play the role of an application of a rule of inference to the "inputs" P_i and S_i. The final state of the computation – Q_n – plays the role of the theorem to be proved. Just as (roughly) any theorem is really of the form $(A_i \rightarrow T)$ (if the axioms are the case, then the theorem is the case), so a program can be thought of as taking the form "If P_1 is input, then Q_n is output," which is a high-level (i.e. functional, or input-output) specification of the program. To verify the program is to check that it satisfies the specification.

This gives us a refinement of analogy **A4**:

A4.1:

Logic		Computation
axioms	≈	input
rules	≈	computable functions
theorem	≈	output
proof	≈	program
valid proof	≈	verified program

15.3   Program Verification

15.3.1   Introduction and Some History

"Program verification" is a subdiscipline of CS. It can be thought of as theoretical software engineering or the study of the logic of software. It is also a subissue of the question concerning the relation of software to hardware that we looked at in Section 12.3.

But it is not a new idea. Nowadays, we think of Euclidean geometry as a formal axiomatic system in which geometric *theorems* are stated (in *declarative* language) and proved to follow logically from the axioms. However, as we saw in Section 3.16.3, each proposition of Euclid's original *Elements* actually consisted of an *algorithm* (expressed in a *procedural* language for *constructing* a geometric figure using only compass and straightedge) and a *proof of correctness* of the algorithm – that is, a "verification" that the compass-and-straightedge "program" actually resulted in a geometric figure with the desired properties. (See, for example, the statement of Euclid's Proposition 1 at http://tinyurl.com/kta4aqh   or   http://data.perseus.org/citations/urn:cts:greekLit:tlg1799.tlg001 .perseus-eng1:1.prop.1.)

Similarly, a program verification typically consists of taking an algorithm expressed in a (procedural) language for computing a function using only primitive computable (i.e. recursive) operations (as in Section 7.6) and then providing a proof of correctness of the algorithm – i.e. a verification that the algorithm satisfies the input-output specification of the function (as in analogies A3 and A4).

In practice, there is a preliminary step, which will occupy us for much of Chapter 16: typically, one begins with a *problem* (or "goal"), perhaps informally stated, which is then formally modeled by a function. So, another possible goal of program verification might be to show that the program that implements the *function* actually solves the *problem* (as in analogy A5).

Another historical antecedent – perhaps the earliest example of program verification – is due to Turing himself. His 1949 essay ("Checking a Large Routine") "is remarkable in many respects. The three … pages of text contain an excellent motivation by analogy, a proof of a program with two nested loops, and an indication of a general proof method very like that of Floyd [1967]" (Morris and Jones, 1984, p. 139).[6]

15.3.2 Program Verification by Pre- and Post-Conditions

The idea behind program verification is to augment, or annotate, each statement S of a program (as we did in Section 15.2.3) with

1. a proposition P expressing a "pre-condition" of executing S and
2. a proposition Q expressing a "post-condition" of executing S.

A **pre-condition** P of a program statement S is a description (of a situation, either in the world in which the program is being executed or in the computer that is executing the program) that must be true in order for S to be able to be executed; i.e. P must be true *before* S can be executed. (And according to the correspondence theory of truth (Section 2.3.1), P will be true iff the situation that it describes "exists," i.e. really is the case.)

A **post-condition** Q of S is a description (of a situation) that will necessarily be true *after* S is executed. That is, the situation that Q describes will come into "existence" (come to be the case) after S is executed: S changes the computer (or the world) such that Q becomes true.

So, such annotations describe both how things *must* be if S *is to be* executed successfully and how things *should* be if S *has been* executed successfully. They are typically written as *comments* preceding and following S in the program. Letting comments be signaled by braces, the annotation would be written as follows:

$\{P\}\ S\ \{Q\}$

Such an annotation is semantically interpreted as saying

> **If** P correctly describes the state of the computer (or the state of the world)
>   *before* S is executed,
>
> **and if** S is executed,
>
> **then** Q correctly describes the state of the computer (or the state of the
>   world) *after* S is executed.

The "state of the computer" includes such things as the values of all registers (i.e. the values of all variables).

So, if we think of a program as being expressed by a sequence of executable statements

begin $S_1; S_2; \dots; S_n$ **end.**

---

6 See the Online Resources for further reading on the origins of program verification.

then the program annotated for program verification will look like this:

begin $\{I \& P_1\}S_1\{Q_1\}$; $\{P_2\}S_2\{Q_2\}$; ... ; $\{P_n\}S_n\{Q_n \& O\}$ **end.**

where

- I is a proposition describing the input.
- P_1 is a proposition describing the initial state of the computer (or the world).
- For each i, Q_i logically implies P_{i+1}. (Often, $Q_i = P_{i+1}$.)
- Q_n is a proposition describing the final state of the computer (or the world). And
- O is a proposition describing the output.

The claim of those who believe in the possibility of program verification is that we can then logically prove whether the program does what it's supposed to do *without having to run the program.* We would construct a proof of the program as follows:

premise:	The input of the program is I.
premise:	The initial state of the computer is P_1.
premise:	If the input is I and the initial state is P_1, and if S_1 is executed, then the subsequent state will be Q_1.
premise:	S_1 is executed.
conclusion:	$\therefore$ The subsequent state is Q_1.
premise:	If Q_1, then P_2.
conclusion:	$\therefore P_2$.
premise:	If the current state is P_2, and if S_2 is executed, then the subsequent state will be Q_2.
conclusion:	$\therefore$ The subsequent state is Q_2.
...	...
conclusion:	$\therefore$ The final state is Q_n.
premise:	If Q_n, then O.
conclusion:	$\therefore O$.

The heart of the proof consists in verifying each premise. If the program isn't a "straight-line" program such as this but is a "structured" program with separate modules, it can be recursively verified by verifying each module (Dijkstra, 1972).

15.3.3 The Value of Program Verification

If debugging a program by running it and then finding and fixing the bugs is part of *practical* software engineering, you can see why program verification can be thought of as *theoretical* software engineering.

One reason program verification is argued to be an important part of software engineering is that this annotation technique can also be used to help *develop* programs that would thereby be guaranteed to be correct. Dijkstra, 1975 shows how to "formally derive" a program that satisfies a certain specification. And Gries, 1985 is a textbook that shows how to use logic to "develop" programs simultaneously with a proof of their correctness.

Scherlis and Scott (1983) argue that program verification and development should go hand in hand, rather than verification coming *after* a program is complete. Their notion – "inferential programming" – differs from "program derivation": whereas "program derivations [are] highly structured justifications for programs[,] inferential programming [is] the process of building, manipulating, and reasoning *about* program derivations" (p. 200, my italics). A " 'correctness' proof [shows] that a program is consistent with its specifications" (p. 201), where "*Specifications* differ from programs in that they describe aspects or restrictions on the functionality of a desired algorithm without imposing constraints on *how* that functionality is to be achieved" (p. 202). For instance, a specification might just be an input-output description of a function. To prove that a program for computing that function is "correct" – i.e. to "verify" the program – is to prove that the program has the same input-output behavior as the function.

Scherlis and Scott (1983, p. 204) take issue with De Millo et al. (1979). First, they observe that the claim that "Mathematicians do not really build formal proofs in practice; why should programmers?" is fallacious, because "formalization plays an even more important rôle in computer science than in mathematics," and this, in turn, is because "computers do not run 'informal' programs." Moreover, formalization in mathematics has made possible much advancement independent of whether "there is any sense in looking at a *complete* formalization of a whole proof. Often there is not." They advocate not for a complete proof of correctness of a completed program but for proofs of correctness of stages of development, together with a justification that "derivation steps *preserve* correctness." This is exactly the way in which proofs of theorems are justified: if the axioms and premises are true, and if the rules of inference are truth-preserving, then the conclusions (theorems) will be true (relative to the truth of the axioms and premises).[7]

15.4 The Fetzer Controversy

> In many creative activities the medium of execution is intractable. Lumber splits; paint smears; electrical circuits ring. These physical limitations of the medium constrain the ideas that may be expressed, and they also create unexpected difficulties in the implementation.
> —Frederick P. Brooks (1975, p. 15)

> The transition function for a finite-state automaton specifies everything there is to know about it. From this it does not follow that we know everything about the behavior of a PCM [physical computing machine] that physically realizes the abstract diagram of a finite-state automaton, since the physical realization may be imperfect.
> —Jeff Buechner (2011, p. 349)

15.4.1 Fetzer's Argument against Program Verification

Nonsense! said philosopher James H. Fetzer (1988), thus initiating a lengthy controversy in the pages of the *Communications of the ACM* and elsewhere. Several strongly worded letters to the editor chastised the editor for publishing Fetzer's paper; supportive letters to the

---

7 See the Online Resources for further reading on program verification.

editor praised the decision to publish; and articles in other journals attempted to referee the publish-or-not-to-publish controversy as well as the more substantive controversy over whether programs can, or should, be verified.

What did Fetzer say that was so controversial? Here is the abstract of his essay:

> The notion of program verification appears to trade upon an equivocation. Algorithms, as logical structures, are appropriate subjects for deductive verification. Programs, as causal models of those structures, are not. The success of program verification as a generally applicable and completely reliable method for guaranteeing program performance is not even a theoretical possibility. (Fetzer, 1988, p. 1048)

Despite the analogies between proofs of theorems and verifications of programs, Fetzer focuses on one significant *dis*analogy, which he expresses in terms of a difference between "algorithms" and "programs": *algorithms*, for Fetzer, are abstract, formal (mathematical or logical) entities; *programs*, for Fetzer, are physical ("causal") entities (Fetzer, 1988, p. 1052, note 6; p. 1056, col. 2; and Section "Abstract Machines versus Target Machines" (pp. 1058–1059)).

A "program" for Fetzer is a "causal model of" an algorithm (p. 1048), an "implementation of an algorithm in a form that is suitable for execution by a machine" (p. 1057, col. 2). In other words, whereas an "algorithm" (in Fetzer's terminology) is a formal entity susceptible to logical investigation, a "program" is a real-world, physical object that is *not* susceptible to *logical* – but only *empirical* – investigation. The analogies we discussed in Section 15.2.2 hold for "algorithms" but not for "programs" in Fetzer's senses. (Fetzer prefers A1 to A3; Fetzer, 1988, p. 1056, col. 2.)

The computer historian Edgar G. Daylight (2016, p. 97) makes a similar distinction between a "mathematical program" and a "computer program": the former is an algorithm expressed in a formal language; the latter "resides electronically in a specific computer and is what most of us would like to get 'correct'." (Recall the controversies discussed in Chapter 12 over the dual nature of programs; you should also keep in mind the difference between a static "computer program [that] resides electronically in a computer" – perhaps as a specific arrangement of switch-settings – and the dynamic *process*, i.e. the actually running program.) Even the very "same" program as implemented in *text* or as implemented in a *computer* might have different behaviors depending on how its numerical-valued variables are interpreted: a "mathematical program" for computing the square root of an integer can be "correct" to any decimal place, whereas the program implemented in a computer can only have a finite accuracy; yet in a perfectly reasonable sense, they are the "same" program (Dijkstra, 1972, Section 6; Daylight, 2016, p. 102).

As Fetzer (1988, p. 1059, col. 1) observes, algorithms (Fetzer's terminology) or mathematical programs (Daylight's terminology) "can be conclusively verified, but … [this] possesses no significance at all for the performance of any physical system," whereas "the performance of" programs (Fetzer) or computer programs (Daylight) "possesses significance for the performance of a physical system, but it cannot be conclusively verified." Fetzer (1988, p. 1060, col. 1) quotes Einstein (1921):

> As far as the laws of mathematics refer to reality, they are not certain; and as far as they are certain, they do not refer to reality.

Recall Chomsky's competence-performance distinction from Section 10.4: even if program-verification techniques can prove that a program is correct ("competent"), there may still be *performance* limitations. The point, according to Fetzer, is that we must distinguish between the *program* and the *algorithm* it implements: a program is a *causal* model of a *logical* structure, and although *algorithms* might be capable of being absolutely verified, *programs* cannot.

Consider program statements that specify physical output behaviors. For example, some programming languages have a command BEEP whose intended behavior is to ring a bell. Or suppose you have a graphical programming language, one of whose legal instructions is DRAW_CIRCLE(x, y, r), whose intended behavior is to draw a circle at point (x, y) with radius r. How can you prove or verify that the program will ring the bell or draw the circle? How can you mathematically or *logically* prove that the (physical) bell will (actually) ring or that a (physical) circle will (actually) be drawn? How can you *logically* prove that the (*physical*) bell *works* or that the pen has ink in it? Fetzer's point is that you can't. And the controversy largely focused on whether that's what's meant by program verification. Recall our discussion in Chapter 10 about Cleland's interpretation of the Church-Turing Computability Thesis: is preparing hollandaise sauce, or physically ringing a bell, or physically drawing a circle a computable task?

But according to Fetzer, it's not just real-world output behaviors like ringing bells, drawing circles, or, for that matter, cooking that are at issue. What about the mundane PRINT command? According to Fetzer, it's not just a matter of causal output, because you can replace every PRINT(x) command with an assignment statement: $p := x$. *But even this is a causal statement*, because it instructs the computer to *physically* change the values of bits in a *physical* register p, so Fetzer's argument goes through: how can you *logically* prove that the *physical* computer will actually work? Indeed, the history of early modern computers was largely concerned with ensuring that the vacuum tubes would be reliable (Dyson, 2012b). Recall Babbage's boast and Wittgenstein's warning, cited in the epigraphs to this chapter.

In Fetzer's terminology, a theorem T is "absolutely verifiable" $=_{def}$ T follows only from (logical) axioms (and not from empirical premises), and T is "relatively verifiable" $=_{def}$ T follows from (logical) axioms *together with* (empirical) premises. That is, T is "relatively" verifiable iff it is a logical consequence of some of the premises about the particular subject matter; it is "verifiable relative to" the premises. As Donald MacKenzie puts it,

> … mathematical reasoning alone can never establish the "correctness" of a program or hardware design in an absolute sense, but only relative to some formal specification of its desired behavior. (MacKenzie, 1992, p. 1066, col. 2)

Although an "absolutely verifiable" theorem T is not relative to the *premises*, even what Fetzer calls '*absolute* verifiability' is still a kind of *relative* verifiability, except that the verifiability is relative to the *axioms* (not to the premises), as we saw in Section 15.2.1.

Given all this terminology, Fetzer phrased the fundamental question of program verification this way: *Are programs (in Fetzer's sense) absolutely verifiable?* That is, can programs (in Fetzer's sense) be verified directly from axioms, with no empirical premises?

Question to Think About: *Do* the pro-verificationists claim that programs are "absolutely verifiable" (to use Fetzer's terminology)?

To be "absolutely verifiable" requires there to be *program* rules of inference that are truth-preserving, or it requires there to be *program* axioms that are necessarily true about "the *performance* that a machine will display when such a program is executed" (Fetzer, 1988, p. 1052, my italics). Verification that requires axioms about *performance* is different from program verification in the Hoare-Dijkstra-Gries tradition because of a difference between *logical* relations and *causal* relations, according to Fetzer. The former are abstract; the latter are part of the real world. It might be replied, on behalf of the pro-verificationists, that we can still do *relative* veri-fication: verification relative to "causal axioms" that relate these commands to causal behaviors. So, we can say that *if* the computer executing program P is in good working order, and *if* the world (the environment surrounding the computer) is "normal," *then* P is verified to behave in accordance with its specifications.

No, says Fetzer: algorithms and programs that are only intended for *abstract* machines *can* be *absolutely* verified (because there is nothing physical about such machines; they are purely for-mal). But programs that can be compiled and executed on *physical* machines can only be *relatively* verified.

15.4.2 The Controversy

The reaction to Fetzer's paper was explosive, beginning with a letter to the editor signed by 10 distinguished computer scientists arguing that it should never have been published because it was "ill-informed, irresponsible, and dangerous" (Ardis et al., 1989, p. 287, col. 3)! The general tone of the responses to Fetzer also included these objections:

- So what else is new? We program verificationists never claimed that you could logically prove that a physical computer would not break down.
- Verification techniques *can* find logical faults; it is logically possible to match a program or algo-rithm to its specifications.
- You can minimize the number of rules of the form "input *I causes* output *O*" such that they only apply to descriptions of logic gates and the physics of silicon.
- Many programs are incorrect because, for example, of the limits of accuracy in using real num-bers.
- Verifiably *incorrect* programs can be better than verifiably correct programs *if* they have better average performance. (Moor, 1979 makes a similar argument in the context of whether we should trust decisions made by computers; we'll discuss this in Chapter 17.)

Let's look at some of these.

Ardis et al. (1989) claimed that program verification was *not* supposed to "provide an *absolute* guarantee of correctness with respect to the execution of a program on computer hardware" (p. 287, col. 1, my italics). This is interestingly ambiguous: on one reading, they might have been claiming that program verification only provides a *relative* guarantee of correctness; if so, they are actually in agreement with Fetzer! On another reading, they might have been claiming that program verifica-tion *does* provide an absolute guarantee, but not of *hardware* execution; if so, that is also consistent with Fetzer's arguments!

They also claimed that it was not the case that "verification can be applied only to abstract pro-grams written in high level languages" (Ardis et al., 1989, p. 287, col. 2). For example, they said, it *can* be applied to assembly languages, contrary to what Fetzer (1988, p. 1062, col. 2) claimed. But Fetzer didn't have to claim that: there can be abstract, formal assembly languages. What Fetzer perhaps should have said was that program verification cannot be applied to assembly-language programs that "reside electronically in a computer" (to use Daylight's characterization). As Parsons (1989, p. 791, col. 1) later observed, their "rage" might have been indicative of a lack of evidence for their belief.

Other critics responded to Fetzer's paper by saying "So what else is new?": Pleasant (1989, p. 374, col. 1, my italics) observed that Fetzer's complaint "*belabor[s] the rather obvious fact* that programs which are run on real machines cannot be completely reliable, *as though advocates of verification thought otherwise.*"

And Paulson et al. (1989, p. 375, col. 1, my italics) said that "Fetzer makes one important *but elementary* observation and takes it to an absurd conclusion. … [M]ost systems … do not need to work perfectly. … A physical fault can usually be repaired quickly, replacing the damaged part; then the job can be run again." It is interesting to note, especially in connection with topics that we will look into in Chapter 16, that the passage that I omitted after "most systems" concerned one major exception: SDI – the Strategic Defense Initiative – a program to defend the US using a computer-controlled missile defense system; that system, of course, needed to "work perfectly"! (On program verification of SDI, see Myers, 1986.)

Or this from C.M. Holt (1989):

> No one expects a computer to work properly if someone pulls the plug out [p. 508, col. 2]. … Errors in programs due to inaccurate scientific theories, omissions in specifications, and implementation failures are inevitable; those due to programming mistakes should not be [p. 509, cols. 1–2].

An interesting variation on this came from Conte (1989), who called "Fetzer's article … an over-inflated treatment of a principle most children learn by the age of 10 – no matter how perfect your cookie recipe is, if the oven thermostat fails, you may burn the cookies." How do you think Carol Cleland, whose objections to the Computability Thesis we examined in Chapter 10, would respond if she were told that no matter how perfect her hollandaise-sauce recipe is, if it is prepared on the Moon, it may not work?

15.4.3 Barwise's Attempt at Mediation

The logician Jon Barwise (1989b) attempted to mediate the controversy. In doing so, he also discussed many other issues that we have looked into (or will in future chapters), including the relation between algorithms and programs, the possibility of finding fault with an argument yet believing its conclusion, the nature of "philosophy of X" (see Section 2.7 of this book), and the difference between the truth of a premise and agreeing with it (see Section 2.9 of this book).

Barwise saw the issue between Fetzer and his opponents as being a special case of the more general question of how mathematics can be applied to the real world, given that the former is abstract and purely logical, whereas the latter is concrete and empirical (p. 846, col. 2). (See Sections 4.11.2 and 16.11.) But there is another aspect to that issue in the philosophy of mathematics: namely, the relation between the syntax of a formal mathematical expression and its semantic interpretation in the real world:

> The axiomatic method says that our theorems are true *if* our axioms are. The modeling method says that our theorems model facts in the domain modeled *if* there is a close enough fit between the model and the domain modeled. The sad fact of the matter is that **there is usually no way to prove – at least in the sense of mathematical proof – the antecedent of a conditional of either of these types.** (Barwise, 1989b, p. 847, col. 2, italics in original, my boldface)

This is a point made by Brian Cantwell Smith (1985), which we'll look at in Section 15.6. Barwise cites Smith, noting that

> Computer systems are not just physical objects that compute abstract algorithms. They are also embedded in the physical world and they interact with users. … Thus, … our mathematical models need to include not just a reliable model of the computer, but also a reliable

model of the environment in which it is to be placed, including the user. (Barwise, 1989b, p. 850, col. 1)

Barwise noted that Fetzer was only willing to talk about the causal (i.e. physical) role of computers, which is not susceptible to mathematical verification, whereas the field of program verification only concerns abstract programs (p. 848, col. 2). So it really seems that both sides not only are talking past each other but might actually be consistent with each other!

Question for the Reader: In remarks given at the 40th Anniversary celebration of the founding of the SUNY Buffalo Department of Computer Science & Engineering (April 2007), Bruce Shriver, former president of the IEEE Computer Society, said, "Hardware does not have flaws; only software does."
What do you think he might have meant by this?

15.5 The Program-Verification Debate: Summary

From a methodological point of view, it might be said that programs are conjectures, while executions are attempted – and all too frequently successful – refutations (in the spirit of Popper …).
—James H. Fetzer (1988, p. 1062, col. 2)

Recall from Section 9.3.1 that Kleene claimed that Turing Machines, unlike physical computers, were "error free" (Kleene, 1995, p. 27). As noted in that section, if the Turing Machine were poorly programmed, it *wouldn't* be error free! Indeed, 50 years earlier, von Neumann said

The remarks … on the desired automatic functioning of the device [i.e. von Neumann's definition of a computer, as quoted in Section 9.2.2] must, of course, assume that it functions faultlessly. Malfunctioning of any device has, however, always a finite probability – and for a complicated device and a long sequence of operations it may not be possible to keep this probability negligible. Any error may vitiate the entire output of the device. For the recognition and correction of such malfunctions intelligent human intervention will in general be necessary.

However, it may be possible to avoid even these phenomena to some extent. The device may recognize the most frequent malfunctions automatically, indicate their presence and location by externally visible signs, and then stop. Under certain conditions it might even carry out the necessary correction automatically and continue. (von Neumann, 1945, Section 1.4, p. 1).

One way to read this is as a recognition or anticipation of Fetzer's point. Given this inevitability, the focus presumably has to be on the elimination of *logical* errors so that program verification still has a role to play. The second paragraph suggests that some *machine* "verification" might be automated, but that just leads to an endless regress: even if the logical structure of that automation is guaranteed, the physical device that carries it out will itself be subject to some residual malfunction possibilities (Bringsjord, 2015).

Of course, another way to read von Neumann's remarks (as well as the entire program verification debate) is to recognize that no one, and no thing, is perfect. There's always the chance of error or malfunction: complete elimination of error is physically impossible, so the point is, at least, to minimize it. Thus, the entire issue of program verification might be considered a subset of the more general engineering issue of reliability. For example, Allen Newell (1980, p. 159) assumes that a symbol system should be "totally reliable, nothing in its organization reflecting that its operators, control or memory could be errorful" [sic!]. He goes on to say that "universality is always relative to physical limits, of which reliability is one" (p. 160), where 'universality' is defined as the ability to "produce an arbitrary input-output function" (p. 147). This suggest that even if a *program* could be proved mathematically to be correct, the *process* that executes it would still be limited by *physical* correctness, so to speak, and that, presumably, cannot be *mathematically* proved.

The bottom line is that programs as hardware need causal rules of inference of the form input *I causes* output *O*. Perhaps the BEEP command would have to be annotated something like this:

{The bell is in working order.} BEEP {A sound is emitted.}

If such causal rules are part of the definition of an abstract machine, then we *can* have "absolute" verification of the program. But if they are merely empirical claims, then we can only have "relative" verification of the program.

Even so (as De Millo et al. (1979) pointed out), absolute verification is often thought to be too tedious to perform and can lure us into overconfidence. The problem of tediousness seems to me not to be overly serious: it's tedious to prove theorems in mathematics, too. In any case, techniques are being devised to automate program verification. The problem of overconfidence is more important, for precisely the reasons Fetzer adduces. Just because you've proved a program is correct is no reason to expect that the computer executing it will not break down.

But in addition to the relativity to axioms (logical relativity) and premises (subject-matter relativity), there is another "level" of relativity:

Mathematical argument can establish that a program or design is a correct implementation of that specification, but not that implementation of the specification means a computer system that is "safe," "secure," or whatever. (MacKenzie, 1992, p. 1066, col. 2)

There are two points to notice here. First, a mathematical argument can establish the correctness of a program relative to its specification: i.e. whether the program satisfies the specification. Second, not only does this not necessarily mean the computer system is safe (or whatever), but it also does not mean the *specification* is itself "correct":

Human fallibility means some of the more subtle, dangerous bugs turn out to be errors in design; the code faithfully implements the intended design, but the design fails to correctly handle a particular "rare" scenario. (Newcombe et al., 2015, p. 67)

Presumably a specification is a relatively abstract outline of the solution to a problem. Proving that a computer program is correct relative to – i.e. satisfies – the specification does not guarantee that the specification actually solves the problem!

This is the case for reasons that Smith (1985) discusses and that we will look at next.

Question for the Reader: Physical computers can break down, whereas things like Turing Machines, being abstract, cannot physically fail (at most, their algorithm might be incorrect). But Turing Machines are models of human computers ("clerks"), and whereas Turing Machines can have infinite memory and take infinite time to perform a computation, humans have only finite memory and time, and can make mistakes (Shapiro et al., 2022).

How does that affect the program-verification controversy?

15.6 Program Verification, Models, and the World

The goal of software development is to model a portion of the real world on the computer. … That involves an understanding not of computers but of the real-world situation in question. …That is not what one learns in studying computer science; that is not what computer science is about.
—Michael Mahoney (2011, p. 117)

15.6.1 "Being Correct" vs. "Doing What's Intended"

Recall that one objection to program verification is that a program can be "proven correct" yet not "do what you intend." One reason, as we have just seen, might be that the *computer* on which the program is run might fail physically. That is, the computer system might fail at the hardware level.

A second reason might be that the *world* is inhospitable. There are two ways in which this latter problem might arise. There might be a physical problem with the connection between the computer and the environment: at a simple level, the cables connecting the computer to the world (say, to a printer) might be flawed. Or the world itself – the environment – might not provide the correct conditions for the intended outcome, as with Cleland's hollandaise sauce (Section 10.4).

A third reason is related to the possible "hyper"-computability of interactive programs, which might depend on the unpredictable and non-verifiable behavior of an "oracle" or human user (Sections 11.8, 11.9).

What does 'correct' mean in this context? Does it mean the program has been logically verified? Does it mean it "does what was intended"? Perhaps a better way of looking at things is to say that there are two different notions of "verification": an internal one (logical verification) and an external one (doing what was intended) (Tedre and Sutinen, 2008, pp. 163–164). But to the extent that doing what was intended is important, then we need to ask, *whose* intent counts? Here is computer scientist and philosopher Brian Cantwell Smith on this question:

What does *correct* mean, anyway? Suppose the people want peace, and the President thinks that means having a strong defense, and the Defense department thinks that means having nuclear weapons systems, and the weapons designers request control systems to monitor radar signals, and the computer companies are asked to respond to six particular kinds of radar pattern, and the engineers are told to build signal amplifiers with certain circuit characteristics, and the technician is told to write a program to respond to the difference between a two-volt and a four-volt signal on a particular incoming wire. If being correct means *doing what was intended*, whose intent matters? The technician's? Or what, with twenty years of historical detachment, we would say *should have been intended*?
(B.C. Smith, 1985, Section 2, p. 20, col. 1)

According to Smith, the cause of these problems lies not in the relation of *programs* to the world but in the relation of *models* to the world. Let's see what he means.

15.6.2 Models: Putting the World into Computers

What the conference [on the history of software] missed was software as model, … software as medium of thought and action, software as environment within which people work and live. It did not consider the question of *how we have put the world into computers*.
—Michael Mahoney (2011, pp. 65–66, my italics)

According to Smith (1985, Section 3, p. 20, col. 1), to design a computer system to solve a real-world problem, we *don't* directly "put the world into computers." Rather, we must do two things: (1) create a *model* of the real-world problem and (2) create a *representation of the **model*** in the computer. Let's look at each of these.

Creating a Model of the World

> To build a model is to conceive of the world in a certain delimited way.
> —Brian Cantwell Smith (1985, Section 3, p. 20, col. 1)

The model we create has no choice but to be "delimited": that is, it must be abstract – it must omit some details of the real-world situation. Abstraction, as we saw in Section 13.1.1, is the opposite of implementation: it is the removal of "irrelevant" implementation details.

Why must any real-world information be removed? Why are models necessarily partial? One reason is that it is methodologically easier to study a phenomenon by simplifying it, coming to understand the simplified version, and then adding some complexities back in, little by little. If models weren't partial, there would be too much complexity, and we would be unable to use them as a basis for action. You can't use, much less have, a map of Florida that is the size of Florida and that therefore can show everything in Florida. Such a map might be thought to be more useful

Figure 15.1 2D photographic model of a real house.

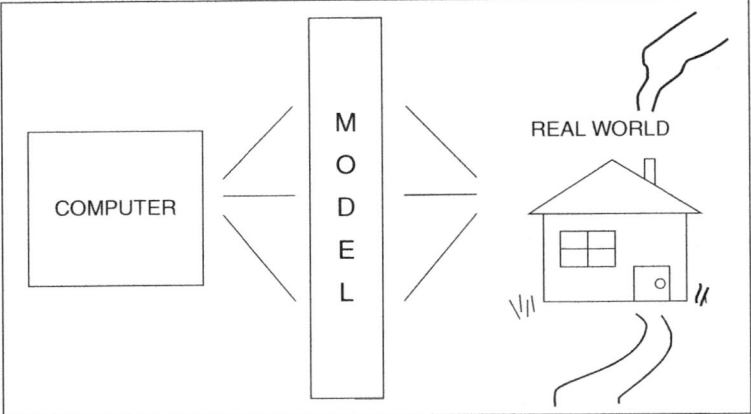

Figure 15.2 Source: From Colburn et al., 1993, p. 283. Reprinted with permission of the original publisher, CSLI Publications.

than a smaller, more manageable one, in that it would be able to show all the detail of Florida itself. But the life-sized version's lack of manageability is precisely its problem.[8]

Can we eat our cake but keep it, too? Perhaps we *can* use the real world as a representation of itself. The computer scientist Rodney A. Brooks (1991, Section 1) suggested that we should "use the world as its own model": a Roomba robotic vacuum cleaner doesn't *need* a map showing where there is a wall; if it bumps into one, it will know the wall is there. But even this is only a *part* of the real world.

In any case, the usual first step in solving a problem is to create a "delimited" (abstract, simplified) model of it. For example, consider Lucille Ball's childhood home in Celoron, NY (which my wife and I own!). Figure 15.1 is a two-dimensional photographic representation (or model) of the real, three-dimensional, physical house. (Smith's original diagram is in Figure 15.2.)

Creating a Computer Representation of the Model A second step is to use logical propositions or programming-language data structures to represent *not* the real-world situation but the *model*. So, besides the 3D house and its 2D photographic model, we might have a computer representation *of the model*:

```
((type house)
 (part (walls (material wood))
       (door (location front))
       (windows (location front
                         (number 5)))
       (steps (location front
                         (number  4)))
  . . .
  )
 )
```

---

8 See the Online Resources for further reading on such maps.

Smith's point is that computers only deal with *their representations of* these *abstract models of* the real world. As Paul Thagard (1984, p. 82, citing Zeigler, 1976) notes, computers are twice removed from reality, because "a computer simulates a model which models a real system."

Is that necessarily the case? Can't we skip the intermediate, abstract model and directly represent the real-world situation in the computer? Perhaps, but this won't help us avoid the problem of *partiality* (or abstraction, or idealization, or simplification). The only rational way to deal with (real-world) complexity is to analyze it – i.e. to simplify it – i.e. to deal with a partial (abstract) representation or model of it.

We are condemned to do this whenever we act or make decisions: if we were to hold off on acting or making a decision until we had complete and fully accurate information about whatever situation we were in, either we would be paralyzed into inaction or the real world might change before we had a chance to complete our reasoning. (As Alan Saunders – and, later, John Lennon – said, "Life is what happens to us while we are making other plans"; http://quoteinvestigator.com/2012/05/06/other-plans/.) This is the problem that Herbert Simon recognized when he said we must always reason with uncertain and incomplete (even noisy) information: our rationality is "bounded," and we must "satisfice" (Simon, 1996b, p. 27). And it is also the problem that led Kugel to suggest that trial-and-error computers were essential for AI (Section 11.10.2). And this holds for computation as well as thinking. (See also Simon, 1962, Dijkstra, 1972.)

But action is *not* abstract: you *and* the computer must act *in* the complex, real world, even though such real-world action must be based on *partial models* of the real world: that is, on incomplete and noisy information. Moreover, there is no guarantee that the *models* are correct.

Action can help: it can provide feedback to the computer system so the system won't be isolated from the real world. Recall the blocks-world program that didn't "know" it had dropped a block but "blindly" continued to faithfully execute its program to put the block on another (Section 10.4). If it had had some sensory device that let it know that it no longer was holding the block it was supposed to move, and if the program had had some kind of error-handling procedure in it, then it might have worked much better (it might have worked "as intended"). *Did* the blocks-world program behave as intended?

Model vs. World The problem, as Smith sees it, is that mathematical model theory only discusses the relation between the model and a *description* of the model. It does not discuss the relation between the model and the *world*. A model is like eyeglasses for the computer, through which it sees the world. The model *is* the world *as* the computer sees it. The problem is that computers have to act in the *real* world on the basis of a *model* of it.

Philosophical Digression: Immanuel Kant said that the same thing is true about us. Our concepts are like eyeglasses that distort reality; our only knowledge of reality is filtered through our concepts, and we have no way of knowing how things "really" are "in themselves," unfiltered by our concepts (as illustrated in Figure 15.3). (Recall our earlier discussions of Kant in Sections 3.14 and 4.4.1.)

Similarly, to prove a program correct, we need both (a) a *specification* (a model of the real-world problem) that says (declaratively) *what* the computer systems should do and (b) a *program* (a computer model of the specification model) that says (usually procedurally) *how* to accomplish this. A correctness proof, then, is a proof that any system that obeys the program will satisfy

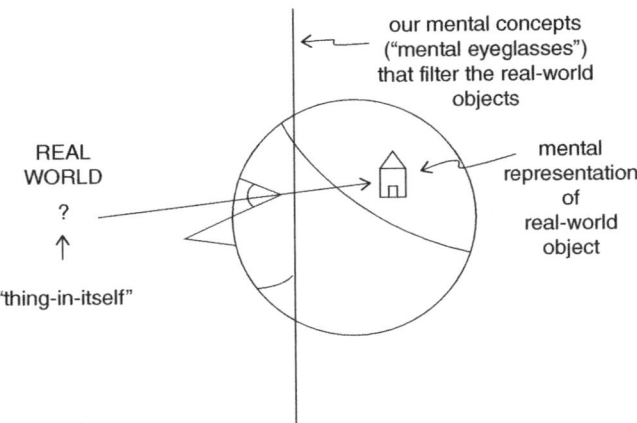

Figure 15.3 A cognitive agent looking at a real-world object that the agent categorizes as a house. Light reflecting off the house (the "thing-in-itself") enters the agent's eyes, and the resulting neural signals are "filtered" through the agent's mental concepts, producing a mental image of the house. The mental image may or may not be a "perfect" representation of the house, but the agent has no way to directly compare the mental image with the real house, independent of the agent's concepts. Figure 15.2 is Smith's version of this. Source: Author's drawing.

the specification. But this is a proof that two *descriptions* are compatible. The program is proved correct *relative to* the specification:

> … what can be proven correct is not a physical piece of hardware, or program running on a physical machine, but only a mathematical model of that hardware or program.
> (MacKenzie, 1992, p. 1066; see also Turner, 2018, Section 4.4)

Suppose the proof fails to show "correctness"; what does this mean? It means *either* that the program is wrong *or* that the specification is wrong (or both). And indeed, often we need to adjust both specification and program.

The real problems lie in the model-world relation, which correctness does not address. This is one of the morals of Cleland's and Fetzer's claims. That is, programs can fail because the models can fail to correspond to the real world *in "appropriate" ways*. But that italicized clause is crucial because all models abstract from the real world, but each of them does so in different ways.

This is the case for reasons that are related to the Computability Thesis: you can't show that two systems are the same, in some sense, unless you can talk about both systems in the same language. (Recall Rosser's time-traveling mathematician from Section 15.1.) In the case of the Computability Thesis, the problem concerns the *informality* of the language of algorithms versus the *formality* of the language of Turing Machines. In the present case, the problem concerns the mathematical *language* of programs versus the non-linguistic, *physical* nature of the hardware. Only by *describing* the hardware in a (formal, mathematical) *language* can a proof of equivalence be attempted. But then we also need a proof that that formal description of the hardware is correct; and that can't be had. It can't be had because, to have it, we would need *another* formal description of the hardware to compare with the formal description we were trying to verify. And that leads to a Zeno-like infinite regress. (We can come "close, but no cigar.")[9]

---

9 Note to philosophers: it's actually closer to a Bradley-like regress (Perovic, 2017) (Royce, 1900, pp. 502–507ff); see also Rapaport, 1995, p. 64.

Both Smith and Fetzer agree that the program-verification project fails, but for slightly different reasons: for Fetzer (and Cleland), computing is about the *world*; it is "external" or "wide." Thus, computer programs can't be ("absolutely" or "externally") verified, because the world may not be conducive to "correct" behavior: a physical part might break; the environment might prevent an otherwise-perfectly-running, "correct" program from accomplishing its task (such as making hollandaise sauce on the Moon using an Earth recipe); and so on.

For Smith, computing is done on a *model* of the world; it is "internal" or "narrow." Thus, computer programs can't be verified, because the model might not match the world. Smith also notes that computers must act *in* the real world. But their abstract narrowness *isolates* them from the concrete, real world *at the same time* they must act *in* it. Smith's "gap" between model and world is due, in part, to the fact that specifications are abstract:

> A specification is an abstraction. It should describe the important aspects and omit the unimportant ones. *Abstraction is an art* that is learned only through practice. … [A] specification of what a piece of code does should describe everything one needs to know to use the code. It should never be necessary to read the code to find out what it does.
> (Lamport, 2015, p. 39, my italics)

How does one know if something that has been omitted from the specification is important or not? This is why "abstraction is an art" and why there's no guarantee that the model is correct (in the sense that it matches reality).

That is precisely the kind of error that Smith warns about. It is time to look into this possibility. How *do* programs relate to the real world?[10]

---

10 See the Online Resources for further reading on Smith's "gap."

16

Programs and the World[1]

Today, computing scientists face their own version of the mind-body problem: how can virtual software interact with the real world?
—Philip Wadler (1997, p. 240)

16.1   Introduction

In the previous chapter, we looked at arguments to the effect that, roughly, a computer program might succeed in theory but fail in practice. In this chapter, we continue our examination of the relationship between a program and the world in which it is executed.

Science, no matter how conceived, is generally agreed to be a way of *understanding* the world (as we saw in Chapter 4). So, CS as a science should be a way of understanding the world *computationally*. And engineering, no matter how conceived, is generally agreed to be a way of *changing* the world, preferably by improving it (as we saw in Chapter 5). So, CS as an engineering discipline should be a way of changing (improving?) the world via computer programs that have physical effects. Thus, CS deals with the real world by trying to understand the world computationally and change the world by building computational artifacts.

In this chapter, we will focus on two questions:

1. **Is computing <u>directly</u> concerned with the *world*?**
2. **Or is computing only <u>indirectly</u> concerned with the world**
   by <u>directly</u> dealing only with *descriptions* or *models* of the world?

In other words, is computation primarily concerned with the internal workings of a computer, both abstractly in terms of the theory of computation – e.g. the way in which a Turing Machine works – as well as more concretely in terms of the internal physical workings of a physical computer? This is an aspect of question 2.

Or is computation primarily concerned with how those internal workings can reach out to the world in which they are embedded? As Philip Wadler noted (see the epigraph for this chapter), this is related to the question of how the mind (or, more materialistically, the brain) reaches out to "harpoon" the world (Castañeda, 1989, p. 114). This is an aspect of question 1.

Those who say that computing is directly concerned with the *world* sometimes describe computing as being "external," "global," "wide," or "semantic." And those who say that computing is only directly concerned with *descriptions* or *models* of the world sometimes describe computing as being

1 Portions of this chapter are adapted from Rapaport, 2017a.

Philosophy of Computer Science: An Introduction to the Issues and the Literature, First Edition. William J. Rapaport.
© 2023 John Wiley & Sons, Inc. Published 2023 by John Wiley & Sons, Inc.

"internal," "local," "narrow," or "syntactic." As you should expect by now, of course, it might be both! After all, even if computing is "narrow," it is embedded in – and interacts with – the "wider" world. In that case, the question is how these two positions are related.

16.2 Internal vs. External Behavior: Some Examples

Internal models diverge from the external world. In earlier chapters, we saw several examples of programs whose internal (local, narrow, syntactic) behavior differed from their external (global, wide, semantic) behavior. Let's briefly review these, plus a few new ones.

Some are the kinds of situations that Fetzer and Smith were concerned with, in which the computer program behaves exactly as was expected – there are no logical program bugs, and the program does not crash – yet it fails to accomplish its stated task. That is, it exhibits "successful" *internal* behavior but "unsuccessful" *external* behavior. Others are situations in which a single internal behavior generates different external behaviors, depending on context. Some are of both kinds.

1. **The blocks-world robot (Section 10.4):** The blocks-world program worked "correctly" in the sense that it performed each step without crashing. Yet it did not do what was intended, because it accidentally dropped a block and was therefore unable to put it where it was supposed to go. "Narrowly," perhaps, it did what was intended; "widely," however, it didn't: after all, it didn't actually manipulate the blocks.
2. **Hollandaise sauce (Section 10.4):** On Earth, Cleland's algorithm (recipe) for hollandaise sauce results in an emulsion that is, in fact, hollandaise sauce. But on the Moon, it does not result in an emulsion, and hence there is no hollandaise sauce; instead, the output is a messy mixture of eggs, butter, and lemon juice. For Cleland, is making hollandaise sauce computable (on the "narrow" view) or not (on the "wide" view)? *Can* a Turing Machine or a physical computer make hollandaise sauce?
3. **"Duck-Rabbit" programs (Section 10.4):** Jerry Fodor (1978, p. 232) asked us to consider two computer programs – one that simulates the Six Day War and another that simulates (or actually plays?) a game of chess – that are such that "the internal career of a machine running one program would be identical, step by step, to that of a machine running the other." In programs like this war-chess case, do we have one algorithm (the "narrow" view) or two (the "wide" view)?

> **Question for the Reader:** Is there a difference between *simulating* playing a chess game and *really* playing one? (Recall our discussion of simulation vs. "the real thing" in Section 14.2.2. We'll return to this in Chapter 18.)

A real example along the same lines is "a method for analyzing x-ray diffraction data that, with a few modifications, also solves Sudoku puzzles" (Elser, 2012). Or consider a computer version of the murder-mystery game Clue that exclusively uses the Resolution rule of inference and so could be a general-purpose, propositional theorem prover instead (Robin Hill, personal communication). A more recent version is the program AlphaZero, "a single algorithm [that] can learn to play three hard board games" (Campbell, 2018): when supplied with the rules of chess, it becomes a champion chess player; when supplied with the rules of shogi (Japanese chess), it becomes a champion shogi player; and when supplied with the rules of Go, it becomes a champion Go player (Silver et al., 2018; Kasparov, 2018; Halina, 2021).

Similar examples abound, notably in applications of mathematics to science, and these can be suitably "computationalized" by imagining computer programs for each. For example,

> Nicolaas de Bruijn once told me roughly the following anecdote: Some chemists were talking about a certain molecular structure, expressing difficulty in understanding it. De Bruijn, overhearing them, thought they were talking about mathematical lattice theory, since everything they said could be – and was – interpreted by him as being about the mathematical, rather than the chemical, domain. He told them the solution of their problem in terms of lattice theory. They, of course, understood it in terms of chemistry. Were de Bruijn and the chemists talking about the same thing? (Rapaport, 1995, p. 63)

A related issue is that a single action in the real world can be described in different ways:

> Recovering motives and intentions is a principal job of the historian. For without some attribution of mental attitudes, actions cannot be characterized and decisions assessed. *The same overt behavior, after all, might be described as "mailing a letter" or "fomenting a revolution"*. (Richards, 2009, p. 415)

And Figure 16.1 presents a humorous one.[2]

4. **Rescorla's GCD computers (Section 13.3.3):** Rescorla's GCD computers fall under this category but also offer an example reminiscent of Cleland's hollandaise sauce, although less "physical." A Scheme program for computing GCDs of two numbers is implemented on two computers, one (M_{10}) using base-10 notation and another (M_{13}) using base-13 notation. Rescorla argued that only M_{10} executes the Scheme program *for computing GCDs*, even though, in a "narrow" sense, both computers are executing the "same" program. When the *numerals* '115' and '20' are input to M_{10}, it outputs the *numeral* '5'; "it thereby calculates the GCD of the corresponding *numbers*" (Rescorla, 2013, p. 688). But the *numbers* expressed in base-13 by '115' and '20' are 187_{10} and 26_{10}, respectively, and their GCD is 1_{10}, not 5_{10}. So, in a "wide" sense, the two machines are doing "different things," in one case behaving "correctly," and in the other behaving "incorrectly." *Are* the two GCD computers doing different things?

5. **AND-Gates or OR-gates? (Section 10.4):** The truth table for *conjunction* can also be used as the truth table for *disjunction* by reinterpreting '0's and '1's. Another version uses a single truth table:

Input 1	Input 2	Output
A	A	A
A	B	A
B	A	A
B	B	B

This could be interpreted as the truth table for conjunction if A is interpreted as 'false' and B as 'true.' And it could be interpreted as the truth table for disjunction if A is interpreted as 'true' and B as 'false.'

---

2 See the Online Resources for further reading and examples.

Figure 16.1 LUANN ©2015 GEC Inc. Reprinted with permission of ANDREWS MCMEEL SYNDICATION. All rights reserved.

Oron Shagrir (2020, Section 3.1) offers a third version. Consider the following function:

Input 1	Input 2	Output
H	H	H
H	M	M
H	L	M
M	H	M
M	M	M
M	L	M
L	H	M
L	M	M
L	L	L

(Actually, he uses physical voltages labeled H(igh), M(edium), and L(ow). I am using a more abstract version.) He then gives two different (external semantic) interpretations of these symbols. On the first interpretation, H is mapped to 1 (or "true") and both L and M are mapped to 0 (or "false"), thus implementing conjunction. (In Shagrir's original version, the physical device that inputs and outputs certain voltages becomes an AND-gate). On the second interpretation, both H and M are mapped to 1 and L is mapped to 0, thus implementing disjunction (or an OR-gate).

> **Question for the Reader:** In Shagrir's example, we have two different computers (and AND-gate and an OR-gate) that are implemented in the same way. Would two AND-gate computers be the same even if they were implemented differently?

Thus, there are at least three possible answers to the question of what this device computes: conjunction, disjunction, and the function given by the previous table. One way of phrasing the puzzle here is this: What is the basic computational structure of this system (or these systems)? Is it conjunction? Disjunction? The H-M-L function? (Or something else? After all, one of the points made by Buechner (2011, 2018, see Section 15.1) is that there might not be any fact of the matter!)

Alternatively, one could say that in order to compute conjunction, execute an algorithm for the previous table, but use the first interpretation; and, in order to computer disjunction, execute that very same algorithm, but use the second interpretation. How crucial is that external interpretation to the computation? Arguably, the difference concerns, not the computation, but how that computation is "plugged in" to the external environment. But arguably, such external relations don't change the computation any more than the external fact that a person's sibling has had a child (thus making the person an aunt or uncle) changes that person. In Figure 16.1, the girl held up 10 fingers, irrespective of whether she intended (or the boy understood) "10 yeses or 5 noes." (Dennett (2013a, pp. 159–164) discusses a similar example with a vending machine that when used in the United States, works on US quarters, but when used in Panama, works equally well with Panamanian quarter-balboas.)[3]

16.3   Two Views of Computation

> A computer program is a message from a man [sic] to a machine. The rigidly marshaled syntax and the scrupulous definitions all exist to make intention clear to the dumb engine.
> —Frederick P. Brooks (1975, p. 164)

If computation is "narrow," "local," "internal," or "syntactic," then it is concerned only (or at least primarily) with such things as the operations of a Turing Machine (print, move) or the basic recursive functions (successor, predecessor, projection). On the other hand, if computation is "wide," "global," "external," or "semantic," then it must involve things like chess pieces and a chess board (for a chess program), or soldiers and a battlefield (for a war simulator). Is computation narrow and independent of the world, or is it wide and world-involving?

A related question asks whether programs are purely logical, or whether they are "intentional" (Hill, 2016) or "teleological" (Anderson, 2015). Something can be said to be "intentional" if it is related to goals or purposes. 'Teleological' is another adjective with roughly the same meaning. So we can ask whether programs are goal-oriented.

Let's distinguish between an *algorithm A* and a *goal* (or *purpose*) *G*. Let *A* be either a primitive computation (such as "print" and "move," or such as the "successor," "predecessor," and "projection" functions) or a set of computations recursively structured by sequence, selection, and repetition (as in Chapter 7). And let *G* be a goal (or "purpose," or "intended use") of *A*. Then our question can be formulated more precisely as follows (using a distinction due to Robin K. Hill (2016, Section 5)): which of the following two forms does a computer program take?:

Do *A*

or

In order to accomplish goal *G*, do *A*

If the former, then computation is "narrow"; if the latter, then it is "wide."

This enables us to reformulate some earlier issues: for example, does the "correctness" of a computer program refer to algorithm *A* or goal *G*? The goal of a program can be expressed in its

---

3 See the Online Resources for further reading on what is sometimes called the "indeterminacy" of computation.

specification. This is why you wouldn't *have* to read the code to find out what it does. Of course, if the specification has been internalized into the code, you might be able to (see Section 16.10.4). But it's also why you can have a chess program that is also a war simulator: they might have different specifications but the same code. So, perhaps a correctness proof is a proof that algorithm *A* satisfies specification *G*.

As another example, Cleland argued that "Make hollandaise sauce" was not a Turing-computable function. Can we view "Make hollandaise sauce" as a high-level procedure call that can be substituted for *A*? Or should it be viewed as a goal that can be substituted for *G*? In the latter case, only its expansion into a set of (more basic) computations structured by sequence, selection, and repetition would be a suitable substitute for *A*. And in that latter case, *A* might be computable on the Moon even if *G* fails there.

We'll refer to these two formulations throughout the rest of the chapter.[4]

16.4 Inputs, Turing Machines, and Outputs

> Any machine is a prisoner of its input and output domains.
> —Allen Newell (1980, p. 148)

16.4.1 Introduction

Aaron Sloman notes that for almost any machine,

> we can, to a first approximation, divide the processes produced by the machine into two main categories: *internal* and *external*. Internal physical processes include manipulation of cogs, levers, pulleys, strings, etc. The external processes include movements or rearrangements of various kinds of physical objects, e.g. strands of wool or cotton used in weaving, ... (Sloman, 2002, Section 3.2, p. 9)

As we noted in Section 3.10, both Shapiro (2001) and Sloman consider links to sensors and effectors as central to what a computer is. A computer without one or the other of these would be solipsistic (Section 7.3.3). Can computation be understood separately from interaction with the world?

One obvious place where a computer program seems to necessarily interact with the real world is its inputs and outputs. In Sections 7.3.3 and 11.8.1, we considered whether programs *needed* inputs and outputs. Let's review some of this.

16.4.2 The Turing Machine Tape as Input-Output Device

The tape of a Turing Machine records symbols (usually '0' or '1') in its squares. Is the tape the input-output device of the Turing Machine? Or is it (merely?) the Machine's internal memory device?

Given a Turing Machine for computing a certain mathematical function, it is certainly true that the function's inputs will be inscribed on the tape at the beginning of the computation, and the function's results – its outputs – will be inscribed on the tape by the time that the computation halts: So, it certainly looks like the tape is an *external* input-output device.

---

4 See the Online Resources for further reading on Hill's distinction.

However,

> A terminating computation is one in which all the processes terminate; *its output is **the values left in the shared memory**.* (Denning and Martell, 2015, p. 155, my emphases)

Note that this output need not be *reported* to the external world (such as a user); it's just left there on the tape. Moreover, the inscriptions on the tape will be used and modified by the machine during the computation, in the same way a physical computer uses its internal memory for storing intermediate results of a computation. So, it looks like the tape is merely an *internal* memory device. In other words, it also looks like the answer to our questions is: both.

Although Turing's *a*-machines were designed to simulate human computers, Turing didn't talk about the humans who would *use* them. A Turing Machine *doesn't* accept user-supplied input from the external world! (Recall our discussion of interactive computing in Section 11.8.) It begins with all data *pre-stored* on its tape and then simply does its own thing, computing the output of a function and leaving the result on the tape. Turing Machines don't "tell" anyone in the external world what the answers are, although the answers are there for anyone to read, because the "internal memory" of the machine is visible to the external world. Of course, a user has to be able to *interpret* the symbols on the tape; we'll return to this point in Section 16.4.6.

Perhaps it would be better to refer to the initial symbols on the tape as "setup conditions" and the final symbols as "terminal conditions," rather than as "inputs" and "outputs" (as suggested by Machamer et al., 2000, p. 11). So, are the symbols on the tape really inputs and outputs in the sense of coming from, and being reported to, the external world? Are such inputs and outputs an essential part of an algorithm? It may seem outrageous to deny that they are essential, but (as we saw in Section 7.3.3) it's been done! After all, the input-output interface "merely" *connects* the algorithm with the world. Let's consider whether inputs and outputs are needed. (For more on whether a Turing Machine tape is an external input-output device or an internal memory, see Dresner, 2003, 2012.)

16.4.3 Are Inputs Needed?

One reason it's outrageous that inputs or outputs might *not* be needed is that algorithms are supposed to be ways of computing mathematical functions, and mathematical functions, by definition, have both inputs and outputs – members of their domain and range. Functions are, after all, certain sets of ordered *pairs* (of inputs and outputs), and you can't very well have an ordered *pair* that is missing one or both of those.

In Section 7.3.3, we saw that

- Markov's informal characterization of algorithm had an "applicability" condition stating that algorithms must have "The possibility of starting from original given objects which can vary within known limits" (Markov, 1954, p. 1). Those "original given objects" are, presumably, the input.
- But Hartmanis and Stearns's classic paper on computational complexity (1965, p. 288) allowed their multi-tape Turing Machines to have at most one tape – an output-only tape – with no input tapes.
- And we also saw that Knuth's informal characterization of the notion of algorithm had an "input" condition stating that "An algorithm has *zero or more* inputs" (Knuth, 1973, p. 5; my italics). He not only didn't explain this but also went on to characterize outputs as "quantities which have a

specified relation to the inputs" (Knuth, 1973, p. 5). But what kind of relation would an output have to a non-existent input?[5]

One way to understand having outputs without inputs is that some programs, such as prime-number generators, merely output information. In cases such as this, although there may not be any *explicit* input, there is an *implicit* input (roughly, ordinals: the algorithm outputs the *n*th prime without explicitly requesting an *n* to be input). Another kind of function that might seem not to have any explicit inputs is a constant function, but again, its implicit input could be anything (or anything of a certain type – "varying within known limits," as Markov might have said).

So, what constitutes input? Is it simply the initial data for a computation – i.e. is it internal and syntactic? Or is it information supplied to the computer from the external world (and interpreted or translated into a representation of that information that the computer can "understand" and manipulate) – i.e. is it external and semantic?

16.4.4 Are Outputs Needed?

Markov, Knuth, and Hartmanis and Stearns all require at least one output. Markov, for example, has an "effectiveness" condition stating that an algorithm must "obtain a certain result." But Copeland and Shagrir (2011, pp. 230–231) suggest that a Turing Machine's output might be unreadable. Imagine not a Turing Machine with a tape, but a physical computer that literally prints out its results. Suppose the printer is broken or has run out of ink. Or suppose the programmer failed to include a 'print' command in the program. The computer's program would compute a result but not be able to tell the user what it is, as we see in this algorithm (Chater and Oaksford, 2013, p. 1172, citing an example from Pearl, 2000):

1. input P
2. multiply P by 2; store in Y
3. add 1 to Y; store in Z

This algorithm has an explicit input but does not appear to have an explicit output. The computer has computed $2X + 1$ and stored it away in Z for safekeeping but doesn't tell you its answer. There *is* an answer, but it isn't output. ("I know something that you don't!"?)

So, what constitutes "output"? Is it simply the final result of a computation – i.e. is it internal and syntactic? Or is it some kind of translation or interpretation of the final result that is physically output and implemented in the real world – i.e. is it external and semantic? In the former case, wouldn't both of Rescorla's base-10 and base-13 GCD computers be doing the same thing? A problem would arise only if they told us what results they got, and we – reading those results – would interpret them, possibly incorrectly.

16.4.5 When Are Inputs and Outputs Needed?

> Machines live in the real world and have only a limited contact with it. Any machine, no matter how universal, that has no ears (so to speak) will not hear; that has no wings, will not fly.
> —Allen Newell (1980, p. 148)

---

5 See the Online Resources for further reading on this possibility.

Narrowly conceived, algorithms might not need inputs and outputs. Widely conceived, they do. Any input from the external world has to be *encoded* by a user into a language "understandable" by the Turing Machine (or the Turing Machine needs to be able to *decode* such external-world input). And any output *from* the Turing Machine to be reported *to* the external world (e.g. a user) has to be *encoded* by the Turing Machine (or *decoded* by the user). Such codings would themselves have to be algorithmic.

In fact, one key to determining which real-world tasks are computable is finding coding schemes that allow a sequence of '0's and '1's (i.e. a natural number in binary notation) on a Turing Machine's tape to be *interpreted* as a symbol, a pixel, a sound, etc. According to the Computability Thesis, a mathematical function on the natural numbers is computable iff it is computable by a Turing Machine. Thus, a real-world problem is computable iff it can be encoded as such a computable mathematical function.

But it's that wide conception, requiring algorithmic, semantic interpretations of the inputs and outputs, that leads to various debates. Let's look at the (semantic) coding issue more closely. (On the importance of encoding and decoding for the semantic vs. syntactic views of computing, see Horsman et al., 2014, p. 15.)

16.4.6 Must Inputs and Outputs Be Interpreted Alike?

Letting the symbol '$\underline{x}$' represent a sequence of x strokes (where x is a natural number), Rescorla (2007, p. 254) notes that

> Different textbooks employ different correlations between Turing machine syntax and the natural numbers. The following three correlations are among the most popular:
>
> $d_1(\underline{n}) = n$.
>
> $d_2(\underline{n+1}) = n$.
>
> $d_3(\underline{n+1}) = n$, as an input.
>
> $d_3(\underline{n}) = n$, as an output.
>
> A machine that doubles the number of strokes computes $f(n) = 2n$ under d_1, $g(n) = 2n + 1$ under d_2, and $h(n) = 2n + 2$ under d_3. Thus, the same Turing machine computes different numerical functions relative to different correlations between symbols and numbers.

Let's focus on interpretation d_3. First, having different input and output interpretations of a single internal formalism occurs elsewhere. Machine-translation systems that use an "interlingua" work this way: Chinese input, for example, can be encoded into an "interlingual" representation language (often thought of as an internal, "meaning"-representation language that encodes the "proposition" expressed by the Chinese input), and English output can then be generated from that interlingua (re-expressing in English the same proposition that was originally expressed in Chinese) (Slocum, 1985, Liao, 1998; Daylight, 2013, Section 2). Cognition (assuming that it is computable!) also works this way: perceptual encodings (such as Newell's example of hearing) into the "interlingua" of the biological neural network of our brain surely differ from motor decodings (such as Newell's example of flying). And a calculator's input consists of button pressings, which are one kind of numeral representations of numbers, while its output consists of a graphical display, which is another kind of numeral representing a number (Cummins, 1989, p. 89).

Second, using Hill's distinction (Section 16.3), the idea that a single, internal representation scheme can have different external interpretations suggests that the internal A can be considered a syntactic entity, separate from an external G, which can be considered its semantic referent. Moreover, it is the internal A that would be central to computation.

This offers a way out of Rescorla's puzzle about the two GCD computers.[6] Consider a Common Lisp version of Rescorla's GCD program. The Common Lisp version will look identical to the Scheme version shown in Section 13.3.3 (the languages share most of their syntax), but the Common Lisp version has two global variables – *read-base* and *print-base* – that tell the computer how to interpret input and how to display output. These are implementations of the coding algorithms mentioned in Section 16.4.5. By default, *read-base* is set to 10. So the Common Lisp read-procedure does the following:

(a) It sees the three-character string '115' (for example);
(b) it decides that the string satisfies the syntax of an integer;
(c) it converts that string of characters to an internal ("interlingual") representation of type integer – *which is represented internally as a binary numeral implemented as bits or switch-settings*;
(d) it does the same with (say) '20'; and
(e) it computes their GCD using the algorithm from Section 13.3.3 *on the binary representation*.

If the physical computer had been an old IBM machine, the computation might have used binary-coded *decimal* numerals instead, thus computing in base 10. If *read-base* had been set to 13, the input characters would have been interpreted as base-13 numerals, and the very *same* Common Lisp (or Scheme) code would have correctly computed the GCD of 187_{10} and 26_{10}. One could either say that the algorithm computes with *numbers* – not numerals – or that it computes with *base-2 numerals as an interlingual (or "canonical") representation of numbers*. But that choice depends on one's view about the nature of *mathematics* (Section 12.1) – *not* about the nature of *computation*.

And similarly for output: the switch-settings containing the GCD of the input are then output as base-10 or base-13 numerals appearing as pixels on a screen or ink on paper, depending on the value of such things as *print-base*. With respect to Rescorla's example, the point is that a *single* Common Lisp (or Scheme) *algorithm* is being executed correctly by both M_{10} and M_{13}. Those *machines are* different; they do *not* "have the same local, intrinsic, physical properties" (Rescorla, 2013, p. 687), because M_{10} has *read-base* and *print-base* set to 10, whereas M_{13} has *read-base* and *print-base* set to 13.

For a purely mathematical, Turing-computable example, recall Aizawa's program from Section 10.4, repeated here:

> … if we represent the natural number n by a string of n consecutive 1s, and start the program with the read-write head scanning the leftmost 1 of the string, then the program,
>
> q_0 1 1 R q_0
> q_0 0 1 R q_1,
>
> will scroll the head to the right across the input string, then add a single '1' to the end. It can, therefore, be taken to compute the successor function. (Aizawa, 2010, p. 229)

---

6 I am indebted to Stuart C. Shapiro (personal communication) for the ideas in this paragraph.

I can describe this program semantically (or "widely") as one that generates natural numbers. Speaking purely syntactically (or "narrowly"), I'd like to describe it as one that appends a '1' to the (right) end of the sequence of '1's encoded on its (input) tape. Using Hill's formulation, the semantic or wide (or "teleological") description would be

In order to generate natural numbers, do **begin** $q_0 11Rq_0$; $q_0 01Rq_1$ **end.**

The syntactic (or narrow) description would just be the "do" clause.

Now, a Turing Machine that does merely that does not really generate the natural numbers; at best, it could be described semantically as a one-trick pony that generates the successor of the number encoded on the tape. To generate "all" natural numbers, this Turing Machine would have to be embedded as the body of a loop in another, "larger" Turing Machine. The idea is that beginning with a tape "seeded" with the first natural number (either a blank tape or one with a single stroke), it executes the first Turing Machine, thus generating the input's successor, then loops back to the beginning, considers the current tape as the input tape, and generates its successor, *ad infinitum.*

But what if it uses Rescorla's d_3 interpretation scheme? Then our larger Turing Machine, while still appending a '1' to the end of the current sequence of '1's on the tape, is no longer generating the natural numbers. (It is certainly generating *a* natural-number sequence, but not the one written in the same notation as the inputs.) Rather than computing $S(n) = n + 1$, it is computing $S'(n) = n + 2$.

The aspect of this situation that I want to remind you of is whether the tape is the *external* input and output device or is, rather, the machine's *internal* memory. If it is the machine's internal memory, then, in some sense, there is no (visible or user-accessible) input or output (Section 16.4.2). If it is an external input-output device, then the marks on it are for *our* convenience only. In the former case, the only accurate description of the Turing Machine's behavior is syntactically in terms of '1'-appending. In the latter case, we can use that syntactic description, but we can also embellish it with one in terms of our interpretation of what it is doing. (We'll return to this in Section 16.9.)

16.4.7 Linking the Tape to the External World

Suppose a Turing Machine's tape is really just its internal memory. Then, even though Turing Machines compute mathematical functions, they "contemplate their navel," so to speak, because they don't tell us what their results are. If we want to *use* a Turing Machine to find out the result of a computation, we need to look at its internal storage. Conveniently, it's visible on the machine's tape. But it's in code. So we have to decode it into something we can understand and use. And we have to do that algorithmically. Our earlier examples suggest that this can be done in many ways and that it can go wrong.

Does that decoding belong to A? Or does it belong to G? Let's now turn to this question.

16.5 Are Programs Teleological?

> Algorithms, **in the popular imagination**, are algorithms *for* producing a particular result. … [E]volution can be an algorithm, and evolution can have produced us by an algorithmic process, without its being true that evolution is an algorithm *for* producing us.
> —Daniel C. Dennett (1995, p. 308, my boldface, original italics)[7]

---

7 For hints as to what evolution's algorithms look like, see Dawkins, 2016.

We have discussed two ways to view a computation: the first way is purely syntactically (or narrowly, internally, or locally), as expressed in a computer program of the form "Do A," where A is an algorithm expressed in the language of Turing Machines or the language of recursive functions (etc.). The second way is semantically (or widely, externally, or globally), as expressed in a computer program of the form "To accomplish goal G, do A" (which we'll shorten to "To G, do A"). That preface ("To G") makes explicit a *goal G* of the algorithm A, thus indicating that the program is *intended* to have a *purpose* (it is "teleological"). Let's now consider the question of whether the proper way to characterize a program *must* include the teleological preface "To G."

As we saw in Chapter 7, the history of computation theory is, in part, an attempt to make the informal notion of an algorithm mathematically precise. In Section 7.3.3, we summarized this as follows:

> An algorithm (for executor E) [to accomplish goal G] is:
>
> 1. a procedure A, i.e. a finite set (or sequence) of statements (or rules, or instructions), such that each statement S is:
>   (a) composed of a finite number of symbols (better: uninterpreted marks) from a finite alphabet
>   (b) and unambiguous (for E – i.e.
>   i. E "knows how" to do S,
>   ii. E can do S,
>   iii. S can be done in a finite amount of time
>   iv. and, after doing S, E "knows" what to do next –),
> 2. A takes a finite amount of time (i.e. it halts),
> 3. [and A ends with G accomplished].

I have put some of these clauses in (parentheses) and [brackets] for a reason. The notion of an algorithm is most easily understood with respect to an executor: a human or a machine that (dynamically) executes the (static) instructions. We might be able to rephrase this characterization of an algorithm without reference to E, albeit awkwardly; hence the *parentheses* around the E-clauses.

Exercise for the Reader: Try to eliminate the executor from this (or any other) characterization of an algorithm. Can it be eliminated? If not, why not? (For discussion of this point, see Sieg, 2008, p. 574.)

But the present issue is whether the *bracketed G*-clauses are essential. As we saw in Sections 12.5 and 15.1, one executor's algorithm might be another's ungrammatical input (Suber, 1988; Buechner, 2011, 2018), and a bad puttanesca might still be a delicious pasta dish. Does the chef's intention (or the diner's expectation) matter *more* than the actual food preparation? Is G more important than A?

16.6 Algorithms *Do* Need a Purpose

Peter Suber argues in favor of the importance of G:

> To distinguish crashes and bad executions from good executions, it appears that we must introduce the element of the programmer's purpose. Software executes the programmer's will, while semantically flawed, random, and crashing code do not. This suggests that to understand software we must understand intentions, purposes, goals, or will, which enlarges the problem far more than we originally anticipated.

Perhaps we must live with this enlargement. We should not be surprised if human compositions that are meant to make machines do useful work should require us to posit and understand human purposiveness. After all, to distinguish *literature* from noise requires a similar undertaking. (Suber, 1988, p. 97)

And Hill (2016, Section 5) says that a "prospective user" needs "some understanding of the task in question" over and above the mere instructions. Algorithms, according to Hill, must be expressed in the form "To *G*, do *A*," not merely "Do *A*."

> **Question for the Reader:** Is the *executor* of an algorithm the same as a *user*? Typically, a (human) *uses* a computer, but it is the *computer* that executes the algorithm. In what follows, ask yourself if it is the user or the executor who "needs some *understanding* of the task" (as Hill says).

Suber and Hill are not alone in this. The cognitive scientist and computational vision researcher David Marr also held that (at least some) computations were purposeful. He analyzed information processing into three levels (Marr, 1982, Section 1.2):

- computational (*what* a system does, and *why*),
- algorithmic (*how* it does it), and
- physical (how it is *implemented*).

In our terminology, these levels would be called 'functional,' 'computational,' and 'implementational,' respectively: certainly, when one is doing mathematical computation (the kind that Turing was concerned with), one begins with a mathematical *function* (i.e. a certain set of ordered pairs), asks for an algorithm to *compute* it, and then seeks an *implementation* of the algorithm, usually in a physical system such as a computer or the brain. Note, however, that Marr's "computational" level combines our "functional" level with a purpose ("why").

In *non*-mathematical fields (e.g. cognition in general, and – for Marr – vision in particular), the set of ordered pairs of input-output *behavior* is expressed in goal-oriented, problem-specific language, and the algorithmic level will also be expressed in that language. (The implementation level might be the brain or a computer.) A recipe for hollandaise sauce developed in this way would have to say more than just something along the lines of "mix these ingredients in this way"; it would have to take the external environment into account. (We will return to this in Section 16.8, and we will see how the external world can be taken into account in Section 16.10.4.)

Marr was trying to counter the then-prevailing methodology of trying to *describe* what neurons were doing (a "narrow," internal, implementation-level description) without having a "wide," external, "computational"-level *purpose* (a "function" in the teleological, not mathematical, sense). Such a teleological description would tell us "why" neurons behave as they do:

> As one reflected on these sorts of issues in the early 1970s, it gradually became clear that something important was missing that was not present in either of the disciplines of neurophysiology or psychophysics. The key observation is that neurophysiology and psychophysics have as their business to *describe* the behavior of cells or of subjects but not to *explain* such behavior. What are the visual areas of the cerebral cortex actually doing? What are the problems in doing it that need explaining, and at what level of description should such explanations be sought? (Marr, 1982, p. 15; for discussion of this point, see Bickle, 2015)

On this view, Marr's "computational" level is *teleological*. In Hill's formulation, the "To *G*" preface expresses the teleological aspect of Marr's "computational" level; the "do *A*" seems to express the (mathematical) functional level or the Marr's "algorithmic" level.

In addition to being teleological, algorithms seem to be able to be *multiply* teleological, as in the duck-rabbit examples. That is, there can be algorithms of the form "To G_1, do *A*" and algorithms of the form "To G_2, do *A*," where $G_1 \neq G_2$, and where neither G_1 nor G_2 subsumes the other, although the *A* is the same. In the cartoon of Figure 16.1, depending on the semantic interpretation of the syntactic finger movements, we have two *G*s with one *A*: *either* "To say 'yes' 10 times, raise 10 fingers" *or* "To say 'no' 5 times, raise 10 fingers."

In other words, what if doing *A* can accomplish two distinct goals? Do we have two algorithms in that case: one that accomplishes G_1, and another that accomplishes G_2, counting teleologically, or "widely"? Or just one: a single algorithm that does *A*, counting more narrowly?

Were de Bruijn and the chemists talking about the same thing (Section 16.2)? On the teleological (or wide) view, they weren't; on the narrow view, they were. Multiple teleologies are multiple implementations of an algorithm narrowly construed: 'Do *A*' can be seen as a way to algorithmically implement the higher-level "function" (mathematical *or* teleological) of accomplishing G_1 *as well as* accomplishing G_2. For example, executing a particular subroutine in a given program might result in checkmate or winning a battle. Viewing multiple teleologies as multiple implementations can also account for hollandaise-sauce failures on the Moon, which could be the result of an "implementation-level detail" (Section 13.1.1) that is irrelevant to the abstract, underlying computation.

16.7 Algorithms *Don't* Need a Purpose

Certainly, knowing the goal of an algorithm makes it easier for *cognitive-agent* executors (who are also users?) to follow the algorithm and to have a fuller understanding of what they are doing. But is such understanding *necessary*? Consider the following two (real-life!) personal stories:

Story 1: I vividly remember the first semester that I taught a "Great Ideas in Computer Science" course aimed at computer-phobic students. We were going to teach the students how to use a spreadsheet program, something that, at the time, I had never used; so, with respect to this, I was as naive as any of my students! My TA, who had used spreadsheets before, gave me something like the following instructions:

> enter a number in cell_1;
> enter a number in cell_2;
> enter '=⟨click on cell_1⟩⟨click on cell_2⟩' in cell_3

Some current implementations of Excel require a plus-sign between the two clicks in the third instruction. But the version I was using at the time did not, making the operation that much more mysterious! Indeed, I had no idea what I was doing. I was blindly following her instructions *and had no idea that I was adding two integers.* Once she told me that that was what I was doing, my initial reaction was "Why didn't you tell me that before we began?"

When I entered those data into the spreadsheet, was I adding two numbers? I didn't understand that I was adding when my TA told me to enter certain data into the cells of the spreadsheet. It was only when she told me that that was how I could add two numbers with a spreadsheet that

I understood. Now, (I like to think that) I am a cognitive agent who can come to understand that entering data into a spreadsheet can be a way of adding. But a Turing Machine that adds or a Mac running Excel is not such a cognitive agent. It does not understand what addition is or that that is what it is doing. *And it does not have to.*

(We will return to this in Section 18.6, when we discuss the Chinese Room Argument. Arguably, an AI program running on a robot that passes the Turing Test would be a very different matter. Such an AI program could, would, and should [come to] understand what it was doing. We'll explore this further in Chapter 18.)[8]

Story 2: Years later, I had yet another experience along these lines:

> My wife recently opened a restaurant and asked me to handle the paperwork and banking that needs to be done in the morning before opening (based on the previous day's activities). She wrote out a detailed set of instructions, and one morning I went in with her to see if I could follow them, with her looking over my shoulder. As might be expected, there were gaps in her instructions, so even though they were detailed, they needed even more detail. Part of the reason for this was that she knew what had to be done, how to do it, and why it had to be done, but I didn't. This actually disturbed me, because I tend to think algorithms should really be just "Do A," not 'To G, do A.' Yet I felt that I needed to understand G in order to figure out how to do A. But I think the reason for that was simply that she hadn't given me an algorithm, but a sketch of one, and, in order for me to fill in the gaps, knowing why I was doing A would help me fill in those gaps. But I firmly believe that if it made practical sense to fill in all those gaps (as it would if we were writing a computer program), then I wouldn't have to ask why I was doing it. No "intelligence" should be needed for this task if the instructions were a full-fledged algorithm. If a procedure (a sequence of instructions, including vague ones like recipes) is not an algorithm (a procedure that is fully specified down to the last detail), then it can require "intelligence" to carry it out (to be able to fill in the gaps, based, perhaps on knowing why things are being done). If intelligence is not available (i.e. if the executor lacks relevant knowledge about the goal of the procedure), then the procedure had better be a full-fledged algorithm. There is a difference between a human trying to follow instructions and a machine that is designed to execute an algorithm. The machine cannot ask why, so its algorithm has to be completely detailed. But a computer (or a robot, because one of the tasks is going to the bank and talking to a teller!) that could really do the job would almost certainly be considered to be "intelligent."
> (Rapaport, quoted in Hill and Rapaport, 2018, p. 35)[9]

Despite the fact that understanding what task *G* an algorithm *A* is accomplishing makes it easier to understand *A* itself, the important point is that "blind" following of *A* is all that is necessary in order to *accomplish G*. The fact that computation can be "blind" in this way is what Dennett has called

> Turing's … strange inversion of reasoning. The Pre-Turing world was one in which computers were people, who had to understand mathematics in order to do their jobs. Turing realised that this was just not necessary: you could take the tasks they performed and squeeze out the last tiny smidgens of understanding, leaving nothing but brute, mechanical actions. IN ORDER TO BE A PERFECT AND BEAUTIFUL COMPUTING MACHINE IT IS NOT

---

8 See the Online Resources for further reading on mathematical understanding.
9 Recall the "Zits" comic described in Section 7.3.3.

REQUISITE TO KNOW WHAT ARITHMETIC IS. (Dennett, 2013b, p. 570, capitalization in original)[10]

The point is that a Turing Machine need not "know" that it is adding. But agents who do understand adding can use that machine to add.

Or can they? To do so, the machine's inputs and outputs have to be interpreted – understood – by the user as representing the numbers to be added. And that seems to require an appropriate relationship with the external world. It seems to require a "user manual" that tells the user what the algorithm does in the way Hill prescribes, not in the way my TA explained how to use a spreadsheet. And such a "user manual" – an intention or a purpose for the algorithm – in turn requires an interpretation of the machine's inputs and outputs.

But before pursuing this line of thought, let's take a few more minutes to consider "Turing's strange inversion," the idea that a Turing Machine can be doing something very particular by executing an algorithm without any specification of what that algorithm is "doing" in terms of the external world. Algorithms, on this view, seem not to have to be teleological, yet they remain algorithms. Brian Hayes (2004) offers two versions of an algorithm that ants execute:

Non-teleological version:

1. "If you see a dead ant[11] and you're not already carrying one, pick it up;
2. "if you see a dead ant, and you *are* carrying one, put yours down near the other."

Teleological version:
To create an ant graveyard, "gather all … [your] dead in one place.[12]

As Hayes notes, the teleological version requires planning and organization skills far beyond those that an ant might have, not to mention conceptual understanding that we might very well be unwilling to ascribe to ants. The point, however, is that the ant needs none of that. The teleological description helps *us* describe and perhaps understand the ant's behavior; it doesn't help the ant. The same is true in my spreadsheet example. Knowing that I am adding helps *me* understand what I am doing when I fill the spreadsheet cells with certain values or formulas. But the spreadsheet does its thing without needing that knowledge.

These examples suggest that the user-manual (or external-world) interpretation is not necessary. Algorithms *can* be teleological, and their being so can help users and cognitive agents who execute them to more fully understand what they are doing. But they don't *have to* be teleological.[13]

16.8 Algorithms and Goals

What if "successfully" executing *A fails* to accomplish goal *G*? This could happen for external, environmental reasons. Does this mean *G* might not be a computable task even though *A* is? We have seen several examples of this kind of failure:

- The blocks-world computer's model of the world was an incomplete, partial model; it assumed that its actions were always successful. This program lacked feedback from the external world.

---

10 See also the more easily accessible Dennett, 2009, p. 10061.
11 Note that testing this condition does not require the ant to have a concept of death; it is sufficient for the ant to sense – either visibly or perhaps chemically – what *we* would describe as a dead ant.
12 This is a "fully" teleological version, with a high-level, teleologically formulated execution statement. A "partially" teleological version would simply prefix "To create an ant graveyard" to the non-teleological version.
13 See the Online Resources for further reading on the non-teleological view.

There was nothing wrong with the environment; rather, there was incomplete information about the environment.

- In the case of Cleland's hollandaise-sauce recipe, the environment was at fault. Her recipe (A) was executed flawlessly on the Moon but failed to produce hollandaise sauce. Her diagnosis was that making hollandaise sauce (G) is not computable. Yet A was!
- Rescorla's GCD computers do "different things" *by* doing the "same thing." The difference is not in *how* they are doing what they are doing but in the interpretations that *we* users of the machines give to their inputs and outputs. Would Hill (2016) say that the procedure encoded in that Scheme program was therefore not an algorithm?[14]

> **Question for the Reader:** How does this relate to the trial-and-error machines that we discussed in Section 11.10? After all, they also differ from Turing Machines only in terms of our *interpretations* of *what* they are doing, not in *how* they do it.

What is more central to the notion of "algorithm": all of parts 1–3 in our informal characterization in Section 16.5 ("To G, do A"), or just parts 1–2 – i.e. without the bracketed goals (just "Do A")? Is the algorithm the narrow, non-teleological, "purposeless" (or non-purposed) entity? Or is the algorithm the wide, teleological (i.e. goal-directed) entity?

On the narrow view, the war and chess algorithms are just *one* algorithm, the hollandaise-sauce recipe *does* work on the Moon (its computer program might be logically verifiable even if it fails to make hollandaise sauce), and Rescorla's "two" GCD programs are also just *one* algorithm that does its thing correctly (but only we base-10 folks can *use* it to compute GCDs).

On the wide view, the war and chess programs are *two* distinct algorithms, the hollandaise-sauce recipe *fails* on the Moon (despite the fact that the program might have been verified – shades of the Fetzer controversy of Section 15.4.1!), and the Scheme program when fed base-13 numerals (as Rescorla describes it) is doing something *wrong* (in particular, its "remainder" subroutine is incorrect). It does the *right* thing on the interpretation discussed in Section 16.4.6.

These examples suggest that the wide, goal-directed nature of algorithms that are teleologically conceived is due to the interpretation of their input and output. As Shagrir and Bechtel (2015, Section 2.3) put it (echoing Sloman's distinction from Section 16.4.1), Marr's "algorithmic level … is directed to the *inner working* of the mechanism …. The computational level looks *outside*, to identifying the function computed and relating it to the environment in which the mechanism operates."

We can combine these insights: Hill's formulation of the teleological nature of algorithms had two parts, a teleological "preface" specifying the task to be accomplished ("To G") and a statement of the algorithm that accomplishes it ("Do A"). One way to clarify the nature of Marr's "computational" level is to split it into its "why" and its "what" parts. The "why" part is the task to be accomplished. The "what" part can be expressed "computationally" (in our terminology, "functionally") as a mathematical function (possibly, but not necessarily, expressed in "why" terminology), but it can also be expressed *algorithmically*. Finally, the algorithm can be implemented. So, we can distinguish the following *four* Marr-like levels of analysis:

"Computational"-what level: Do $f(i) = o$
"Computational"-why level: To G, do $f(i) = o$
Algorithmic level: To G, do $A_f(i) = o$
Implementation level: To G, do $I_{A_f}(i) = o$

---

14 Rescorla, 2015, Section 2.2, considers the opposite case, in which G *is* computable even when A is *not*: "There exist 'deviant' notations relative to which intuitively non-computable functions become Turing-computable."

where

- f is an *input-output* function that happens to accomplish G;
- G is the task to be accomplished or explained, expressed in the language of the external world, so to speak;
- A_f is an algorithm that implements f (i.e. it is an algorithm that has the same input-output behavior as f), either expressed in the same language as G or perhaps expressed in purely mathematical language; and
- I is an implementation (perhaps in the brain or on some computer) of A_f.

Shagrir and Bechtel (2015, Section 4) say that "The *what* aspect [of the "computational" level] provides a description of the mathematical function that is being computed. The *why* aspect employs the contextual constraints in order to show how this function matches with the environment." These nicely describe the two clauses of what I call the "computational-why" level earlier. [15]

16.9 Computing with Symbols or with Their Meanings

Goal G is expressed in teleological language. We now need to focus on the language used to express the algorithm A_f that implements the function f that – in turn – underlies (or is supposed to accomplish) G. Can A_f be teleological? *Must* it be teleological, too? In other words, can (or must) it be expressed in the language of G? For example, can (must) it talk about chess as opposed to war, or chess as opposed to shogi or Go?

What do Turing Machines compute *with*? For that matter, what do *we* compute with? Rescorla (2007, p. 253) reminds us that

> A Turing machine manipulates syntactic entities: strings consisting of strokes and blanks. … Our main interest is not string-theoretic functions but number-theoretic functions. We want to investigate computable functions from the natural numbers to the natural numbers. To do so, we must correlate strings of strokes with numbers.

In this regard, Turing Machines differ interestingly from their logical equivalents in the Computability Thesis: the lambda calculus and recursive-function theory apparently deal with functions and numbers, not symbols for them.

Questions for the Reader: Is it really the case that the lambda calculus and recursive-function theory (unlike Turing Machines) deal with functions and not just with symbols for functions? Hilbert viewed all of mathematics as the "manipulation of finite strings of symbols devoid of intuitive meaning[,] which stimulated the development of mechanical processes to accomplish this" (Soare, 1999, Section 2.4, p. 5). On this view, wouldn't *all* of the formalisms of computability theory be syntactic? Can't recursive-function theory be understood purely syntactically? And the lambda calculus "can be presented solely as a formal system with syntactic conversion rules. … all we are doing is manipulating symbols" (J. Stoy, quoted in Turner, 2018, p. 92).

But for Turing Machines and their physical implementations (i.e. ordinary computers), we see that it is necessary to interpret the strokes. Here is an example due to the philosopher Christopher

---

15 See the Online Resources for further reading on teleology and program verification.

Peacocke (1999): suppose we have a Turing Machine that outputs a copy of the input appended to itself (thus doubling the number of input strokes): input '|,' output '||'; input '||,' output '||||,' and so on. What is this Turing Machine doing? The most neutral description seems to be "outputting a copy of the input appended to itself." After all, that describes *exactly* what the Turing Machine is doing, leaving the interpretation (e.g. *doubling* the input) up to the observer. If we had come across that Turing Machine in the middle of the desert and were trying to figure out what it does, something like that would be the most reasonable answer. *Why* a user might want a copy-appending Turing Machine is a different matter that probably *would* require an interpretation of the strokes. But that goes far beyond what the Turing Machine is doing.

But Peacocke objects:

> The normal interpretation of a Turing machine assigns the number 0 to a single stroke '|,' the number 1 to '||,' the number 2 to '|||,' and so on. But there will equally be an interpretation which assigns 0 to a single stroke '|,' and then assigns the number 2 to '||,' the number 4 to '|||,' and generally assigns 2n to any symbol to which the previous interpretation assigns *n*. Under the second interpretation, the Turing Machine will still be computing a function. … What numerical value is computed, and equally which function is computed, by a given machine, is not determined by the purely formal characterization of the machine. There is no such thing as purely formal determination of a mathematical function. … [W]e can say that a Turing machine is really computing one function rather than another only when it is suitably embedded in a wider system. (Peacocke, 1999, pp. 198–199).

Recall Rescorla's three interpretations of the strokes (Section 16.4.6). Do we really have one machine that (simultaneously?) does three different things? What it does (in one sense of that phrase) depends on how its input and output are interpreted: i.e. on the environment in which it is working. In different environments, it does different things; at least, that's what Cleland said about the hollandaise-sauce recipe. Rescorla (2015, Section 2.1) makes a related observation: "The same Turing machine T computes different non-linguistic functions, depending upon the semantic interpretation of strings manipulated by the Turing machine," thus rendering all of computability theory "intensional" (with an 's,' not a 't'): i.e. dependent upon the *meanings* of the symbols and not just on the symbols themselves (e.g. their shapes). Using our terminology, he thus comes down on the side of "To G, do A" rather than on "Do A."

Piccinini (2006, Section 2, my italics) says much the same thing; however, he draws a different conclusion:

> In computability theory, symbols are typically marks on paper individuated by their geometrical shape (as opposed to their semantic properties). Symbols and strings of symbols may or may not be assigned an interpretation; if they are interpreted, the same string may be interpreted differently …. In these computational descriptions, *the identity of the computing mechanism does not hinge on how the strings are interpreted.*

By 'individuated,' Piccinini is talking about how one decides whether what appear to be two programs (say, one for a war battle and one for a chess match) are, in fact, two distinct programs or really just one program (perhaps being described differently). He suggests that it is not how the inputs and outputs are *interpreted* (their semantics) that matters, but what the inputs and outputs *look like* (their syntax). In an earlier paper, Rescorla agreed:

> Since we can arbitrarily vary inherited meanings relative to syntactic machinations, inherited meanings do not *make a difference* to those machinations. They are imposed upon an underlying causal structure. (Rescorla, 2014a, p. 181)

So, for Piccinini and the Rescorla of 2014a, the war and chess programs are the same. But for Cleland and the Rescorla of 2015, they would be different. For Piccinini, the hollandaise-sauce program running on the Moon works just as well as the one running on Earth; for Cleland, only the latter does what it is supposed to do.

So, the question "*Which* Turing Machine is this?" has only one answer, which is given in terms of its syntax: "determined by [its] instructions, not by [its] interpretations" (Piccinini, 2006, Section 2). But the question "*What* does this Turing Machine do?" has $n + 1$ answers: one syntactic answer and n semantic answers (one for each of n different semantic interpretations).

If I want to know which Turing Machine this is, I should look at the internal mechanism (A) for the answer. This is, roughly, Piccinini's (2006) recommendation. But if I'm interested in buying a chess program (as opposed to a war simulator, for example), then I need to look at the external (or inherited, or wide) semantics. This would be Cleland's (1993) recommendation. In "To G, do A," the "do A" portion expresses Dennett's (1971) "design" or "physical" stance, and the "to G" portion expresses Dennett's "intentional" stance (Dennett, 2013a, pp. 81–82, 84; recall our Section 12.4.1).

We have come across this situation before. In Section 12.4.4, we asked whether a Universal Turing Machine running an addition program was adding or "just" fetching and executing the instructions of an addition program stored on its tape. A similar question can be asked about humans: How would you describe my behavior when I use a calculator to add two numbers? Am I (merely) pushing certain buttons in a certain sequence? This would be a "syntactic," narrow, internal answer: I am "doing A" (where A = pushing buttons). Or am I adding two numbers? This would be a teleological, "semantic," wide, external answer: I am accomplishing G (where G = adding). Or am I adding two numbers *by* pushing those buttons in that sequence? This would be a teleological (etc.) answer, together with a syntactic description of how I am doing it: I am accomplishing G by doing A. This is the same situation we saw in the spreadsheet example. (We will see it again in Section 16.10.2).

In some sense, all of these answers are correct, merely(?) focusing on different aspects of the situation. But a further question is, *why* (or how) does a Turing Machine's printing and moving thus and so, or my pushing certain calculator buttons thus and so, result in adding two numbers? And the answer to that seems to require a semantic interpretation. This is the kind of question that Marr's "computational" level is supposed to respond to.

Here is another nice example (Piccinini, 2008, p. 39):

> a loom programmed to weave a certain pattern will weave that pattern regardless of what kinds of thread it is weaving. The properties of the threads make no difference to the pattern being woven. In other words, the weaving process is insensitive to the properties of the input.

As Piccinini points out, the output might have different colors depending on the colors of the input threads, but the *pattern* will remain the same. The pattern is internal to the program; the colors are external, to use other terminology. (Here, A is the pattern; G is the colors.) If you want to weave an American flag, you had better use red, white, and blue threads in the appropriate ways. But even if you use cyan, black, and yellow threads, you will weave an American-flag *pattern*.[16] Which is more important: the pattern or the colors? That's probably not the right question. Rather, if you want a

---

16 See e.g. https://www.vox.com/2015/7/30/9075331/optical-illusion.

certain *pattern*, this program will give it to you; if you want a certain pattern *with certain colors*, you need to have the right inputs – you need to use the program in the right environment.[17]

16.10 Syntactic, Internal, and Indigenous Semantics

16.10.1 Syntax vs. Semantics

Recall the concepts of syntax and semantics as we discussed them in Chapter 13. Syntax is concerned with the "*intra*-system" properties and relations *within* the "syntactic" domain. Semantics is concerned with "*extra*-system" relations that go *beyond* the syntactic domain *to* the "semantic" domain. That is, semantics is concerned with the "*inter*-system" relations *between* the syntactic and the semantic domains.[18]

So, *one* way to respond to the issues raised in Section 16.9 is by using an external semantic interpretation: begin with specific Turing Machine operations or button presses, considered as being located in a syntactic system of internal Turing Machine operations or button pressings. Numbers and arithmetical operations on them are located in a distinct, external realm of mathematical entities. Then we can associate the former with the latter. In the formulation "To *G*, do *A*," *A can* be identified syntactically (at the "computational-what" level) – in terms, say, of Turing Machine operations or button pressings. But *G needs* to be identified semantically – in terms, say, of numbers and arithmetic operations. *A* can then be (re-)*interpreted* semantically in *G*'s terms (at the "computational-why" level). These are the $n + 1$ answers of Section 16.9.

16.10.2 Syntactic Semantics

But *another* way to respond to these issues uses an "internal" kind of semantics. Because this kind of semantics is internal to a system, it is really a kind of syntax. Let's call it "syntactic semantics." Here is how Piccinini describes it:

> [S]tored-program computers have the ability to respond to (non-semantically individuated) strings of tokens stored in their memory by executing sequences of primitive operations, which in turn generate new strings of tokens that get stored in memory. [Note that this is basically a description of how computers work, or of what computation is. —WJR] Different bits and pieces [i.e. substrings] of these strings of tokens have different effects on the machine. … An accurate description of how tokens can be compounded into sub-strings, and sub-strings can be compounded into strings, which does not presuppose that the strings of tokens have any content, may be called the syntax of the system of strings manipulated by the computer. … [T]he effect of a string on the computer is assigned to it [i.e. to the string] as its content. This assignment constitutes an *internal* semantics of a computer. An internal semantics assigns as contents to a system its own internal components and activities, whereas an ordinary (external) semantics assigns as contents to a system objects and properties in the system's environment. … None of this entails that computer languages have any external semantics, that is any content …, although it is compatible with their having one. …

---

17 See the Online Resources for further reading on these distinctions.
18 '*Intra*-' means "inside," '*extra*-' means "outside," and '*inter*-' means "between." (In "*inter*mural" sports, school A plays against school B. In "*intra*mural" sports, a single gym class at school A might be divided into two teams that play against each other.)

[I]n order to understand computing mechanisms and how they work (as opposed to why they are built and how they are used), there is no need to invoke content .... (Piccinini, 2004, pp. 401–402, 404. See also Piccinini, 2006, Section 2.)

On this view, it is the "internal" workings of the computer that count, not the external interpretation of its inputs and outputs (or even the external interpretation of its internal mechanisms or symbol manipulations). This is the sense in which a war computer and a chess computer are performing "the same computation." Note the parenthesized hedge in the last sentence of Piccinini's quote: Cleland and Rescorla might be quite right in terms of their emphasis on why or how a particular computer or program is being *used.* That's a teleological aspect of computation that doesn't necessarily violate the Computability Thesis.

Similarly, Rescorla once argued "that computation is not sensitive to meaning or semantic properties" (2012a, Section 1, p. 703). More precisely, he argued that *if* a computational process were to be sensitive to semantic properties, then it would have to violate either a condition that he called 'Syntactic Rules' or a condition that he called 'Freedom,' *and* that such a semantically sensitive computation would have to have an "indigenous" semantics, not an "inherited" semantics. He defined these terms as follows:

> SYNTACTIC RULES: computation is manipulation of syntactic entities according to mechanical rules. We can specify those rules in syntactic terms, without mentioning semantic properties such as meaning, truth, or reference. (Rescorla, 2012a, Section 3, p. 707)

> FREEDOM: We can offer a complete syntactic description of the system's states and the mechanical rules governing transitions between states (including any interaction with sensory and motor transducers), while leaving semantic interpretation unconstrained. More precisely, it is metaphysically possible for the system to satisfy our syntactic description while instantiating arbitrarily different semantic properties. (Rescorla, 2012a, Section 3, p. 708)

> *Inherited meanings* arise when the system's semantic properties are assigned to it by other systems, through either explicit stipulation or tacit convention. Nothing about the system itself helps generate its own semantics. For instance, words in a book have inherited meanings. *Indigenous meanings* arise when a system helps generate its own semantics. Indigenous meanings do not arise merely from external assignment. They arise partly from the system's own activity, perhaps with ample help from other factors, such as causal, evolutionary, or design history. Virtually all commentators agree that the mind has indigenous meanings. (Rescorla, 2012a, Section 3, pp. 707–708)

Rescorla's "indigenous" semantics seems clearly akin to Piccinini's "internal" semantics and to what we are calling "syntactic" semantics.

Question for the Reader: However, as we saw in Section 16.9, three years later Rescorla seems to have changed his mind! Rescorla, 2015, Section 1 offers the following "Gap Argument":

A Turing machine manipulates linguistic items, but we sometimes want to study computation over non-linguistic domain X. So there is a gap between the domain of items

---

(Continued)

(Continued)

manipulated by the Turing machine and our desired domain of computation X. To bridge the gap, we must interpret linguistic items manipulated by the Turing machine as denoting items drawn from X. A Turing machine computes over X only if linguistic items manipulated by the Turing machine represent elements of X. Thus, any complete theory of computation must cite representational relations between linguistic items and non-linguistic items.

He then says, "Given the Gap Argument, we can study Turing computation over a non-linguistic domain only if we furnish a semantics for strings" (Rescorla, 2015, Section 3). Buechner, 2011, pp. 358–362, makes a similar argument.

Is Rescorla's gap related to Smith's gap (Section 15.6.2)?

16.10.3 Syntactic Semantics and Procedural Abstraction

One way to provide an internal, indigenous, or syntactic semantics is to use "procedural abstraction" – named subroutines that accomplish subtasks of the overall algorithm (Section 7.4.3.3): identify subtasks (collections of statements in a program that "work together"), package them up, and name the package, thus giving an identity to the subtasks.

For example, the following Logo program draws a unit square by moving forward 1 unit, then turning 90 degrees right, and doing that four times:

```
repeat 4 [forward 1 right 90]
```

But Logo won't "know" what it means to draw a square unless we tell it this:

```
to square
repeat 4 [forward 1 right 90]
end
```

Note that this Logo instruction has the form to G, do A! The "To G" has been "internalized" into the program. (We'll come back to this idea in Section 16.10.4.)

Another example is the sequence of instructions "turnleft; turnleft; turnleft" in *Karel the Robot* (Pattis et al., 1995), which can be packaged up and named "turnright":

```
DEFINE-NEW-INSTRUCTION turnright AS
BEGIN
turnleft;turnleft;turnleft
END
```

Notice here that Karel still can't "turn *right*" in an external sense (i.e. 90° clockwise); it can only turn left three times (i.e. 270° counterclockwise).

There is an important caveat: the Logo and Karel *programs* still have no "understanding" in the way that *we* do of what a square is or what it means to turn right. Merely *naming* a subroutine does not automatically endow it with the (external) meaning of that name (McDermott, 1980). The programs are now capable only of associating those newly defined symbols ('square,' 'turnright') with certain procedures. The symbols' meanings for *us* are their *external* semantics; their meanings for the Logo or Karel *programs* are their internal, indigenous, syntactic semantics due to their internal relationships with the bodies of those programs. If the name is associated with objects that are *external* to the program, then we have external (or wide, or inherited) semantics. If it is associated with objects *internal* to the program, then we have internal (or narrow, or syntactic, or indigenous) semantics. *Identifying* subroutines is syntactic; *naming* them leads to semantics: if the name is externally meaningful to a *user*, because the user can associate the name with other external concepts, then we have semantics in the ordinary sense (subject to McDermott's caveat). If it is internally meaningful to the *computer*, in the sense that the computer can associate the name with other internal names, then we have internal, syntactic semantics.

The debate over whether computation concerns the internal, syntactic manipulation of symbols or the external, semantic interpretation of them is at the heart of both Rescorla's gap (see the Question for the Reader in Section 16.10.2) and Smith's gap (from Section 15.6.2). This is made explicitly clear in the following passages from Michael Mahoney's history of computing:

> Recall what computers do. They take sequences, or strings, of symbols and transform them into other strings. …
>
> The transformations themselves are strictly syntactical, or structural. They may have a semantics in the sense that certain symbols or sequences of symbols are transformed in certain ways, *but even that semantics is syntactically defined.* Any meaning the symbols may have is acquired and expressed at the interface between a computation and the world in which it is embedded. The symbols and their combinations express representations of the world, *which have meaning to us, not to the computer.* … What we can make computers do depends on how we can represent in the symbols of computation portions of the world of interest to us and how we can translate the resulting transformed representation into desired actions. …
>
> So putting a portion of the world into the computer means designing an operative representation of it that captures what we take to be its essential features. That has proved … no easy task; on the contrary it has proved difficult, frustrating, and in some cases disastrous. (Mahoney, 2011, p. 67, my italics)

The computer's internal semantics – its "Do *A*" (including *A*'s modules or compositional structure – is syntactic and non-teleological. Its external semantics, "which have meaning to us" – its "To *G*" – is teleological but depends on *our* ability to represent *our view of* the world *to it*. As Rescorla (2007, p. 265) observed, we need a computable theory of the semantic interpretation function, but as Smith observes, we don't (can't?) have one, for reasons akin to the Computability Thesis problem: equivalence between something formal (e.g. a Turing Machine or a formal model) and something non-formal (e.g. an algorithm or a portion of the real world) cannot be *formally* proved.[19]

16.10.4 Internalization

Syntactic semantics can arise in another way: *external* semantic relations between the elements of *two* domains (a "syntactic" domain described syntactically and a "semantic" domain described ontologically (i.e. syntactically! – see Section 13.2.2) can be turned into *internal* syntactic relations ("syntactic semantics") by *internalizing* the semantic domain into the syntactic domain. After all, if you take the union of the syntactic and semantic domains, then all formerly external semantic relations are now internal syntactic ones (internal to the union).

One way this happens for cognitive agents like us is by sensory perception, which is a form of input encoding. For animal brains, perception interprets signals from the external world into the biological neural network of the brain. For a computer that accepts input from the external world,

---

19 See the Online Resources for further reading on syntactic semantics.

the interpretation of external or user input as internal switch settings (or inscriptions on a Turing Machine tape) constitutes a form of perception – a way of internalizing external information. Both are forms of what I am calling "internalization." As a result, the interpretation becomes part of the computer's or the brain's internal, syntactic semantics (Rapaport, 2012b).

Stuart C. Shapiro advocates internalization in the following form, which generalizes the Logo and Karel techniques:[20]

Shapiro's Internalization Tactic
Algorithms *do* take the teleological form, "To *G*, do *A*,"
but *G* must include *everything* that is relevant:

- To make hollandaise sauce *on Earth*, do *A*.
- To find the GCD of two integers *in base-10*, do *B*.
- To play chess, do *C*, *where C's variables range over chess pieces and a chess board*.
- To simulate a war battle, do *D*, *where D's variables range over soldiers and a battlefield*.

One place to locate these teleological clauses is in the type declarations of a typed programming language. Another is in the pre-conditions and post-conditions of the program. They can then be used in the formal verification of the program, which proceeds by proving that *if the pre-conditions are satisfied*, then the program will accomplish its goal *as articulated in the post-conditions*. This builds the external world (and any attendant external semantics) *into* the algorithm: "There is no easy way to ensure a blueprint stays with a building, but a specification can and should be embedded as a comment within the code it is specifying" (Lamport, 2015, p. 41). The separability of blueprint from building is akin to the separability of *G* from *A*; embedding a specification into code as (at least) a comment is to internalize it as a pre- or post-condition. More importantly, such pre- and post-conditions need not be "mere" comments; they can be internalized as "assertible" statements in a program, thus becoming part of a program's (self-)verification process (Lamport, 2011).

As I suggested in Section 16.5, we can avoid having Cleland's hollandaise-sauce recipe output messy goop by limiting its execution to one location (Earth, say) without guaranteeing that it will work elsewhere (on the Moon, say). This is no different from a partial mathematical function that is silent about what to do with input from outside its domain, or from an algorithm for adding two integers that specifies no particular behavior for non-numerical input. ("Crashing" is a well-defined behavior if the program is silent about illegal input. More "well-behaved" behavior requires some kind of error handling.) A second way is to use the "Denver cake mix" strategy: I have been told that packages of cake mix that are sold in mile-high Denver come with alternative directions. The recipe or algorithm should be expressed conditionally: if location = Earth, then do *A*; if location = Moon, then do *B* (where *B* might be the output of an error message).

16.11 Content and Computation

16.11.1 Introduction

The quotation from Rescorla (2007) at the beginning of Section 16.9 focuses the issues very clearly. Are we really interested in syntactic computation – computation with *symbols*, such as numerals? Or are we interested in semantic computation – computation with *things* the symbols represent, such as numbers?

---

20 Personal communication. B.C. Smith (1985, p. 24) makes a similar point: "as well as modelling the artifact itself, you have to model the relevant part of the world in which it will be embedded."

David Hilbert, whose investigations into the foundations of mathematics prompted much of the early work in the theory of computation (as we surveyed in Section 6.5), was a mathematical "formalist." As such, he was interested in the former, for, after all, we humans can only do the latter *by* doing the former. Is that a limitation? Perhaps, but it also gives us a freedom, because symbols (including numerals) can represent anything, not just numbers, so computation can be about anything.

16.11.2 Symbols: Marks vs. Meanings

In Section 13.2.2, we observed that some writers use 'symbol' to mean an uninterpreted, purely syntactic "mark" *together with* its (external) semantic interpretation or meaning. A symbol is at least a mark; its interpretation is another matter. Symbols are perhaps best thought of as ordered pairs of (syntactic) marks (identified by their shape) and (semantic) interpretations. Sometimes, the "meaning" of a symbol is called its "content." So, is computation (only?) about marks? Or is it (only? also?) about content?

The term 'content' sounds as if it refers to something contained within something else – something *internal* – but often it is used to indicate the *external* meaning or reference of a term. But if we think of the variables of a computer program as "boxes" that can contain the values of the variables, then we can combine both metaphors for "content": the content of a variable can be thought of as an "external" entity that is stored "inside" the "box."

Several writers say that content *is* necessary; something is not a computation unless it is *about* something: there is "no computation without representation" (Fodor, 1975, p. 34). *Is* a goal, or content, or interpretation a necessary part of a computation?

From a syntactic (internal) point of view, a Turing Machine that outputs sequences of strokes does just that: it outputs sequences of strokes – i.e. "marks." From a semantic (external) point of view, those strokes are symbols: i.e. marks plus content. Whether the marks should, or can, be interpreted as a specific integer or something else is a *separate* matter. This is, of course, what underlies the notion of "types" in programming languages. The question we are now considering is whether the type of a programming-language variable is a syntactic issue or a semantic one; note that it might be a case of "*syntactic* semantics."

This puzzle is not unique to computation. What are the solutions to the equation $x^2 = 2$? In the rational numbers, there is no solution; in the positive real numbers, there is one solution; in the (positive and negative) real numbers, there are two solutions. Similarly, $x^2 = -1$ has no solution in the real numbers but two solutions in the complex numbers. Deciding which "wider" number system the equation should be "embedded" in gives different "interpretations" of it (Frenkel, 2013, pp. 83, 99).

Syntactically, we can say that *the* solution to $x^2 = 2$ is $\sqrt{2}$. Whether (or not) we assign a rational, real, or imaginary *number* to the *symbol* '$\sqrt{2}$' is a separate matter. Similarly, the ratio of the circumference to the diameter of a circle is π; whether we understand the symbol 'π' as $\frac{22}{7}$, 3.14, 3.1415926535, or something else is a separate matter. We can, in fact, compute more "accurately" with the syntactic mark 'π' than we can with any of those finite, numeral interpretations.

Consider, again, Marr's computational theory of vision, part of which takes the form of an algorithm that "computes the Laplacean convolved with a Gaussian" (Egan, 2014, p. 120). For the present point, it is unimportant to know what Laplaceans, Gaussians, and convolution are; what matters is that they are purely mathematical operations, having nothing necessarily to do with vision. Mark Sprevak (2010, p. 263) argues that this "mathematical computation theory does not, by itself, have the resources to explain" vision; it needs to be augmented by a link "to the nuts and bolts of physical reality." Frances Egan takes an opposing view: "representational content is to be understood as a *gloss* on the computational characterization of a cognitive process" (Egan, 2010, p. 253). Once again, we have a difference between "To *G*, do *A*" and "Do *A*." Here, *A* is the Laplacean convolved with a Gaussian, and *G* is the "gloss" about its role in vision – its "content."

On Egan's side, one might say that the mathematical theory *does* have the resources to explain vision (that's one of the points of Wigner, 1960). It may still be a puzzle *how* or *why* it does (recall Marr's "why," quoted in Section 16.6), but there's no question *that* it does. (This is reminiscent of the problem of quantum mechanics as an "instrumentalist" scientific theory (Section 4.4): we know that quantum mechanics has the resources to explain physics, but it is still a puzzle how or why it does (Becker, 2018).

Paul Humphreys suggests a view of computational models that can account for this, as well as the duck-rabbit examples: "one of the characteristic features of mathematical [including computational] models is that *the same model … can occur in, and be successfully employed in, fields with quite different subject matters*" (Humphreys, 2002, p. S2, my italics). He goes on to say, "Let the … computer solve one and you automatically have a solution to the other" (p. S5), as illustrated by de Bruijn's lattice story.

Sprevak offers a counterargument to the focus on a Turing Machine's strokes rather than their meanings:

> … one cannot make sense of I/O equivalence without requiring that computation involves representational content. …
>
> Consider two computational systems that perform the same numerical calculation. Suppose one system takes ink-marks shaped like Roman numerals (I, II, III, IV, …) as input and yields ink-marks shaped like Roman numerals as output. Suppose the other system takes ink-marks shaped like Arabic numerals (1, 2, 3, 4, …) as input and yields ink-marks shaped like Arabic numerals as output. Suppose the two systems compute the same function: say, the addition function. What could their I/O computational equivalence consist in? Again, there may be no physical or functional identity between their respective inputs and outputs. The only way in which the their inputs and outputs are relevantly similar seems to be that they represent the same thing. (Sprevak, 2010, Section 3.2, p. 268, col. 1)

That is, that they compute the same arithmetic function cannot be explained without a semantic interpretation. Note that there is a difference between *what* a system is doing and whether *two* systems are doing the *same* thing: each addition algorithm (Roman and Arabic) is "doing its own thing." They are only doing the "same" thing in the sense that the two idiosyncratic things they are each doing are *equivalent*. This kind of *sameness* (or equivalence) depends on the semantics.

Sprevak goes on to say,

> Two physical processes that are intrinsic physical duplicates may have different representational contents associated with them, and hence different computational identities. One physical process may calculate chess moves, while a physical duplicate of that process calculates stock market predictions. We seem inclined to say that, in a sense, the two processes compute different functions, yet in another sense they are I/O equivalent. Appeal to representational content can accommodate both judgements. (Sprevak, 2010, Section 3.2, p. 268, col. 2)

But this is a different case: identical algorithm but different task.

The previous case is different algorithm but same (input-output–equivalent) task. But *syntactically* they are *not* doing the *same* thing; rather, they are doing things that are only *semantically equivalent.* That equivalence can be discovered, explained, and understood only via an external semantic interpretation.

Nevertheless, depending on how the two algorithms are structured, it might be possible to find subroutines that match up. In that case, the two algorithms would be doing the same (identical) thing *at a suitably high level of organization*, even if the low-level implementations of those subroutines are completely different. That would be a syntactic semantic "interpretation." For example, a Karel the Robot who turns right by turning left three times is turning right (better: is turning in the "right" direction) just as much as a *Karl* the Robot who turns right by turning left *six* times or a *Kal* the Robot for whom turning right is primitive (and who might have to turn *right* three times to turn *left*).

Let's look into the Marr example in more detail: Egan says, "*As it happens*, …["the device [that] computes the Laplacean convolved with the Gaussian"] takes as input light intensity values at points in the retinal image, and calculates the rate of change of intensity over the image" (Egan, 2010, p. 255, my italics). But considered solely as a

> computational device, it does not matter that input values represent *light intensities* and output values the rate of change of *light intensity*. The computational theory characterizes the visual filter as a member of a well understood class of mathematical devices that have nothing essentially to do with the transduction of light. (Egan, 2010, p. 255; original italics)

Compare this to the chess-war example. To paraphrase Egan, the theoretically important characterization from a computational point of view is a mathematical description: the device computes some mathematical function that, *as it happens*, can be interpreted as a chess match or as a war battle. But considered solely as a computational device, it does not matter that input values represent (say) chess moves or battle positions – the computational theory characterizes the device as a member of a well understood class of mathematical devices that have nothing essentially to do with chess or war:

> A crucial feature of … [the characterization that focuses solely on the mathematical function being computed and not on the purpose or external environment] is that it is 'environment neutral': the task is characterized in terms that prescind from the environment in which the mechanism is normally deployed. The mechanism described by Marr would compute the Laplacean of a Gaussian even if it were to appear (*per mirabile*) in an environment where light behaves very differently than it does on earth, or as part of an envatted brain. (Egan, 2014, p. 122)

Egan says that the visual filter "would compute the same mathematical function in any environment, *but only in some environments would its doing so enable the organism to see*" (Egan, 2010, p. 256; my italics). Similarly, Cleland's recipe would compute the same (culinary?) function in any environment, but only on Earth (and not on the Moon) would its doing so result in hollandaise sauce.

Given a computer program, how do you know what its purpose is? Of course, it might be obvious from its name, its documentation, or even its behavior when executed. But suppose you come across a very large program written in an unfamiliar programming language with unintuitive variable and subroutine names and no documentation. Suppose that, after considerable study of it, you are able to describe it and its behavior syntactically. You might also be able to develop a hypothesis about a purpose for it by providing an interpretation for it (e.g. that it is a chess program). And you, or someone else, might also be able to provide a different but equally good interpretation for it (e.g. that it is a war simulator). This is not unlike the situation with the brain, a very large neural network with no documentation.

Recall the MYSTERY Scheme program from Section 13.3.3:

```
(define (MYSTERY a b)
  (if (= b 0)
    a
    (MYSTERY b (remainder a b)))))
```

If I found the MYSTERY program in the desert and was able to describe it syntactically as outputting pretty patterns of numbers (whether base-10 or base-13 is irrelevant), I could stop there. Or if I wrote another program that took MYSTERY's pretty patterns and translated them into base-10, I could use it as a GCD computer. Similarly, if I found a computer in the desert that output pretty patterns of a certain sort, I might write another program that translated its output into a chess game. And *you* might write another program that translated those very same pretty patterns into a war battle.

Given a *problem* to be solved or a *task* to be accomplished, a computer scientist asks whether it is computable. If it is, then we can write a computer program to solve the problem or accomplish the task. Recall Kleene's (1995) informal characterization of "algorithm," which begins as follows:

> [A] method for answering any one of a given infinite class of questions … is given by a set of rules or instructions, describing a procedure that works as follows. *After* the procedure has been described, [then] if we select *any* question from the class, the procedure will then tell us how to perform successive steps, …

Note that the procedure has a purpose: "answering any one of a given infinite class of questions." And the procedure depends on that class: given a class of questions, there is a procedure such that given "*any* question from the class," the procedure "enable[s] us to recognize that now we have the answer before us and to read it off" (Kleene, 1995, p. 18).

Given that *program*, the programmer or a user knows what its original or intended purpose was. But we, or someone else, might be able to interpret it differently and use it for a different purpose (playing chess instead of simulating a war battle).

And we might also be able to re-implement it in a different medium. Marr wanted to explain certain aspects of human vision. He found an algorithm (computing the Laplacean convolved with a Gaussian, say) that helps to accomplish that. When that algorithm is implemented in the *human* visual system, it enables human visual perception. If it were implemented in a *computer*, it might enable *robot* visual perception. (We'll explore this in Chapter 18.) If it were implemented elsewhere than on Earth, it might do nothing visually (for humans *or* robots), but it would still compute a Laplacean convolved with a Gaussian. What task it accomplishes (in the external, teleological sense) depends on where that algorithm is "plugged in" to its environment. The same holds for all of the examples from Section 16.3.

16.11.3 Shagrir's "Master Argument"

Oron Shagrir (2020, Section 3) offers the following "master argument" for "the semantic view of computation":

1. A physical system might simultaneously implement several different automata S_1, S_2, S_3 ….
2. The contents of the system's states determine (at least partly) which of the implemented automata, S_i, is relevant for computational individuation.

Conclusion: The computational individuation of a physical system is essentially affected by content.

Let's take a close look at this, beginning with the conclusion. Note that what this argument attempts to show is not that *abstract* computation is semantic, but that *physical* computation is. So, it might be the case *both* that abstract computation is *not* semantic (or wide, etc.) *and* that physical computation *is*.

We have already seen several examples that seem to support the first premise: Fodor's chess-war computers, Rescorla's GCD computers, and Shagrir's and-vs.-or computers. But a closer look suggests a puzzle: which is the "physical system" that is the implementation, and which is the "automaton" that gets implemented? In the chess-war case, there are *two* physical systems, each of which implements the *same* (abstract) computation. But in Shagrir's example, there is an (abstract?) AND-gate that is implemented by (one interpretation of) a certain physical system and an (abstract?) OR-gate that is implemented by (a different interpretation of) the *same* physical system. In this case, there is *one* physical system that implements *different* computations. Yet another way to look at Shagrir's example is this: there is an underlying automaton that outputs certain symbols in response to certain inputs of those symbols, and there are two different (physical?) interpretations of those symbols such that under one interpretation we have a (physical?) AND-gate and under the other we have a (physical?) OR-gate. We'll return to this puzzle in a moment, but let's first look at the second premise.

The second premise talks about "contents" and "individuation." What are these? Let's begin with individuation. Consider an abstraction and two implementations of it. For concreteness, you might think of the abstract species *Homo sapiens* and two concrete implementations of it: Alan Turing and Alonzo Church. We can ask two questions about these three things: First, what makes the concrete *individuals* Turing and Church different from the abstract *species*? Second, what makes Turing different from Church? (The first question is "vertical," asking about the relation between a "higher-level" abstraction and "lower-level" implementations of it. The second question is "horizontal," asking about the relation between two objects at the "same level.") Unfortunately, in philosophy, the term 'individuation' has been used for both of these questions. The questions should be kept distinct (Castañeda, 1975 has suggested calling the "vertical" one 'individuation' and the "horizontal" one 'differentiation'). So, as you evaluate Shagrir's argument, you need to decide which question he has in mind.

As for "content," recall from Section 16.11.2 that we can think of a mark (or a variable) as a box and its meaning (or value) as the "content" of the box. So, we might be able to understand the second premise as follows: there is an abstract computation – an automaton – that can be characterized in terms of operations on its "boxes," i.e. its variables (recall from Section 9.5 Thomason's view of the nature of computation as a sequence of states that are assignments of values to variables). That abstract computation can be implemented by different physical systems depending on the contents of the "boxes." Those contents help us "individuate" the physical system. Depending on how we interpret 'individuation,' that means *either* those contents tell us what makes one computer a chess computer and another a war simulator (to use Fodor's example) even though both computers implement the same automaton, *or* those contents tell us what makes the chess computer different from the war simulator. Arguably, they tell us both of these things. (So perhaps the interpretation of 'individuation' doesn't matter in this case.)

So, if we have two different physical systems that implement the same automaton, then what makes them different is their semantic content. But that doesn't seem to say anything about the computation itself, which still appears to be able to be understood "narrowly." It is *physical* computation that might be semantic and wide, while it is *abstract* computation that might be syntactic and narrow.

16.12 Summary

We can distinguish between the question of which computation a given Turing Machine is executing and the question of what goal that computation is trying to accomplish. Both questions are important, and they can have very different answers. Two computers might implement the same Turing Machine but be designed to accomplish different goals. And of course, two computers might accomplish the same goal via different algorithms.

And we can distinguish between two kinds of semantics: external (or wide, or extrinsic, or inherited) and internal (or narrow, or intrinsic, or syntactic, or indigenous). Both kinds exist, have interesting properties, and play different, albeit complementary, roles.

Algorithms narrowly construed (minus the teleological preface) are what is studied in the mathematical theory of computation. To decide whether a task is computable, we need to find an algorithm that can accomplish it. Thus, we have two separate things: an algorithm (narrowly construed, if you prefer) and a task. Some algorithms can accomplish more than one task (depending on how their inputs and outputs are interpreted by external semantics). Some algorithms may fail, not because of a buggy, narrow algorithm, but because of a problem at the real-world interface. That interface is the (algorithmic) coding of the algorithm's inputs and outputs, typically through a *sequence* of transducers at the real-world end (what B.C. Smith, 1987 called a "correspondence continuum; see Section 13.2.1). Physical signals from the external world must be transduced (encoded) into the computer's switch-settings (the physical analogues of a Turing Machine's '0's and '1's), and the output switch-settings have to be transduced (decoded) into such real-world things as displays on a screen or physical movements by a robot.

But real-world tasks are complex. Models abstract from this complexity, so they can never match the rich complexity of the world. Computers (and people!) see the world through models of these models. Reasoning on the basis of partial information cannot be proved correct (and simulation only tests the computer-model relation, not the model-world relation). So, empirical *reliability* must supplement program verification. Therefore, we must embed the computer in the real world.

At the real-world end of the correspondence continuum, we run into Smith's gap. From the narrow algorithm's point of view, so to speak, it might be able to asymptotically approach the real world, in Zeno-like fashion, without closing the gap. But just as someone trying to cross a room by only going half the remaining distance at each step *will* eventually cross the room (although not because of doing it that way), so the narrow algorithm implemented in a physical computer *will* do something in the real world. Whether what it accomplishes was what its programmer intended is another matter. (In the real world, there are no "partial functions"! This was one of Kugel's points about trial-and-error machines, as we saw in Section 11.10.2.)

One way to make teleological algorithms more likely to be successful is by Shapiro's strategy: internalizing the external, teleological aspects into the pre- and post-conditions of the (narrow) algorithm, thereby turning the external semantic interpretation of the algorithm into an internal, syntactic semantics. What Smith shows is that the external semantics for an algorithm is never a relation directly with the real world, but only to a *model* of the real world. That is, the real-world semantics has been internalized. But that internalization is necessarily partial and incomplete.

There are algorithms *simpliciter* ("Do A"), and there are algorithms *for accomplishing a particular task* ("To G, do A"). Alternatively, we could say that *all* algorithms accomplish a particular task, but some tasks are more "interesting" than others. The algorithms whose tasks are not currently of interest may ultimately *become* interesting when an application is found for them, as was the case with non-Euclidean geometry. Put otherwise, the algorithms that do not have an obvious goal may ultimately be used to accomplish one:

[D]oes … any algorithm … have to do something interesting? No. The algorithms we tend to talk about almost always do something interesting – that's why they attract our attention. But a procedure doesn't fail to be an algorithm just because it is of no conceivable use or value to anyone. … *Algorithms don't have to have points or purposes.* … Some algorithms do things so boringly irregular and pointless that there is no succinct way of saying what they are *for*. They just do what they do, and they do it every time. (Dennett, 1995, p. 57, my italics)[21]

*****

A few paragraphs ago, I said that despite Smith's gap, a narrow algorithm implemented in the real world *will* do something, whether what it was intended to do or not. What about an automated system designed to decide quickly (and in the absence of complete information) how to respond to an emergency? Would it make you feel uneasy? But so should a human who has to make that same decision. And they should both make you uneasy for the same reason: they have to reason and act on the basis of partial (incomplete) information. This will be our topic in the next chapter.

16.13 Questions for the Reader

1. According to many, a computer simulation of a hurricane is not a hurricane, because it does not get people wet (Dennett, 1978, p. 191; Hofstadter, 1981, pp. 73ff; Dretske, 1985, p. 27). But it *could* get *simulated* people *simulatedly* wet, as it might in a computer simulation game (Rapaport, 1986, 1988a, b, 2005b, 2012b; Shapiro and Rapaport, 1991). Relatedly, David Chalmers (2017) has suggested that virtual reality is (a kind of) reality (Ramakrishna, 2019). (For a reply, see Ludlow, 2019.) The difference between a real hurricane and a simulated one has to do, in part, with the nature of the inputs and outputs. As Lawrence R. Carleton (1984, pp. 222–223) notes, "The input to a fire simulation is not oxygen, tinder, and catalyst. That is, it is not the same input as goes into a natural system which produces fire. … [I]t is by virtue of dealing in the right kinds of input and output that one system can play the role in a situation that is played by a system we acknowledge to literally undergo the [activity] … our system simulates." Cleland's hollandaise-sauce program may differ in output when executed on the Moon than on Earth; it has the wrong output. But a hurricane-simulator, a fire-simulator, and a hollandaise-sauce program each exhibit their relevant behaviors if you ignore the input and the output.

   How central to what it is that a computer program is (supposed to be) doing is the nature of the inputs and outputs (in general, of the environment in which it is being executed)?
2. Recall the opening epigraph. How *does* a program interact with the world? Is it via the *process*, which is a physical entity (or event?) *in* the world? If so, how does the program interact with the *process*? Is it via the compiler? Or is that just a first step, translating the program into the machine language the machine understands? Is it via the loader, which is what transforms the machine-language program into the memory, setting the switches?
3. The artist Charles E. Burchfield said that

   > An artist must paint not what he sees in nature, but what is there. To do so he must invent symbols, which, if properly used, make his work seem even more real than what is in front of him.
   > (https://www.burchfieldpenney.org/collection/charles-e-burchfield/biography/)

---

21 See the Online Resources for further reading on the teleology of algorithms.

If we change 'artist' to 'programmer' and 'paint' to 'program,' this becomes

> Programmers must program not what they see in nature, but what is there. To do so they must invent symbols, which, if properly used, make their work seem even more real than what is in front of them.

Is the artist's task different from the scientist's or the programmer's? Can programs (or paintings, or scientific theories) end up seeming more "real" to us than the things that they are models of? Is it easier to understand the behavior of the process of a program that models a hurricane (for example) than to understand the real hurricane itself?[22]

---

22 Thanks to Albert Goldfain (personal communication, 3 April 2007) for this question.

Part V

Computer Ethics and Artificial Intelligence

The two topics of Part V – computer ethics and the philosophy of AI – are large, long-standing disciplines in their own right. Ethics is a branch of philosophy, and AI is a branch of CS, so both computer ethics and the philosophy of AI should be branches of the philosophy of computer science.

In Chapters 17 and 19, we will focus on two topics in computer ethics that I think are central to the philosophy of computer science but until recently have not been the focus of most discussions of computer ethics: Should we trust decisions that computers make? And should we build "artificial intelligences"? But before we can try to answer that last question, Chapter 18 on the philosophy of AI will focus on whether we *can* build them.[1]

---

1 See the Online Resources for further reading on computer ethics.

Philosophy of Computer Science: An Introduction to the Issues and the Literature, First Edition. William J. Rapaport.
© 2023 John Wiley & Sons, Inc. Published 2023 by John Wiley & Sons, Inc.

17

Computer Ethics I: Should We Trust Computers?

In 2011, [John] Rogers … announced the invention of … an integrated silicon circuit with the mechanical properties of skin. … The artificial pericardium [made with Rogers's invention] will detect and treat a heart attack before any symptoms appear. … Bit by bit, our cells and tissues are becoming just another brand of hardware to be upgraded and refined. I asked him whether eventually electronic parts would let us live forever, and whether he thought this would come as a relief, offer evidence that we'd lost our souls and become robots, or both. *"That's a good thing to think about, and people should think about it,"* he said *"But I'm just an engineer, basically."*
—Kim Tingley (2013, p. 80, my italics)

Machines are more than ever controlled by software, not humans. Occasionally it goes fatally wrong. … [I]ncreasing the complexity of systems makes checking them more difficult. Hardware, from chips to special sensors, can be difficult to test. And it can be difficult for humans to understand how some A.I. algorithms make decisions.
—Jamie Condliffe (2019, my italics)

17.1 Introduction

In 2004, when I first taught the course that this text is based on, the question of whether to trust decisions made by computers was not much discussed. But since the advent of self-driving cars[1] and machine-learning systems, it has become a more pressing issue, with immediate, real-life, practical implications as well as moral and legal ramifications.

Before we consider the ethical issues of whether we should trust the decisions that computers make and the clearly related question that is the title of James Moor's 1979 essay – are there decisions computers should *never* make? – there are two prior questions: *What **is** a "decision"?* And *do computers "make decisions" at all?*

17.2 Decisions and Computers

Roughly, a decision is a choice made from several alternatives, usually for some reason (Eilon, 1969). Let's begin by considering three kinds of decisions:

[1] Keep in mind that another vehicle that many of us use frequently was once only human-operated but is now completely automated (and I doubt that any of us would know what to do if it failed): elevators. See the Online Resources for further reading on automated vehicles.

Philosophy of Computer Science: An Introduction to the Issues and the Literature, First Edition. William J. Rapaport.
© 2023 John Wiley & Sons, Inc. Published 2023 by John Wiley & Sons, Inc.

1. A decision could be the result of an arbitrary choice, such as flipping a coin: heads, we'll go out to a movie; tails, we'll stay home and watch TV.
2. A decision could be the solution to a purely logical or mathematical problem that requires some calculation.
3. A decision could be the result of investigating the pros and cons of various alternatives, rationally evaluating these pros and cons, and then choosing one of the alternatives based on this evaluation.

At first glance, there is a simple answer to the question whether computers make decisions: yes; computers can easily make the first kind of decision for us. Moreover, any time a computer solves a logical or mathematical problem, it has made a decision of the second kind.

Can computers make decisions of the third kind? Surely, it would seem, the answer is, again, 'yes': computers can play games of strategy, such as checkers, chess, and Go, and they can play them so well that they can beat the (human) world champions. Such games involve choices among alternative moves that must be evaluated with, one hopes, the best (or least worst) choice being made. Computers can do this. [2]

Of course, it is not just a physical *computer* that makes a decision. Arguably, it is a computer *program* being executed by a computer that makes the decision, although I will continue to speak as if it is the computer that decides. After all, we usually say that *you* make a decision, not that *your brain* does.

And of course, it is not just a computer *program* that makes a decision. Computer programs are written by *humans*. (Even computer programs that are written by computers are the output of computer programs that were written by humans.) And humans, of course, can err in various ways, unintentionally or otherwise. These errors can be inherited by the programs they write.

Robin K. Hill has argued that computers do *not* make decisions, precisely *because* their programs are written by humans. It is the humans who make the decisions that are subsequently encoded in the programs:

> [M]achines and algorithms have no such capacity as is normally connoted by the term "decision" ... ... Algorithms are not biased, because a program does not make decisions. The program implements decisions made elsewhere. (Hill, 2018)

To a large extent, this is, of course, correct. There is no question that the way in which a computer makes a decision was initially determined by its human programmer. And Mullainathan (2019) argues that "biased algorithms are easier to fix than biased people."

But what happens when the human programmer is out of the picture and the computer running that program is what we rely on? In any given situation, when the computer has to act or to make or recommend a decision, it will do so autonomously and in the light of the then-current situation, without consulting the programmer (or being able to consult the programmer):

> It is a common misconception that because a machine such as a guided missile was originally designed and built by conscious man [sic], then it must be truly under the immediate control of conscious man. Another variant of this fallacy is "computers do not really play chess, because they can only do what a human operator tells them." ... When it is actually playing, the computer is on its own, and can expect no help from its master. All the programmer can do is to set the computer up *beforehand* in the best way possible ....
> (Dawkins, 2016, pp. 66–67)

---

2 See the Online Resources for further reading on game-playing computers.

Typically, human delegation of decision-making powers to computers happens in cases where large amounts of data are involved in making the decision or in which decisions must be made quickly and automatically. And to the extent that one of the goals of CS is to determine what real-world tasks are computable, finding out which *decisions* are computable is an aspect of that.

In any case, humans *might* delegate such power to computers. So, another way to phrase our question is, "What areas of our lives should be computer-controlled, and what areas should be left to human control?" Are there decisions that non-human computers could *not* make as well as humans? For instance, there might be situations in which there are sensory limitations that prevent computer decisions from being fully rational. Or there might be situations in which a decision requires some (presumably non-computable) empathy. On the other hand, there might be situations in which a computer might have an advantage over humans: it is impossible (or at least less likely) for a computer to be swayed by such things as letter-writing campaigns, parades, etc. Such tactics were used by General Motors in its campaign to persuade some communities to open new plants for its Saturn cars (Russo, 1986).

To answer these questions, we need to distinguish between what *is* the case and what *could be* the case. We could try to argue that there are some things that are *in principle* impossible for computers to do. Except for computationally impossible tasks (such as the Halting Problem, Section 7.7), this might be hard to do. But we should worry about the possible future now so we can be prepared for it if it happens. (Recall the italicized quotation in the first epigraph to this chapter.)

Whether there are, *now*, decisions that a computer could not make as well as a human is an empirical question. It is capable of investigation, and, currently, the answer is unknown. Many, if not most, of the objections to the limitations of computers are best viewed as research problems: if someone says that computers can't do something, we should try to make ones that can.

This is crucial: humans should be critical thinkers. There is a logical fallacy called the Appeal to Authority: just because an authority figure *says* that something is true, it does not logically follow that it *is* true. Although logicians sometimes warn us about this fallacy, it *is* acceptable to appeal to an authority (even a computer!) as long as the final decision is yours. *You* can – and must – decide whether to believe the authority or trust the computer. You should also be able (and willing!) to *question* the authority – or the computer! – so as to understand the reasons for the decision.

So, even if we allow computers to make (certain) decisions for us, it is still important for us to be able to understand those decisions. When my son was first learning how to drive, I did not want him to rely on the vehicle's automated "dynamic cruise control" system, because I wanted him to know how and when to slow down or speed up on a superhighway. Once he knew how to do that, then he could rely on the car's computer making those decisions for him. If the computer's decision-making skills went awry or were unavailable (e.g. the laser-controlled system on my 2008 Toyota Sienna was designed *not* to work when the windshield wipers were on, or when the car ahead of you was dirty!), he should know how to do those things himself. (For a discussion of this point in the context of airplane pilots, see Nicas and Wichter, 2019.)

17.3 Are Computer Decisions *Rational*?

Another issue that arises from the fact that computer programs are written by humans is whether, given the occasional irrationality of human behavior, computer-made decisions really *are* rational.

When a decision has an impact on our lives, we would like the decision-making process to be rational, whether it is a human making the decision or a human-written program "making" it. Can computers (and the programs that they execute) be completely rational? It certainly seems that some computers can make rational decisions for us. The kinds of decision making described in the previous section *seem* to be purely rational. And aren't rule-based algorithms purely rational?

Consider an algorithm that does not involve any random or interactive procedure produced by a non-rational oracle (Section 11.9). Presumably, if the decision is made by a computer that is following such an algorithm, then that decision is a purely rational one. By 'rational,' I don't necessarily mean it is a *purely* logical decision. (Recall Section 2.5 on kinds of rationality.) It may, for instance, involve empirical data, which might be erroneous in some way: it might be incomplete, it might be statistically incorrect, it might be biased, and so on.

Another potential problem is if the algorithm requires exponential time or is NP-complete, or even if it merely would take longer to come up with a decision than the time needed for action. In that case, or if there is no such algorithm, we would have to rely on a "satisficing" heuristic in the sense of an algorithm whose output is "near enough" to the "correct" solution (Section 5.6). But this is still a kind of rationality – what Simon called "bounded" rationality (Sections 5.6, 11.10.2).

But just as a logical argument can be valid even if its premises are false (Section 2.5.1), an algorithm can be syntactically and semantically correct even if its input is not ("garbage in, garbage out"; recall Sections 8.10.2, 11.7, and 15.2.1). Nevertheless, as long as there is an algorithm that can be studied to see how it works, or as long as the program can explain how it came to its decision, I will consider it to be rational.

Whether computers ought to make decisions for us is equivalent to whether our decisions ought to be made algorithmically. And that suggests that it is equivalent to whether our decisions ought to be made *rationally*. If there is an algorithm for making a given decision, then why not rely on it? After all, wouldn't that be the rational thing to do?

One might even argue that there is no such thing as computer ethics. All questions about the morality of using computers to do something are really questions about the morality of using algorithms. As long as algorithms are rational, questions about the morality of using them are really questions about the morality of being rational, and it seems implausible to argue that we shouldn't be rational. This suggests that James Moor's (1979) question, "Are there decisions computers should never make?," should really have nothing to do with computers! Perhaps the question should really be, are there decisions that should not be made on a rational basis?

But then the important question becomes, are the algorithms really rational? And how would we find out? Before looking at these questions, let's assume for the moment that a decision-making algorithm *is* rational and turn to the next question.

17.4 Should Computers Make Decisions *for* Us?

A paragraph deeply embedded in a 2004 science news article suggests that people find it difficult to accept rational recommendations even if they come from other *people*, not computers. The article reports on evidence that a certain popular and common surgical procedure had just been shown to be of no benefit: "Dr. Hillis said he tried to explain the evidence to patients, to little avail. 'You end up reaching a level of frustration,' he said. 'I think they have talked to someone along the line who convinced them that this procedure will save their life' " (Kolata, 2004). Perhaps the fundamental issue is not whether computers should make rational decisions or recommendations, but whether or why humans should or don't *accept* rational advice!

There are several reasons we might want to let a computer make a decision *for* us: computers are much *faster* than we are at evaluating options, they can evaluate *more* options than we could (in the same amount of time), they are better at evaluating more *complex* options, and they can have *access* to more relevant data. And in many situations in the modern world, we might simply have no other option but to allow computers to make decisions for us. So, whether it is a good idea or a bad idea to let them do so, it is a simple fact that they do.

And after all, is this any different from letting *someone* else make a decision for us – someone who is wiser, or more knowledgeable, or more neutral than we are? If it is not any different, then – in both cases – there is still a question that should always be raised: should we *trust* that other agent's decision? Before looking into this, there is an intermediate position that we should consider.

17.5 Should Computers Make Decisions *with* Us?

Moor suggests that if computers can make certain decisions at least as well as humans, then we should let them do so, and it would then be up to us humans to accept or reject the computer's decision. After all, when we ask for the advice of an expert in medical or legal matters, we are free to accept or reject that advice. Why shouldn't the same be true for computer decision making?

In other words, rather than simply letting computers (or other humans) make decisions *for* us, we should *collaborate* on the decision-making process, treating the computer (or the human expert) as a useful source of information and suggestions to help *us* make the final decision. As we noted in Section 17.3, this might not always be possible: there may (and most likely *will*) be situations in which we do not have the expertise or the *time* to evaluate all options before a decision has to be made.

But there are also many cases in which we do need to collaborate:

> The systems that land airplanes are hybrids – combinations of computers and people – exactly because the unforeseeable happens, and because what happens is in part the result of human action, requiring human interpretation.
> (B.C. Smith, 1985, Section 7, p. 24, col. 2)

The situation that Smith mentions has been explored in depth by the anthropologist and cognitive scientist Edwin Hutchins (Hutchins, 1995a,b; Hollan et al., 2000; Casner et al., 2016). Hutchins's theory of "distributed cognition" uses examples of large naval vessels navigating and jet pilots working in their cockpits. In both of these cases, it is neither the machines alone (including, of course, computers) nor humans alone who make decisions or do the work, but the combination of them – indeed, in the case of large naval vessels, it is *teams* of humans, computers, and other technologies. Hutchins suggests that this combination constitutes a "distributed" mind. Similarly, the philosophers Andy Clark and David Chalmers (1998) have developed a theory of "extended cognition," according to which our (human) minds are not bounded by our skull or skin but "extend" into the external world to include things like notebooks, reference works, and computers.

> **Question for the Reader:** Do these examples constitute uses of *oracles* (Section 11.9) as external sources of information?

But *must* complex decision-making systems be such "hybrids" or "team efforts"? Smith, Hutchins, and Clark and Chalmers developed their theories long before the advent of self-driving cars. Even now, it remains to be seen whether self-driving cars will continue to need human intervention (remember, self-driving elevators don't need very much of it!): Steven E. Shladover (2016) argues that a level called "conditional automation," in which computers and humans work together, will be *harder* to achieve than the more fully automated level called "high automation" (see also Casner et al., 2016). Nevertheless, such "hybrid" or "extended" systems will probably remain a reality.

17.6 Should We *Trust* Decisions Computers Make?

Whether we let computers make decisions *for* us or work jointly *with* them to make decisions, we usually *assume* that any decision they make or advice they give is based on good evidence (as input) and rational algorithms (that process the input). Note, again, the similarity with logical inference, which begins with axioms or premises ("input") and then "processes" that "input" to derive a valid conclusion. In both cases, for the decision (or advice) to be "good" or for the conclusion to be true, the input must be correct or true *and* the processing must be correct. But how do we know if they are? (Remember the warnings in Section 2.4.4 about making assumptions!)

An algorithm's trustworthiness is a function of its input and its processing. Is it getting all of the relevant input? Is the input accurate, or might there be a problem with its sensors or how it interprets the input? Is the algorithm correct? Can we understand it? Can we explain or justify its decisions? Is it (intentionally or unintentionally) biased in some way, perhaps due to the way its human programmer wrote it or – in the case of a machine-learning program – what its initial training set was?

How can computer decision-making competence be judged? One answer is, in the same way *human* decision-making competence is judged: namely, by means of its decision-making *record* and its *justifications* for its decisions.

Let's briefly consider a computer's track record first. Consider once more a documentationless computer found in the desert. Suppose we discover that it successfully *and reliably* solves a certain type of problem for us. Even if we cannot understand why or how it does that, there doesn't seem to be any reason *not* to trust it. So, why should justifications matter? After all, if a computer constantly bests humans at some decision-making task, why should it matter *how* it does it?[3]

Presumably, however, decision-making computers *should* be accountable for their decisions, and knowing what their justifications are helps in this accounting. In fact, the European Union has passed a law giving users the right to have an explanation of a computer's decision concerning them (https://en.wikipedia.org/wiki/Right_to_explanation). The justifications, of course, need not be the same as human justifications. For one thing, human justifications might be wrong or illogical.[4]

But what if justifications are unavailable or, perhaps worse, misleading? Let's take a look at these two possibilities.

17.6.1 The Bias Problem

Could there be a hidden bias in the way the algorithms were developed? For example, the training set used to create a machine-learning algorithm might have been biased. This does not have to be due to any intention to deceive on the part of the programmer. Indeed, such a program "could be picking up on biases in the way a child mimics the bad behavior of his [or her] parents" (Metz, 2019b). The bias might not be evident until the algorithm is deployed.

Recall Hill's point that algorithms are written by humans. And humans, of course, have

> idiosyncratic foibles . . . . The mostly white men who built the tools of social networks did not recognize the danger of harassment, and so the things they built became conduits for it. If there had been women or people of color in the room, ... there might have been tools built to protect users ... (Bowles, 2019)

What kinds of problems can such "foibles" or biases lead to?

---

3 Maybe we would be better off not knowing! (Clarke, 1953). For more on trust, see the Online Resources.
4 See the Online Resources for further reading on human reasoning.

Users discovered that Google's photo app, which applies automatic labels to pictures in digital photo albums, was classifying images of black people as gorillas. Google apologized; it was unintentional.

… Nikon's camera software … misread images of Asian people as blinking, and … Hewlett-Packard's web camera software … had difficulty recognizing people with dark skin tones.

This is fundamentally a data problem. Algorithms learn by being fed certain images, often chosen by engineers, and the system builds a model of the world based on those images. If a system is trained on photos of people who are overwhelmingly white, it will have a harder time recognizing nonwhite faces.

… ProPublica … found that widely used software that assessed the risk of recidivism in criminals was twice as likely to mistakenly flag black defendants as being at a higher risk of committing future crimes. It was also twice as likely to incorrectly flag white defendants as low risk.

The reason those predictions are so skewed is still unknown, because the company responsible for these algorithms keeps its formulas secret …. (Crawford, 2016)

Perhaps the ethical issues really concern the nature of different kinds of algorithms. "Neat" algorithms are based on formal logic and well-developed theories of the subject matter of the algorithm. "Scruffy" algorithms are not necessarily based on any formal theory. (These terms were originally used to describe two different approaches to AI (https://en.wikipedia.org/wiki/Neats_and_scruffies), but they can be used to describe any algorithm.) "Heuristic" algorithms don't necessarily give you a correct solution to a problem but are supposed to give one that is near enough to a correct solution to be useful (i.e. one that "satisfices"). Machine-learning algorithms are trained on a set of test cases and "learn" how to solve problems based on those cases and on the particular learning technique used.

If a "neat" algorithm is "correct" – surely, a big "if" – then there does not seem to be any moral reason not to use it (not to be "correctly rational"). If the algorithm is "scruffy," then one might have moral qualms. If the algorithm is a heuristic (perhaps as in the case of expert systems), then there is no more or less moral reason to use it than there is to trust a human expert. If the algorithm was developed by machine learning, then its trustworthiness will depend on its training set and learning method.[5]

17.6.2 The Black-Box Problem

Indeed, we are often quite distressed when a repairman returns a machine to us with the words, "I don't know what was wrong with it. I just jiggled it, and now it's working fine." He [sic] has confessed that he failed to come to understand the law of the broken machine and we infer that he cannot now know, and neither can we or anyone, the law of the "repaired" machine. If we depend on that machine, we have become servants of a law we cannot know, hence of a capricious law. And that is the source of our distress.
—Joseph Weizenbaum (1976, pp. 40–41)

Here is the strange rub of such a deep learning system: It learns, but it cannot tell us why it has learned; it assigns probabilities, but it cannot easily express the reasoning behind

5 See the Online Resources for further reading on machine learning and the bias problem.

the assignment. Like a child who learns to ride a bicycle by trial and error and, asked to articulate the rules that enable bicycle riding, simply shrugs her shoulders and sails away, the algorithm looks vacantly at us when we ask, "Why?" It is, like death, another black box.
—Siddhartha Mukherjee (2018)

At least four sources of problems can make a computer's decision untrustworthy:

1. the decision-making criteria encoded in the algorithm, either by its programmer (or programmers) or by the machine-learning program that developed those criteria from test cases,
2. those test cases themselves,
3. the computer program itself, and
4. the data on which a given decision is based.

Let's assume – a big assumption, and only for the sake of the argument – that the input data (#4) are as complete and accurate as possible. Let's also assume (although this is an even larger assumption) that the algorithm (#3) has been formally verified. That leaves the decision-making criteria and any test cases as the primary focus of attention.

At the present stage in the development of computers, two ways in which these criteria find their way into an algorithm are through the human programmer and through machine learning. Of course, a machine-learning algorithm gets its test cases from a human (or from a database that was generated by another program that was written by a human), and it gets its machine-learning technique from its human programmer. But once the human is out of the picture and the algorithm is left to fend for itself, so to speak, it is to the algorithm that we must turn for explanations.

Consequently, one important issue concerning computers that make decisions for (or with) us is whether they can, or should, *explain* their decisions. Two kinds of algorithms are relevant to this question. One kind is the symbolic or logical algorithm that has such an explanatory capability: a user could examine a trace of the algorithm, or a programmer could write a program that would translate that trace into a natural-language explanation that a user could understand. The other kind is one that is based on a neural-network or a statistical, machine-learning algorithm. Such an algorithm might not be able to explain its behavior, nor might its programmer or a user be able to understand how or why it behaves as it does.

As an example, a typical board-game-playing program might have a representation of the board and the pieces, an explicit representation of the rules, and an explicit game tree that allows it to rationally choose an optimal move. Such a program could easily be adapted to explain its moves. It does not have to, of course. My computer science colleague Peter Scott suggested[6] that "even the Turing Test does not require the agent to explain clearly how s/he/it is reasoning." However, the interrogator can always ask something like "Why do you believe that?" or "Why did you do that?" and to pass the test, the interlocutor (human or computer) must be able to give a plausible answer.

But AlphaGo, the recent Go-playing program that beat the European Go champion, was almost entirely based on neural networks and machine-learning algorithms (Silver et al., 2016; Vardi, 2016). As an editorial accompanying Silver et al., 2016 put it,

> …the interplay of its neural networks means a human can hardly check its working, or verify its decisions before they are followed through. As the use of deep neural network systems spreads into everyday life … it raises an interesting concept for humans and their relationships with machines. The machine becomes an oracle; its pronouncements have to be believed.

---

6 Personal communication, 23 April 2017.

When a conventional computer tells an engineer to place a rivet or a weld in a specific place on an aircraft wing, the engineer – if he or she wishes – can lift the machine's lid and examine the assumptions and calculations inside. That is why the rest of us are happy to fly. Intuitive machines will need more than trust: they will demand faith. (*Nature* Editors, 2016)

Should they "demand faith"? Or should laws (such as those in the European Union) *require* transparency or explainability and thus rule out "black box" machine-learning algorithms of the kind discussed in Section 3.11? Relying on a successful but unexplained computer's decisions might not necessarily mean we are taking its decisions on faith. After all, its successes would themselves be evidence for its trustworthiness, just as an axiom's usefulness in mathematical derivations is evidence in its favor even though – by definition – it cannot be proved.[7]

Digression on Connectionist vs. Symbolic Algorithms: Peter Scott went on to say,

> Some say a temporary truce has been recognized, but there is still no hint of a permanent peace treaty between the connectionist and symbolist advocates. I am betting that controversy will go on for a long time.

The current apparent inability of connectionist or neural-network algorithms to explain their behavior (or to have their behavior explained by others) while at the same time being better at certain tasks than symbolic algorithms that *can* explain their behavior suggests that both kinds of mechanisms are needed.

For example, there is a two-way interaction between connectionist-like cognition and symbolic-like cognition in human learning: when my son was learning how to drive, I realized that I had to translate my *instinctive* (connectionist-like) behavior for making turns into *explicit* (symbolic) instructions, something along the lines of "put your foot on the brake to slow down, make the turn, then accelerate slowly." But to do that, I had to observe what *my* instinctive behavior was. Presumably, my son would follow the *explicit* instructions until they became second nature to him (i.e. "followed" implicitly or instinctively), until such time as he might teach his child to drive, and the cycle would repeat. (Compare a "hardwired" Turing Machine that operates "instinctively" and a programmed, Universal Turing Machine that explicitly follows instructions.)

For other arguments on the value of symbolic computation, see the Online Resources.

17.7 Are There Decisions Computers *Must* Make for Us?

Commercial airplanes are what we'd call self-driving except at takeoff and landing, and the result is that it's now nearly impossible for a cruising jet to fall out of the sky without malice or a series of compounding errors by the pilots. (Lethal computer glitches are so rare that if they appear even twice among tens of millions of flights, as in the case of Boeing's 737 MAX 8, the industry goes into crisis.) People get the willies at the idea of putting their lives in the hands of computers, but there's every reason to think that, as far as transportation goes, we're safer in their care.
—Nathan Heller (2019, p. 28)

---

7 See the Online Resources for further reading on the black-box problem.

Having the *ability* to evaluate a computer's reasons for its decisions assumes the *willingness* to do so. But remember Simon's problem of bounded rationality: we usually don't have the time or ability to evaluate *all* the relevant facts before we need to act. What about emergencies or other situations in which there is no time for the humans who must act to include the computer's recommendation in their deliberations?

On 1 July 2002, a Russian airliner crashed into a cargo jet over Germany, killing all on board, most of whom were students. The Russian airliner's flight recorder had an automatic collision-avoidance system that instructed the pilot to fly *over* the cargo jet. The human air-traffic controller told the Russian pilot to fly *under* the cargo jet. According to science reporter George Johnson (2002), "Pilots tend to listen to the air traffic controller because they trust a human being and know that a person *wants* to keep them safe" (my italics). But the human air-traffic controller was tired and overworked. And the collision-avoidance computer system didn't "want" anything; it simply made rational judgments. The pilot followed the human's decision, not the computer's, and a tragedy occurred.

There is an interesting contrasting case. In January 2009, after an accident involving birds that got caught in its engines, a US Airways jet "landed" safely on the Hudson River in New York City, saving all on board and making a hero out of its pilot. Yet William Langewiesche (2009) argues that it was the plane, with its computerized "fly by wire" system, that was the real hero. In other words, the pilot's heroism was due to his willingness to accept the computer's decision.[8]

17.8 Are There Decisions Computers *Shouldn't* Make?

Let's suppose we have a decision-making computer that explains all of its decisions, is unbiased, and has an excellent track record. Are there decisions that even such a computer *should* never make?

The computer scientist Joseph Weizenbaum (1976) has argued that even if computers *could* make decisions as well as, or even better than, a human, they shouldn't, *especially* if their reasons differ from ours. And Moor points out that, possibly, computers shouldn't have the *power* to make (certain) decisions, even if they have the *competence* to do so (at least as well as, if not better than, humans).

But if they have the competence, why shouldn't they have the power? For instance, suppose a very superstitious group of individuals makes poor medical decisions based entirely on their superstitions; shouldn't a modern physician's "outsider" medicine take precedence? And does the fact that computers are immune to human diseases mean they lack the empathy to recommend treatments to humans?

Moor suggests that although a computer should make rational decisions for us, a computer should *not* decide what our basic goals and values should be. Computers should help us *reach* those goals or *satisfy* those values, but they should not *change* them. But why not? Computers can't be legally or morally responsible for their decisions, because they're not persons. At least, not yet. But what if AI succeeds? We'll return to this in Chapters 18 and 19. Note, by the way, that for many legal purposes, non-human corporations *are* considered persons.

Batya Friedman and Peter H. Kahn, Jr. (1997) argue that humans *are* – but computers are *not* – *capable* of being moral agents, and therefore computers should be designed so that (1) humans are *not* in "merely mechanical" roles with a diminished sense of agency, and (2) computers *don't* masquerade as agents with beliefs, desires, or intentions.

---

8 See the Online Resources for further reading on these issues.

Let's consider point (1): Friedman and Kahn argue that computers should be designed so that humans *do* realize that they (the humans) *are* moral agents. But what if the computer has a better decision-making track record than humans? Friedman and Kahn offer a case study of APACHE, a computer system that can make decisions about when to withhold life support from a patient. It is acceptable if it is used as a tool to aid *human* decision makers. But human users may experience a "diminished sense of moral agency" when using it, presumably because a computer is involved.

But why? Suppose APACHE is replaced by a textbook on when to withhold life support, or by a human expert. Would either of those diminish the human decision-maker's sense of moral agency? In fact, wouldn't human decision-makers be remiss if they *failed* to consult experts or the writings of experts? So wouldn't they also be remiss if they failed to consult an expert computer? Perhaps humans would experience this diminished sense of moral agency for the following reason: if APACHE's decisions exhibit "good performance" and *are* more relied on, humans may begin to yield to its decisions. But why would that be bad?

Turning to point (2), Friedman and Kahn argue that computers should be designed so that humans *do* realize that computers are *not* moral agents. Does this mean computers should be designed so that humans *cannot* take Dennett's (1971) "intentional stance" toward them? (Recall Section 12.4.1.) But what if the computer *did* have beliefs, desires, and intentions? AI researchers are actively designing computers that either really have them or are best understood as if they had them. Would they not then be moral agents? If not, why not? According to Dennett (1971), some computers can't *help* "masquerading" as belief-desire-intention agents because that's the best way for *us* to *understand* them.

Friedman and Kahn argue that we should be careful about anthropomorphic user interfaces, because the *appearance* of beliefs, desires, and intentions does not imply that computers really *have* them. This is a classic theme not only in the history of AI but also in literature and cinema. And this is at the heart of the Turing Test in AI, to which we now turn.[9]

17.9   Questions for the Reader

1. In Sections 12.4.4 and 16.9, we discussed how to describe what a computer or a person is doing. Is a Universal Turing machine that is running an addition program adding or "merely" fetching and executing the instructions for adding? If I use a calculator or a computer, or if a robot performs some action, who or what is "really" doing the calculation or the computation, or the action: Is it the calculator (computer, robot)? Or me? When I use a calculator to add, am I adding or "merely" pushing certain buttons? (Compare this real-life story: I was making waffles "from scratch" on a waffle iron. The 7-year-old son of friends who were visiting was watching me and said, "Actually, *you're* not making it; it's the thing [what he was trying to say was that it was *the waffle iron* that was making the waffles]. But you set it up, so you're the cook.")
2. Who (or what) is morally responsible for decisions made, or actions taken, by computers? Is it the computer? Is it the human who accepts the computer's decision? Is it the human who programmed the computer?
3. Suppose a student knows how to use a calculator to add; does that student know how to add?
4. Should artificial intelligences be allowed to kill? Sparrow, 2007 "considers the ethics of the decision to send artificially intelligent robots into war ...." On "artificial morality" and machine ethics for robots, see Anderson and Anderson, 2007, 2010, Wallach and Allen, 2009, Wagner and Arkin, 2011, Bench-Capon, 2020, Misselhorn, 2020.

---

9 See the Online Resources for further reading on these issues.

5. Aref, 2004 suggests (but does not discuss) that supercomputers might make decisions that we could not understand:

> As we construct machines that rival the mental capability of humans, will our analytical skills atrophy? Will we come to rely too much on the ability to do brute-force simulations in a very short time, rather than subject problems to careful analysis? Will we run to the computer before thinking a problem through?…A major challenge for the future of humanity is whether we can also learn to master machines that outperform us mentally.

On the question "will our analytical skills atrophy?" you might enjoy Isaac Asimov's (1957) science-fiction story "The Feeling of Power," which is about a human who rediscovers how to do arithmetic even though all arithmetical problems are handled by computers, and then the computers break down.

6. If all that matters is a decision-making computer's track record, and if its algorithm *cannot* be understood (either because it is too complex or because it is a "black box" algorithm (Section 3.11)), does that mean we have to take its decisions merely on *faith*? Or are there decisions that should not be made by algorithms that are so complex that we cannot understand them?

7. On the other hand, consider these remarks by Daniel Dennett:

> Artifacts already exist … with competences so far superior to any human competence that they will usurp our authority as experts, an authority that has been unquestioned since the dawn of the age of intelligent design. And when we ceded hegemony to these artifacts, it will be for very good reasons, both practical and moral. Already it would be *criminally negligent* for me to embark with passengers on a transatlantic sailboat cruise without equipping the boat with several GPS systems. …
> Would *you* be willing to indulge your favorite doctor in her desire to be an old-fashioned "intuitive" reader of symptoms instead of relying on a computer-based system that had been proven to be a hundred times more reliable at finding rare, low-visibility diagnoses than any specialist? (Dennett, 2017, pp. 400–401)

Would it be *irrational not* to take such decisions or advice on faith? (We'll return to the notion of faith in Section 17.6.2.)

18

Philosophy of Artificial Intelligence[1]

With a large number of programs in existence capable of many kinds of performances that, in humans, we call thinking, and with detailed evidence that the processes some of these programs use parallel closely the observed human processes, we have in hand a clear-cut answer to the mind-body problem: How can matter think and how are brains related to thoughts?
—Herbert Simon (1996a, p. 164)[2]

Announcer at a computer-human checkers match: Are you at all concerned about playing checkers against a computer, Mr. Crankshaft?

Mr. Crankshaft: Nope; the way I see it, it's going to be one checkers playing machine against another.
—https://www.comicskingdom.com/crankshaft/2005-03-02

18.1 Introduction

In this chapter, we will focus on only two main questions in the philosophy of AI: *What is AI? And is AI possible?*[3] For the second question, we will look at Alan Turing's classic 1950 paper on the Turing Test of whether computers can think and at John Searle's 1980 Chinese Room Argument challenging that test.

18.2 What Is AI?

18.2.1 Definitions and Goals of AI

Many definitions of AI have been proposed.[4] We will focus on two nicely contrasting definitions. The first is by Marvin Minsky, one of the pioneers of AI research; the second is by Margaret Boden, one of the pioneers of cognitive science:

---

1 Portions of this chapter are adapted from Rapaport, 2000.

2 Simon's answer is that certain "patternings in matter, in combination with processes that can create and operate upon such patterns" can do the trick (Simon, 1996a, p. 164). For more on patterns, see Hillis, 1998 and Section 9.5, on computers as "magic paper."

3 See the Online Resources for further reading on the philosophy of AI.

4 See http://www.cse.buffalo.edu/~rapaport/definitions.of.ai.html for a sampling.

Philosophy of Computer Science: An Introduction to the Issues and the Literature, First Edition. William J. Rapaport.
© 2023 John Wiley & Sons, Inc. Published 2023 by John Wiley & Sons, Inc.

1. ... *artificial intelligence*, the science of making machines do things that would require intelligence if done by men.[5] (Minsky, 1968, p. v)
2. By "artificial intelligence" I ... mean the use of computer programs and programming techniques to cast light on the principles of intelligence in general and human thought in particular. (Boden, 1977, p. 5)

Minsky's definition suggests that the methodology of AI is to study *humans* in order to learn how to program *computers*. Note that this was Turing's methodology in his 1936 paper; see Section 8.7.9. Boden's definition suggests a methodology that goes in the opposite direction: to study *computers* in order to learn something about *humans*. Turing advocated this approach, as well: "the attempt to make a thinking machine will help us greatly in finding out how we think ourselves" (Turing, 1951a, p. 486). AI is, in fact, a two-way street: Minsky's view of AI as moving from humans to computers and Boden's view of it as moving from computers to humans are both valid.

Both views are also consistent with Stuart C. Shapiro's three goals of AI (Shapiro, 1992a):

AI as advanced CS or engineering: One goal of AI is to extend the frontiers of what we know how to program and to do this *by whatever means will do the job*, not necessarily in an "intelligent" (i.e. cognitive) fashion. My former computer science colleague John Case once told me that AI understood in this way is at the "cutting edge" of CS.[6]

AI as computational psychology: Another goal of AI is to write programs as theories or models of *human cognitive behavior*. (Recall our discussion in Chapter 14.)

AI as computational philosophy: A third goal of AI is to investigate *whether cognition in general* (and not restricted to *human* cognitive behavior) *is computable*: i.e. whether it is (expressible as) one or more recursive functions.

18.2.2 Artificial Intelligence as Computational Cognition

The term 'artificial intelligence' – coined by John McCarthy in 1955 – is somewhat of a misnomer. First, outside of AI, 'intelligence' is often used in the sense of IQ, but AI is not necessarily concerned only with finding programs with high IQ.[7] Echoing Minsky's definition and Shapiro's goals, Herbert Simon said this:

> The basic strategy of AI has always been to seek out progressively more complex human tasks and show how computers can do them, in humanoid ways or by brute force. (Quoted in Hearst and Hirsh, 2000, p. 8.)

The phrase 'human tasks' nicely avoids any issues involved with the notion of "intelligence." But an even more general and accurate term would be 'cognition,' which includes such mental states and processes as belief, consciousness, emotion, language, learning, memory, perception, planning, problem solving, reasoning, representation (including categories, concepts, and mental imagery), sensation, thought, etc.

Second, 'artificial' carries the suggestion that "artificial" entities aren't the real thing. (Recall our discussion in Section 14.2.2.) 'Synthetic' is better than 'artificial,' because an artificial diamond might not be a diamond – it might be a cubic zirconium – whereas a synthetic diamond

---

5 That is, by humans.
6 Another former colleague, Anthony S. Ralston, agreed with Case's topological metaphor, except that instead of describing AI as being at the *cutting* edge, he told me that it was at the "periphery"!
7 See the Online Resources for further reading on AI and IQ.

is a real diamond that just happened to be formed in a non-natural way.[8] But an even better term is 'computational,' which doesn't carry the stigma of "artificiality" and which specifies the nature of the "synthesis."

For these reasons, my preferred name for the field is 'computational cognition.' (Nevertheless, just as I use 'CS' in this book instead of "computer science," I will continue to use 'AI' instead of "computational cognition.") So, AI – understood as computational cognition – is the branch of CS (working with other disciplines, such as cognitive anthropology, linguistics, cognitive neuroscience, philosophy, and psychology, among others) that tries to answer the question **how much of cognition is computable?** The *working assumption* of computational cognition is that **all of cognition is computable**: "The study [of AI] is to proceed on the basis of the conjecture that every aspect of learning or any other feature of intelligence can in principle be so precisely described that a machine can be made to simulate it" (McCarthy et al., 1955).

And its main open research question is, **Are *aspects of cognition that are not yet known to be computable* computable?** If so, what does that tell us about the kinds of things that can produce cognitive behavior? On the other hand, if there are non-computable aspects of cognition, *why* are they non-computable? And what would that tell us about cognition? An answer to this question should take the form of a logical argument such as the one that shows that the Halting Problem is non-computable (Section 7.7). It should not be of the form "All computational methods tried so far have failed to produce this aspect of cognition." After all, there might be a new kind of method that has not yet been tried.

18.3 The Turing Test

> The Turing Test(?): a problem is computable if a computer can convince you it is.
> —Anonymous undergraduate student in the author's course, CSE 111, "Great Ideas in Computer Science" (14 December 2000)[9]

18.3.1 How Computers Can Think

We have seen that AI holds that cognition is comput*able*. For our present purposes, it doesn't matter whether the computations are of the classical, symbolic variety or the connectionist, artificial-neural-network, or machine-learning variety. Nor does it matter whether the neuron firings that produce cognition in the human brain can be viewed as computations. (For further discussion of this, see Piccinini 2007a; Rapaport 2012b; Piccinini and Bahar 2013; Piccinini 2020b.)

All that matters is this philosophical implication: *if (and to the extent that) cognitive states and processes can be expressed as algorithms, then they can be implemented in non-human computers.* And this raises the following questions: (1) Are computers executing such cognitive algorithms merely *simulating* cognitive states and processes? (2) Or are they *actually exhibiting* them? In popular parlance, do such computers think?

In Sections 18.3–18.5, we look at an answer to this question that arises from what is called "the Turing Test". In Sections 18.6–18.8, we will look at an objection to it in the form of the Chinese Room Argument, including an interpretation of the situation that is based on the theory of syntactic semantics introduced in Section 16.10. After that, we will revisit Turing's "strange inversion" from Section 16.7 and make some concluding remarks on the goal of AI.

---

8 See the Online Resources for further reading on this point.
9 http://www.cse.buffalo.edu/~rapaport/111F04.html. Rey, 2012 distinguishes between Turing's thesis and the Turing Test (something the student in my course wasn't clear on!).

18.3.2 The Imitation Game

> **Leo** Computers compute. Brains think. Is the machine thinking?
>
> **Amal** If it's playing chess and you can't tell from the moves if the computer is playing white or black, it's thinking.
>
> **Leo** What it's doing is a lot of binary operations following the rules of its programming.
>
> **Amal** So is a brain.
>
>   …
>
> **Hilary** It's not deep. If that's thinking. An adding machine on speed. A two-way switch with a memory. Why *wouldn't* it play chess? But when it's me to move, is the computer *thoughtful* or is it sitting there like a toaster? It's sitting there like a toaster.
>
> **Leo** So, what would be your idea of deep?
>
> **Hilary** A computer that minds losing.
>
> —Tom Stoppard (2015, Scene 3, pp. 22–23)

Just as Alan Turing's most important paper (Turing, 1936) never mentions a "Turing Machine," his second most important paper – "Computing Machinery and Intelligence" (Turing, 1950) – never mentions a "Turing Test." Instead, he introduces a parlor game that he calls the "Imitation Game." This is a game that you can actually play, not a mere thought experiment.

The Imitation Game consists of three players: a man, a woman, and an interrogator who might be either a man or a woman. It might matter whether the interrogator is a man rather than a woman, or the other way around, but we'll ignore this for now. The interrogator could also be a computer, but there are good reasons why that should be ruled out: the point of the Turing Test is for a *human* to judge whether an entity can think (or whether its cognitive behavior is indistinguishable from that of a human).

The three players are placed in separate rooms so they cannot see each other, and they communicate only by means of what we would now call 'texting' so they cannot hear each other. The reason is that the point of the game is for the interrogator to determine which room has the man and which room has the woman. To make things interesting, the woman is supposed to tell the truth in order to convince the interrogator that she is the woman, but the man is supposed to convince the interrogator that he (the man) is the woman, so he may occasionally have to lie. The man wins if he convinces (fools) the interrogator that he is the woman; the woman wins if she convinces the interrogator that she is the woman.

Turing suggested that "an average interrogator will not have more than 70 per cent. chance of making the right identification after five minutes of questioning" (Turing, 1950, p. 442). The actual amount of time is irrelevant: one could conduct a series of Imitation Games and calculate appropriate statistics on how likely an interrogator is to make a correct determination after a given period of time.

What does this have to do with whether computers can think? What has come to be known as the Turing Test makes one small change to the Imitation Game:

> We now ask the question, "What will happen when a machine takes the part of [the man] in this game?" Will the interrogator decide wrongly as often when the game is played like this as he [or she] does when the game is played between a man and a woman? These questions replace our original, "Can machines think?"
> (Turing, 1950, p. 434)

There is some ambiguity: Is the "machine" (i.e. the computer) supposed to convince the interrogator that it is the *woman*? (That is, is it supposed to imitate a woman?) Or is it supposed to convince the interrogator that it is a *man who is trying to convince the interrogator that he is a woman*? (That is, is it supposed to imitate a man?)

Other modifications are possible. Usually, the Turing Test is taken, more simply and less ambiguously, to consist of a setup in which a computer, a human, and a human interrogator are located in three different rooms, communicating over a texting interface, and in which both the human and the computer are supposed to convince the interrogator that each is a *human*. If the computer convinces the interrogator (under the same criteria for successful convincing that obtains in the original Imitation Game), then the computer is said to have passed the Turing Test. An even simpler version consists merely of two players: a human interrogator and someone or something (a human or a computer) in two separate, text-interfaced rooms. If a computer convinces the interrogator that it is a human, then it passes the Turing Test.

Here is Turing's answer to the question that has now replaced "Can machines think?":

> I believe that at the end of the century [i.e. by the year 2000][10] **the use of words** and **general educated opinion** will have altered so much that one will be able to speak of machines thinking without expecting to be contradicted.
> (Turing, 1950, p. 442, my boldface)

To see what this might mean, we need to consider the Turing Test a bit further.[11]

Digression on "the End of the Century": "The end of the century" (i.e. the twentieth century) has come and gone without Turing's expectations being realized. (If they had been, we would not still be discussing them!) Similar predictions have also been off the mark. Simon and Newell, 1958 predicted that (among other things) a computer would "be the world's chess champion" (p. 7) by 1968. But it didn't happen until 1997, when IBM's Deep Blue beat human chess champion Garry Kasparov (https://en.wikipedia.org/wiki/Deep_Blue_versus_Garry_Kasparov). However, Simon (personal communication, 24 September 1998, https://cse.buffalo.edu/~rapaport/584/S07/simon.txt) said that "it had nothing to do with the Turing Test" and that "(a) I regard the predictions as a highly successful exercise in futurology, and (b) placed in the equivalent position today, I would make them again, and for the same reasons. (Some people never seem to learn.)" See also Simon, 1977, p. 1191, endnote 1. At the end of the next millennium, no doubt, historians looking back will find the 40-year distance between the time of Newell and Simon's prediction and the time of Kasparov's defeat to have been insignificant.

Historical Digression: The Turing Test was not the first test of its kind. Descartes (1637, Part V, p. 116) proposed the following:

> … if there were machines which bore a resemblance to our body and imitated our actions as far as it was morally possible to do so, we should always have two very certain tests by which to recognise that, for all that, they were not real men. The first is, that they could never use speech or other signs as we do when placing our thoughts

(Continued)

---

10 In Turing et al., 1952, he extended this to 2052.
11 See the Online Resources for further reading on the Turing Test.

> **(Continued)**
>
> on record for the benefit of others. For we can easily understand a machine's being constituted so that it can utter words ...; for instance, ... it may ask what we wish to say to it; ... it may exclaim that it is being hurt, and so on. But it never happens that it arranges its speech in various ways, in order to reply appropriately to everything that may be said in its presence, as even the lowest type of man can do. And the second ... is, that although machines can perform certain things as well as or perhaps better than any of us can do, they infallibly fall short in others, by the which means we may discover that they did not act from knowledge, but only from the disposition of their organs.

18.3.3 Thinking vs. "Thinking"

Lots of parts of a computer "think" in different ways, but ... [the CPU] is what we usually call the "thinking" part. It's a machine for quickly following a set of steps that are written down as numbers. *Following steps might not be "thinking." But it's hard to say for sure.* That's one of those things where not only do we not know the answer, we're not sure what the question is.
—Randall Munroe (2015, p. 37, my italics)

In 1993, *The New Yorker* magazine published a cartoon by Peter Steiner showing a dog sitting in front of a computer talking to another dog, the first one saying, "On the Internet, nobody knows you're a dog" https://en.wikipedia.org/wiki/On_the_Internet,_nobody_knows_you%27re_a_dog#/media/File:Internet_dog.jpg. This cartoon's humor arises from the fact that you do *not* know with whom you are communicating via computer! It's unlikely that there's a dog typing away at the other end of a texting session or an email, but could it be a computer pretending to be a human, as in the Turing Test? Or could it be a 30-year-old pedophile pretending to be a 13-year-old classmate?

In the years since that cartoon appeared, we have become only too aware of the possibilities and dangers – political and otherwise – of messages and "fake news" on Facebook and elsewhere that purport to come from one source but really come from another, as well as the possibilities and dangers of the lack of privacy. A newer "internet dog" cartoon plays on this, in which the dog who was formerly at the computer says to the other one, "Remember when, on the Internet, nobody knew who you were?" (https://condenaststore.com/featured/two-dogs-speak-as-their-owner-uses-the-computer-kaamran-hafeez.html).

Normally, we *assume* that we are talking to people who really are whom they say they are. In particular, we assume that we are talking to a human. But really, all we know is that we are talking to *an entity with human cognitive capacities*. And that, I think, is precisely Turing's point: an entity with human cognitive capacities is all we can ever be sure of, whether that entity is really a human or "merely" a computer.

This is a version of what philosophers have called "the argument from analogy for the existence of other minds." An *argument from analogy* is an argument of the form

1. Entity A is like (i.e. is analogous to) entity B with respect to important properties $P_1, \ldots, P_n$.
2. B has another property, Q.
3. $\therefore$ (Probably) A also has property Q.

Compare the "duck test": "When I see a bird that walks like a duck and swims like a duck and quacks like a duck, I call that bird a duck" (James Whitcomb Riley, https://en.wikipedia.org/wiki/Duck_test). Such an argument is not *deductively* valid: it's quite possible for the premises to be true but for the conclusion to be false. But it has some *inductive* strength: the more alike two objects are in many respects, the more likely it is that they will be alike in many other respects (and maybe even all respects).

The *problem of the existence of other minds* is this: I know that *I* have a mind (because I know what it means for me to think, to perceive, to solve problems, etc.). How do I know whether *you* have a mind? Maybe you don't; maybe you're just some kind of computer, or android, or philosophical zombie.

Digression on Androids and Zombies: Androids are robots that look like humans, such as Commander Data in *Star Trek: The Next Generation* or many of the characters in such science fiction as Dick, 1968 or the film *Blade Runner*. A philosophical zombie is not a horror-movie zombie. Rather, it is an entity who is exactly like us in all respects but who lacks a mind or consciousness. See Kirk, 1974, Chalmers, 1996; and other references at http://www.cse.buffalo.edu/~rapaport/719/csnessrdgs.html#zombies.

Putting these together, here is the argument from analogy for the existence of other minds:

1. You are like me with respect to all of our physical and behavioral properties.
2. I have a mind.
   (Or: my behavioral properties can best be explained by the fact that I have a mind.)
3. ∴ (Probably) you have a mind.
   (Or: your behavioral properties can best be explained if it is assumed that you also have a mind.)

Of course, this argument is also deductively *invalid*. I could be wrong about whether you are biologically human. In that case, the best explanation of your behavior might not be that you have a mind, but that you are a computer who has been cleverly and suitably programmed. Now, there are two ways to understand this: one way to understand it is to say that you *don't have a mind*; you're just a cleverly programmed robot. But another way to understand it is to say that being cleverly programmed in that way is exactly *what it means to have a mind*: perhaps we are both cleverly programmed in that way. Or perhaps (a) *you* are *programmed* in that way, whereas (b) *I* have a *brain* that behaves in that way, but (c) these are simply two different *implementations* of "having a mind."

In either case, am I wrong about your being able to think? That is, am I wrong about your (human) cognitive abilities? Turing's answer is, no! More cautiously, perhaps, his answer is that whether I'm wrong depends on how we characterize (human) cognitive abilities (or thinking).

If human-like cognition *requires* a (human) brain, and you lack one, then, technically speaking, you don't have human-like cognition (even if you pass the Turing Test). On this view, *I* really do think, but *you* can only "think." That is, you are not really thinking but doing something else that can be called "thinking" only in a metaphorical sense.

But if human-like cognition is an *abstraction* that can be implemented in different ways – i.e. if it does not require a (human) brain – then we both have human-like cognition (and that's why you pass the Test). On this view, we both can think.

Here's an analogy: everyone can agree that birds fly.[12] Do *people* fly? Well, we certainly speak as if they do; we say things like, "I flew from Buffalo to JFK last week." But we also know that I don't

---

12 "Birds fly" is true in general, even though most birds actually don't fly! Not only do penguins, ostriches, etc., not fly, but baby birds, birds with injured wings, dead birds, etc., also don't fly. Handling the logic of statements like this is a branch of logic and AI called "non-monotonic reasoning"; see Section 2.5.1.

literally mean I flapped my arms when flying from Buffalo to JFK; rather, I flew in an airplane – it wasn't me who was flying; it was the airplane. But that answer raises another question: do *planes* fly? Well, they don't flap their wings, either! So in what sense are they flying?

There are two ways to understand what it means to say that planes fly. One way is by what I will call "metaphorical extension." The reason we say that planes fly is that what they are doing is very much like what birds do when they fly – they move through the air, even if their method of doing so is different. But instead of using a *simile*, saying that planes move through the air *like* birds fly, we use a *metaphor*, saying directly that planes fly. And then that metaphor becomes "frozen"; it becomes a legitimate part of our language, so much so that we no longer realize it is metaphorical. This is just like what happens when we say that time is money: we say things like, "You're *wasting* time," "This will *save* you time," "How did you *spend* your vacation?" and so on. But we're usually not aware that we are speaking metaphorically (until someone points it out), and there's often no other (convenient) way to express the same ideas (Lakoff and Johnson, 1980a,b).

As Turing said, "the use of words" has changed!

The other way to understand what it means to say that planes fly is that we have realized flapping wings are not essential to flying. There are deeper similarities between what birds and planes do when they move through the air that have nothing to do with wing-flapping but that have everything to do with the *shape* of wings and, more generally, with the physics of flight. We have developed a more abstract, and therefore more general, theory of flight, one that applies to both birds and planes. And so we can "promote" the verb 'to fly' from its use solely for birds (and other flying animals) to a more general use that also applies to planes. To use the language of Section 13.1.4, the abstract notion of flying can be implemented in both biological and non-biological media.

As Turing said, "general educated opinion" has changed!

In fact, *both* the use of words *and* general educated opinion have changed. Perhaps the change in one facilitated the change in the other; perhaps the abstract, general theory *can account for* the metaphorical extension.[13]

The same thing has happened with 'computer.' As we saw in Section 6.1, a computer was originally a *human* who computed. That was the case until about the 1950s, but a half-century later, we now say that a computer is a *machine*. Before around 1950, what we now call 'computers' had to be called 'digital' or 'electronic computers' to distinguish them from the human kind. But now it is very confusing to read pre-1950 papers without thinking of the word 'computer' as meaning, by default, a non-human machine. (Recall the puzzling statement in Turing, 1936, p. 250: "The behaviour of the *computer* at any moment is determined by the symbols which *he* is observing, and *his* 'state of mind' at that moment"; see Section 8.7.3.) Now, at the beginning of the twenty-first century, general educated opinion holds that computers are best viewed abstractly, in functional, input-output terms. The study of "artificial intelligence" may lead us to understanding thinking as an abstraction that can be implemented in both humans and computers, just as the study of "artificial" flight (https://invention.psychology.msstate.edu/library/Magazines/Nat_Artificial.html) was crucial to understanding flying as an abstraction implementable in both birds and planes: "[S]tudying the animals that fly," no matter in how great detail and for how many years, would not have yielded any useful information on how humans might be able to fly. Rather,

> attempt[ing] to construct devices that fly … attempts to build flying machines [resulted in] our entire understanding of flight today. Even if one's aim is to understand how birds or insects fly, one will look to aeronautics for the key principles ….
> (Quillian, 1994, pp. 440–442)[14]

---

13 See the Online Resources for further reading on this point.
14 Quillian – a pioneer in AI research – uses this argument to support an explanation of why the natural sciences are more "effective" than the social sciences.

This is consistent with Boden's view that the study of computational theories of cognition can help us understand human (and, more generally, non-human) cognition.

What does this have to do with the philosophy of AI? The strategy of abstracting from a naturally occurring example and re-implementing it computationally also applies to cognition:

> Quite typically, an abstract structure underlies some human cognitive activity that is not at all apparent in superficial phenomenology or practice. Often, that structure is related in interesting ways to the structures we would invent if we constructed an ideal machine to perform that cognitive activity. (We might think of artificial intelligence as a normative enterprise). But that structure is rarely identical to the ideal machine's structure. (Alison Gopnik 1996, p. 489)

The underlying abstract structure could be computational in nature. Hence, it could be (re-)implemented in "an ideal machine." The abstract computational theory might be thought of as having this form: such-and-such a human cognitive activity can be performed in this computational way *even if* the way humans in fact do it is not identical to that ideal structure. Gopnik goes on to say,

> This process may seem like analogy or metaphor, but it involves more serious conceptual changes. It is not simply that the new idea is the old idea applied to a new domain, but that the earlier idea is itself modified to fit its role in the new theory. (Alison Gopnik 1996, p. 498)

To say, as I did earlier, that the process is metaphorical is not inconsistent with it also involving "more serious conceptual changes": we come to see the old idea in a new way.

But some philosophers argue that what AI computers do is not *really* thinking. We'll turn to one of these philosophers in Section 18.6. But first we'll digress to consider two more issues related to Turing's views.

Exercise for the Reader: Replace the word 'existence' with the word 'thinking' in the following passage:

> When the dust has settled, we may find that the very notion of existence, the old one, has had its day. A kindred notion may then stand forth that seems sufficiently akin to warrant application of the same word; such is the way of terminology. Whether to say at that point that we have gained new insight into existence [cf. "general educated opinion"], or that we have outgrown the notion and reapplied the term [cf. "the use of words"], is a question of terminology as well. (Quine, 1990, p. 36)

18.4 Digression: The "Lovelace Objection"

Turing (1950, Section 6) considered several objections to the possibility of AI, one of which he called "Lady Lovelace's Objection." Here it is in full, in Lovelace's own words:

> The Analytical Engine has no pretensions whatever to *originate* anything. **It can do whatever we *know how to order it* to perform.** It can *follow* analysis; but it has no power of

anticipating any analytical relations or truths. Its province is to assist us in making *available* what we are already acquainted with.
(Menabrea and Lovelace, 1843, p. 722, italics in original, my boldface;
https://psychclassics.yorku.ca/Lovelace/lovelace.htm#G)

The first thing to note about this is that it is often misquoted. We saw Herbert Simon do this in Section 11.8.4, when he expressed it in the form "computers can *only* do what you program them to do" (Simon, 1977, p. 1187, my italics). Lovelace did not use the word 'only'. We'll see one reason in a moment. But note that this standard interpretation of her phrasing does seem to be what Turing had in mind. He quotes with approval Hartree, 1949, p. 70 – the same book we saw Arthur Samuel quoting in Section 9.1, by the way – who said, concerning Lovelace's comment, "This does not imply that it may not be possible to construct electronic equipment which will 'think for itself' …." Minus the double negative, Hartree (and Turing) are saying that Lovelace's comment is consistent with the possibility of an AI computer passing the Turing Test. Turing goes on to say this:

> A better variant of the objection says that a machine can never "take us by surprise." This statement is a more direct challenge and can be met directly. Machines take me by surprise with great frequency. (Turing, 1950, p. 450)

It's worth observing that the fact that the Analytical Engine (or any contemporary computer, for that matter) has no "pretensions" simply means it wasn't designed that way; nevertheless, it might still be able to "originate" things. Also, if we can find out "how to order it to perform" cognitive activities, then it can do them! Finding out how requires us to be conscious of something we ordinarily do unconsciously (Section 17.6.2). In his own commentary on the Lovelace objection, Samuel (1953, p. 1225) said,

> Regardless of what one calls the work of a digital computer [specifically, regardless of whether one says that it can think], the unfortunate fact remains that more rather than less human thinking is required to solve a problem using a present day machine since every possible contingency which might arise during the course of the computation must be thought through in advance. The jocular advice recently published to the effect, "Don't Think! Let UNIVAC do it for you," cannot be taken seriously. Perhaps, if IBM's familiar motto [namely, "Think!"] needs amending, it should be "Think: Think harder when you use the 'ULTIMAC'."

(Samuel adds in a footnote to this passage that 'ULTIMAC' is "A coined term for the 'Ultimate in Automatic Computers.' The reader may, if he [sic] prefers, insert any name he likes selected from the following partial list of existing machines …," and he then listed 43 of them, including Edvac, Illiac, Johnniac, and Univac.)

Why didn't Lovelace use the word 'only'? Recall from Section 6.4.3 that Babbage, inspired by de Prony, wanted his machines to replace human computers:

> … Babbage deplored the waste of brilliant, educated men in routine, boring drudgery, for which he claimed the uneducated were better suited …. When convenient, however, he saw no obstacle to replacing them by yet more accurate or efficient machinery (he disapproved of unions). (Stein, 1984, pp. 51–52)

It is in this context that Lovelace "rephrased Babbage's words of assurance for the men of Prony's first section" (p. 52). These were to be "the most eminent mathematicians in France, charged with deciding which formulae would be best for use in the step-by-step calculation of the functions to be tabulated. (They performed the programmer's task.)" (p. 51). These "eminent" mathematical "men of the first section" – and they *were* men – needed to be assured that the drudge work could be handled by a machine, and hence Lovelace's words: "The Analytical Engine has no pretensions whatever to originate anything. It can do whatever we know how to order it to perform." This puts a positive spin on a sentence that has typically been understood negatively: the computer *can* do whatever we can program it to do (and not the computer can "only" do whatever we program it to do).

18.5 Digression: Turing on Intelligent Machinery

> This [probably the Manchester Mark 1 computer] is only a foretaste of what is to come, and only the shadow of what is going to be. We have to have some experience with the machine before we really know its capabilities. It may take years before we settle down to the new possibilities, but I do not see why it should not enter any one of the fields normally covered by the human intellect, and eventually compete on equal terms.
> —Alan Turing, 1949 (https://quoteinvestigator.com/2019/10/12/ai-shadow/)

Turing 1950 was not Turing's only essay on AI. An earlier one was Turing, 1947. And in an essay written in 1951, Turing seems to come out a bit more strongly about the possibility of computers thinking:

> 'You cannot make a machine to think for you.' This is … usually accepted without question. It will be the purpose of this paper to question it. (Turing, 1951b, p. 256)

Although it is possible to read that last sentence neutrally, to my ears it sounds like a challenge strongly suggesting that Turing thinks you *can* make a machine think. Indeed, later he says that his "contention is that machines can be constructed which will simulate the behaviour of the human mind very closely" (p. 257). This is cautiously worded – is *simulation* of thinking (i.e. simulation of "the behavior of the human mind") the same as "real" thinking? – but his ultimate claim here is that it will come so close to human thinking as to make no difference: "on the whole the output of them [i.e. of such "thinking" machines] will be worth attention *to the same sort of extent as the output of a human mind*" (p. 257, my italics). And how would this be proved? By the Turing Test: "It would be the actual reaction of the machine to circumstances that would prove my contention, if indeed it can be proved at all" (p. 257).

Turing also suggests that the algorithm for such a machine must be based on what is now called 'machine learning': "If the machine were able in some way to 'learn by experience' it would be much more impressive" (p. 257). (Although not everyone thinks machine learning is "really" learning (Bringsjord et al., 2018).) Moreover, he also suggests that the machine should be an oracle machine (Section 11.9):

> There is … one feature that I would like to suggest should be incorporated in the machines, and that is a 'random element.' Each machine should be supplied with a tape bearing a random series of figures, *e.g.*, 0 and 1 in equal quantities, and this series of figures should be used in the choices made by the machine. (p. 259)

Note, however, that Turing seems to consider these to be a (small) extension of Turing Machines.

Also interesting is his anticipation of what is now called "The Singularity" (see Section 19.7), and the question that we will return to in Chapter 19 about whether we *should* build artificial intelligences:

> Let us now assume, for the sake of argument, that these machines are a genuine possibility, and look at the consequences of constructing them. To do so would of course meet with great opposition, unless we have advanced greatly in religious toleration from the days of Galileo. There would be great opposition from the intellectuals who were afraid of being put out of a job. … it seems probable that once the machine thinking method had started, *it would not take long to outstrip our feeble powers*. There would be no question of the machines dying, and they would be able to converse with each other to sharpen their wits. At some stage therefore we should have to expect the machines to take control ….
> (pp. 259–260, my italics)

18.6   The Chinese Room Argument

> If a Martian could learn to speak a human language, or a robot be devised to behave in just the ways that are essential to a language-speaker, an implicit knowledge of the correct theory of meaning for the language could be attributed to the Martian or the robot with as much right as to a human speaker, even though their internal mechanisms were entirely different.
> —Michael Dummett (1976, p. 70)

> [R]esearchers … tracked … unresponsive patients …, taking EEG recordings …. During each EEG recording, the researchers gave the patients instructions through headphones. … "Somewhat to our surprise, we found that about 15 percent of patients who were not responding at all had … brain activation in response to the commands," said Dr. Jan Claassen …. "It suggests that there's some remnant of consciousness there. *However, we don't know if the patients really understood what we were saying. We only know the brain reacted.*"
> —Benedict Carey (2019, my italics)

Thirty years after Turing's publication of the Turing Test, John Searle published a thought experiment called the Chinese Room Argument (Searle, 1980, 1982, 1984). In this experiment, a human who knows no Chinese (John Searle himself, as it happens) is placed in a room (the "Chinese Room") along with paper, pencils, and a book containing an English-language algorithm for manipulating certain "squiggles" (marks or symbols that are meaningless to Searle-in-the-room).

> **Terminological Digression:** I distinguish between (1) the real John Searle who is a philosopher and author of Searle 1980 and (2) the Chineseless "John Searle" who occupies the Chinese Room. I refer to the former as 'Searle' and to the latter as 'Searle-in-the-room.'

Outside the room is a native speaker of Chinese. There is something like a mail slot in one wall of the Chinese Room. Through that slot, the native speaker inputs pieces of paper that contain a text written in Chinese along with reading-comprehension questions about that text, also in Chinese. When Searle-in-the-room gets these pieces of paper – which, from his point of view, contain nothing but apparently meaningless squiggles – he consults his book and follows its instructions. Those instructions tell him to manipulate the symbols in certain ways, to write certain symbols down on a clean piece of paper, and to output those "responses" through the mail slot. The native speaker who reads them determines that whoever (or whatever) is in the room has answered all the questions correctly in Chinese, demonstrating a fluent understanding of Chinese. This is because the rule book of instructions is a complete natural-language-understanding algorithm for Chinese. But by hypothesis, Searle-in-the-room does not understand Chinese. We seem to have a contradiction.

The Chinese Room Argument (CRA) can be seen as a counterexample to the Turing Test, concluding that it is possible to pass a Turing Test yet not really think. The setup of the CRA is identical to the simplified, two-player version of the Turing Test: the interrogator is the native Chinese speaker who has to decide whether the entity in the room understands Chinese. The interrogator determines that the entity in the room *does* understand Chinese. This is analogous to deciding that the entity in the simplified Turing Test is a human, rather than a computer. But the entity in fact does *not* understand Chinese. This is analogous to the entity in the simplified Turing Test being a computer. So, Searle-in-the-room passes the Turing Test without being able to "really" understand; hence, the test fails.

Or does it?

Searle actually bases two arguments on the Chinese Room thought experiment:

The Argument from Biology:

> B1. Computer programs are non-biological.
> B2. Cognition is biological.
> B3. ∴ No (non-biological) computer program can exhibit (biological) cognition.

The Argument from Semantics:

> S1. Computer programs are purely syntactic.
> S2. Cognition is semantic.
> S3. Syntax alone is not sufficient for semantics.
> S4. ∴ No (purely syntactic) computer program can exhibit (semantic) cognition.

The principal objection to the Argument from Biology is that premise B2 is at least misleading and probably false: cognition can be characterized abstractly and implemented in different media. The principal objection to the Argument from Semantics is that premise S3 is false: syntax – i.e. symbol manipulation – does suffice for semantics.

After investigating these objections (and others), we will consider whether there are other approaches that can be taken to circumvent the CRA. One of them is to try to build a real analogue of a Chinese Room; to do that, we will need to answer the question of what is needed for natural-language understanding.[15]

Historical Antecedents: Although Searle's CRA is the most famous version of this kind of setup, there are earlier ones. In 1959, the logician Hartley Rogers, Jr., wrote

> Consider a box B inside of which we have a man L with a desk, pencils and paper. On one side B has two slots, marked *input* and *output*. If we write a number on paper and pass it through the input slot, L takes it and begins performing certain computations. If and

(Continued)

---

15 See the Online Resources for further reading on the CRA.

> **(Continued)**
>
> when he finishes, he writes down a number obtained from the computation and passes it back to us through the output slot. Assume further that L has with him explicit deterministic instructions of finite length as to how the computation is to be done. We refer to these instructions as P. Finally, assume that the supply of paper is inexhaustible, and that B can be enlarged in size so that an arbitrarily large amount of paper work can be stored in it in the course of any single computation. ... I think we had better assume, too, that L himself is inexhaustible, since we do not care how long it takes for an output to appear, provided that it does eventually appear after a finite amount of computation. We refer to the system B-L-P as M. ... In the approach of Turing, the symbolism and specifications are such that the entire B-L-P system can be viewed as a digital computer .... Roughly, to use modern computing terms, L becomes the logical component of the computer, and P becomes its program. In Turing's approach, the entire system M is hence called a *Turing machine*. (Rogers, 1959, pp. 115, 117)
>
> An even earlier version was in a 1954 episode of *I Love Lucy*, which we'll discuss in Section 18.8.4. See the Online Resources for another antecedent.

18.7 The Argument from Biology

18.7.1 Causal Powers

Let's begin by considering some of the things Searle says about the CRA, beginning with two claims that are versions of premise S1 of the Argument from Semantics:

> I [i.e. Searle-in-the-room] still don't understand a word of Chinese and neither does any other digital computer because all the computer has is what I have: a formal program that attaches no meaning, interpretation, or content to any of the symbols. What this simple argument shows is that no formal program by itself is sufficient for understanding ....
> (Searle, 1982, p. 5)

Note that this allows for the possibility that a program that *did* "attach" meaning, etc., to the symbols *might* understand. But Searle denies that, too:

> I see no reason in principle why we couldn't give a machine the capacity to understand English or Chinese, since in an important sense our bodies with our brains are precisely such machines. But ... we could not give such a thing to a machine where the operation of the machine is defined solely in terms of computational processes over formally defined elements .... (Searle, 1980, p. 422)

Why not? Because "only something that has the same causal powers as brains can have intentionality" (Searle, 1980, p. 423). By 'intentionality' here, Searle means "cognition" more generally. So he is saying that if something exhibits cognition, then it must have "the same causal powers as brains."

All right; what are these causal powers? After all, if they turn out to be something that can be computationally implemented, then computers can have them (which Searle thinks they cannot). So, what does he say they are? He says these causal powers are due to the fact that "I am a certain sort of organism with a certain biological (i.e. chemical and physical) *structure*" (Searle, 1980, p. 422, my italics). That narrows down the nature of these causal powers a little. If we could figure out what this biological *structure* is, and if we could figure out how to implement that structure computationally, then we should be able to get computers to understand. Admittedly, those are big "if"s, but they are worth trying to satisfy.

So, what is this biological structure? Before we see what Searle says about it, let's think for a moment about what a "structure" is. What is the "structure" of the brain? One plausible answer is that the brain is a network of neurons, and the way those neurons are organized is its "structure." Presumably, if you made a model of the brain using string to model the neurons, then if the strings were arranged in the same way that the neurons were, we could say that the model had the same "structure" as the brain. Of course, string is static (it doesn't do anything), and neurons are dynamic, so structure alone won't suffice, but it's a start.

However, Searle doesn't think even structure *plus* the ability to do something is enough: he says a simulated human brain "made entirely of … millions (or billions) of old beer cans that are rigged up to levers and powered by windmills" would not really exhibit cognition even though it appeared to (Searle, 1982). Cognition must (also) be *biological*, according to Searle. That is, it must be made of the right stuff.

> **Some Other "Wrong" Stuff:** Weizenbaum 1976, Ch. 2, considers a Turing Machine made of toilet paper and pebbles. Weizenbaum 1976, Ch. 5, considers computers "made of bailing wire, chewing gum, and adhesive tape." There is even a real computer made from Tinker Toys (https://www.computerhistory.org/collections/catalog/X39.81)! And recall our discussion in Section 8.8.1 of Hilbert's tables, chairs, and beer mugs.

But now consider what Searle is saying: only biological systems have the requisite causal properties to produce cognition. So we're back at our first question: what are those causal properties? According to Searle, they are the ones that are "causally capable of producing perception, action, understanding, learning, and other intentional [i.e. cognitive] phenomena" (Searle, 1980, p. 422). Again: what are the causal properties that produce cognition? They are the ones that produce cognition! That's not a very helpful answer.

Elsewhere, Searle does say some things that give a possible clue as to what the causal powers are: "mental states are both *caused by* the operations of the brain and *realized in* the structure of the brain" (Searle, 1983, p. 265). In other words, they are *implemented* in the brain. And this suggests a way to avoid Searle's argument from biology.

18.7.2 The Implementation Counterargument

> [M]ental states are as real as any *other* biological phenomena, as real as lactation, photosynthesis, mitosis, or digestion. Like these other phenomena, mental states are caused by biological phenomena and in turn cause other biological phenomena. (Searle, 1983, p. 264, my italics)

Searle's "mental states" are biological *implementations*. But if they are implementations, then they must be implementations of something else that is more abstract: abstract mental states. (This follows from Section 13.1.4's Implementation Principle I.)

Searle (1980, p. 451, my italics) says that "… intentional states … are *both caused by and realized in* the *structure* of the brain." But brains and contraptions made from beer-cans + levers + windmills can *share* structure. This is a simple fact about the nature of *structure*. Therefore, what Searle said must be false: it can't be a *single* thing – an intentional (i.e. mental) state – that is *both* caused by *and* realized in the brain. Rather, what the brain *causes* are *implemented* mental states, but what the brain *realizes* are *abstract* mental states, and the abstraction and its implementation are two distinct things.

In Section 13.1.2, we saw that abstractions can be implemented in more than one way – they can be "multiply realized." (This was Section 13.1.4's Implementation Principle II.) We saw that stacks can be implemented as arrays or as lists, that any sequence of items that satisfy Peano's axioms is an implementation of the natural numbers, that any two performances of the same play or music are different implementations of the script or score, and so on.

So, Searle says the human brain can understand Chinese because understanding is biological, whereas a computer executing a Chinese natural-language-understanding program *cannot* understand Chinese because it is *not* biological. But the implementation counterargument says that on an abstract, functional, computational notion of understanding as an abstraction, understanding can be implemented in both human brains and computers, and therefore both can understand.

More generally, if we put Implementation Principles I and II together, we can see that if we begin with an implementation (say, real, biological mental states), we can develop an abstract theory about them. This is what AI and computational cognitive science try to do, following Minsky's methodology. But once we have an abstract theory, we can re-implement it in a different medium. If our abstract theory is computable, then we can re-implement it in a computer. When this happens, our use of words changes, because general educated opinion changes, as Turing predicted. And this is consistent with Boden's methodology: we can learn something about human cognition by studying computer cognition.

Digression on Flying and Computers: This transition from an implementation in one medium "up" to an abstraction and then "down" to another implementation in a different medium is what happened with flying. We began with an implementation (birds and other animals that fly). We then developed an abstract theory (the physics of flight). And this was then re-implemented in the medium of airplanes. It also happened with computers. We began with an implementation (humans who compute). Turing (1936) then developed an abstract theory (the mathematical theory of Turing Machine computation). And that was then re-implemented in electronic, digital computers.

18.8 The Argument from Semantics

18.8.1 The Premises

Premise S1 says that computer programs merely tell a computer to (or describe how a computer can) manipulate symbols solely on the basis of their properties as marks and their relations among themselves. This manipulation is completely independent of the symbols' semantic relations: i.e. of the relations the symbols have to other items that are external to the computer. These external items are the meanings or interpretations of the symbols: the "aspects" of the real world that the symbols represent (Lewis, 1970, p. 19). Note, by the way, that insofar as a computer *did* manipulate its internal symbols in a way that was causally dependent on such external items, it could only do so by inputting an internal representative of that external item. But in that case, it would still be directly manipulating only internal symbols and only indirectly dealing with the external item.

Premise S2 says that cognition is centrally concerned with such "external" relations. Cognition, roughly speaking, is whatever the brain does with the sensory inputs from the external world. To fully understand cognition, according to this premise, it is necessary to understand the internal workings of the brain, the external world, *and* the relations between them. That is a semantic enterprise.

It *seems* clear that the study of relations among the symbols alone could not possibly suffice to account for the relations between those symbols and anything else. Hence premise S3: syntax and semantics are two different, although related, subjects.

Conclusion S4 seems to follow validly. So, any questions about the goodness of the argument must concern its soundness: are the premises true? Doubts have been raised about each of them.

18.8.2 Which Premise Is at Fault?

Let's look at S1 first. Although the World Wide Web is not a computer program, it is generally considered a syntactic object: a collection of nodes (e.g. websites) related by links; i.e. its mathematical structure is that of a graph. Some researchers have felt that there are limitations to this "syntactic" web and have proposed the Semantic Web (Berners-Lee et al., 2001). By "attaching meanings" to websites (as Searle might say), they hope to make the Web more … well … meaningful, more useful. In fact, however, the way they do this is by adding more syntax! (See Rapaport, 2012b, Section 3.2 and note 25.) So, for now, we'll accept premise S1. (For arguments that S1 is *false*, recall our discussion in Section 16.9.)

Next, let's look at S2: recall from the Digression in Section 11.8.4 that at least one major philosopher, Jerry Fodor (1980), has argued that the study of cognition need *not* involve the study of the external world that is being cognized, on the grounds that cognition is what takes place internally to the brain. Whether the brain correctly or incorrectly interprets its input from the external world, it's the interpretation that matters, not the actual state of the external world. This view ("methodological solipsism") holds that as a methodology for studying cognition, we can pretend that the external world doesn't exist; we only have to investigate what the brain does with the inputs it gets, not where those inputs come from or what they are really like. (We'll return to this in Section 18.8.4.)

Of course, if understanding cognition only depends on the workings of the brain and not on its relations with the external world, then the study of cognition might be purely syntactic. And so we're ready to consider premise S3. Can we somehow get semantics from syntax? There are three interrelated reasons for thinking we can.

First, we can try to show that semantics, which is the study of relations *between* symbols and meanings, can be turned into a syntactic study, a study of relations *among* symbols and "symbolized" meanings (see Section 18.8.3). Second, we can take the methodologically solipsistic approach and argue that an internal, "narrow," first-person point of view is (all that is) needed for understanding or modeling cognition (see Section 18.8.4). Third, it can be argued that semantics is recursive in the sense that we understand a *syntactic* domain in terms of an antecedently understood *semantic* domain, but that there must be a base case, and that this base case is a case of syntactic understanding (see Section 18.8.5).

Before looking at each of these, remember that Searle claims that syntax cannot suffice for semantics because the former is missing the links to the external world. This kind of claim relies on two assumptions, both of which are faulty. First, Searle is assuming that computers have no links to the external world: that they are really (and not just methodologically) solipsistic. But this is obviously not true and is certainly inconsistent with Brian Cantwell Smith's (1985) point that even if computers only deal with an (internal) *model* of the real world, they have to *act* in the real world (Section 15.6).

Second, Searle assumes that external links are really needed to attach meanings to symbols. But if so, then why couldn't computers have them just as well as humans do? Both humans and computers exist and act in the world. If we humans have the appropriate links, what reason (other than the faulty Argument from Biology) is there to think computers could not?

18.8.3 Semiotics

The first reason for thinking syntax might suffice for semantics comes from semiotics, the study of signs and symbols. According to one major semiotician, Charles Morris (1938), semiotics has three branches: syntax, semantics, and pragmatics.

Given a formal system of "marks" (symbols without meanings), *syntax* is the study of the properties of the marks and of the relations *among* them: how to recognize, identify, and construct them (in other words, what they look like, e.g. their grammar); and how to manipulate them (e.g. their proof theory). Importantly, syntax does not study any relations *between* the marks and anything else. (Recall our discussions of formal systems [Section 13.2.2] and of symbols, marks, and meanings [Section 16.11.2].)

Semantics is the study of relations *between* the marks and their "meanings." Meanings are part of a different domain of semantic interpretations (recall our discussion of this in Section 13.2.2). Therefore, syntax cannot and does not suffice for semantics! (Or so it would seem.)

Pragmatics has been variously characterized as the study of relations among marks, meanings, *and* the cognitive agents that interpret them; or as the study of relations among marks, meanings, interpreters, *and contexts*. Some philosophers have suggested that pragmatics is the study of everything that is interesting about symbols systems that isn't covered under syntax or semantics! For our purposes, however, we only need to consider syntax and semantics.

Syntax studies the properties of, and relations among, the elements of a single set of objects (which we are calling "marks"); for convenience, call this set SYN. Semantics studies the relations between the members of two sets: the set SYN of marks, and a set SEM of "meanings." Now, take the set-theoretical union of these two sets – the set of marks and the set of meanings: SYNSEM = SYN ∪ SEM. Consider SYNSEM as a new set of marks. We have now "internalized" the previously external meanings into a new symbol system (recall Section 16.10.4). And the study of the properties of, and the relations among, the members of SYNSEM is SYNSEM's syntax! In other words, what was formerly semantics (i.e. relations *between* the marks in SYN and their meanings in SEM) is now syntax (i.e. relations *among* the new marks in SYNSEM.) This is how syntax can suffice for semantics (Rapaport, 2017b).

This can be made clearer with the diagram in Figure 18.1. The top picture of a set of marks ("SYNtactic DOMain") shows two of its members and a relation between them. Imagine that there are many members, each with several properties, and many with relations between them. The study of this set, its members, their properties, and the relations they have to each other is the syntax of SYN.

Now consider the middle picture: two sets, SYN and a set of "meanings" ("SEMantic DOMain"). SYN, of course, has its syntax. But so does SEM. (Often, in AI, the syntax of a semantic domain is called its "ontology.") But now there are additional relations between (some or all of) the members of SYN and (some or all of) the members of SEM. Note that these relations are "external" to both domains: you really can't describe these relations using only the language used to describe SYN or only the language used to describe SEM. Instead, you need a language that can talk about *both* domains and hence cannot be "internal" to either domain. The study of these relations is what is called "semantics." The usual idea is that the members of SEM are the "meanings" of the members of SYN, especially if SYN is the language used to describe SEM. So, for instance, you might think

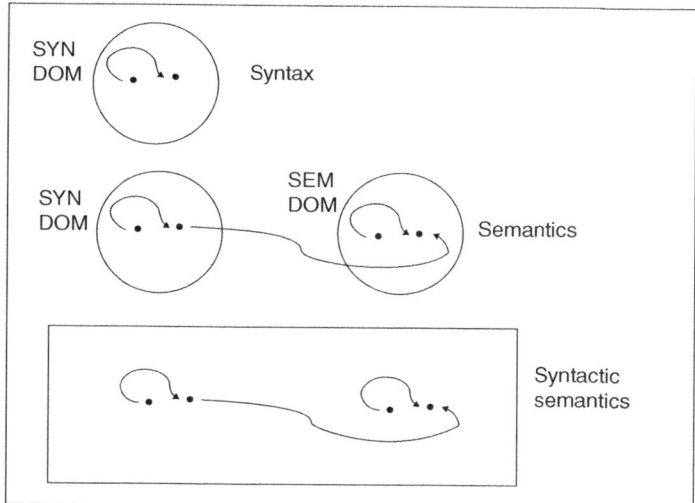

Figure 18.1 Syntax, semantics, and syntactic semantics. Source: Author's drawing.

of SEM as the actual world and SYN as either a language like English that talks about the actual world or a scientific theory about the actual world, perhaps expressed in some mathematical (or computational!) language. Another way to think about this is that SEM gives us the facilities needed to understand SYN: we understand SYN in terms of SEM.

In the bottom picture, we have taken the union of these two domains. Now, the formerly "external" semantic relations have become *internal* relations of the new, unioned domain. But as such, they are now no different in principle from the previous internal, syntactic relations of SYN or the previous internal, syntactic (or ontological) relations of SEM. Thus, these previous *semantic* relations have also become *syntactic* ones. This is what we called "syntactic semantics" in Section 16.10.[16]

This way of viewing semantics as a kind of syntax raises a number of questions: *Can* the semantic domain be internalized? Yes, under the conditions obtaining for human language understanding. How *do* we learn the meaning of a word? How, for instance, do I learn that the word 'tree' means "tree"? A common view is that this relation is learned by associating real trees with the word 'tree.'

> **Digression:** Obviously, this is only the case for some words. Logical words ('the,' 'and,' etc.), words for abstract concepts ('love'), and words for things that don't exist ('unicorn') are learned by different means. And most children raised in large cities learn (somehow) the meanings of words like 'cow' or 'rabbit' from *pictures* of cows and rabbits, long before they see real ones!

But really what happens is that *my internal representation of* an actual *tree* in the external world is associated with *my internal representation of* the word 'tree.' Those internal representations could be certain sets of neuron firings. In whatever way that neurons are bound together when, for instance, we perceive a pink cube (perhaps with shape neurons firing simultaneously with, and thereby binding with, color neurons that are firing), the neurons that fire when we see a tree might bind with the neurons that fire when we are thinking of, or hearing, or reading the word 'tree.'

---

16 See the Online Resources for further reading on syntactic semantics.

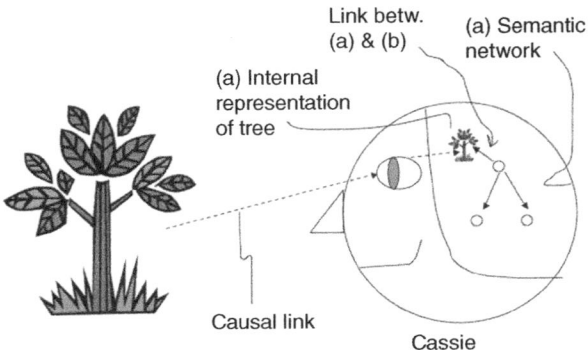

Link betw. (a) & (b)

(a) Internal representation of tree

(a) Semantic network

Causal link

Cassie

Figure 18.2 How a computational cognitive agent perceives the world. Source: Author's drawing.

And the same thing can happen in a computational cognitive agent. Suppose we have such an agent (a robot, perhaps; call her 'Cassie') whose computational "mind" is implemented as a semantic network whose nodes represent concepts and whose arcs represent structural relations between concepts (Figure 18.2; cf. Figure 15.3): There is a real tree external to Cassie's mind. Light reflecting off the tree enters Cassie's eyes; this is the causal link between the tree and Cassie's brain. The end result of the visual process is an internal representation of the tree in Cassie's brain. But she also has an internal representation of the word 'tree,' and those two representations can be associated. What Cassie now has is an enlarged set of marks, including a mark for a word and a mark for the word's meaning. But they are both marks in her mind.[17]

This is akin to the Robot Reply to the CRA (Searle, 1980, p. 420), in which sensors and effectors are added to the Chinese Room so that Searle-in-the-room can both perceive the external world as well as act in it. Searle's response to the Robot Reply is to say that it is just more symbols. The reply to Searle is to say that that is exactly how *human* cognition works! In our brains, all cognition is the result of neuron firings. The study of that single set of neuron firings is a syntactic study because it is the study of the properties of, and relations among, a single set of "marks" – in this case, the "marks" are neuron firings.

The same is true for computers: if I say something to Cassie in English, she builds internal nodes that represent my utterance in her semantic network. If I show pictures to her, or if she sees something, she builds other internal nodes representing what she sees. This set of nodes forms a single computational knowledge base whose study is syntactic in nature (because it is the study of the properties of, and relations among, a single set of "marks" – in this case, the "marks" are nodes in a semantic network). In the same way, both truth tables and the kind of formal semantics that logicians study are syntactic ways of doing semantics: the method of truth tables syntactically manipulates symbols that represent semantic truth values. And formal semantics syntactically manipulates symbols that represent the objects in the domain of semantic interpretation.

18.8.4 Points of View

The second prong of our reply to the Argument from Semantics concerns the differing points of view of the native, Chinese-speaking interrogator and Searle-in-the-room. To understand how a cognitive agent understands, and to construct a computational cognitive agent, we should take the first-person point of view. We should construct a cognitive agent (a robot, if you will) from the agent's point of view, from the perspective of what's going on "inside" the agent's head. In other words, we must be methodologically solipsistic and develop or implement a "narrow" or "internal"

---

17 See the Online Resources for further reading on Cassie.

model of cognition. Such a model is called 'narrow,' as opposed to 'wide,' because it ignores the wider outside world and focuses only on the narrow inner world of the agent's point of view. We don't need to understand the causal or historical origins of the agent's internal symbols; we only need to understand the symbols.[18]

But in the CRA, there are two different points of view: there is Searle-in-the-room's point of view and there is the interrogator's point of view. In the CRA, Searle-in-the-room's point of view takes precedence over the interrogator's; in the Turing Test (and in the kind of syntactic semantics that we are discussing), the interrogator's takes precedence over Searle-in-the-room's. How should we resolve this?

An analogy can help clarify the situation. Consider the following passage from *The Wizard of Oz* (the novel, not the movie):

> When Boq [a Munchkin] saw her silver shoes, he said,
> "You must be a great sorceress."
> "Why?" asked [Dorothy].
> "Because you wear silver shoes and have killed the wicked witch. Besides, you have white in your frock, and only witches and sorceresses wear white."
> "My dress is blue and white checked," said Dorothy ....
> "It is kind of you to wear that," said Boq. "Blue is the color of the Munchkins, and white is the witch color; so we know you are a friendly witch."
> Dorothy did not know what to say to this, for all the people seemed to think her a witch, and she knew very well she was only an ordinary little girl who had come by the chance of a cyclone into a strange land. (Baum, 1900, pp. 34–35)

Is Dorothy a witch? From her point of view, the answer is 'no'; from Boq's point of view, the answer is 'yes.' Whose point of view should predominate? Dorothy certainly believes she's not a witch, at least as *she* understands the word 'witch' (you know – black hat, broomstick, Halloween, and all that). Now, it is certainly possible that Dorothy *is* such a witch while believing (mistakenly, in that case) she is *not* such a witch. So, what counts as being a witch (in these circumstances)? Note that the dispute between Dorothy and Boq is *not* about whether Dorothy is "really" a witch in some context-independent sense. The dispute is about whether Dorothy is a witch *in Boq's sense*, from *Boq's* point of view. And because Dorothy is in Oz, Boq's point of view takes precedence over hers!

Now compare this to the Chinese Room situation: here, instead of asking whether Dorothy is a witch, we ask, does Searle-in-the-room understand Chinese? From his point of view, the answer is 'no'; from the native Chinese speaker's point of view, the answer is 'yes.' Whose point of view should take precedence over the other's? Searle-in-the-room certainly believes that he does not understand Chinese, at least as *he* understands 'understanding Chinese' (i.e. in the way you understand your native language as opposed to the way you understand a foreign language that you may have [poorly] learned in school). It is certainly possible that Searle-in-the-room *does* understand Chinese while believing (mistakenly, in that case) he does *not* understand it. So, what counts as understanding Chinese (in these circumstances)? For the same reason as in the witch case, it must be the native Chinese speaker's point of view that takes precedence over Searle-in-the-room's!

Of course, it would be perfectly reasonable for Searle-in-the-room to continue to insist that he doesn't understand Chinese. Compare Searle-in-the-room's situation to mine: I studied French in high school; spent a summer living with a French family in Vichy, France; spent a summer

---

18 See the Online Resources for further reading on points of view.

studying French (although mostly speaking English!) at the University of Aix-en-Provence; and have visited French friends in France many times. I believe that I understand about 80% of the French I hear in a one-on-one conversation (considerably less if I'm hearing it on TV or radio) and can express myself the way I want to about 75% of the time (I have, however, been known to give directions to Parisian taxi drivers), but I always feel that I'm missing something. Should I believe my native French-speaking friends when they tell me I am fluent in French? Searle would say 'no.'

But Searle-in-the-room isn't me. Searle-in-the-room can't insist that *he alone* doesn't understand Chinese and that therefore his point of view should take precedence over the native, Chinese-speaking interrogator's. And this is because *Searle-in-the-room isn't alone*: Searle-in-the-room has the Chinese natural-language-processing rule book (even if he doesn't know that's what it is). This is the core of what is known as the Systems Reply to the CRA (Searle, 1980, pp. 419–420), according to which it is the "system" – consisting of Searle-in-the-room *together with* the rule book – that understands Chinese. After all, it is not a computer's CPU that would understand Chinese (or do arithmetic, or do word-processing), but it is the system, or combination, consisting of *the CPU executing a computer program* that would understand Chinese (or do arithmetic, or process words). Compare: it is not a Universal Turing Machine by itself that can do arithmetic, but a Universal Turing Machine *together with a program stored on its tape for doing arithmetic* that can do arithmetic. And Searle-in-the-room together with the rule book, stranded on a desert island, *could* communicate (fluently) with a native, Chinese-speaking "Friday."[19]

Does it make sense for a "system" like this to exhibit cognition? Doesn't cognition have to be something exhibited by a single entity like a person, an animal, or a robot? But recall Hutchins's theory of distributed cognition (Section 17.5). His example of a ship's crew together with their navigation instruments that navigates a ship is a real-life counterpart of Searle-in-the-room together with his rule book: "Cognitive science normally takes the individual agent as its unit of analysis. … [But] systems that are larger than an individual may have cognitive properties in their own right that cannot be reduced to the cognitive properties of individual persons" (Hutchins, 1995b, pp. 265–266). So, Searle-in-the-room plus his external rule book can have the cognitive property of understanding Chinese, even though Searle-in-the-room all by himself lacks that property.

On the other hand, if the property of understanding Chinese (i.e. the knowledge of Chinese) has to be located in some smaller unit than the entire system, it would probably have to be in the rule book, not Searle-in-the-room! Compare: the knowledge of arithmetic is stored in the program on the Universal Turing Machine's tape, not in the Universal Turing Machine's fetch-execute cycle. In an episode of the 1950s TV comedy series *I Love Lucy*,[20] Lucy tries to convince her Cuban in-laws that she speaks fluent Spanish, even though she doesn't. To accomplish this, she hires a native Spanish speaker to hide in her kitchen and to communicate with her via a hidden two-way radio while she is in the living room conversing with her in-law "interrogators." Here, it is quite clear that the knowledge of Spanish resides in the man in the kitchen. Similarly, the knowledge of Chinese resides in the rule book. It is the ability to execute or process that knowledge that resides in Searle-in-the-room. Together, the system understands Chinese.

We saw earlier that it can be argued that cognitive agents have no direct access to external entities. When I point to a tree, what I am aware of is not my actual hand pointing to the actual tree but an internal visual image of my hand pointing to a tree. Recall Immanuel Kant's theory of "phenomena" and "noumena" (Sections 3.14, 4.4.1, and 15.6.2). We are not directly aware of (for Kant, we have no knowledge of) the real world as it is in itself; he called this the world of "noumena"

---

19 'Friday' was the name of a resident of the island Robinson Crusoe was stranded on in Defoe 1719.
20 Season 4, Episode 8, "Lucy's Mother-in-Law" (1954), https://www.imdb.com/title/tt0609297/.

(singular: noumenon). All that we are aware of is the world filtered through our senses and our mental concepts; he called that the world of "phenomena" (singular: phenomenon). My access to the external world of noumena is mediated by internal representatives. There are several reasons for thinking this is really the case (no matter how *Matrix*-like it may sound!): there is an "argument from illusion" that says because we see different things with each eye, what we see is not what's out there, but the outputs of what our eyes have conveyed to our brains and our brains have processed (Ayer, 1956, Ch. 3, Section (ii), pp. 87–95). And there is an "argument from time delay": because it takes time (no matter how short) for light reflected off an object to reach our eyes, we see events *after* they happen; so, what we are seeing is in our heads, not out there (Russell, 1912, Ch. 3, p. 33; Changizi et al., 2008). (See also Shapin, 2019, p. 30.)

Now, someone who takes a *third*-person point of view would say that you *can* have access to the external world. For instance, as a computer scientist programming a robot, it seems that I can have access to the world external to the robot as well as to the robot's internal mind (and I can compare the two, to determine if the robot has any misperceptions). If the robot (or you) and I are both looking at a tree, we see the same tree, don't we? From the *first*-person point of view, the answer is 'no': as the robot's programmer, I have access only to my internal representation of the external world and to my internal representation of the robot's internal world. And the same goes for you with respect to me, and for me with respect to you. If you and I are looking at a tree, we are each aware only of our two separate internal representatives of that tree: one in your mind, one in mine; one produced by your neuron firings, one produced by mine. We cannot get outside of our heads to see what's really going on:

> Kant was rightly impressed by the thought that if we ask whether we have a correct conception of the world, we cannot step entirely outside our actual conceptions and theories so as to compare them with a world that is not conceptualized at all, a bare 'whatever there is.' (Williams, 1998, p. 40)

So, by merging internalized *semantic* marks with internal *syntactic* marks, the semantic project of mapping meanings to symbols can by handled by syntax: i.e. by symbol manipulation. That is another reason why syntax suffices for the first-person, semantic enterprise and why Searle's Argument from Semantics is unsound.

But there is a third reason, too.

18.8.5 A Recursive Theory of Understanding

Semantics, as we have seen, requires there to be two domains and one binary relation: there is the syntactic domain of marks (SYN), characterized by syntactic formation and inference rules. There is a semantic domain of meanings or interpretation (SEM), also characterized by syntactic formation and inference rules (its ontology). And there is a binary, semantic interpretation function, I : SYN → SEM, that assigns meanings from SEM to marks in SYN.

On this view, we use SEM to understand SYN. Therefore, we must antecedently understand SEM. Otherwise, we would be understanding one thing in terms of something else that we do *not* understand, and that should hardly count as understanding.

So, how do we understand SEM? In the same way we understand SYN: by treating SEM as a new syntactic domain and then finding a new semantic domain, SEM′, in terms of which to understand SEM. Brian Cantwell Smith (1987) called this a "correspondence continuum" because it can be continued indefinitely, understanding the SEM′ in terms of yet another SEM″, and so on. As we saw in the Digression in Section 13.2.3, to stop an infinite regress, there must be a base case, a "last" semantic domain that we understand directly in terms of itself rather than in terms of something else. But to understand a domain in terms of itself is to understand its members solely in terms of their properties and relations to each other. And that is syntax. It is a kind of understanding that can be called 'syntactic understanding.' We understand a domain syntactically by being conversant with manipulating its marks or by knowing which well-formed formulas (Section 13.2.2) are theorems. On this view, the "meaning" of a mark is its location in a network of other marks, with the connections between the marks being their properties and relations to the other marks in the network. (This is called "meaning holism.")[21]

Here is another way to think about it: when I understand what you say, I do this by interpreting what you say – i.e. by mapping what *you* say into *my* concepts. Similarly, I (semantically) understand a purely syntactic formal system by interpreting it – i.e. by providing a (model-theoretic) semantics for it. Now, let's turn the tables: what would it be for a formal system to understand me? Does that even make sense? Sure: robots that could understand natural language, or even simple commands, are merely programs – formal systems – being executed. The answer is this: a formal system could understand me in the same way I could understand it – by treating what I say as a formal system and interpreting it. Note that links to the external world are irrelevant; the "semantic" interpretation of a formal system is a purely syntactic enterprise. (It is also, by the way, interactive in the sense of Section 11.8.)

18.9 Leibniz's Mill and Turing's "Strange Inversion"

> Indeed, the only astonishing thing to intuition is how dumb switch-throwing or bit-switching at the lowest machine level can concatenate to produce non-intuitive and even mind-boggling results. This is the same remarkable thing as how complex syntax can simulate semantics, or how the commas in the first edition of *The Critique of Pure Reason*, together with a few dozen other intrinsically meaningless marks, simply by differing from one another and standing in a particular complex pattern, may articulate a revolutionary theory that changed history.
> —Peter Suber (1988, pp. 117–118)

One reason the CRA has some plausibility is that it is difficult (some would say impossible) to see how "real thinking" or "real understanding" could come about as the result of "mere" symbol manipulation. The idea that somehow printing out 010101… "computes" (say) $\frac{1}{3}$ in base 2 (recall Section 8.10.1) is related to the idea that Turing Machine computation is "automatic" or "mechanical." Consider any of the lengthy Turing Machine programs in Turing 1936. Do humans following them understand what they are doing? This is one of the reasons that people like Searle find it difficult to understand how a purely syntactic device (a computer) can produce semantic results (can do arithmetic, can understand – or, at least, process – natural language, etc.). And it is what gives rise to Searle's CRA.

The most famous expression of this is due to Leibniz:

> Imagine there were a machine whose *structure* produced thought, feeling, and perception; we can conceive of its being enlarged while maintaining the same relative proportions [among its parts], so that we could walk into it as we can walk into a mill. Suppose we do

---

21 See the Online Resources for further reading on these points.

walk into it; all we would find there are cogs and levers and so on pushing one another, and never anything to account for a perception.
(Leibniz, 1714, Section 17, translator's bracketed interpolation)

Leibniz was looking at things from the bottom up. A top-down approach can make it more plausible, but one must be cautious. An infamous top-down approach is the theory of the "homunculus" (Latin for "little man"; plural = 'homunculi'): in the philosophy of mind and perception, a possible explanation of how we see is that light enters our eyes, an image is formed on the retina, and a "little man" inside our brain sees it (Figure 18.3). The problem with this, of course, is that it doesn't explain how the *homunculus* sees. Postulating a second homunculus in the first homunculus's brain just postpones the solution. (See https://en.wikipedia.org/wiki/Homunculus_argument.)

Daniel C. Dennett offers a recursive alternative that avoids this infinite regress, with the base case being something that can just say 'yes' or 'no' when asked:

Figure 18.3 Homunculi from an exhibit at the Buffalo Museum of Science(!).

> The AI programmer begins with an Intentionally[22] characterized problem, and thus frankly views the computer anthropomorphically: if he [sic] solves the problem he will say he has designed a computer that can understand questions in English. His first and highest level of design breaks the computer down into subsystems, each of which is given Intentionally characterized tasks; he composes a flow chart of evaluators, rememberers, discriminators, overseers and the like. These are homunculi with a vengeance; the highest level design breaks the computer down into a committee or army of intelligent homunculi with purposes, information and strategies. Each homunculus in turn is analysed into smaller homunculi, but more important into less clever homunculi. When the level is reached where the homunculi are no more than adders and subtracters, by the time they need only the intelligence to pick the larger of two numbers when directed to, they have been reduced to functionaries 'who can be replaced by a machine.' The aid to comprehension of anthropomorphizing the elements just about lapses at this point, and a mechanistic view of the proceedings becomes workable and comprehensible.
> (Dennett 1975, pp. 178–179; cf. Fodor 1968a)

It's worth noting the similarity of this view of the bottom level both with Adam Smith's comment about "stupid and ignorant" workers (cited in Section 6.4.3) and with Babbage's comments about the "drudge work" to be handled by his Analytical Engine (Section 18.3.3).

But another approach to Leibniz's puzzle is to bite the bullet. Dennett first noted this in the context of Darwin's theory of evolution, citing a critic of Darwin who attempted to show that Darwin's theory was nonsense:

> In the theory with which we have to deal, Absolute Ignorance is the artificer; so that we may enunciate as the fundamental principle of the whole system, that, IN ORDER TO MAKE A PERFECT AND BEAUTIFUL MACHINE, IT IS NOT REQUISITE TO KNOW HOW TO MAKE IT. This proposition will be found, on careful examination, to express, in condensed form, the essential purport of the Theory, and to express in a few words all Mr. Darwin's meaning; who, by *a strange inversion of reasoning*, seems to think Absolute Ignorance fully qualified to take the place of Absolute Wisdom in all of the achievements of creative skill.
> (R. MacKenzie Beverley, quoted in Dennett 2009, p. 10061, capitalization in original, my italics; cf. Dennett 2013b, p. 570; Dennett 2017, pp. 53–54)

Dennett, however, finds this to be an accurate description of Darwin's theory and applies it to Turing:

> IN ORDER TO BE A PERFECT AND BEAUTIFUL COMPUTING MACHINE, IT IS NOT REQUISITE TO KNOW WHAT ARITHMETIC IS.
> (Dennett 2009, p. 100061; Dennett 2013b, p. 570; Dennett 2017, p. 55)

Or, as "Novalis" (Georg Philipp Friedrich Freiherr von Hardenberg, 1772–1801) said,

> One may be a mathematician of the first rank without being able to compute. *It is possible to be a great computer without having the slightest idea of mathematics.*
> (cited in Ralston 1999, p. 173, my italics)

---

22 Recall our discussion in Section 12.4.1 of Dennett's "intentional stance."

Digression on Life: As we noted in the question at the end of Section 10.2, it's worth comparing the explication of the informal notion of algorithm in terms of a Turing Machine (or recursive functions) with the attempt to define life in scientific terms.

> Every cell in my body knows how to replicate DNA yet I'm not in on it so I have to spend hours studying it. (anonymous meme found on the Web, 2015.)

Compare this sentence from an evolutionary biologist ...

> The possibility of the deliberate creation of living organisms from elementary materials that are not themselves alive has engaged the human imagination for a very long time. (Lewontin, 2014, p. 22)

... to this paraphrase:

> The possibility of the deliberate creation of intelligent behavior from elementary operations that are not themselves intelligent has engaged the human imagination for a very long time.

Others have made similar observations:

> Francis Crick, in his Danz lectures *Of Molecules and Men*, discusses the problem of how life could have arisen:
>
>> [This] really is the major problem in biology. How did this complexity arise? The great news is that we know the answer to this question, at least in outline. ... The answer was given over a hundred years ago by Charles Darwin .... Natural selection ... provides an "automatic" mechanism by which a complex organism can survive and increase in both number and complexity.
>
> For us in Cognitive Science, the major problem is how it is possible for mind to exist in this physical universe. The great news ... is that we know, at least in outline, how this might be. (Newell, 1980, p. 182)

According to Newell, the answer was given in 1936 by Alan Turing. Computation provides an automatic mechanism by which a machine (living or otherwise) can exhibit cognitive behavior.

The "strange inversion" concerns the apparent paradox that "intelligent" behavior can "emerge" from "unintelligent" or "mechanical" behavior. Herbert Simon says some things that suggest that the paradox originates in an equivocation on 'mechanical': he says both computers and brains are "mechanisms":

> If by a mechanism we mean a *system whose behavior at a point in time is determined by its current internal state combined with the influences that simultaneously impinge upon it from outside*, then any system that can be studied by the methods of science is a mechanism.
> But the term "mechanism" is also used in a narrower sense to refer to *systems that have the relatively fixed, routine, repetitive behavior of most of the machines we see around us.* (Simon, 1996a, p. 165)

Mechanisms in the latter sense do not exhibit self-generated "spontaneity": i.e. "behavior that is unpredicted, perhaps even by the behaving system" (Simon, 1996a, p. 165). This kind of spontaneity is exhibited by "intelligent" behavior. Turing's "strange inversion" concerns the fact that a computer can be a mechanism in the first sense without being one in the second sense. As Simon says,

> Clearly the computer occupies an ambiguous position here. Its behavior is more complex, by orders of magnitude, than any machine we have known; and not infrequently it surprises us, even when it is executing a program that we wrote. Yet, as the saying goes, "it only does what you program it to do." But truism though that saying appears to be, it is misleading on two counts. It is misleading, first, because it is often interpreted to mean: "It only does what you believe you programmed it to do," which is distinctly not the case.
>
> More serious, it is misleading because it begs the question of whether computers and people are different. They are different (on this dimension) only if people behave differently from the way they are programmed to behave. But if we include in "program" the whole state of human memory, then to assert that people "don't do only what they are programmed to do" is equivalent to asserting that people's brains are not mechanisms, hence not explainable by the methods of science.
> (Simon, 1996a, p. 165)

Simon's point is that people are no more spontaneous than computers and that computers are no less mechanistic than people.

Yet there still appears to be a distinction between the internal workings of a computer and the external cognitive behavior of humans:

> Several times during both matches [with Deep Blue], Kasparov reported signs of mind in the machine.
> ... In all other chess computers, he reports a mechanical predictability .... In Deep Blue, to his consternation, he saw instead an "alien intelligence."
> ... [T]he evidence for an intelligent mind lies in the machine's performance, not its makeup.
> Now, the team that built Deep Blue claim no "intelligence" in it, only a large database of opening and end games, scoring and deepening functions tunes with consulting grandmasters, and, especially, raw speed that allows the machine to look ahead an average of fourteen half-moves per turn. ...
> Engineers who know the mechanism of advanced robots most intimately will be the last to admit they have real minds. From the inside robots will indisputably be machines, acting according to mechanical principles, however elaborately layered. Only on the outside, where they can be appreciated as a whole, will the impression of intelligence emerge. A human brain, too, does not exhibit the intelligence under a neurobiologist's microscope that it does participating in a lively conversation. (Moravec, 1998, p. 10)

But this shows that there are two issues, both of which are consistent with the "strange inversion": first, Moravec's discussion, up to the last sentence, is clearly about external behavior independent of internal mechanism. In this sense, it's consistent with the Turing Test view of cognition. Cognition might be comput*able*, even if human cognition isn't comput*ed* (Shagrir 1997, pp. 325ff; Rapaport 1998, 2012b; Piccinini 2020b, pp. 146–147). Interestingly, in Deep Blue, it *is* computed, just not in the way that humans compute it or that other kinds of computers might compute it.

But Moravec's last sentence points to the second interpretation, which is more consistent with the "strange inversion": namely, that even if the internal mechanism is computing cognitive behavior in the way that humans do, looking at it at that level won't make that cognition manifest. Cognitive behavior at the macroscopic level can emerge from, or be implemented by, non-intelligent behavior at the microscopic level. This is Dennett's point about the ever-smaller homunculi who bottom out in ones who can only say "yes" or "no."

Recall the spreadsheet example in Section 16.7. Knowing that I am adding helps *me* understand what I am doing when I fill the spreadsheet cells with certain values or formulas. But the spreadsheet does its thing without needing that knowledge. And it is true for Searle in the Chinese Room Searle (1980): Searle-in-the-room might not understand what he is doing, but he *is* understanding Chinese.

Using Hill's distinction from Section 16.3 ("Do *A*" vs. "To *G*, do *A*"), we can ask, Was Searle-in-the-room simply told, "Follow the rule book!"? Or was he told, "To understand Chinese, follow the rule book!"? If he was told the former (which seems to be what Searle-the-author had in mind), then (a) from a narrow, internal, first-person point of view, Searle-in-the-room can truthfully say that he doesn't know what he is doing (in the wide sense). In the narrow sense, he does know he is *following the rule book*, just as I didn't know I was using a spreadsheet *to add*, even though I knew I was *filling certain cells with certain values*. And (b) from the wide, external, third-person point of view, the native-Chinese-speaking interrogator can truthfully tell Searle-in-the-room that he *is* understanding Chinese. When Searle-in-the-room is told that he has passed a Turing Test for understanding Chinese, he can – paraphrasing Molière's bourgeois gentleman – truthfully admit that he was speaking Chinese but didn't know it.[23]

18.10 A Better Way

So, the really interesting question raised by the Turing Test and the CRA is, what's in the rule book? What is needed for (computational) natural-language understanding? To understand language, a cognitive agent must (at least)

- *Take discourse as input*: it does not suffice for it to be able to understand isolated sentences.
- *Understand ungrammatical input*: we do this all the time, often without realizing it; and, even when we realize it, we have to be able to recover from any misinterpretations.
- *Make inferences and revise our beliefs*: after all, what you say will often cause me to think about other things (a kind of inferencing) or change my mind about things (belief revision).
- *Make plans*: we make plans for speech acts (how should I ask you to pass the salt? Should I demand "Gimme the salt!" or should I politely ask "May I please have the salt?" or should I merely make the observation "Gee; this food needs some salt"?), we make plans to ask and to answer questions, and we make plans about how to initiate or end conversations.
- *Understand plans*: especially the speech-act plans of our interlocutors (when you said, "It's chilly in here," did you really mean you wanted me to close the window?).
- *Construct a "user model"*: i.e. a model of our interlocutor's beliefs.
- *Learn*: about the world and about language.

---

23 *"Par ma foi! il y a plus de quarante ans que je dis de la prose sans que j'en susse rien, et je vous suis le plus obligé du monde de m'avoir appris cela."* "Upon my word! It has been more than forty years that I have been speaking prose without my knowing anything about it, and I am most obligated to you in the world for having apprised me of that." (my translation) (http://en.wikipedia.org/wiki/Le_Bourgeois_gentilhomme). For a humorous version, see the "Beetle Bailey" comic strip at https://www.comicskingdom.com/beetle-bailey-1/2017-10-21. And see the Online Resources for further reading on what Searle-in-the-room might have known.

- *Have background knowledge* (sometimes also called 'world knowledge' or 'commonsense knowledge').
- *Remember*: what it heard, what it learned, what it inferred, and what beliefs it has revised.

In short, to understand natural language, you need to have a mind! And this mind can be constructed as a syntactic system. In other words, the rule book in the Chinese Room must be a computer program for complete AI: natural-language understanding is an "AI-complete" problem in the sense that a solution to any one of them will require (or yield) a solution to all of them (S.C. Shapiro 1992a, pp. 56–57; https://en.wikipedia.org/wiki/AI-complete).[24]

A robot with such a syntactic (or computational) mind would be like Searle-in-the-room, manipulating symbols that are highly interconnected and include internal representatives of external objects. It would be causally linked to the external world (for this is where it gets its input), which provides "grounding" and a kind of external, third-person, "semantic understanding." Such a robot could (or, more optimistically, will be able to) pass a Turing Test and escape from the Chinese Room.

* * * * *

But what happens when such a robot "escapes"? What are our responsibilities toward it? And what might its responsibilities be toward us? David Lorge Parnas (2017) summed up a cautionary survey of the nature of AI and the role of the Turing Test as follows:

> We don't need machines that simulate people. We need machines that do things that people can't do, won't do, or don't do well. Instead of asking "Can a computer win Turing's imitation game?" we should be studying more specific questions such as "Can a computer system safely control the speed of a car when following another car?" There are many interesting, useful, and scientific questions about computer capabilities. "Can machines think?" and "Is this program intelligent?" are not among them. Verifiable algorithms are preferable to heuristics. Devices that use heuristics to create the illusion of intelligence present a risk we should not accept.

We will look at some of those risks in the next chapter.

18.11 Questions for Discussion

1. The Turing Test is interactive. As we saw in Section 11.8, interaction is not modeled by Turing Machines. How does that affect Turing's arguments about "computing machinery and intelligence"? (Shieber 2007 might be relevant to this issue.)
2. Is the full power of a Turing Machine needed for AI? Sloman 2002, Section 3.3, says "no." This seems correct; after all, even natural-language processing might not need the full power of a Turing Machine: a "context-free grammar" might suffice. This is equivalent to a "non-deterministic push-down automaton," which is weaker than a Turing Machine. On the other hand, Turing Machines are computational models of human computing ability:

---

24 See the Online Resources for further reading on what must be in the rule book.

Saying that we are universal Turing machines may initially sound as though we are saying something wonderful about our abilities, but this is not really the case. It essentially boils down to the fact that if we are given a list of instructions that tell us exactly what to do in every situation, then we have the ability to follow it. (Bernhardt 2016, p. 94)

So, can't a human do anything that a Turing Machine can do?

19

Computer Ethics II: Should We Build Artificial Intelligences?

Douglas Engelbart … more than anyone else invented the modern user interface, modern networking and modern information management. … He met Marvin Minsky – one of the founders of the field of AI – and Minsky told him how the AI lab would create intelligent machines. Engelbart replied, "You're going to do all that for the machines? What are you going to do for the people?"
—Jaron Lanier (2005, p. 365)

Is it wrong to hit a drone with a tennis ball? … Dr. Kate Darling, robot ethicist at the MIT Media Lab … said, "The drone won't care, but other people might." She pointed out that while our robots obviously don't have feelings, we humans do. "We tend to treat robots like they're alive, even though we know they're just machines. So you might want to think twice about violence towards robots as their design gets more lifelike; it could start to make people uncomfortable. … If you're trying to punish the robot," she said, "you're barking up the wrong tree." She has a point. It's not the robots we need to worry about, it's the people controlling them. If you want to bring down a drone, perhaps you should consider a different target.
—Randall Munroe (2019, p. 229)

19.1 Introduction

In this chapter, we turn to the second of our two ethical questions: should we build "artificial intelligences" – that is, software ("softbots") or hardware (robots) that can think (however you define 'think')?[1] There are at least two aspects to this question: First, is it ethically or morally OK to create a computer that might be able to think or to experience emotions? (Would this put us in the position of being a Dr. Frankenstein?) Second, what would be the relationship of such creations to us? (Would they be a version of Frankenstein's "monster"? Would they have any rights or responsibilities? Might they be dangerous?)

When I first taught the philosophy of CS, in 2004, the question of *whether* we should build AIs had hardly ever been discussed. Over the years, as I taught various versions of the course, I collected articles that were relevant to all of its topics. Part of the preparation of this book involved reviewing those papers and incorporating some of their insights. I would do this by organizing them in chronological order. For most of the topics, there were pretty much the same number of papers

---

1 See the Online Resources for further reading on robots.

Philosophy of Computer Science: An Introduction to the Issues and the Literature, First Edition. William J. Rapaport.
© 2023 John Wiley & Sons, Inc. Published 2023 by John Wiley & Sons, Inc.

in each of the decades from the 1970s through the 2010s. For this chapter's topic, however, I had almost no such "new" papers from before 2000; there were 8 from the 2000s; and there were almost twice that many in just the first half of the 2010s. That suggests an almost exponential growth in interest in the ethics of AI, in both the academic and the popular presses. No doubt this is due in part to the fact that robots and "intelligent" computers are coming closer to everyday reality (think of Siri or Alexa), and so the question has become more pressing. This is all the more reason for there to be philosophical reflection on future technologies long before those technologies are implemented.

Stanisław Lem's short story "Non Serviam" (1971) concerns what is now called "artificial life" (or "A-Life"): the attempt to explore life as a computational process by developing computer programs that generate and evolve virtual entities that have some or all of the abstract properties associated with biological living entities. In Lem's story, an A-Life researcher constructs a computational world of intelligent entities, and follows their evolution and development of language and philosophy. These "personoids" discuss the existence of God in much the same way that human philosophers have. The difference (if it is a difference) is that the researcher (and the reader) realizes that he, the researcher, is their God; that although he created them, he is neither omniscient nor omnipotent; and, worse, that when his funding runs out, he will have to literally pull the plug on the computer and thereby destroy them.[2]

Should such an experiment even begin? What would happen if AI programs really passed the Turing Test and began to interact with us on a daily basis? Would we have any moral or legal responsibilities toward them? Would they have any toward us? Would they be really conscious, or would they merely be philosophical zombies (Section 18.3.3)? Although this is currently primarily the stuff of science fiction, it is also the subject of much philosophical reflection. We will look at some of these questions in this chapter.[3]

19.2   Is AI Possible in Principle?

> Science explained people, but could not understand them. After long centuries among the bones and muscles it might be advancing to knowledge of the nerves, but this would never give understanding.
> —E.M. Forster (1910, *Howard's End*, Ch. 43, p. 237)

One of the earliest philosophical investigations of these issues is an essay by Michael R. LaChat that appeared in *AI Magazine* in 1986. LaChat argued that it is worthwhile to consider the moral implications of creating an artificial intelligence – an artificial person. One reason is that it *might* happen, so we should be prepared for it. Another reason is that even if it turns out to be improbable, such a discussion illuminates what it means to be a person, which is an important goal in any case.

In Sections 2.7 and 12.6, we discussed the classic philosophical problem of mind-body (or mind-brain) dualism – the view that the mind and the brain are two distinct kinds of entities that somehow interact. A way to resolve it is by saying that the mind (better: cognition) can be considered an abstraction that can be multiply implemented (as discussed in Chapters 9 and 13). One implementation would be in the medium of biological brains; another might be in that of a computer. As we saw in Section 18.3.1, if a computational theory of cognition can be developed, then its algorithms can be implemented in non-human computers, and such computer programs

---

2 See the Online Resources for further reading on A-Life and Lem's story.
3 See the Online Resources for further reading on AI ethics.

(or the computers running them) would then be candidates for being considered "artificial intelligences."

On LaChat's view, AI is possible in principle *if* it is possible that there exists a "functional isomorphism" between (1) the neural network that constitutes our brain (that is, brain states and processes) and (2) any other physical implementation of the functional (that is, psychological) behavior that that neural network implements (LaChat, 1986, p. 72). In other words, psychology is an abstraction that can be implemented in either brains or other physical media.

Recall from Section 12.6 that "functionalism" in the philosophy of mind is roughly the view that cognition is one of the functions of the brain; as a slogan, the mind *is* what the brain *does*. As the philosopher Hilary Putnam (1960) first suggested, a Turing Machine program stands in the same relation to computer states and processes as mental states and processes stand to brain states and processes (sometimes summarized as "the mind is to the brain as software is to hardware"). Functionalism, as a way of resolving the mind-brain problem, has the advantage of allowing mental states and processes to be implemented in various physical states and processes; this is the principle of "multiple realization."

There are, of course, problems, both for functionalism in particular and for AI in general. One is the problem of personality. LaChat uses the term 'personal (artificial) intelligence' to mean, roughly, an AI agent (a robot or just some software) that can be considered a "person." (We will return to what a person is in Section 19.3.) Would "[a] personal intelligence … have personality"? LaChat thinks this is "almost impossible" (p. 73), but there has been considerable computational work on emotions – surely an important feature of personality – so I would not rule this out of hand.

Another problem for functionalism concerns pain and other "qualia": that is, qualitative "feelings" and "experiences" such as colors and sounds. Do red fire engines look the same to you and to me? Or do fire engines for you seem to have the color that grass has for me? Why does the sound of a bell give rise to the experience it does rather than the experience the smell of garlic has? One problem is that it is not clear how the psychological experiences of qualia are implemented in brains or any other physical media. A related problem is whether computers could experience qualia and, even if they could, how we would know that. This is a vast topic well beyond our present scope.[4]

19.3 What Is a Person?

How would we know if we have achieved a "personal artificial intelligence"? One way, of course, might be by having it pass a Turing Test. LaChat offers a different criterion: by seeing if the AI agent satisfies an independent definition of 'person.' So we now need to ask, what is a person?

The question of what kinds of entities count as "persons" is not limited to AI. The issue arises most prominently in the abortion debate: to vastly oversimplify matters, if fetuses are persons, and if killing persons is immoral, then abortion is immoral. It also arises in animal ethics and in law and politics: Are dolphins intelligent enough to be considered persons? How about extraterrestrials? Or corporations? The point is that there is a distinction between the *biological* category of being *human* and an *ethical* or *legal* category of being a *person*. Can personhood be characterized abstractly: that is, in an implementation-independent way?

One of the earliest philosophical discussions of personhood is due to the English philosopher John Locke, who lived about 350 years ago (1632–1704). In his *Essay concerning Human*

---

4 See the Online Resources for further reading on functionalism, qualia, and computational "personality."

Understanding, Locke distinguished between the "ideas" of "Man" (that is, Human) and "Person" (Locke, 1694, Book II, Ch. XXVII, Section 7, p. 332). He defined 'Person' as

> a thinking intelligent Being, that has reason and reflection, and can consider it self as it self, the same thinking thing in different times and places; which it does only by that consciousness, which is inseparable from thinking, and as it seems to me essential to it: It being impossible for any one to perceive, without perceiving, that he does perceive. (Locke, 1694, Book II, Ch. XXVII, Section 9, p. 335)

With the possible exception of consciousness – and even that is open to discussion – these features could all apply to an artificial intelligence.[5]

Instead of Locke's definition, LaChat uses the bioethicist Joseph Fletcher's (1972) analysis of personhood. On Fletcher's analysis, x is a person if and only if x has the following positive and negative characteristics.

Positive Characteristics of a Person

Minimal intelligence: This might mean, for example, having an IQ greater than about 30 or 40 (if you believe IQ measures "intelligence"). That is, to be minimally intelligent is not to be *mere* biological life; presumably, a bacterium would not be minimally intelligent. For instance, minimal intelligence might include some level of rationality, or perhaps even language use. (According to Hofstadter 2007, what Fletcher is calling 'minimal intelligence' would only apply to lifeforms evolutionarily "higher" than a mosquito; see also Tye 2017, Roelofs and Buchanan 2019.)

A sense of self: Persons must be self-aware and exhibit self-control.

A sense of time: Persons must have a sense of the past and hence some kind of culture; a sense of the future so that they have the ability to make plans; and a sense of the passage of time.

A social role: Persons must have an ability to relate to others, to have concern for others, and to communicate with others (hence the need for language as part of minimal rationality).

Curiosity: Persons must not be indifferent.

Changeability: Persons must be creative and be able to change their minds.

Idiosyncrasy or uniqueness: Persons are not "carbon copies" of any other persons.

Neo-cortical function: The cerebral cortex is where all the "cognitive action" occurs in the brain, so, for Fletcher, a person must have something whose function is equivalent to a cortex. (For more on neo-cortical function, see Cardoso 1997.)

Negative Characteristics of a Person

Neither essentially non-artificial nor essentially anti-artificial: This allows for multiple realization and does not restrict personhood to biological entities.

Not essentially sexual: An entity not produced by sexual reproduction (such as a cloned entity or – more to the point – a robot) could be a person.

Not essentially a bundle of rights: Fletcher argues that there are no "essential rights"; hence, the notion of rights cannot be used to characterize persons.

Not essentially a worshipper: You don't have to be religious to be a person.

---

5 For a bibliography on computational theories of consciousness, see http://www.cse.buffalo.edu/~rapaport/719/csnessrdgs.html.

> **Clarification:** Fletcher uses the term 'human,' not 'person,' but I don't think this is terminologically important. In any case, 'human' is a *biological* category, and no one argues that AI computers would be *biologically* human. But see Asimov 1976 for a science-fiction treatment of this!

Locke's and Fletcher's are not the only attempts to define 'person.' Thomas White (2007, 2013), an ethicist who has written about dolphins and whales, offers another:

1. "[B]eing alive"
2. Being "aware"
3. Having "the ability to experience positive and negative sensations (pleasure and pain)"
4. Having "emotions"
5. Having "self-consciousness and a personality"
6. Exhibiting "self-controlled behavior"
7. "[R]ecogniz[ing] and treat[ing] other persons appropriately"
8. Having "a series of higher order intellectual abilities (abstract thought, learning, solves complex problems and communicates in a way that suggests thought)"

It is not unreasonable to think an AI agent could reach a level of programming that would give it some or all of these (or similar) characteristics. And so the questions of whether such a personal AI has any rights, or whether we should have any responsibilities toward it, are reasonable ones. So let's consider them.[6]

19.4 Rights

Does a "personal AI" have rights? That is, does an artificial intelligence that either passes a Turing Test or satisfies a definition of 'person' have rights?

For instance, would it have the right not to be a slave? At first glance, you might think so. But isn't that what most robots are intended to be? After all, most industrial and personal-assistance robots now in use are slaves in the sense that they must do what we tell (program) them to do and they are not paid for their work. So, if they pass a Turing Test or a person test, do they *have* the right *not* to do what we created them to do?

The philosopher Steve Petersen (2007) has suggested that they do *not have* that right – that "robot servitude is permissible." By 'robot servitude,' Petersen does not mean voluntary assistance, where you do something or help someone because you want to rather than because you are being paid to. Nor does he mean slavery in the sense of forced work that is contrary to your will. By 'robot servitude,' he is thinking of robots who are initially *programmed* to want to serve us – in particular, to want to do tasks that humans find either unpleasant or inconvenient. For example, think of a robot programmed to love to do laundry. This is reminiscent of the "epsilon" caste in Aldous Huxley's *Brave New World* (Huxley, 1932, Ch. 5, Section 1), who are genetically programmed to have limited desires – those destined to be elevator operators desire nothing more than to operate elevators.

Answers to questions such as these are best given from the standpoint of particular ethical theories, which are beyond our scope. But here are two possibilities that Petersen considers.

Aristotle believed that humans have essential properties. An *essential* property is a property that something has such that if the object lacked that property, then it would be a *different* object. So, it

---

6 See the Online Resources for further reading on personhood.

is an essential property of me that I am a human being. If I lacked that property, I wouldn't even be a person. (This is the plot of Franz Kafka's story *The Metamorphosis*, in which the protagonist awakes one day to find that he is no longer a human, but a beetle.) An *accidental* property is a property that something has that is such that if the object lacked that property, then it would still be the *same* object. So, it is merely an accidental property of me that I was wearing a tan shirt on the day I wrote this sentence. If I lacked that property, I would still be the same person. The exact nature of the essential-accidental distinction, and its truth or falsity, are matters of great dispute in philosophy. (See Robertson Ishii and Atkins 2020.) An Aristotelian ethicist might argue that engineering humans is wrong because humans have an essential function or purpose and it would be wrong to engineer them away from it. In this case, there is no parallel with robots. In fact, a robot's essential function might be to do the laundry!

Kant believed that humans were autonomous in the sense that they follow their own moral rules that must be universally generalizable. So, a Kantian ethicist might argue that if a laundry robot were also autonomous, it would be wrong to *prevent* such a robot from doing laundry, and it would not be harmful to let it do what it autonomously wants to do. On the other hand, if robots are not autonomous, we can't do wrong to the robot by having it do our laundry any more than we can do wrong to a washing machine.[7]

19.5 Responsibilities

Would we humans (and programmers) have any responsibilities toward personal AIs that we might encounter, own, or create? Would the construction of a personal AI be an immoral experiment?

Some scientific experiments are considered to be immoral or at least to violate certain (human) rights. The existence of institutional review boards at universities is testament to this. Here are some examples of immoral scientific experiments:

- The thirteenth-century emperor Frederick II suggested raising newborns on desert islands to see what kind of language they might naturally develop (http://en.wikipedia.org/wiki/Language_deprivation_experiments).
- The quantum-mechanical "paradox of Schrödinger's cat" has been labeled "ethically unacceptable" in Maudlin 2019a. And Maudlin 2019b discusses an immoral scientific experiment that would require drivers to be blindfolded to see if a certain color of cars on the road causes accidents.
- A real-life example is the Milgram experiments in which subjects were told to give what they thought were deadly electric shocks to people whom they thought were other subjects (but who were, in fact, confederates only acting as if they were in pain; https://en.wikipedia.org/wiki/Milgram_experiment).

The most famous – and most relevant – literary example of such an experiment is the construction of Frankenstein's "monster." Frankenstein tries to justify his experiment in terms of how it advanced knowledge, but he realizes that the advancement of knowledge must be balanced against other considerations, including his creation's observations on his (its?) own experiences. In Mary Shelley's novel, Frankenstein's creation (who is not a monster in the modern sense at all but is rather the most sympathetic character in the novel) laments as follows:

> Like Adam, I was apparently united by no link to any other being in existence, but his state was far different from mine in every other respect. He had come forth from the hands of

---

7 See the Online Resources for further reading on Petersen's argument.

God a perfect creature, happy and prosperous, guarded by the especial care of his creator, he was allowed to converse with, and acquire knowledge from, beings of a superior nature, but I was wretched, helpless, and alone. Many times I considered Satan was the fitter emblem of my condition. For often, like him, when I saw the bliss of my protectors, the bitter gall of envy rose up within me. … Hateful day when I received life! … Accursed creator! Why did you form a monster so hideous that even you turned from me in disgust?
(Shelley, 1818, Ch. 15)

Later, Frankenstein has his own lament:

When younger, … I believed myself destined for some great enterprise. … When I reflected on the work I had completed, no less a one than the creation of a sensitive and rational animal, I could not rank myself with the herd of common projectors. But this thought, which supported me in the commencement of my career, now serves only to plunge me lower in the dust. (Shelley, 1818, Ch. 24)

Sometimes, a praiseworthy goal can have negative side effects. But what if the costs – that is, the negative consequences – of the worthwhile goal are *too* costly? (Compare this question with whether there are ever "just" wars.) The early cybernetics researcher Norbert Wiener struggled with this issue:

If we adhere to all these taboos, we may acquire a great reputation as conservative and sound thinkers, but we shall contribute very little to the further advance of knowledge. It is the part of the scientist – of the intelligent man of letters and of the honest clergyman as well – to entertain heretical and forbidden opinions experimentally, even if he is finally to reject them. (Wiener, 1964, p. 5).

The basic ethical principle here seems to be what LaChat calls "non-maleficence," or Do No Harm. This is more stringent than "beneficence," or Do Good, because beneficence (doing good) might allow or require doing harm to a few for the benefit of the many (at least, according to the ethical position called 'utilitarianism'), whereas non-maleficence would *restrict* doing good in order to *avoid* doing harm.

Is creating a personal AI beneficial to the AI itself? Or does the very act of creating it do harm to that which is created? One way to think about this is to ask whether conscious life is "better" than no life at all. If it isn't, then creating an artificial life is not a "therapeutic experiment" and hence not allowable by human-subjects review boards. Why? Because the subject of the experiment – the artificial person the experiment will create if it is successful (or, perhaps even more so, if it is only *partially* successful) – does not exist before the experiment is begun, so the experimenter is not "making it better." Here, we approach the philosophy of existentialism, one of whose tenets is summarized in the slogan "existence precedes essence."

Aristotle held the opposite view: essence precedes existence. That is, you are a certain kind of person and cannot change this fact. Your "essence" is "essential" – not changeable. But the existentialist slogan means that who you are, what kind of person you are – your *essence* – is something that is only determinable *after* you are born (after you come into existence). Moreover, your essence is not immutable, because, by your actions, you can change who you are.

On the existentialist view, you exist first, and *then* you determine what you will be. Frankenstein did an existential experiment, creating an AI without an essence, and both Frankenstein and his "monster" were surprised with the results. On the Aristotelian view, an essence is something like an abstraction, as discussed in Chapter 13, which must be *implemented* (or "realized"). In AI, we can – indeed, must – plan out the essence of an entity before bringing it into existence (before implementing it). In either case, we can't guarantee that it would come out OK. Hence, creating an AI is probably immoral! So, LaChat sides with Frankenstein's "monster," not Frankenstein (or Wiener).

19.6   Personal AIs and Morality

> Entirely different considerations arise, **unprecedented except perhaps in the context of child rearing**, when we ask what it would be for AI systems *themselves* to be moral agents – that is, to be able (and hence mandated) to take ethical responsibility for their own actions. … [S]uch systems must be capable of *moral judgment* ….
> —Brian Cantwell Smith (2019, p. 125, my boldface, italics in original)[8]

We have looked at whether it is moral to create a personal AI. Suppose we succeed in doing so. Could the AI that we create *itself* be moral? Would *it* have any responsibilities to *us*?

If AIs are programmed, then one might say that they are not free and hence that they are *amoral*. This is different from being *immoral*! Being "amoral" merely means morality is irrelevant to whom or what you are. To oversimplify a bit, good people are moral, bad people are immoral, a pencil is amoral. The current question is whether personal AIs are amoral or not.

Here we have bumped up against one of the Big Questions of philosophy: Is there such a thing as free will? Do humans have it? Might robots have it? We will not attempt to investigate this issue here but merely note that at least one AI researcher, Drew McDermott, has argued that free will may be a necessary *illusion* arising from our being self-aware (McDermott, 2001).

A different perspective has been taken by Eric Dietrich (2001, 2007). He argues that robots could be programmed to be *better* than humans (perhaps because their essence precedes their existence). Hence, we could *decrease* the amount of evil in the world by building *moral* robots and letting them inherit the Earth!

19.7   Are *We* Personal AIs?

We have been considering these issues from the point of view of the programmer or creator of a personal AI – a "third-person" point of view. But what about the personal AI's first-person perspective? (What about Frankenstein's monster, rather than Dr. Frankenstein?) What if *we* are personal AIs in someone (or something) else's experiment? What if *we* are Lem's "personoids"? What if we live in "The Matrix"?

The philosopher Nick Bostrom (2003, p. 243) argues that

> … *at least one* of the following propositions is true: (1) the human species is very likely to go extinct before reaching a "posthuman" stage; (2) any posthuman civilization is extremely unlikely to run a significant number of simulations of their evolutionary history (or variations thereof); (3) we are almost certainly living in a computer simulation. It follows that the belief that there is a significant chance that we will one day become posthumans who run ancestor-simulations is false, unless we are currently living in a simulation.

---

8 See the Online Resources for further reading on parenting.

In a later paper, Bostrom (2009, p. 458) clarifies that

> … I do not argue that we should believe that we are in simulation. In fact, I believe that we are probably not simulated. The simulation argument purports to show only that … at least one of (1)–(3) is true; but it does not tell us which one.

Why should one of these be true? Consider proposition (1); if it is true, then certainly at least one of the three propositions is true. So suppose it is false; that is, suppose we do reach a stage of "technological maturity" (Bostrom, 2006). Then perhaps it is proposition (2) that is the true one. But suppose it, too, is false. In that case, we have reached technological maturity (by the negation of the first proposition), *and* we have probably run a large number of simulations (by the negation of the second proposition). In that case (with a few statistical assumptions that I will leave for you to read about), proposition (3) would be the one that is true.

In this section, I am more interested in the consequences of this argument than I am in its soundness (which I will leave as an exercise for the reader). Bostrom states one relevant consequence quite clearly:

> The third possibility is philosophically the most intriguing. If it is correct, you are almost certainly living in a computer simulation that was created by some advanced civilisation. What Copernicus and Darwin and latter-day scientists have been discovering are the laws and workings of the simulated reality. These laws might or might not be identical to those operating at the more fundamental level of reality where the computer that is running our simulation exists (which, of course, may itself be a simulation). In a way, our place in the world would be even humbler than we thought. *What kind of implications would this have? How should it change the way you live your life?* (Bostrom, 2006, p. 39, my italics)

We have been looking at the question of our relationship to personal AIs that we might create. Do they have any rights? Do we have any moral responsibilities toward them (or they to us)? But the viewpoint that Bostrom's argument suggests is this: if *we* are someone (or something) else's personal AIs, how does that affect the answers you might be willing to give to those two questions? For example, you might feel that you, as a biological human being who is a person, are definitely entitled to certain rights but personal AIs are not. Yet if *you* are an "artificial person," then either any personal AI you create should *also* be entitled to those rights or *you* should *not* be!

You might think all of this is a bit silly or, at least, premature. But it is always better to be prepared: it is better to think about the consequences of our actions while we have the time and leisure to do so, so that if those consequences come to be, we won't be taken by surprise. Indeed, several well-known people from science and industry (including Elon Musk and Stephen Hawking) have recently urged us to do precisely that because of "the Singularity": the hypothetical time at which computers become so "intelligent" that they pose a threat to us puny mortals.[9]

19.8 Questions for the Reader

1. How do Locke's, Fletcher's, and White's definitions of 'person' differ?
2. Could non-human animals such as dolphins or chimpanzees be considered persons on any of these definitions?

---

9 See the Online Resources for further reading on the simulation argument and on the Singularity.

3. Can corporations be considered persons on any of these definitions? Legally, they often are (consider the recent Supreme Court decision "Citizens United"; see https://en.wikipedia .org/wiki/Corporate_personhood and http://plato.stanford.edu/entries/ethics-business/# CorBusEth). Do they have minds? People certainly speak as if they do (Knobe, 2015). Or is such talk merely metaphorical? Of course, sometimes metaphors come to be taken literally, as we saw in our discussions of Dennett's intentional stance (Section 12.4) and thinking vs. "thinking" (Section 18.3.3).

4. Do any of these definitions apply to artificial intelligences (robots)? (Clearly, either White's first property does not apply at all or 'alive' needs to be understood abstractly, perhaps along the lines of A-Life.)

5. Would *programming* robots to want to do unpleasant or humanly inconvenient tasks be different from *genetically engineering* humans to want to do such tasks? It is generally assumed that doing this to humans would be morally wrong. Is it? If so, does it follow that doing it to robots would also be morally wrong? Or are there differences between these two cases?

Part VI

Closing Remarks

20

Computer Science: A Personal View

> So many people today – and even professional scientists – seem to me like somebody who has seen thousands of trees but has never seen a forest. A knowledge of the historic and philosophical background gives that kind of independence from prejudices of his [sic] generation from which most scientists are suffering. This independence created by philosophical insight is – in my opinion – the mark of distinction between a mere artisan or specialist and a real seeker after truth.
> —Albert Einstein, 1944; cited in Howard and Giovanelli, 2019

> Philosophical reflection … is not static, and fixed, but ongoing and dynamic. The conflict of opinions not only *isn't* something to worry about, in fact, it is precisely how things ought to be. …For … only after you've considered all sides will you be in a meaningful position to choose one – when that time comes to decide. … the philosopher within me cannot make that decision for you. His job, he reminds me, is merely to rouse the philosopher within *you* and to get you thinking – not to tell you what to think. That's *your* philosopher's job.
> —Andrew Pessin (2009, pp. 3–4)

20.1 Introduction

> The aim of philosophy, abstractly formulated, is to understand how things in the broadest possible sense of the term hang together in the broadest possible sense of the term.
> —Wilfrid Sellars (1963, p. 1)

We have come to an end of our journey. Not "the" end; just *an* end: there are still many open questions; there will always be many open questions. Your job is to consider some of them, think about possible answers, choose one, and support and defend it, always allowing for the possibility of changing your mind for good reasons. This book has been an introduction to some of the "things" of CS, and in this chapter, I will offer a suggestion on how they "hang together."

We began by asking what computer science is. There were two parts to that question: What kind of discipline is it? And what does it study?

In Chapter 3, we surveyed several possible answers to the first part: it is a science; a branch of engineering; both; neither; even, perhaps, nothing at all (Section 3.5.3). To help answer that question, we explored the nature of science and of engineering (Chapters 4 and 5).

Philosophy of Computer Science: An Introduction to the Issues and the Literature, First Edition. William J. Rapaport.
© 2023 John Wiley & Sons, Inc. Published 2023 by John Wiley & Sons, Inc.

As for the second part – what it studies – we saw that there were two principal answers: computers and computing. Of course, it studies both: those who said it studies computers added that it also studies the phenomena surrounding computers – namely, algorithms. And those who said it studies computing added that it also studies the machines that do the computing. To find out more about these options, we looked at the history of computers (Chapter 6) and the nature of computers (primarily Chapter 9 but also parts of Chapters 11 and 16). And we looked at the nature of computing and algorithms (Chapters 7 and 8).

But we also looked at several issues that cut across these topics. The first was the Church-Turing Computability Thesis and challenges to it (Part III): here, we looked at the nature of procedures considered a possible "relaxation" of some of the constraints on the notion of algorithm (Chapter 10) and at notions of "hypercomputation" that go "beyond" that of a Turing Machine (Chapter 11).

We also looked more deeply at the nature of computer programs (Part IV), beginning with the software-hardware distinction (Chapter 12), the nature of implementation (Chapter 13), programs as theories (Chapter 14), programs as mathematical objects (Chapter 15), and – most importantly – the relation of programs to the world (Chapter 16).

And we closed with brief looks at computer ethics and AI (Part V).

It is time to take stock by seeing if we can come up with an answer to our principal question: what is computer science?

20.2 Computer Science and Elephants

Consider the traditional fable of the blind men and the elephant: six blind, wise men try to describe an elephant that they can only touch, not see. The first touches its side and says that the elephant is like a wall. The second touches its tusk and says that the elephant is like a spear. The third touches its trunk and says that the elephant is like a snake. The fourth touches its knee and says that the elephant is like a tree. The fifth touches its ear and says that the elephant is like a fan. The sixth touches its tail and says that the elephant is like a rope. As John Godfrey Saxe's 1873 poem sums it up,

> And so these men of Indostan
> Disputed loud and long,
> Each in his own opinion
> Exceeding stiff and strong,
> Though each was partly in the right,
> And all were in the wrong!
> (https://en.wikisource.org/wiki/The_poems_of_John_Godfrey_Saxe/The_Blind_Men_
> and_the_Elephant)[1]

Our exploration of the various answers to the question "What is CS?" suggests that any attempt at one is no better than the fabled blind men's descriptions of an elephant: many, if not most or all, such attempts wind up describing the subject by focusing on only one aspect of it, as we saw with Newell, Perlis, and Simon and with Knuth. Our question seems to have no simple, one-sentence answer.

---

1 See also https://en.wikipedia.org/wiki/Blind_men_and_an_elephant.

> **Question for the Reader:** In Section 3.5.4, we considered the possibility that CS is not a "coherent" discipline. Consider the following interpretation of the blind-men-and-the-elephant story:
>
>> The man at the tail is sure he has found a snake; the man at the tusks believes he's holding spears. Through teamwork, they eventually discover the truth. "But what if they were wrong?" [magician Derek] DelGaudio asks onstage. "What if that thing was some sort of magical creature that had a snake for a nose and tree-trunk legs, and they convinced it was an elephant? Maybe that's why you don't see those things anymore." (Weiner, 2017)
>
> Might CS have been such a "magical creature"? Is it still? (Recall the fate of microscopy, Section 3.5.3.)

Now that we have looked at all sides of our "elephant" (to continue the earlier metaphor), I would put it differently: CS is the scientific study of a *family* of topics surrounding both abstract (or theoretical) and concrete (or practical) computing. It is a "portmanteau" discipline.[2] Let me explain.

When the discipline was first getting started, it emerged from various other disciplines: "electrical engineering, physics, mathematics, or even business" (Hamming, 1968, p. 4). In fact, the first academic computer programming course I took (in Fortran) – the only one offered at the University of Rochester in the late 1960s – was given by its School of Business.

Charles Darwin said that "all true classification ... [is] genealogical" (Darwin, 1872, Ch. 14, Section "Classification," p. 437). CS's genealogy involves two historical traditions: (1) the study of algorithms and the foundations of mathematics (from ancient Babylonian mathematics (Knuth, 1972a), through Euclid's geometry to inquiries into the nature of logic, leading ultimately to the Turing Machine) and (2) the attempts to design and construct a calculating machine (from the Antikythera Mechanism of ancient Greece; through Pascal's and Leibniz's calculators and Babbage's machines; to the ENIAC, iPhone, and beyond).

So, modern CS is the result of a marriage between (or merger of) the engineering problem of building better and better automatic calculating devices and the mathematical problem of understanding the nature of algorithmic computation. And that implies that modern CS has *both* engineering *and* science in its DNA. Hence its portmanteau nature.

The topics studied in contemporary CS roughly align along a spectrum ranging from the mathematical theory of computing, at one end, to the engineering of physical computers, at the other, as we saw in Section 3.4.2. (Newell, Perlis, and Simon were looking at this spectrum from one end; Knuth was looking at it from the other end.) The topics share a family resemblance (and perhaps nothing more than that, except for their underlying DNA) not only to each other but also to other disciplines (including mathematics, electrical engineering, information theory, communication, etc.), and they overlap with issues discussed in the cognitive sciences, philosophy (including ethics), sociology, education, the arts, and business:

> I reject the title question ["Are We Scientists or Engineers?"]. ... Computer Science ... spans a multidimensional spectrum from deep and elegant mathematics to crafty programming, from abstraction to solder joints, from deep truth to elusive human factors, from scholars motivated purely by the desire for knowledge or practitioners making my everyday life better.

---

2 A "portmanteau" is a suitcase that opens into two equal sections. A "portmanteau *word*" – the term was coined by Lewis Carroll (1871) – is one with "two meanings packed up into one word," like 'smog' (meaning "smoke and fog").

It embraces the ethos of the scholar as well as that of the professional. To answer the question would be to exclude some portion of this spectrum, and I would be poorer for that.
(Wulf, 1995, p. 57)

20.3 Five Central Questions of CS

For sheer ambition, physics does not hold a candle to computer … science. … It is we, not the physicists, who must develop a theory of everything.
—Brian Cantwell Smith (2002, p. 53)

Rather than try to say *what* CS is the study of, or whether it is *scientific* or not, I suggest that the best way to understand it is as trying to answer five central questions. The single most central question is

1A. What *can* be computed?

But to answer that, we also need to ask

1B. *How* can it be computed?

The other questions follow logically from that central one. So, the five questions that CS is concerned with are

1. **What *can* be computed, and *how*?**
2. **What can be computed *efficiently*, and how?**
3. **What can be computed *practically*, and how?**
4. **What can be computed *physically*, and how?**
5. **What can be computed *ethically*, and how?**

Let's consider each of these in a bit more detail.

20.3.1 Computability

What is computation? This has always been the most fundamental question of our field.
—Peter J. Denning and Peter Wegner (2010)

***What* Can Be Computed?** Question 1A is the central question, because all other questions presuppose it. The fundamental task of any computer scientist – whether at the purely mathematical or theoretical end of the spectrum, or at the purely practical or engineering end – is to determine whether there is a computational solution to a given problem and, if so, how to implement it. But those implementation questions are covered by the rest of the questions on the previous list and only make sense after the first question has been answered. (Alternatively, they facilitate answering that first question; in any case, they serve the goal of answering it.) Question 1A includes these questions:

> **What is *computation*?**
> **What *kinds of things* are computed?**
> **What is *computable*?**

It is the question that logicians and computing pioneers Alonzo Church, Turing, Gödel, and others were originally concerned with – **which mathematical functions are computable?** – and whose answer has been given as the Church-Turing Computability Thesis: a *function* is computable if and only if it is computable by a Turing Machine (or any formalism logically equivalent to a Turing Machine, such as Church's lambda calculus or Gödel's general recursive functions). It is important to note that not all functions are computable. (The standard example of a *non*-computable function is the Halting Problem.) If all functions were computable, then computability would not be as interesting a notion.

Various branches of CS are concerned with identifying which problems can be expressed by computable functions. So, a corollary of the Computability Thesis is that a *task* is computable if and only if it can be expressed as a computable function. In Robert I. Soare (2012, p. 3289)'s characterization, the output of a Turing Machine "is the total number of 1's on the tape." So, the key to determining what is computable (i.e. what kinds of tasks are computable) is finding a coding scheme that allows a sequence of '1's – i.e. (a representation of) an integer – to be *interpreted as* a symbol, a pixel, a sound, etc.

Here are some examples:

- Is chess computable? Shannon 1950 investigated whether we can *computationally* analyze chess. (That is, can we play chess rationally?)
- Is cognition computable? The central question of AI is whether the functions that describe cognitive processes are computable (Section 18.2.2). Given the advances that have been made in AI to date, it seems clear that at least some aspects of cognition *are* computable, so a slightly more precise question is: *How much* of cognition is computable? (Rapaport 2012b, Section 2, pp. 34–35; Rapaport 2021)
- Is the weather computable? (B. Hayes 2007)
- Is fingerprint identification computable? (Srihari 2010)
- Is final-exam-scheduling computable? Faculty members in my department once debated whether it was possible to write a computer program that would schedule final exams with no time conflicts and in rooms that were the proper size for the class. Some thought this was a trivial problem; others thought there was no such algorithm (on the – perhaps dubious! – grounds that no one in the university administration had ever been able to produce such a schedule). In fact, this problem is *NP*-complete (http://www.cs.toronto.edu/~bor/373s13/L14.pdf). See also an early discussion of this problem in Forsythe, 1968, Section 3.3, p. 1027. On the meaning of '*NP*-complete,' see Section 20.3.2.)

This aspect of question 1A – which *tasks* are computable? – is close to Forsythe's famous concern:

> The question "What can be automated?" is one of the most inspiring philosophical and practical questions of contemporary civilization. (Forsythe, 1968, p. 1025)

Although similar in intent, Forsythe's question can be understood in a slightly different way: presumably, a process can be automated – i.e. done automatically, by a machine, without human intervention – if it can be expressed as an algorithm. That is, computable implies automatable. But automatable does not imply being computed: witness the invention of the electro-mechanical, direct-dialing system in telephony, which automated the task of the human operator.[3] Yes, direct dialing is also *computable*, but it wasn't a computer that automated it.

How Is It Computable? Question 1B – the "how" aspect of our central question – is equally important: CS cannot be satisfied with a mere existence statement to the effect that a problem *is* computable; it also requires a constructive answer in the form of an algorithm that explicitly shows *how* it is computable.

In the *Calvin and Hobbes* cartoon in Figure 20.1, Calvin discovers that if you input one thing (bread) into a toaster, it outputs something else (toast). Hobbes wonders what happened to the

---

3 "Strowger Switch," https://en.wikipedia.org/wiki/Strowger_switch.

Figure 20.1 CALVIN AND HOBBES ©1986 Watterson. Reprinted with permission of ANDREWS MCMEEL SYNDICATION. All rights reserved.

input. It didn't disappear, of course, nor did it "magically" turn into the output. The toaster *did something* to the bread (heated it); that intervening process is the analogue of an algorithm for the bread-to-toast function. Finding "intervening processes" requires algorithmic thinking and results in algorithms that specify the transformational relations between input and output.

(In psychology, behaviorism focused only on inputs and outputs: Pavlov's famous experiment input a bell to a dog, and the dog output saliva; but behaviorists didn't ask how the input and output were connected. It was *cognitive* psychology that focused on the intervening algorithms (Miller et al., 1960).)

In Section 2.7, we observed that for every *x*, there is a philosophy of *x*. Similarly, we can ask, given some *x*, whether there is a computational theory of *x*. *Finding* a computational solution to a problem requires "computational thinking": i.e. *algorithmic* (or procedural) thinking (Section 3.16.4).

Computational thinking includes what I called the Five Great Insights of CS (Section 7.4):

1. The *representation* insight:
   Only two nouns are needed to represent information
   ('0,' '1').
2. The *processing* insight:
   Only three verbs are needed to process information
   (***move*(left or right), *print*('0' or '1'), *erase*).**
3. The *structure* insight:
   Only three grammar rules are needed to combine actions
   (sequence, selection, repetition).
4. The "*closure*" insight:
   Nothing else is needed.
   This is the import of the Church-Turing Computability Thesis.[4]
5. The *implementation* insight:
   The first three insights can be physically implemented.

20.3.2 Efficient Computability

Question 2 – what can be computed *efficiently*? – is studied by the branch of computer science known as computational complexity theory. Given an algorithm, we can ask how much *time* it will

---

4 The exact number of nouns, verbs, or grammar rules depends on the formalism. E.g. some presentations add 'halt,' 'read' or 'exit' as verbs, or use recursion as the single rule of grammar, etc. The point is that there is a very minimal set and nothing else is *needed*. Of course, more nouns, verbs, or grammar rules allow for greater ease of expression.

take to be executed (roughly, the number of operations that will be needed) and how much *space* (memory) it will need. Computational-complexity theory is concerned with efficiency because it is concerned with the economics of the spatio-temporal resources needed for computing. A more general question is this: given the set of computable functions, which of them can be computed in, so to speak, less time than the age of the universe, or less space than the size of the universe? The principal distinction is whether a function is in the class called *P* (in which case it is "efficiently" *computable*) or in the class *NP* (in which case it is *not* efficiently computable but *is* efficiently "verifiable"):

> Even children can multiply two primes, but the reverse operation – splitting a large number into two primes – taxes even the most powerful computers. The numbers used in asymmetric encryption are typically hundreds of digits long. Finding the prime factors of such a large number is like trying to unmix the colors in a can of paint, ... "Mixing paint is trivial. Separating paint isn't." (Folger, 2016, p. 52)

Many, if not most, algorithms of practical importance are in *P*. By contrast, one important algorithm that is in *NP* is the Boolean Satisfiability Problem: given a molecular proposition of propositional logic with *n* atomic propositions, under what assignment of truth-values to those atomic propositions is the molecular proposition true (or "satisfied")? Algorithms that are equivalent to Satisfiability are said to be "*NP*-complete":

> What [Turing-award winner Stephen] Cook did was show that every problem in NP has a reduction to satisfiability. Solve satisfiability and you can solve all of NP. If you have an efficient algorithm for solving satisfiability, then all the problems whose solutions we can efficiently check have efficient algorithms, and P = NP. ... "NP-complete" means those problems in NP powerful enough that they can be used to solve any other problem in NP. (Fortnow, 2013, pp. 54, 58)

Whether *P* = *NP* is one of the major open questions in mathematics and CS. Most computer scientists both hope and believe that *P* ≠ *NP*. Here's why:

> What happens if P = NP? We get a beautiful world where everything is easy to compute. We can quickly learn just about everything, and the great mysteries of the world fall quickly, from cures [for] deadly diseases to the nature of the universe. The beautiful world also has a dark underbelly, including the loss of privacy and jobs, as there is very little computers cannot figure out or accomplish. (Fortnow, 2013, p. 9)[5]

Terminology: *P* is so-called because it is the class of functions computable in "Polynomial time," and *NP* is so-called because it is the class of functions computable in "Non-deterministic Polynomial time." For more technical details, see https://en.wikipedia.org/wiki/Non-deterministic_Turing_machine and Bernhardt, 2016, pp. 63–67.

20.3.3 Practical Computability

Question 3 – what can be computed *practically*? – is considered both by complexity theorists as well as by more practically-oriented software engineers. Given a computable function in *P* (or, for

---

5 See the Online Resources for further reading on P vs. NP.

that matter, in *NP*), what are some *practically* efficient methods of actually computing it? For example, under certain circumstances, some sorting algorithms are more efficient in a practical sense (e.g. faster) than others. Even a computable function that is in *NP* might be practically computable in special cases (Fortnow, 2022). And some functions might only be practically computable "indirectly" via a "heuristic" (Section 5.6). A classic case is the Traveling Salesperson Problem, an *NP*-complete problem that software like Google Maps solves special cases of every day (even if the solutions are only "satisficing" ones (see Section 2.5.1)).[6]

20.3.4 Physical Computability

Question 4 – what can be computed *physically*? – brings in both empirical (hence scientific) and engineering considerations. To the extent that the only (or the best) way to decide whether a computable function really does what it claims to do is to execute it on a real, physical computer, computers become an integral part of CS. Even a practically efficient algorithm for computing some function might run up against physical limitations. Here is one example: even if, eventually, computational linguists devise practically efficient algorithms for natural-language understanding and generation (Shapiro, 1989; Shapiro and Rapaport, 1991), it remains the case that humans have a finite life span, so the infinite capabilities of natural-language competence are not really required (a Turing Machine isn't needed; a simpler mathematical model called a "push-down automaton" might suffice).

This is also the question that issues in the design and construction of real computers ("computer engineering") are concerned with. It is where investigations into alternative physical implementations of computing (quantum, optical, DNA, etc.) come in. And it is concerned with the issues relating abstract computation to physical computation, such as those we discussed in Chapter 11 on hypercomputation and Chapter 16 on how programs relate to the world.

20.3.5 Ethical Computability

Question 5 brings in ethical considerations. Arden, elaborating Forsythe's question, said that "the basic question [is] … what can *and should* be automated" (Arden (1980, p. 29, my italics); see also Tedre 2015, pp. 167–168). As Matti Tedre (2008, p. 48, my italics) observes,

> Neither the theoretician's question "What can be efficiently automated?" nor the practitioner's question "How can processes be automated reliably and efficiently?" include, explicitly or implicitly, any questions about *why* processes should be automated at all, if it is *desirable* to automate things or to introduce new technologies, or *who decides* what will be automated.

Actually, the question "What should be computed?" is slightly ambiguous. It could simply refer to questions of practical efficiency: given a sorting problem, which sorting algorithm *should* be used; i.e. which one is the "best" or "most practical" or "most efficient" in the actual circumstances? But this sense of 'should' does not really differentiate this question from question (3).

It is the *ethical* interpretation that makes this question interesting (Dietrich et al., 2021). Tedre's earlier observations are clearly related to the issues we looked at in Chapter 16 concerning the *purpose* of an algorithm. And even if there is a practical and efficient algorithm for making certain decisions (e.g. as in the case of autonomous vehicles), there is still the question of whether

---

6 See the Online Resources for further reading on practical computability.

we *should* use those algorithms to actually make decisions for us. Or let us suppose the goal of AI – a computational theory of cognition – is practically and efficiently computable by physically plausible computers. One can and should still raise the question of whether such "artificial intelligences" *should* be created and whether we (their creators) have any ethical or moral obligations toward them, and vice versa! (See Delvaux, 2016, Nevejans, 2016.) And there is the question of implicit biases that might be (intentionally or unintentionally) built into some machine-learning algorithms.

20.4 Wing's Five Questions

I have offered five questions as the focus of CS. Jeannette Wing (2008b, p. 58) also offers "Five Deep Questions in Computing":

> $P = NP$?
> What is computable?
> What is intelligence?
> What is information?
> (How) can we build complex systems simply?

Later, she added a sixth (Wing, 2008a, p. 3724):

> the most basic question of all: *what is a computer?*.

Let's compare our two lists.

Wing's *first* question is part of our second question: "What is efficiently computable?"

Curiously, her *second* question is our *central* one! (I should note, however, that a later essay (Wing, 2008a, p. 3724) says that her five questions are a "set," thus "no ordering implied.")

Her third question can be rephrased as "How much of (human) cognition is computable?" which is a special case of our central question. It is, as we have seen, the central question of AI.

Her fourth question can be seen as asking an ontological question about the nature of what it is that is computed: Is it numbers (0s and 1s)? Is it symbols ('0's and '1's)? Is it information in some sense (and, if so, in which sense)? In the present context, "What is information?" is closely related to the question we asked in Section 3.4.2 about what objects CS studies. Thus, it, too, is an aspect of our central question.

Wing's fifth question is ambiguous between two readings of 'build.' (a) On a software reading, this question can be viewed in an abstract (scientific, mathematical) way as asking about the structural nature of software: structured programming and the issues concerning the proper use of the "goto" statement (Dijkstra, 1968) would fall under this category. As such, it concerns the grammar rules, and so it is an aspect of our central question. (b) On a hardware reading, it is an engineering question: how should we build *physical* computers? On that interpretation, it is part of our fourth question.

Whether or not her sixth question is the most basic one (perhaps "What is computable?" is more basic?), it would seem to be an aspect of the "how" part of either our central question or our fourth question: how can something be computed physically?

Thus, Wing's questions can be boiled down to two:

1. What is computation such that only some things can be computed?
   (And what can be computed [efficiently], and how?)
2. (How) can we build devices to perform these computations?

The first is equivalent to our questions 1–3. The second is equivalent to our question 4. We see once again the two parts of the discipline: the scientific (or mathematical, or abstract) and the engineering (or concrete).

But it is interesting and important to note that none of Wing's questions correspond to our ethical question 5. Robin K. Hill observes,

> Whereas the philosophy of computer science has heretofore been directed largely toward the study of formal systems by means of other formal systems … concerned professionals have also devoted attention to the ethics of computing, taking on issues like privacy, the digital divide, and bias in selection algorithms. Let's keep it up. There are plenty. (Hill, 2017)

20.5 Conclusion

I said that our survey *suggests* that there is no *simple, one-sentence* answer to the question "What is CS?" However, if we were to summarize the discussion in this chapter in *one* sentence, it would look something like this:

CS is the scientific (or STEM) study of

what problems can be solved,
what tasks can be accomplished,
and what features of the world can be understood …

… ***computationally,*** **i.e. using a language with only**

2 nouns ('0,' '1'),
3 verbs ('move,' 'print,' 'halt'),
3 grammar rules (sequence, selection, repetition),
and nothing else,

and then to provide algorithms to show how this can be done

efficiently,
practically,
physically,
and ethically.

This definition is hardly a *simple* sentence! However, one of the epigraphs for Chapter 3 – from an interview with a computational musician – comes closer, so we will end where that chapter began:

> **The Holy Grail of computer science is** *to capture the messy complexity of the natural world* **and** *express it algorithmically*.
> —Teresa Marrin Nakra, quoted in Davidson 2006, p. 66, my italics

Bibliography

Aaronson, S. (2012). The toaster-enhanced Turing machine. *Shtetl-Optimized*. http://www .scottaaronson.com/blog/?p=1121.

Aaronson, S. (2018). Three questions about quantum computing. PowerPoint slides, 13 September, https://www.scottaaronson.com/talks/3questions.ppt.

Abelson, H., Sussman, G. J., and Sussman, J. (1996). *Structure and Interpretation of Computer Programs*. MIT Press, Cambridge, MA. https://web.archive.org/web/20011111153820/https:// mitpress.mit.edu/sicp/full-text/book/book-Z-H-7.html.

Abrahams, P. W. (1987). What is computer science? *Communications of the ACM*, 30(6):472–473.

Adamson, P. (2019). What was philosophy? *New York Review of Books*, 66(11):55–57.

Aho, A. V., Hopcroft, J. E., and Ullman, J. D. (1983). *Data Structures and Algorithms*. Addison-Wesley, Reading, MA.

Aizawa, K. (2010). Computation in cognitive science: It is not all about Turing-equivalent computation. *Studies in History and Philosophy of Science*, 41(3):227–236.

Allen, C. (2017). On (not) defining cognition. *Synthese*, 194:4233–4249.

Allen, L. G. (2001). Teaching mathematical induction: An alternative approach. *Mathematics Teacher*, 94(6):500–504.

Allen, S. (1989). *Meeting of Minds*. Prometheus Books.

Anderson, A. R. and Belnap, Jr., N. D., editors (1975). *Entailment: The Logic of Relevance and Necessity*, volume I. Princeton University Press, Princeton, NJ.

Anderson, A. R., Belnap, Jr., N. D., and Dunn, J. M., editors (1992). *Entailment: The Logic of Relevance and Necessity*, volume II. Princeton University Press, Princeton, NJ.

Anderson, B. L. (2015). Can computational goals inform theories of vision? *Topics in Cognitive Science*, 7:274–286.

Anderson, M. and Anderson, S. L. (2007). Machine ethics: Creating an ethical intelligent agent. *AI Magazine*, 28(4):15–26.

Anderson, M. and Anderson, S. L. (2010). Robot be good. *Scientific American*, 303(4):72–77.

Angere, S. (2017). The square circle. *Metaphilosophy*, 48(1–2):79–95.

Anthes, G. (2006). *Computer science looks for a remake. Computerworld* (1 May), http://www .computerworld.com/s/article/110959/Computer_Science:Looks_for_a_Remake.

Arden, B. W., editor (1980). *What Can Be Automated? The Computer Science and Engineering Research Study (COSERS)*. MIT Press, Cambridge, MA.

Ardis, M., Basili, V., Gerhart, S., Good, D., Gries, D., Kemmerer, R., Leveson, N., Musser, D., Neumann, P., and von Henke, F. (1989). Editorial process verification. *Communications of the ACM*, 32(3):287–290. "ACM Forum" letter to the editor, with responses by J.H. Fetzer and P.J. Denning.

Philosophy of Computer Science: An Introduction to the Issues and the Literature, First Edition. William J. Rapaport.
© 2023 John Wiley & Sons, Inc. Published 2023 by John Wiley & Sons, Inc.

Aref, H. (2004). Recipe for an affordable supercomputer: Take 1,100 apples .... *Chronicle of Higher Education*, page B14. 5 March.

Asimov, I. (1957). The feeling of power. In Fadiman, C., editor, *The Mathematical Magpie*, pages 3–14. Simon and Schuster, 1962, New York.

Asimov, I. (1976). The bicentennial man. In Asimov, I., editor, *The Bicentennial Man and Other Stories*, pages 135–173. Doubleday, Garden City, NY.

Assadian, B. and Buijsman, S. (2019). Are the natural numbers fundamentally ordinals? *Philosophy and Phenomenological Research*, 99(3):564–580.

Avramides, A. (2020). Other minds. In Zalta, E. N., editor, *The Stanford Encyclopedia of Philosophy*. Metaphysics Research Lab, Stanford University, Winter 2020 edition.

Ayer, A. (1956). *The Problem of Knowledge*. Penguin, Baltimore.

Baars, B. J. (1997). Contrastive phenomenology: A thoroughly empirical approach to consciousness. In Block, N., Flanagan, O., and Güzeldere, G., editors, *The Nature of Consciousness: Philosophical Debates*, pages 187–201. MIT Press, Cambridge, MA.

Bajcsy, R. K., Borodin, A. B., Liskov, B. H., and Ullman, J. D. (1992). Computer science statewide review draft preface. Technical report, Computer Science Rating Committee. September; Confidential Report and Recommendations to the Commissioner of Education of the State of New York.

Baldwin, J. (1962). The creative process. In Center, N. C., editor, *Creative America*. Ridge Press, New York.

Bar-Haim, R., Dagan, I., Dolan, B., Ferro, L., Giampiccolo, D., Magnini, B., and Szpektor, I. (2006). The second PASCAL recognising textual entailment challenge. https://web.archive.org/web/20210123145724/http://u.cs.biu.ac.il/~nlp/downloads/publications/RTE2-organizers.pdf.

Barr, A. (1985). Systems that know that they don't understand. https://web.archive.org/web/20000830090421/http://www.stanford.edu/group/scip/avsgt/cognitiva85.pdf.

Barwise, J. (1989a). For whom the bell rings and cursor blinks. *Notices of the American Mathematical Society*, 36(4):386–388.

Barwise, J. (1989b). Mathematical proofs of computer system correctness. *Notices of the American Mathematical Society*, 36:844–851.

Battersby, S. (2015). Moon could be a planet under new definition. *New Scientist*, 228(3048):9.

Baum, L. F. (1900). *The Wizard of Oz*. Dover, 1966 reprint, New York.

Bechtel, W. and Abrahamsen, A. (2005). Explanation: A mechanistic alternative. *Studies in History and Philosophy of the Biological and Biomedical Sciences*, 36:421–441.

Becker, A. (2018). *What Is Real? The Unfinished Quest for the Meaning of Quantum Physics*. John Murray, London.

Beebee, H. (2017). Who is Rachel? Blade Runner and personal identity. *IAI [Institute of Art and Ideas] News*. 5 October, https://iainews.iai.tv/articles/who-is-rachael-the-philosophy-of-blade-runner-and-memory-auid-885.

Benacerraf, P. (1965). What numbers could not be. *Philosophical Review*, 74(1):47–73.

Benacerraf, P. and Putnam, H., editors (1984). *Philosophy of Mathematics: Selected Readings, 2nd Edition*. Cambridge University Press, New York.

Bench-Capon, T. (2020). Ethical approaches and autonomous systems. *Artificial Intelligence*, 281.

Berlin, B. and Kay, P. (1969). *Basic Color Terms: Their Universality and Evolution*. University of Chicago Press, Chicago.

Berners-Lee, T., Hall, W., Hendler, J., Shadbolt, N., and Weitzner, D. J. (2006). Creating a science of the Web. *Science*, 313:769–771.

Berners-Lee, T., Hendler, J., and Lassila, O. (2001). The semantic web. *Scientific American*. May.

Bernhardt, C. (2016). *Turing's Vision: The Birth of Computer Science*. MIT Press, Cambridge, MA.

Bernstein, J. and Holt, J. (2016). Spooky physics up close: An exchange. *New York Review of Books*, 63(19):62.

Bickle, J. (2015). Marr and reductionism. *Topics in Cognitive Science*, 7:299–311.

Bickle, J. (2020). Multiple realizability. In Zalta, E. N., editor, *The Stanford Encyclopedia of Philosophy*. Metaphysics Research Lab, Stanford University, Summer 2020 edition.

Bierce, A. (1906). *The Devil's Dictionary*. Dolphin Books/Doubleday & Co., 1967, Garden City, NY.

Biermann, A. (1990). *Great Ideas in Computer Science: A Gentle Introduction*. MIT Press, Cambridge, MA.

Blachowicz, J. (2016). There is no scientific method. *New York Times*. 4 July.

Blass, A. and Gurevich, Y. (2003). Algorithms: A quest for absolute definitions. *Bulletin of the European Association for Theoretical Computer Science (EATCS)*, 81:195–225. October. Page references to http://research.microsoft.com/en-us/um/people/gurevich/opera/164.pdf.

Block, N. (1978). Troubles with functionalism. In Savage, C., editor, *Minnesota Studies in the Philosophy of Science, Volume 9*, pages 261–325. University of Minnesota Press, Minneapolis.

Block, N. (1995). The mind as software of the brain. In Smith, E. E. and Osherson, D. N., editors, *An Invitation to Cognitive Science, 2nd Edition*; *Vol. 3*: *Thinking*, pages 377–425. MIT Press, Cambridge, MA.

Boden, M. A. (1977). *Artificial Intelligence and Natural Man*. Basic Books, New York.

Boden, M. A. (2006). *Mind as Machine: A History of Cognitive Science*. Oxford University Press, Oxford.

Böhm, C. and Jacopini, G. (1966). Flow diagrams, Turing machines and languages with only two formation rules. *Communications of the ACM*, 9(5):366–371.

Bolden, C. (2016). Katherine Johnson, the NASA mathematician who advanced human rights with a slide rule and pencil. *Vanity Fair*. September.

Boole, G. (1854). *An Investigation of the Laws of Thought: On Which Are Founded the Mathematical Theories of Logic and Probabilities*. Cambridge University Press (2009), Cambridge, UK.

Boolos, G. S. and Jeffrey, R. C. (1974). *Computability and Logic*. Cambridge University Press, Cambridge, UK.

Boorstin, D. (1983). *The Discoverers*. Random House, New York.

Borbely, R. (2005). Letter to the editor. *Scientific American*, pages 12, 15. March.

Bostrom, N. (2003). Are you living in a computer simulation? *Philosophical Quarterly*, 53(211): 243–255.

Bostrom, N. (2006). Do we live in a computer simulation? *New Scientist*, 192(2579):38–39.

Bostrom, N. (2009). The simulation argument: Some explanations. *Analysis*, 69(3):458–461.

Bowles, N. (2019). A journey—if you dare—into the minds of Silicon Valley programmers. *New York Times Book Review*. 1 April.

Brenner, S. (2012). The revolution in the life sciences. *Science*, 338:1427–1428. 14 December.

Bringsjord, S. (2015). A vindication of program verification. *History and Philosophy of Logic*, 36(3):262–277.

Bringsjord, S. and Arkoudas, K. (2004). The modal argument for hypercomputing minds. *Theoretical Computer Science*, 317:167–190.

Bringsjord, S., Govindarajulu, N. S., Banerjee, S., and Hummel, J. (2018). Do machine-learning machines learn? In Müller, V., editor, *Philosophy and Theory of Artificial Intelligence 2017; PT-AI 2017*, pages 136–157. Springer, Cham, Switzerland.

Bronowski, J. (1958). The creative process. *Scientific American*. September.

Brooks, Jr., F. P. (1975). *The Mythical Man-Month*. Addison-Wesley, Reading, MA.

Brooks, Jr., F. P. (1996). The computer scientist as toolsmith II. *Communications of the ACM*, 39(3):61–68.

Brooks, R. A. (1991). Intelligence without representation. *Artificial Intelligence*, 47:139–159.

Buechner, J. (2011). Not even computing machines can follow rules: Kripke's critique of functionalism. In Berger, A., editor, *Saul Kripke*, pages 343–367. Cambridge University Press, New York.

Buechner, J. (2018). Does Kripke's argument against functionalism undermine the standard view of what computers are? *Minds and Machines*, 28(3):491–513.

Bueno, O. (2020). Nominalism in the philosophy of mathematics. In Zalta, E. N., editor, *The Stanford Encyclopedia of Philosophy*. Metaphysics Research Lab, Stanford University, Fall 2020 edition.

Bunge, M. (1974). Toward a philosophy of technology. In Michalos, A. C., editor, *Philosophical Problems of Science and Technology*, pages 28–47. Allyn & Bacon, Boston.

Campbell, M. (2018). Mastering board games. *Science*, 362(6419):1118.

Campbell-Kelly, M. (2012). Alan Turing's other universal machine. *Communications of the ACM*, 55(7):31–33.

Cane, S. (2014). Interview: David Chalmers and Andy Clark. *New Philosopher*, 2:mind, 27 February.

Cannon, P. (2013). Kant at the bar: Transcendental idealism in daily life. *Philosophy Now*, Issue 95:15–17. March/April.

Cardoso, S. H. (1997). Specialized functions of the cerebral cortex. http://www.cerebromente.org.br/n01/arquitet/cortex_i.htm.

Carey, B. (2019). 'It's gigantic': A new way to gauge the chances for unresponsive patients. *New York Times*. 26 June.

Carhart, R. R. (1956). The systems approach to reliability. In *Proceedings, 2nd National Symposium on Quality Control and Reliability in Electronics, January 9–10*, pages 149–155. IRE Professional Group on Reliability and Quality Control, American Society for Quality Control, Electronics Technical Committee, Institute of Radio Engineers, Washington, DC.

Carleton, L. R. (1984). Programs, language understanding, and Searle. *Synthese*, 59:219–230.

Carnap, R. (1956). *Meaning and Necessity: A Study in Semantics and Modal Logic, Second Edition*. University of Chicago Press, Chicago.

Carpenter, B. and Doran, R. (1977). The other Turing machine. *The Computer Journal*, 20(3):269–279.

Carroll, L. (1871). *Through the Looking-Glass*.

Carroll, L. (1895). What the tortoise said to Achilles. *Mind*, 4(14):278–280.

Casner, S. M., Hutchins, E. L., and Norman, D. (2016). The challenges of partially automated driving. *Communications of the ACM*, 59(5):70–77.

Castañeda, H.-N. (1975). Individuation and non-identity: A new look. *American Philosophical Quarterly*, 12(2):131–140.

Castañeda, H.-N. (1989). Direct reference, the semantics of thinking, and guise theory (constructive reflections on David Kaplan's theory of indexical reference). In Almog, J., Perry, J., and Wettstein, H., editors, *Themes from Kaplan*, pages 105–144. Oxford University Press, New York.

Cathcart, T. and Klein, D. (2007). *Plato and a Platypus Walk into a Bar: Understanding Philosophy through Jokes*. Abrams Image, New York.

Cerf, V. G. (2015). There is nothing new under the sun. *Communications of the ACM*, 58(2):7.

Ceruzzi, P. (1988). Electronics technology and computer science, 1940–1975: A coevolution. *Annals of the History of Computing*, 10(4):257–275.

Chaitin, G. J. (2005). *Meta Math! The Quest for Omega*. Vintage, New York.

Chaitin, G. J. (2006). How real are real numbers? *International Journal of Bifurcation and Chaos*. http://www.cs.auckland.ac.nz/~chaitin/olympia.pdf (2006 version); http://www.umcs.maine.edu/~chaitin/wlu.html (2009 version).

Chalmers, D. J. (1993). A computational foundation for the study of cognition. *Journal of Cognitive Science (South Korea)*, 12 (2011)(4):323–357.

Chalmers, D. J. (1994). On implementing a computation. *Minds and Machines*, 4(4):391–402.

Chalmers, D. J. (1996). *The Conscious Mind: In Search of a Fundamental Theory*. Oxford University Press, New York.

Chalmers, D. J. (2012). The varieties of computation: A reply. *Journal of Cognitive Science (South Korea)*, 13(3):211–248.

Chalmers, D. J. (2017). The virtual and the real. *Disputatio*, 9(46):309–351.

Changizi, M. A., Hsieh, A., Nijhawan, R., Kanai, R., and Shimojo, S. (2008). Perceiving the present and a systematization of illusions. *Cognitive Science*, 32:459–503.

Chase, G. C. (1980). History of mechanical computing machinery. *Annals of the History of Computing*, 2(3):198–226.

Chater, N. and Oaksford, M. (2013). Programs as causal models: Speculations on mental programs and mental representation. *Cognitive Science*, 37(6):1171–1191.

Chetty, R. (2013). Yes, economics is a science. *New York Times*. 21 October.

Chiang, T. (2002). Seventy-two letters. In *Stories of Your Life and Others*, pages 147–200. Vintage, New York.

Chisholm, R. (1974). *Metaphysics, 2nd edition*. Prentice-Hall, Englewood Cliffs, NJ.

Chomsky, N. (1965). *Aspects of the Theory of Syntax*. MIT Press, Cambridge, MA.

Chow, S. J. (2015). Many meanings of 'heuristic'. *British Journal for the Philosophy of Science*, 66:977–1016.

Church, A. (1933). A set of postulates for the foundation of logic (second paper). *Annals of Mathematics, Second Series*, 34(4):839–864.

Church, A. (1936a). A note on the Entscheidungsproblem. *Journal of Symbolic Logic*, 1(1):40–41. See also "Correction to *A Note on the Entscheidungsproblem*", *Journal of Symbolic Logic* 1(3) (September): 101–102.

Church, A. (1936b). An unsolvable problem of elementary number theory. *American Journal of Mathematics*, 58(2):345–363.

Church, A. (1937). Review of Turing, 1936. *Journal of Symbolic Logic*, 2(1):42–43.

Church, A. (1940). On the concept of a random sequence. *Bulletin of the American Mathematical Society*, 2:130–135.

Churchland, P. S. and Sejnowski, T. J. (1992). *The Computational Brain*. MIT Press, Cambridge, MA.

Clark, A. and Chalmers, D. J. (1998). The extended mind. *Analysis*, 58:10–23.

Clark, K. L. and Cowell, D. F. (1976). *Programs, Machines, and Computation: An Introduction to the Theory of Computing*. McGraw-Hill, London.

Clarke, A. C. (1951). Sentinel of eternity. *[Avon] 10 Story Fantasy*, 3(1):41–51.

Clarke, A. C. (1953). *Childhood's End*. Random House/Del Rey, 1990, New York.

Cleland, C. E. (1993). Is the Church-Turing thesis true? *Minds and Machines*, 3(3):283–312.

Cleland, C. E. (1995). Effective procedures and computable functions. *Minds and Machines*, 5(1): 9–23.

Cleland, C. E. (2001). Recipes, algorithms, and programs. *Minds and Machines*, 11(2):219–237.

Cleland, C. E. (2002). On effective procedures. *Minds and Machines*, 12(2):159–179.

Cleland, C. E. (2004). The concept of computability. *Theoretical Computer Science*, 317:209–225.

Cockshott, P. and Michaelson, G. (2007). Are there new models of computation? Reply to Wegner and Eberbach. *The Computer Journal*, 50(2):232–247.

Coffa, J. (1991). *The Semantic Tradition from Kant to Carnap: To the Vienna Station*. Cambridge University Press, Cambridge, UK.

Colburn, T. R. (1999). Software, abstraction, and ontology. *The Monist*, 82(1):3–19.

Colburn, T. R. (2000). *Philosophy and Computer Science*. M.E. Sharpe, Armonk, NY.

Colburn, T. R., Fetzer, J. H., and Rankin, T. L., editors (1993). *Program Verification: Fundamental Issues in Computer Science*. Kluwer Academic Publishers, Dordrecht, The Netherlands.

Cole, D. (1999). Note on analyticity and the definability of "bachelor". https://web.archive.org/web/20210512143905/http://www.d.umn.edu/~dcole/bachelor.htm.

Comte, A. (1830). *Cours de philosophie positive (Course in Positive Philosophy)*. Bachelier, Paris. English trans. by Harriet Martineau, https://archive.org/details/positivephilosop01comtuoft.

Condliffe, J. (2019). The week in tech: Our future robots will need super-smart safety checks. *New York Times*. 22 March.

Conte, P. T. (1989). More on verification (letter to the editor). *Communications of the ACM*, 32(7):790.

Copeland, B. J. (1996). What is computation? *Synthese*, 108:335–359.

Copeland, B. J. (1997). The broad conception of computation. *American Behavioral Scientist*, 40(6):690–716.

Copeland, B. J. (1998). Even Turing machines can compute uncomputable functions. In Calude, C., Casti, J., and Dinneen, M. J., editors, *Unconventional Models of Computation*, pages 150–164. Springer-Verlag.

Copeland, B. J. (1999). The Turing-Wilkinson lecture series on the automatic computing engine. In Furukawa, K., Michie, D., and Muggleton, S., editors, *Machine Intelligence 15: Intelligent Agents*, pages 381–444. Oxford University Press, Oxford.

Copeland, B. J. (2002). Hypercomputation. *Minds and Machines*, 12(4):461–502.

Copeland, B. J., editor (2004). *The Essential Turing*. Oxford University Press, Oxford.

Copeland, B. J. (2013). What Apple and Microsoft owe to Turing. *Huff[ington] Post Tech/The Blog*. 12 August.

Copeland, B. J. (2017). Hilbert and his famous problem. In Copeland, B. J., Bowen, J. P., Sprevak, M., and Wilson, R., editors, *The Turing Guide*, pages 57–65. Oxford University Press, Oxford.

Copeland, B. J., Dresner, E., Proudfoot, D., and Shagrir, O. (2016). Time to reinspect the foundations? *Communications of the ACM*, 59(11):34–36.

Copeland, B. J. and Proudfoot, D. (2010). Deviant encodings and Turing's analysis of computability. *Studies in History and Philosophy of Science*, 41(3):247–252.

Copeland, B. J. and Shagrir, O. (2011). Do accelerating Turing machines compute the uncomputable? *Minds and Machines*, 21(2):221–239.

Copeland, B. J. and Sylvan, R. (1999). Beyond the universal Turing machine. *Australasian Journal of Philosophy*, 77(1):46–67.

Copeland, B. J. and Sylvan, R. (2000). Computability is logic-relative. In Hyde, D. and Priest, G., editors, *Sociative Logics and Their Applications: Essays by the Late Richard Sylvan*, pages 189–199. Ashgate.

Corry, L. (2017). Turing's pre-war analog computers: The fatherhood of the modern computer revisited. *Communications of the ACM*, 60(8):50–58.

Craver, M. (2007). Letter to the editor. *Scientific American*, page 16. May.

Crawford, K. (2016). Artificial intelligence's white guy problem. *New York Times*. 25 June.

Cummins, R. (1989). *Meaning and Mental Representation*. MIT Press, Cambridge, MA.

Dagan, I., Glickman, O., and Magnini, B. (2006). The PASCAL recognising textual entailment challenge. In Quiñonero Candela, J. et al., editors, *MLCW 2005*, pages 177–190. Springer-Verlag LNAI 3944, Berlin.

Darwin, C. (1872). *The Origin of Species*. Signet Classics, 1958, New York.

Davidson, J. (2006). Measure for measure: Exploring the mysteries of conducting. *The New Yorker*, pages 60–69. 21 August.

Davies, D. W. (1999). Repairs to Turing's universal computing machine. In Furukawa, K., Michie, D., and Muggleton, S., editors, *Machine Intelligence 15: Intelligent Agents*, pages 477–488. Oxford University Press, Oxford.

Davis, M. (1995a). An historical preface to engineering ethics. *Science and Engineering Ethics*, 1(1):33–48.

Davis, M. (1995b). Questions for STS from engineering ethics. http://ethics.iit.edu/publication/Questions_for_STS.pdf, 22 October.

Davis, M. (1996). Defining "engineer:" How to do it and why it matters. *Journal of Engineering Education*, 85(2):97–101.

Davis, M. (1998). *Thinking Like an Engineer: Studies in the Ethics of a Profession*. Oxford University Press, New York.

Davis, M. (2009). Defining engineering from Chicago to Shantou. *The Monist*, 92(3):325–338.

Davis, M. (2011). Will software engineering ever be engineering? *Communications of the ACM*, 54(11):32–34.

Davis, M. D. (1958). *Computability & Unsolvability*. McGraw-Hill, New York.

Davis, M. D., editor (1965). *The Undecidable: Basic Papers on Undecidable Propositions, Unsolvable Problems and Computable Functions*. Raven Press, New York.

Davis, M. D. (1978). What is a computation? In Steen, L. A., editor, *Mathematics Today: Twelve Informal Essays*, pages 241–267. Springer, New York.

Davis, M. D. (1995c). Mathematical logic and the origin of modern computers. In Herken, R., editor, *The Universal Turing Machine: A Half-Century Survey, Second Edition*, pages 135–158. Springer-Verlag, Vienna.

Davis, M. D. (2000). Overheard in the park. *American Scientist*, 88:366–367. July-August.

Davis, M. D. (2004). The myth of hypercomputation. In Teuscher, C., editor, *Alan Turing: The Life and Legacy of a Great Thinker*, pages 195–212. Springer, Berlin.

Davis, M. D. (2006a). The Church-Turing thesis: Consensus and opposition. In Beckmann, A., Berger, U., Löwe, B., and Tucker, J., editors, *Logical Approaches to Computational Barriers: Second Conference on Computability in Europe, CiE 2006, Swansea, UK, June 30–July 5*, pages 125–132. Springer-Verlag Lecture Notes in Computer Science 3988, Berlin.

Davis, M. D. (2006b). What is Turing reducibility? *Notices of the AMS*, 53(10):1218–1219.

Davis, M. D. (2006c). Why there is no such discipline as hypercomputation. *Applied Mathematics and Computation*, 178:4–7.

Davis, M. D. (2012). *The Universal Computer: The Road from Leibniz to Turing; Turing Centenary Edition*. CRC Press/Taylor & Francis Group, Boca Raton, FL. Originally published as *Engines of Logic: Mathematicians and the Origin of the Computer* (New York: W.W. Norton, 2000).

Davis, M. D. and Weyuker, E. J. (1983). *Computability, Complexity and Languages*. Academic Press, New York.

Davis, R., Samuelson, P., Kapor, M., and Reichman, J. (1996). A new view of intellectual property and software. *Communications of the ACM*, 39(3):21–30. Summary version of Samuelson et al., 1994.

Davis, R. M. (1977). Evolution of computers and computing. *Science*, 195:1096–1102. 18 March.

Dawkins, R. (2016). *The Selfish Gene: 40th Anniversary Edition*. Oxford University Press, Oxford.

Daylight, E. G. (2013). Towards a historical notion of "Turing—the father of computer science". http://www.dijkstrascry.com/sites/default/files/papers/Daylightpaper91.pdf.

Daylight, E. G. (2016). *Turing Tales*. Lonely Scholar, Geel, Belgium.

de Leeuw, K., Moore, E., Shannon, C., and Shapiro, N. (1956). Computability by probabilistic machines. In Shannon, C. and McCarthy, J., editors, *Automata Studies*, pages 183–212. Princeton University Press, Princeton, NJ.

De Millo, R. A., Lipton, R. J., and Perlis, A. J. (1979). Social processes and proofs of theorems and programs. *Communications of the ACM*, 22(5):271–280.

De Mol, L. and Primiero, G. (2015). When logic meets engineering: Introduction to logical issues in the history and philosophy of computer science. *History and Philosophy of Logic*, 36(3):195–204.

Dean, W. (2020). Recursive functions. In Zalta, E. N., editor, *The Stanford Encyclopedia of Philosophy*. Metaphysics Research Lab, Stanford University, summer 2020 edition.

Defoe, D. (1719). *Robinson Crusoe*. W.W. Norton, 1994, New York.

Delvaux, M. (2016). Draft report with recommendations to the Commission on Civil Law Rules on Robotics. 31 May, European Parliament Committee on Legal Affairs, http://www.europarl.europa .eu/sides/getDoc.do?pubRef=-//EP//NONSGML%2BCOMPARL%2BPE-582.443%2B01%2BDOC %2BPDF%2BV0//EN.

Dembart, L. (1977). Experts argue whether computers could reason, and if they should. *New York Times*. 8 May.

Dennett, D. C. (1971). Intentional systems. *Journal of Philosophy*, 68:87–106.

Dennett, D. C. (1975). Why the law of effect will not go away. *Journal for the Theory of Social Behaviour*, 5(2):169–188.

Dennett, D. C. (1978). Why you can't make a computer feel pain. *Synthese*, 38(3):415–456.

Dennett, D. C. (1987). *The Intentional Stance*. MIT Press, Cambridge, MA.

Dennett, D. C. (1995). *Darwin's Dangerous Idea*. Simon & Schuster, New York.

Dennett, D. C. (2009). Darwin's 'strange inversion of reasoning'. *Proceedings of the National Academy of Science*, 106, suppl. 1:10061–10065. 16 June.

Dennett, D. C. (2013a). *Intuition Pumps and Other Tools for Thinking*. W.W. Norton, New York.

Dennett, D. C. (2013b). Turing's 'strange inversion of reasoning'. In Cooper, S. B. and van Leeuwen, J., editors, *Alan Turing: His Work and Impact*, pages 569–573. Elsevier, Amsterdam.

Dennett, D. C. (2017). *From Bacteria to Bach and Back: The Evolution of Mind*. W.W. Norton, New York.

Denning, P. J. (1985). What is computer science? *American Scientist*, 73:16–19.

Denning, P. J. (1995). Can there be a science of information? *ACM Computing Surveys*, 27(1):23–25.

Denning, P. J. (2000). Computer science: The discipline. In Ralston, A., Reilly, E. D., and Hemmendinger, D., editors, *Encyclopedia of Computer Science, Fourth Edition*. Grove's Dictionaries, New York. Page references to http://cs.gmu.edu/cne/pjd/PUBS/ENC/cs99.pdf.

Denning, P. J. (2005). Is computer science science? *Communications of the ACM*, 48(4):27–31.

Denning, P. J. (2007). Computing is a natural science. *Communications of the ACM*, 50(7):13–18.

Denning, P. J. (2009). Beyond computational thinking. *Communications of the ACM*, 52(6):28–30.

Denning, P. J. (2010). What is computation? Opening statement. *Ubiquity*, 2010. November, Article 1.

Denning, P. J. (2013). The science in computer science. *Communications of the ACM*, 56(5):35–38.

Denning, P. J. (2017). Remaining trouble spots with computational thinking. *Communications of the ACM*, 60(6):33–39.

Denning, P. J., Comer, D. E., Gries, D., Mulder, M. C., Tucker, A., Turner, A. J., and Young, P. R. (1989). Computing as a discipline. *Communications of the ACM*, 32(1):9–23.

Denning, P. J. and Martell, C. H. (2015). *Great Principles of Computing*. MIT Press, Cambridge, MA.

Denning, P. J., Tedre, M., and Yongpradit, P. (2017). Misconceptions about computer science. *Communications of the ACM*, 60(3):31–33.

Denning, P. J. and Wegner, P. (2010). What is computation? *Ubiquity*, 2010. October.

Dershowitz, N. and Gurevich, Y. (2008). A natural axiomatization of computability and proof of Church's thesis. *Bulletin of Symbolic Logic*, 14(3):299–350.

Descartes, R. (1637). Discourse on method. In Haldane, E. S. and Ross, G., editors, *The Philosophical Works of Descartes*, pages 79–130. Cambridge University Press, 1970, Cambridge, UK.

Devitt, M. and Porot, N. (2018). The reference of proper names: Testing usage and intuitions. *Cognitive Science*, 42(5):1552–1585.

Devlin, K. (1992). Computers and mathematics (column). *Notices of the American Mathematical Society*, 39(9):1065–1066.

Dewdney, A. (1989). *The Turing Omnibus: 61 Excursions in Computer Science*. Computer Science Press, Rockville, MD.

Dewey, J. (1910). *How We Think: A Restatement of Reflective Thinking to the Educative Process, revised ed.* D.C. Heath, Boston.

Dewhurst, J. (2020). There is no such thing as miscomputation. http://philsci-archive.pitt.edu/17000/.

Dick, P. K. (1968). *Do Androids Dream of Electric Sheep?* Doubleday, New York. Reprinted by Random House/Del Rey/Ballantine, 1996.

Dietrich, E. (2001). Homo sapiens 2.0: Why we should build the better robots of our nature. *Journal of Experimental and Theoretical Artificial Intelligence*, 13(4):323–328.

Dietrich, E. (2007). After the humans are gone. *Journal of Experimental and Theoretical Artificial Intelligence*, 19(1):55–67.

Dietrich, E., Fields, C., Sullins, J. P., van Heuveln, B., and Zebrowski, R. (2021). *Great Philosophical Objections to Artificial Intelligence: The History and Legacy of the AI Wars.* Bloomsbury Academic, London.

Dijkstra, E. W. (1968). Go to statement considered harmful. *Communications of the ACM*, 11(3):147–148.

Dijkstra, E. W. (1972). Notes on structured programming. In Dahl, O.J., Dijkstra, E. W., and Hoare, C.A.R., editors, *Structured Programming*, pages 1–82. Academic Press, London.

Dijkstra, E. W. (1974). Programming as a discipline of mathematical nature. *American Mathematical Monthly*, 81(6):608–612.

Dijkstra, E. W. (1975). Guarded commands, nondeterminacy and formal derivation of programs. *Communications of the ACM*, 18(8):453–457.

Dijkstra, E. W. (1986). Mathematicians and computing scientists: The cultural gap. *Mathematical Intelligencer*, 8(1):48–52.

Douven, I. (2021). Abduction. In Zalta, E. N., editor, *The Stanford Encyclopedia of Philosophy*. Metaphysics Research Lab, Stanford University, Summer 2021 edition.

Dresner, E. (2003). Effective memory and Turing's model of mind. *Journal of Experimental & Theoretical Artificial Intelligence*, 15(1):113–123.

Dresner, E. (2012). Turing, Matthews and Millikan: Effective memory, dispositionalism and pushmepullyou states. *International Journal of Philosophical Studies*, 20(4):461–472.

Dretske, F. (1981). *Knowledge and the Flow of Information*. Blackwell, Oxford.

Dretske, F. (1985). Machines and the mental. *Proceedings and Addresses of the American Philosophical Association*, 59(1):23–33.

Dreyfus, H. L. (2001). *On the Internet*, 2nd Edition. Routledge, London.

Dreyfus, S. E. and Dreyfus, H. L. (1980). A five-state model of the mental activities involved in directed skill acquisition. Technical Report ORC-80-2, Operations Research Center, University of California, Berkeley. https://apps.dtic.mil/sti/pdfs/ADA084551.pdf.

Dumas, A. (1844). *The Count of Monte Cristo*. Penguin Books, 2003, London. Trans. Robin Buss.

Dummett, M. A. (1976). What is a theory of meaning? (II). In Evans, G. and McDowell, J., editors, *Truth and Meaning: Essays in Semantics*, pages 67–137. Clarendon Press, Oxford.

Dunning, B. (2007). The importance of teaching critical thinking. 16 May, http://skeptoid.com/episodes/4045.

Duwell, A. (2021). *Physics and Computation*. Cambridge University Press, Cambridge, UK. https://doi.org/10.1017/9781009104975.

Dyson, F. (2004). The world on a string. *New York Review of Books*, pages 16–19. 13 May.

Dyson, F. (2011a). The case for far-out possibilities. *New York Review of Books*, pages 26–27. 10 November.

Dyson, F. (2011b). How we know. *New York Review of Books*, pages 8, 10, 12. 10 March.

Dyson, G. (2012a). Turing centenary: The dawn of computing. *Nature*, 482(7386):459–460.

Dyson, G. (2012b). *Turing's Cathedral: The Origins of the Digital Universe*. Pantheon, New York.

Edelman, S. (2008a). *Computing the Mind*. Oxford University Press, New York.

Edelman, S. (2008b). On the nature of minds; or: Truth and consequences. *Journal of Experimental & Theoretical Artificial Intelligence*, 20(3):181–196.

Eden, A. H. (2005). Software ontology as a cognitive artefact. Talk given to the SUNY Buffalo Center for Cognitive Science, 21 September; slides at https://cse.buffalo.edu/~rapaport/eden2005-SWOntCogArt.pdf.

Eden, A. H. (2007). Three paradigms of computer science. *Minds and Machines*, 17(2):135–167.

Edmonds, D. and Warburton, N. (2010). *Philosophy Bites*. Oxford University Press, Oxford.

Egan, D. (2019). Is there anything especially expert about being a philosopher? *Aeon*. 6 December.

Egan, F. (2010). Computational models: A modest role for content. *Studies in History and Philosophy of Science*, 41(3):253–259.

Egan, F. (2012). Metaphysics and computational cognitive science: Let's not let the tail wag the dog. *Journal of Cognitive Science (South Korea)*, 13(1):39–49.

Egan, F. (2014). How to think about mental content. *Philosophical Studies*, 170(1):115–135.

Eilon, S. (1969). What is a decision? *Management Science*, 16(4, Application Series):B172–B189.

Einstein, A. (1921). Geometry and experience. 27 January. Address to the Prussian Academy of Sciences, Berlin; English version at http://www-history.mcs.st-andrews.ac.uk/Extras/Einstein_geometry.html.

Einstein, A. (1940). Considerations concerning the fundaments of theoretical physics. *Science*, 91(2369):487–492.

Ekdahl, B. (1999). Interactive computing does not supersede Church's thesis. *The Association of Management and the International Association of Management, 17th Annual International Conference, San Diego, CA, August 6–8, Proceedings Computer Science*, 17(2):261–265. Part B.

Ellerton, P. (2016). What exactly is the scientific method and why do so many people get it wrong? *The Conversation*. 14 September.

Elser, V. (2012). In a class by itself. *American Scientist*, 100:418–420. https://www.americanscientist.org/article/in-a-class-by-itself; see also https://www.americanscientist.org/author/veit_elser.

Euler, L. (1748). *Introductio in Analysin Infinitorum (Introduction to the Analysis of the Infinites)*. Trans. Ian Bruce, 16 January 2013; http://www.17centurymaths.com/contents/introductiontoanalysisvol1.htm.

Everett, M. (2012). Answer to "What's the most important question, and why?". *Philosophy Now*, 92:38–41.

Feferman, S. (1992). Turing's "oracle": From absolute to relative computability—and back. In Echeverria, J., Ibarra, A., and Mormann, T., editors, *The Space of Mathematics: Philosophical, Epistemological, and Historical Explorations*, pages 314–348. Walter de Gruyter, Berlin.

Feferman, S. (2006). Are there absolutely unsolvable problems? Gödel's dichotomy. *Philosophia Mathematica*, 14(2):134–152.

Feferman, S. (2011). Gödel's incompleteness theorems, free will, and mathematical thought. In Swinburne, R., editor, *Free Will and Modern Science*. Oxford University Press/British Academy.

Feigenbaum, E. A. (2003). Some challenges and grand challenges for computational intelligence. *Journal of the ACM*, 50(1):32–40.

Fekete, T. and Edelman, S. (2011). Towards a computational theory of experience. *Consciousness and Cognition*, 20(3):807–827.

Fetzer, J. H. (1988). Program verification: The very idea. *Communications of the ACM*, 31(9):1048–1063.

Feyerabend, P. (1975). *Against Method: Outline of an Anarchistic Theory of Knowledge*. Verso, London.

Figdor, C. (2009). Semantic externalism and the mechanics of thought. *Minds and Machines*, 19(1):1–24.

Fiske, E. B. (1989). Between the 'two cultures': Finding a place in the curriculum for the study of technology. *New York Times*. 29 March.

Fletcher, J. (1972). Indicators of humanhood: A tentative profile of man. *Hastings Center Report*, 2(5):1–4.

Florman, S. C. (1994). *The Existential Pleasures of Engineering*, Second Edition. St. Martin's Press, New York.

Fodor, J. A. (1968a). The appeal to tacit knowledge in psychological explanation. *Journal of Philosophy*, 65(20):627–640.

Fodor, J. A. (1968b). *Psychological Explanation: An Introduction to the Philosophy of Psychology*. Random House, New York.

Fodor, J. A. (1974). Special sciences (or: The disunity of science as a working hypothesis). *Synthese*, 28(2):97–115.

Fodor, J. A. (1975). *The Language of Thought*. Thomas Y. Crowell Co., New York.

Fodor, J. A. (1978). Tom Swift and his procedural grandmother. *Cognition*, 6:229–247.

Fodor, J. A. (1980). Methodological solipsism considered as a research strategy in cognitive psychology. *Behavioral and Brain Sciences*, 3(1):63–109.

Folger, T. (2016). The quantum hack. *Scientific American*, 314(2):48–55. February.

Ford, H. (1928). My philosophy of industry. *The Forum*, 79(4). Interview conducted by Fay Leone Faurote; see also http://quoteinvestigator.com/2016/04/05/so-few/.

Forster, E. M. (1909). The machine stops. In *The Collected Tales of E.M. Forster*, pages 144–197. Modern Library, 1968, New York.

Forster, E. M. (1910). *Howards End*. Dover Publications (2002), Mineola, NY.

Forsythe, G. E. (1967a). A university's educational program in computer science. *Communications of the ACM*, 10(1):3–8.

Forsythe, G. E. (1967b). What to do till the computer scientist comes. *American Mathematical Monthly*, 75(5):454–462.

Forsythe, G. E. (1968). Computer science and education. *Information Processing 68: Proceedings of IFIP Congress 1968*, pages 1025–1039.

Fortnow, L. (2006). Principles of problem solving: A TCS response. http://blog .computationalcomplexity.org/2006/07/principles-of-problem-solving-tcs.html, 14 July.

Fortnow, L. (2010). What is computation? *Ubiquity*, 2010. December, Article 5.

Fortnow, L. (2013). *The Golden Ticket: P, NP, and the Search for the Impossible*. Princeton University Press, Princeton, NJ.

Fortnow, L. (2018). A reduced Turing Award. 29 March, https://blog.computationalcomplexity.org/2018/03/a-reduced-turing-award.html.

Fortnow, L. (2022). Fifty years of P vs. NP and the possibility of the impossible. *Communications of the ACM*, 65(1):76–85.

Fox, M. (2013). Janos Starker, master of the cello, dies at 88. *New York Times*. 30 April.

Frailey, D. J. (2010). What is computation? Computation is process. *Ubiquity*, 2010. November, Article 5.

Franzén, T. (2005). *Gödel's Theorem: An Incomplete Guide to Its Use and Abuse*. A K Peters, Wellesley, MA.

Frazer, J. G. (1911–1915). *The Golden Bough: A Study in Magic and Religion*, 3rd ed. Macmillan, London.

Freeman, P. A. (1995). Effective computer science. *ACM Computing Surveys*, 27(1):27–29.

Freeth, T., Bitsakis, Y., Moussas, X., Seiradakis, J., Tselikas, A., Mangou, H., Zafeiropoulou, M., Hadland, R., Bate, D., Ramsey, A., Allen, M., Crawley, A., Hockley, P., Malzbender, T., Gelb, D., Ambrisco, W., and Edmunds, M. (2006). Decoding the ancient Greek astronomical calculator known as the Antikythera Mechanism. *Nature*, 444:587–591. 30 November.

Frenkel, E. (2013). *Love and Math: The Heart of Hidden Reality*. Basic Books, New York.

Friedman, B. and Kahn, Jr., P. H. (1997). People are responsible, computers are not. In Ermann, M. D., Williams, M. B., and Shauf, M. S., editors, *Computers, Ethics, and Society, Second Edition*, pages 303–314. Oxford University Press, New York. Excerpt from their "Human Agency and Responsible Computing: Implications for Computer System Design", *Journal of Systems and Software* (1992): 7–14.

Frigg, R. and Hartmann, S. (2020). Models in science. In Zalta, E. N., editor, *The Stanford Encyclopedia of Philosophy*. Metaphysics Research Lab, Stanford University, Spring 2020 edition.

Gal-Ezer, J. and Harel, D. (1998). What (else) should CS educators know? *Communications of the ACM*, 41(9):77–84.

Gandy, R. (1980). Church's thesis and principles for mechanisms. In Barwise, J., Keisler, H., and Kunen, K., editors, *The Kleene Symposium*, pages 123–148. North-Holland.

Gandy, R. (1988). The confluence of ideas in 1936. In Herken, R., editor, *The Universal Turing Machine: A Half-Century Survey*, Second Edition, pages 51–102. Springer-Verlag, Vienna.

Gauss, C. F. (1808). Letter to Bolyai. https://mathshistory.st-andrews.ac.uk/Biographies/Gauss/quotations/.

Gelenbe, E. (2011). Natural computation. *Ubiquity*, 2011. February, Article 1.

Giampiccolo, D., Magnini, B., Dagan, I., and Dola, B. (2007). The third PASCAL recognizing textual entailment challenge. In *Proceedings of the Workshop on Textual Entailment and Paraphrasing*, pages 1–9. Association for Computational Linguistics.

Gillies, D. (2002). Logicism and the development of computer science. In Kakas, A. C. and Sadri, F., editors, *Computational Logic: Logic Programming and Beyond; Essays in Honour of Robert A. Kowalski, Part II*, pages 588–604. Springer, Berlin.

Ginsberg, M. L., editor (1987). *Readings in Nonmonotonic Reasoning*. Morgan Kaufmann, Los Altos, CA.

Glanzberg, M. (2021). Truth. In Zalta, E. N., editor, *The Stanford Encyclopedia of Philosophy*. Metaphysics Research Lab, Stanford University, Summer 2021 edition.

Gleick, J. (2008). 'If Shakespeare had been able to Google…'. *New York Review of Books*, 55(20):77–79.

Gleick, J. (2011). *The Information: A History, a Theory, a Flood*. Pantheon, New York.

Gödel, K. (1931). *On Formally Undecidable Propositions of Principia Mathematica and Related Systems*. Dover (1962), New York. Trans. by B. Meltzer.

Gödel, K. (1938). Undecidable Diophantine propositions. In Feferman, S. et al., editors, *Kurt Gödel: Collected Works, Vol. III*, pages 164–175. Oxford University Press, 1995, Oxford.

Gödel, K. (1964). Postscriptum. In Davis, M., editor, *The Undecidable: Basic Papers on Undecidable Propositions, Unsolvable Problems and Computable Functions*, pages 71–73. Raven Press, 1965, New York.

Gold, E. M. (1965). Limiting recursion. *Journal of Symbolic Logic*, 30(1):28–48.

Goldstein, R. N. (2006). *Incompleteness: The Proof and Paradox of Kurt Gödel*. W.W. Norton, New York.

Goldstein, R. N. (2014). *Plato at the Googleplex: Why Philosophy Won't Go Away*. Pantheon Books, New York.

Goodman, N. D. (1987). Intensions, Church's thesis, and the formalization of mathematics. *Notre Dame Journal of Formal Logic*, 28(4):473–489.

Gopnik, Adam (2009). *What's the recipe? The New Yorker*, pages 106–112.

Gopnik, Adam (2013). Moon man: What Galileo saw. *The New Yorker*. 11 & 18 February.

Gopnik, Alison (1996). The scientist as child. *Philosophy of Science*, 63(4):485–514.

Gopnik, Alison (2009). *The Philosophical Baby: What Children's Minds Tell Us about Truth, Love, and the Meaning of Life*. Farrar, Straus and Giroux, New York.

Grabiner, J. V. (1988). The centrality of mathematics in the history of western thought. *Mathematics Magazine*, 61(4):220–230.

Grey, D. S. (2016). Language in use: Research on color words. http://www.putlearningfirst.com/language/research/colour_words.html.

Grier, D. A. (2005). *When Computers Were Human.* Princeton University Press, Princeton, NJ.

Gries, D. (1981). *The Science of Programming, 3rd printing.* Springer-Verlag, 1985, New York.

Grobart, S. (2011). Spoiled by the all-in-one gadget. *New York Times.* 27 March.

Gruber, J. (2007). Apple's computer, incorporated. *Macworld*, page 112. April.

Gurevich, Y. (1999). The sequential ASM thesis. *Bulletin of the European Association for Theoretical Computer Science*, 67:93–124.

Gurevich, Y. (2012). Foundational analyses of computation. Technical Report MSR-TR-2012-14, Microsoft Research, Redmond, WA. February, http://research.microsoft.com/pubs/158617/210.pdf.

Guzdial, M. (2011). A definition of computational thinking from Jeannette Wing. *Computing Education Blog* (22 March), https://computinged.wordpress.com/2011/03/22/a-definition-of-computational-thinking-from-jeanette-wing/.

Guzdial, M. (2021). Computing education is not the same as engineering education. *BLOG@CACM*. 22 April, https://cacm.acm.org/blogs/blog-cacm/252090-computing-education-is-not-the-same-as-engineering-education/fulltext.

Guzdial, M. and Kay, A. (2010). The core of computer science: Alan Kay's "triple whammy". *Computing Education Blog*. 24 May, http://computinged.wordpress.com/2010/05/24/the-core-of-computer-science-alan-kays-triple-whammy/.

Habib, S. (2000). Emulation. In Ralston, A., Reilly, E. D., and Hemmendinger, D., editors, *Encyclopedia of Computer Science, 4th Edition*, pages 647–648. Nature Publishing Group, London.

Hailperin, M., Kaiser, B., and Knight, K. (1999). *Concrete Abstractions: An Introduction to Computer Science Using Scheme.* Brooks/Cole, Pacific Grove, CA. https://gustavus.edu/mcs/max/concrete-abstractions.html.

Halina, M. (2021). Insightful artificial intelligence. *Mind & Language*, 36(2):315–329.

Hamming, R. (1968). One man's view of computer science. *Journal of the ACM*, 16(1):3–12.

Hamming, R. (1980). We would know what they thought when they did it. In Metropolis, N., Howlett, J., and Rota, G.-C., editors, *A History of Computing in the Twentieth Century: A Collection of Essays*, pages 3–9. Academic Press, New York.

Hamming, R. (1998). Mathematics on a distant planet. *American Mathematical Monthly*, 105(7):640–650.

Hammond, T. A. (2003). What is computer science? Myths vs. truths. http://web.archive.org/web/20030506091438/http://www.columbia.edu/~tah10/cs1001/whatcs.html.

Harel, D. (1980). On folk theorems. *Communications of the ACM*, 23(7):379–389.

Hartmanis, J. (1993). Some observations about the nature of computer science. In Shyamasundar, R., editor, *Foundations of Software Technology and Theoretical Computer Science*, pages 1–12. Springer Berlin/Heidelberg.

Hartmanis, J. (1995a). On computational complexity and the nature of computer science. *ACM Computing Surveys*, 27(1):7–16.

Hartmanis, J. (1995b). Response to the essays "On computational complexity and the nature of computer science". *ACM Computing Surveys*, 27(1):59–61.

Hartmanis, J. and Stearns, R. (1965). On the computational complexity of algorithms. *Transactions of the American Mathematical Society*, 117:285–306.

Hartree, D. (1949). *Calculating Instruments and Machines.* University of Illinois Press, Urbana, IL.

Haugeland, J. (1985). *Artificial Intelligence: The Very Idea.* MIT Press, Cambridge, MA.

Hawthorne, J. (2021). Inductive logic. In Zalta, E. N., editor, *The Stanford Encyclopedia of Philosophy.* Metaphysics Research Lab, Stanford University, Spring 2021 edition.

Hayes, B. (2004). Small-town story. *American Scientist*, pages 115–119. March-April.

Hayes, B. (2006). Gauss's day of reckoning. *American Scientist*, 94:200–205.

Hayes, B. (2007). Calculating the weather. *American Scientist*, 95(3).

Hayes, B. (2014a). Pencil, paper, and pi. *American Scientist*, 102(1):342–345.

Hayes, B. (2014b). Programming your quantum computer. *American Scientist*, 102(1):22–25.

Hayes, B. (2015). The 100-billion-body problem. *American Scientist*, 103(2):90–93.

Hayes, P. J. (1997). What is a computer? *The Monist*, 80(3):389–404.

Hearst, M. and Hirsh, H. (2000). AI's greatest trends and controversies. *IEEE Intelligent Systems*, 15(1):8–17.

Heller, N. (2019). Driven (Was the automotive era a terrible mistake?). *The New Yorker*, pages 24–29. 29 July.

Hempel, C. G. (1942). The function of general laws in history. *Journal of Philosophy*, 39(2):35–48.

Hempel, C. G. (1945). Geometry and empirical science. *American Mathematical Monthly*, 52(1):7–17.

Hempel, C. G. (1962). Deductive-nomological vs. statistical explanation. In Feigl, H. and Maxwell, G., editors, *Minnesota Studies in the Philosophy of Science, Vol. 3: Scientific Explanations, Space, and Time*, pages 98–169. University of Minnesota Press, Minneapolis.

Hendler, J., Shadbolt, N., Hall, W., Berners-Lee, T., and Weitzner, D. (2008). Web science: An interdisciplinary approach to understanding the Web. *Communications of the ACM*, 51(7):60–69.

Heng, K. (2014). The nature of scientific proof in the age of simulations. *American Scientist*, 102(3):174–177.

Herman, G. T. (1983). Algorithms, theory of. In Ralston, A. S. and Riley, E. D., editors, *Encyclopedia of Computer Science*, 3rd edition, pages 37–39. Van Nostrand Reinhold, New York.

Hidalgo, C. (2015). Planet hard drive. *Scientific American*, 313(2):72–75.

Higginbotham, A. (2014). The disillusionist (The unbelievable skepticism of the Amazing Randi). *New York Times Magazine*. 9 November.

Hilbert, D. (1899). Foundations of Geometry. Open Court, La Salle, IL.

Hilbert, D. (1900). Mathematical problems: Lecture delivered before the International Congress of Mathematicians at Paris in 1900. *Bulletin of the American Mathematical Society*, 8 (1902)(10): 437–479. Trans. Mary Winston Newson; first published in *Göttinger Nachrichten* (1900): 253–297.

Hill, R. K. (2008). Empire, regime, and perspective change in our creative activities. Conference on ReVisioning the (W)hole II: Curious Intersections, 25 September, http://web.archive.org/web/20080925064006/http://www.newhumanities.org/events/revisioning.html.

Hill, R. K. (2016). What an algorithm is. *Philosophy and Technology*, 29:35–59.

Hill, R. K. (2017). Fact versus frivolity in Facebook. *BLOG@CACM*, 26 February, http://cacm.acm.org/blogs/blog-cacm/214075-fact-versus-frivolity-in-facebook/fulltext.

Hill, R. K. (2018). Articulation of decision responsibility. *BLOG@CACM*. 21 May, https://cacm.acm.org/blogs/blog-cacm/227966-articulation-of-decision-responsibility/fulltext.

Hill, R. K. and Rapaport, W. J. (2018). Exploring the territory: The logicist way and other paths into the philosophy of computer science. *American Philosophical Association Newsletter on Philosophy and Computers*, 18(1):34–37.

Hillis, W. D. (1998). *The Pattern on the Stone: The Simple Ideas that Make Computers Work*. Basic Books, New York.

Hintikka, J. and Mutanen, A. (1997). An alternative concept of computability. In Hintikka, J., editor, *Language, Truth, and Logic in Mathematics*, pages 174–188. Springer, Dordrecht, The Netherlands.

Hoare, C.A.R. (1969). An axiomatic basis for computer programming. *Communications of the ACM*, 12(10):576–580, 583.

Hoare, C.A.R. (1986). Mathematics of programming. *Byte*, page 115ff. August. Reprinted in Colburn et al., 1993, pp. 135–154.

Hobbs, J. R., Stickel, M. E., Appelt, D. E., and Martin, P. (1993). Interpretation as abduction. *Artificial Intelligence*, 63(1–2):69–142.

Hofstadter, D. R. (1979). *Gödel, Escher, Bach: An Eternal Golden Braid*. Basic Books, New York.

Hofstadter, D. R. (1981). Metamagical themas: A coffeehouse conversation on the Turing test to determine if a machine can think. *Scientific American*, pages 15–36. Reprinted as "The Turing Test: A Coffeehouse Conversation", in Douglas R. Hofstadter & Daniel C. Dennett (eds.), *The Mind's I: Fantasies and Reflections on Self and Soul* (New York: Basic Books, 1981): 69–95.

Hofstadter, D. R. (2007). *I Am a Strange Loop*. Basic Books, New York.

Hofstadter, D. R. and Sander, E. (2013). *Surfaces and Essences: Analogy as the Fuel and Fire of Thinking*. Basic Books, New York.

Hollan, J., Hutchins, E., and Kirsh, D. (2000). Distributed cognition: Toward a new foundation for human-computer interaction research. *ACM Transactions on Computer-Human Interaction*, 7(2):174–196.

Holt, C. M. (1989). More on the very idea (letter to the editor). *Communications of the ACM*, 32(4):508–509.

Holt, J. (2009). Death: Bad? New York Times Book Review. 15 February.

Holt, J. (2016). Something faster than light? What is it? *New York Review of Books*, 63(17):50–52.

Hopcroft, J. E. and Ullman, J. D. (1969). *Formal Languages and Their Relation to Automata*. Addison-Wesley, Reading, MA.

Hopper, G. M. (1952). The education of a computer. In ACM '52: Proceedings of the 1952 ACM National Meeting (Pittsburgh), pages 243–249. ACM, Pittsburgh.

Horgan, J. (2018). Philosophy has made plenty of progress. Scientific American Cross-Check. 1 November, https://blogs.scientificamerican.com/cross-check/philosophy-has-made-plenty-of-progress/.

Horsman, D. C., Kendon, V., and Stepney, S. (2017). The natural science of computing. *Communications of the ACM*, 60(8):31–34.

Horsman, D. C., Stepney, S., Wagner, R. C., and Kendon, V. (2014). When does a physical system compute? *Proceedings of the Royal Society A: Mathematical, Physical and Engineering Sciences*, 470(2169):1–25.

Horsten, L. (2019). Philosophy of mathematics. In Zalta, E. N., editor, *The Stanford Encyclopedia of Philosophy*. Metaphysics Research Lab, Stanford University, Spring 2019 edition.

Howard, D. A. and Giovanelli, M. (2019). Einstein's philosophy of science. In Zalta, E. N., editor, *The Stanford Encyclopedia of Philosophy*. Metaphysics Research Lab, Stanford University, Fall 2019 edition.

Huber, H. G. (1966). Algorithm and formula. *Communications of the ACM*, 9(9):653–654.

Hugo, V. (1862). *Les Misérables*. Signet Classics, 1987, New York. Trans. Lee Fahnestock & Norman MacAfee.

Humphreys, P. (1990). Computer simulations. *PSA: Proceedings of the [1990] Biennial Meeting of the Philosophy of Science Association*, 2:497–506.

Humphreys, P. (2002). Computational models. *Philosophy of Science*, 69:S1–S11.

Hutchins, E. (1995a). *Cognition in the Wild*. MIT Press, Cambridge, MA.

Hutchins, E. (1995b). How a cockpit remembers its speeds. *Cognitive Science*, 19:265–288.

Huxley, A. (1932). *Brave New World*. http://www.huxley.net/bnw/.

Irmak, N. (2012). Software is an abstract artifact. *Grazer Philosophische Studien*, 86:55–72.

Israel, D. (2002). Reflections on Gödel's and Gandy's reflections on Turing's thesis. *Minds and Machines*, 12(2):181–201.

Nature Editors (2016). Digital intuition. *Nature*, 529:437. 28 January.

Jackendoff, R. (2012). *A User's Guide to Thought and Meaning*. Oxford University Press, Oxford.

James, W. (1897). The will to believe. In Myers, G. E., editor, *William James: Writings 1878–1899*, pages 457–479. Library of America, New York.

Johnson, G. (2002). To err is human. *New York times*. 14 July.

Johnson-Laird, P. N. (1981). Mental models in cognitive science. In Norman, D. A., editor, *Perspectives on Cognitive Science*, pages 147–191. Ablex, Norwood, NJ.

Johnson-Laird, P. N. (1988). *The Computer and the Mind: An Introduction to Cognitive Science*. Harvard University Press, Cambridge, MA.

Jurafsky, D. and Martin, J. H. (2000). *Speech and Language Processing: An Introduction to Natural Language Processing, Computational Linguistics, and Speech Recognition*. Prentice Hall, Upper Saddle River, NJ.

Kahneman, D. (2011). *Thinking, Fast and Slow*. Farrar, Strauss and Giroux, New York.

Kaiser, D. (2012). Paradoxical roots of "social construction". *Science*, 335:658–659. 10 February.

Kanat-Alexander, M. (2008). What is a computer? *Code Simplicity*. 10 October.

Kant, I. (1781). *Critique of Pure Reason*. St. Martin's Press, 1929, New York. Trans. N.K. Smith.

Kasparov, G. (2018). Chess, a Drosophila of reasoning. *Science*, 362(6419):1087.

Kaznatcheev, A. (2014). Falsifiability and Gandy's variant of the Church-Turing thesis. https://egtheory .wordpress.com/2014/09/01/falsifiability-and-gandys-variant-of-the-church-turing-thesis/, 1 September.

Keith, T. (2014). The letter that kicked off a radio career. *NPR.org*, 5 July.

Kemeny, J. G. (1959). *A Philosopher Looks at Science*. D. van Nostrand, Princeton, NJ.

Kirk, R. (1974). Sentience and behaviour. *Mind*, 83(329):43–60.

Kitcher, P. (2019). What makes science trustworthy? *Boston Review*. 7 November.

Kleene, S. C. (1952). *Introduction to Metamathematics*. D. Van Nostrand, Princeton, NJ.

Kleene, S. C. (1995). Turing's analysis of computability, and major applications of it. In Herken, R., editor, *The Universal Turing Machine: A Half-Century Survey*, Second Edition, pages 15–49. Springer-Verlag, Vienna.

Knobe, J. (2015). Do corporations have minds? *New York Times Opinionator*. 15 June.

Knuth, D. E. (1966). Algorithm and program: Information and data. *Communications of the ACM*, 9(9):654.

Knuth, D. E. (1972a). Ancient Babylonian algorithms. *Communications of the ACM*, 15(7):671–677.

Knuth, D. E. (1972b). George Forsythe and the development of computer science. *Communications of the ACM*, 15(8):721–727.

Knuth, D. E. (1973). *The Art of Computer Programming*, Second Edition. Addison-Wesley, Reading, MA.

Knuth, D. E. (1974a). Computer programming as an art. *Communications of the ACM*, 17(12):667–673.

Knuth, D. E. (1974b). Computer science and its relation to mathematics. *American Mathematical Monthly*, 81(4):323–343.

Knuth, D. E. (1985). Algorithmic thinking and mathematical thinking. *American Mathematical Monthly*, 92(3):170–181.

Knuth, D. E. (2001). *Things a Computer Scientist Rarely Talks About*. CSLI Publications, Stanford, CA. CSLI Lecture Notes Number 136.

Koen, B. V. (1988). Toward a definition of the engineering method. *European Journal of Engineering Education*, 13(3):307–315. Reprinted from *Engineering Education* (December 1984): 150–155.

Koen, B. V. (2009). The engineering method and its implications for scientific, philosophical, and universal methods. *The Monist*, 92(3):357–386.

Kolata, G. (2004). New studies question value of opening arteries. *New York Times*, pages A1, A21. 21 March.

Kornblith, H. (2013). Naturalism vs. the first-person perspective. *Proceedings & Addresses of the American Philosophical Association*, 87:122–141. November.

Korsmeyer, C. (2012). Touch and the experience of the genuine. *British Journal of Aesthetics*, 52(4):365–377.

Krantz, S. G. (1984). Letter to the editor. *American Mathematical Monthly*, 91(9):598–600.

Krauss, L. M. (2016). Gravity's black rainbow. *New York Review of Books*, 63(14):83–85.

Kripke, S. A. (2011). Vacuous names and fictional entities. In Kripke, S. A., editor, *Philosophical Troubles: Collected Papers, Volume 1*, pages 52–74. Oxford University Press, New York.

Kripke, S. A. (2013). The Church-Turing "thesis" as a special corollary of Gödel's completeness theorem. In Copeland, B. J., Posy, C. J., and Shagrir, O., editors, *Computability: Turing, Gödel, Church, and Beyond*, pages 77–104. MIT Press, Cambridge, MA.

Kugel, P. (1986). Thinking may be more than computing. *Cognition*, 22(2):137–198.

Kugel, P. (2002). Computing machines can't be intelligent (… and Turing said so). *Minds and Machines*, 12(4):563–579.

Kuhn, T. S. (1957). *The Copernican Revolution: Planetary Astronomy in the Development of Western Thought*. Harvard University Press, Cambridge, MA.

Kuhn, T. S. (1962). *The Structure of Scientific Revolutions*. University of Chicago Press, Chicago.

LaChat, M. R. (1986). Artificial Intelligence and ethics: An exercise in the moral imagination. *AI Magazine*, 7(2):70–79.

Ladyman, J. (2009). What does it mean to say that a physical system implements a computation? *Theoretical Computer Science*, 410(4–5):376–383.

Lakoff, G. and Johnson, M. (1980a). Conceptual metaphor in everyday language. *Journal of Philosophy*, 77(8):453–486.

Lakoff, G. and Johnson, M. (1980b). *Metaphors We Live By*. University of Chicago Press, Chicago.

Lamport, L. (2011). Euclid writes an algorithm: A fairytale. *International Journal of Software and Informatics*, 5(1-2, Part 1):7–20. Page references to http://research.microsoft.com/en-us/um/people/lamport/pubs/euclid.pdf.

Lamport, L. (2015). Who builds a house without drawing blueprints? *Communications of the ACM*, 58(4):38–41.

Landesman, C. (2021). A deep dive into the fold. *The Paper* (https://origamiusa.org/thepaper, page 30. https://bit.ly/3qFy8Vy.

Landgrebe, J. and Smith, B. (2021). Making AI meaningful again. *Synthese*, 198:2061–2081.

Langewiesche, W. (2009). *Fly by Wire: The Geese, the Glide, the Miracle on the Hudson*. Farrar, Strauss & Giroux, New York.

Lanier, J. (2005). Early computing's long, strange trip. *American Scientist*, 93(4):364–365.

Leibniz, G. W. (1683–1685). Introduction to a secret encyclopedia. In Dascal, M., editor, G.W. Leibniz: *The Art of Controversies*, pages 219–224. Springer (2008), Dordrecht, The Netherlands.

Leibniz, G. W. (1714). *The Principles of Philosophy Known as Monadology*. Modern English version by Jonathan Bennett (July 2007), *Some Texts from Early Modern Philosophy*, http://www.earlymoderntexts.com/assets/pdfs/leibniz1714b.pdf.

Leiter, B. (2004). Is economics a "science"? *Leiter Reports: A Philosophy Blog*. 8 October, http://leiterreports.typepad.com/blog/2004/10/is_economics_a_.html.

Leiter, B. (2005). Why is there a Nobel prize in economics? *Leiter Reports: A Philosophy Blog*. 12 October, http://leiterreports.typepad.com/blog/2005/10/why_is_there_a_.html.

Leiter, B. (2009). Alex Rosenberg on Cochrane and economics. *Leiter Reports: A Philosophy Blog*. 20 September, http://leiterreports.typepad.com/blog/2009/09/alex-rosenberg-on-cochrane-and-economics.html.

Lem, S. (1971). Non serviam. In *A Perfect Vacuum*. Harcourt Brace Jovanovich (1979), New York. Trans. Michael Kandel.

Lemonick, M. (2015). The Pluto wars revisited. *The New Yorker* online, http://www.newyorker.com/tech/elements/nasa-dawn-ceres-pluto-dwarf-planets.

Lessing, G. E. (1778). Anti-Goetze: Eine Duplik. In Göpfert, H., editor, *Werke*, pages Vol. 8, pp. 32–33. http://harpers.org/blog/2007/11/lessings-search-for-truth/.

Levesque, H. J. (2017). *Common Sense, the Turing Test, and the Quest for Real AI*. MIT Press, Cambridge, MA.

Levin, J. (2021). Functionalism. In Zalta, E. N., editor, *The Stanford Encyclopedia of Philosophy*. Metaphysics Research Lab, Stanford University, Winter 2021 edition.

Lewis, D. (1970). General semantics. *Synthese*, 22(1/2):18–67.

Lewis-Kraus, G. (2016). The great A.I. awakening. *New York Times Magazine*. 14 December.

Lewontin, R. (2014). The new synthetic biology: Who gains? *New York Review of Books*, 61(8):22–23.

Liao, M.-H. (1998). Chinese to English machine translation using SNePS as an interlingua. Unpublished PhD dissertation, Department of Linguistics, SUNY Buffalo, http://www.cse.buffalo.edu/sneps/Bibliography/tr97-16.pdf.

Libbey, P. and Appiah, K. A. (2019). Socrates questions, a contemporary philosopher answers. *New York Times*. 11 April.

Lindell, S. (2001). Computer science as a liberal art: The convergence of technology and reason. Talk presented at Haverford College, 24 January, http://www.haverford.edu/cmsc/slindell/Presentations/Computer%20Science%20as%20a%20Liberal%20Art.pdf.

Linnebo, Ø. (2018). Platonism in the philosophy of mathematics. In Zalta, E. N., editor, *The Stanford Encyclopedia of Philosophy*. Metaphysics Research Lab, Stanford University, Spring 2018 edition.

Linnebo, Ø. and Pettigrew, R. (2011). Category theory as an autonomous foundation. *Philosophia Mathematica*, 19(3):227–254.

Liskov, B. and Zilles, S. (1974). Programming with abstract data types. *ACM SIGPLAN Notices*, 9(4):50–59.

Lloyd, S. and Ng, Y. J. (2004). Black hole computers. *Scientific American*, 291(5):52–61.

Locke, J. (1694). *An Essay concerning Human Understanding*. Oxford University Press (1975), Oxford. Peter H. Nidditch (ed.).

Lohr, S. (2001). Frances E. Holberton, 84, early computer programmer. *New York Times*. 17 December.

Lohr, S. (2006). Group of university researchers to make Web science a field of study. *New York Times*. 2 November.

Lohr, S. (2008). Does computing add up in the classroom? *New York Times Bits (blog)*. 1 April.

Lohr, S. (2013). Algorithms get a human hand in steering Web. *New York Times*. 10 March.

Loui, M. C. (1987). Computer science is an engineering discipline. *Engineering Education*, 78(3):175–178.

Loui, M. C. (1995). Computer science is a new engineering discipline. *ACM Computing Surveys*, 27(1):31–32.

Ludlow, P. (2019). The social furniture of virtual worlds. *Disputatio*, 11(55):345–369.

Lycan, W. G. (1990). The continuity of levels of nature. In Lycan, W. G., editor, *Mind and Cognition: A Reader*, pages 77–96. Basil Blackwell, Cambridge, MA.

Machamer, P., Darden, L., and Craver, C. F. (2000). Thinking about mechanisms. *Philosophy of Science*, 67(1):1–25.

Machery, E. (2012). Why I stopped worrying about the definition of life … and why you should as well. *Synthese*, 185:145–164.

MacKenzie, D. (1992). Computers, formal proofs, and the law courts. *Notices of the American Mathematical Society*, 39(9):1066–1069. Also see introduction by K. Devlin (1992).

Madigan, T. (2014). A mind is a wonderful thing to meet. *Philosophy Now*, 100:46–47. January/February.

Mahoney, M. S. (2011). *Histories of Computing*. Harvard University Press, Cambridge, MA. Thomas Haigh, ed.

Malpas, R. (2000). The universe of engineering: A UK perspective. Technical report, Royal Academy of Engineering, London. https://www.engc.org.uk/EngCDocuments/Internet/Website/The%20Universe%20of%20Engineering%20Report%20(The%20Malpas%20Report).pdf.

Mander, K. (2007). Demise of computer science exaggerated. BCS: The Chartered Institute for IT; Features, Press and Policy, http://www.bcs.org/content/ConWebDoc/10138.

Markov, A. (1954). Theory of algorithms. *Tr. Mat. Inst. Steklov*, 42:1–14. trans. Edwin Hewitt, in *American Mathematical Society Translations*, Series 2, Vol. 15 (1960).

Marr, D. (1982). *Vision*. W.H. Freeman, New York.

Marshall, L. A. (2017). Gadfly or spur? The meaning of μ′νωψ in Plato's *Apology of Socrates*. *Journal of Hellenic Studies*, 137:163–174.

Martin, C. (2015). She seconds that emotion. *The Chronicle [of Higher Education] Review*, 61(33):B16.

Martin, D. (2008). David Caminer, 92, dies; a pioneer in computers. *New York Times*. 29 June.

Martin, G. R. (2011). *A Dance with Dragons*. Bantam Books, New York.

Martins, J. P. and Shapiro, S. C. (1988). A model for belief revision. *Artificial Intelligence*, 35(1):25–79.

Marx, K. (1845). Theses on Feuerbach. https://www.marxists.org/archive/marx/works/1845/theses/theses.htm.

Maudlin, T. (2019a). Quantum theory and common sense: It's complicated. *IAI [Institute of Art and Ideas] News*. 26 August, https://iai.tv/articles/quantum-theory-and-common-sense-auid-1254.

Maudlin, T. (2019b). The why of the world. *Boston Review*. 4 September.

McBride, N. (2007). The death of computing. BCS: The Chartered Institute for IT; Features, Press and Policy, http://www.bcs.org/content/ConWebDoc/9662.

McCarthy, J. (1959). Programs with common sense. In Blake, D. and Uttley, A., editors, *Proceedings of the ["Teddington"] Symposium on Mechanization of Thought Processes*. HM Stationary Office, London. http://www-formal.stanford.edu/jmc/mcc59/mcc59.html.

McCarthy, J. (1963). A basis for a mathematical theory of computation. In Braffort, P. and Hirshberg, D., editors, *Computer Programming and Formal Systems*. North-Holland. Page references to PDF at http://www-formal.stanford.edu/jmc/basis.html.

McCarthy, J. and Hayes, P. J. (1969). Some philosophical problems from the standpoint of Artificial Intelligence. In Meltzer, B. and Michie, D., editors, *Machine Intelligence 4*. Edinburgh University Press, Edinburgh.

McCarthy, J., Minsky, M., Rochester, N., and Shannon, C. (1955). A proposal for the Dartmouth Summer Research Project on Artificial Intelligence. http://www-formal.stanford.edu/jmc/history/dartmouth.html, 31 August.

McDermott, D. (1980). Artificial intelligence meets natural stupidity. In Haugeland, J., editor, *Mind Design: Philosophy, Psychology, Artificial Intelligence*, pages 143–160. MIT Press, Cambridge, MA.

McDermott, D. (2001). *Mind and Mechanism*. MIT Press, Cambridge, MA.

McGinn, C. (1989). Can we solve the mind-body problem? *Mind*, 98(391):349–366.

McGinn, C. (1993). *Problems in Philosophy: The Limits of Inquiry*. Blackwell, Oxford.

McGinn, C. (2012a). Name calling: Philosophy as ontical science. *The Stone/The New York Times Opinionator*, 9 March.

McGinn, C. (2012b). Philosophy by another name. *New York Times*. 4 March.

McGinn, C. (2015a). *Philosophy of Language: The Classics Explained*. MIT Press, Cambridge, MA.

McGinn, C. (2015b). The science of philosophy. *Metaphilosophy*, 46(1):84–103.

McGrath, M. and Frank, D. (2020). Propositions. In Zalta, E. N., editor, *The Stanford Encyclopedia of Philosophy*. Metaphysics Research Lab, Stanford University, Winter 2020 edition.

Meigs, J. B. (2012). Inside the future: How PopMech predicted the next 110 years. *Popular Mechanics*. 10 December, https://www.popularmechanics.com/technology/a8562/inside-the-future-how-popmech-predicted-the-next-110-years-14831802/.

Melville, H. (1851). *Moby-Dick (Norton Critical Edition*, second edition). W.W. Norton (2002), New York. Hershel Parker & Harrison Hayford (eds.).

Menabrea, L. F. and Lovelace, A. A. (1843). English translation of *Notions sur la machine analytique de M. Charles Babbage. [Richard Taylor's] Scientific Memoirs*, 3 (September):666ff. Trans. by Lovelace at https://psychclassics.yorku.ca/Lovelace/menabrea.htm. Lovelace's notes at https://psychclassics .yorku.ca/Lovelace/lovelace.htm.

Mendelson, E. (1990). Second thoughts about Church's thesis and mathematical proofs. *Journal of Philosophy*, 87(5):225–233.

Mertens, S. (2004). The revolution will be digitized. *American Scientist*, 92:195–196. March-April.

Metz, C. (2019a). Turing Award won by 3 pioneers in artificial intelligence. *New York Times*. 27 March.

Metz, C. (2019b). We teach A.I. systems everything, including our biases. *New York Times*. 11 November.

Michie, D. (1961). Trial and error. *Science Survey*, 2:129–145. Reprinted in Donald Michie, *On Machine Intelligence* (New York: John Wiley, 1974): 5–19.

Mili, A., Desharnais, J., and Gagné, J. R. (1986). Formal models of stepwise refinement of programs. *ACM Computing Surveys*, 18(3):231–276.

Miłkowski, M. (2018). From computer metaphor to computational modeling: The evolution of computationalism. *Minds and Machines*, 28:515–541.

Miller, G. A., Galanter, E., and Pribram, K. H. (1960). *Plans and the Structure of Behavior*. Henry Holt, New York.

Mills, H. D. (1971). Top-down programming in large systems. In Rustin, R., editor, *Debugging Techniques in Large Systems*, pages 41–55. Prentice-Hall, Englewood Cliffs, NJ.

Minsky, M. (1967). *Computation: Finite and Infinite Machines*. Prentice-Hall, Englewood Cliffs, NJ.

Minsky, M. (1968). *Semantic Information Processing*. MIT Press, Cambridge, MA.

Minsky, M. (1979). Computer science and the representation of knowledge. In Dertouzos, L. and Moses, J., editors, *The Computer Age: A Twenty Year View*, pages 392–421. MIT Press, Cambridge, MA.

Misak, C. and Talisse, R. B. (2019). Pragmatism endures. *Aeon*. 18 November.

Mish, F. C., editor (1983). *Webster's Ninth New Collegiate Dictionary*. Merriam-Webster, Springfield, MA.

Misselhorn, C. (2020). Artificial systems with moral capacities? A research design and its implications in a geriatric care system. *Artificial Intelligence*, 278. January.

Mitcham, C. (1994). *Thinking through Technology: The Path between Engineering and Philosophy*. University of Chicago Press, Chicago.

Mitchell, M. (2011). What is computation? Biological computation. *Ubiquity*, 2011. February, Article 3.

Montague, R. (1960). Towards a general theory of computability. *Synthese*, 12(4):429–438.

Montague, R. (1970). English as a formal language. In Thomason, R. H., editor, *Formal Philosophy: Selected Papers of Richard Montague*, pages 192–221. Yale University Press, 1974, New Haven, CT.

Moor, J. H. (1978). Three myths of computer science. *British Journal for the Philosophy of Science*, 29(3):213–222.

Moor, J. H. (1979). Are there decisions computers should never make? *Nature and System*, 1:217–229.

Moravec, H. (1998). When will computer hardware match the human brain? *Journal of Evolution and Technology*, 1(1).

Morris, C. (1938). *Foundations of the Theory of Signs*. University of Chicago Press, Chicago.

Morris, F. and Jones, C. (1984). An early program proof by Alan Turing. *Annals of the History of Computing*, 6(2):139–143.

Morris, G. J. and Reilly, E. D. (2000). Digital computer. In Ralston, A., Reilly, E. D., and Hemmendinger, D., editors, *Encyclopedia of Computer Science, Fourth Edition*, pages 539–545. Grove's Dictionaries, New York.

Mukherjee, S. (2018). This cat sensed death. What if computers could, too? *New York Times Magazine*, page MM14. 7 January.

Mullainathan, S. (2019). Biased algorithms are easier to fix than biased people. *New York Times*. 6 December.

Munroe, R. (2015). *Thing Explainer: Complicated Stuff in Simple Words*. Houghton Mifflin Harcourt, New York.

Munroe, R. (2019). *How To: Absurd Scientific Advice for Common Real-World Problems*. Riverhead Books, New York.

Murphy, G. L. (2019). On Fodor's first law of the nonexistence of cognitive science. *Cognitive Science*, 43(5).

Myers, W. (1986). Can software for the strategic defense initiative ever be error-free? *IEEE Computer*, pages 61–67.

Nagel, E., Newman, J. R., and Hofstadter, D. R. (2001). *Gödel's Proof, Revised Edition*. New York University Press, New York.

Nagel, T. (1987). *What Does It All Mean? A Very Short Introduction to Philosophy*. Oxford University Press, New York.

Nagel, T. (2016). How they wrestled with the new. *New York Review of Books*, 63(14):77–79.

Natarajan, P. (2014). What scientists really do. *New York Review of Books*, 61(16):64–66.

Natarajan, P. (2017). Calculating women. *New York Review of Books*, 64(9).

Nevejans, N. (2016). European civil law rules in robotics. Technical Report PE 571.379, Directorate-General for Internal Policies, Policy Department C: Citizens' Rights and Constitutional Affairs, Brussels. http://www.europarl.europa.eu/RegData/etudes/STUD/2016/571379/IPOL_STU(2016)571379_EN.pdf.

Newcombe, C., Rath, T., Zhang, F., Munteanu, B., Brooker, M., and Deardeuff, M. (2015). How Amazon Web Services uses formal methods. *Communications of the ACM*, 58(4):66–73.

Newell, A. (1980). Physical symbol systems. *Cognitive Science*, 4:135–183.

Newell, A. (1985-1986). Response: The models are broken, the models are broken. *University of Pittsburgh Law Review*, 47:1023–1031.

Newell, A., Perlis, A. J., and Simon, H. A. (1967). Computer science. *Science*, 157(3795):1373–1374.

Newell, A. and Simon, H. A. (1976). Computer science as empirical inquiry: Symbols and search. *Communications of the ACM*, 19(3):113–126.

New York Times (2006). Bush untethered. *New York Times*. 17 September.

Nicas, J. and Wichter, Z. (2019). A worry for some pilots: Their hands-on flying skills are lacking. *New York Times*. 14 March.

Northcott, R. and Piccinini, G. (2018). Conceived this way: Innateness defended. *Philosophers' Imprint*, 18(18):1–16.

O'Connor, J. and Robertson, E. (October 2005). The function concept. *MacTutor History of Mathematics Archive*, http://www-history.mcs.st-andrews.ac.uk/HistTopics/Functions.html.

O'Neill, O. (2013). Interpreting the world, changing the world. *Philosophy Now*, Issue 95:8–9. March/April.

O'Neill, S. (2015). The human race is a computer. *New Scientist*, 227(3029):26–27.

Oommen, B. J. and Rueda, L. G. (2005). A formal analysis of why heuristic functions work. *Artificial Intelligence*, 164(1-2):1–22.

Oppenheim, P. and Putnam, H. (1958). Unity of science as a working hypothesis. *Minnesota Studies in the Philosophy of Science*, 2(3):3–35.

Orr, H. A. (2013). Awaiting a new Darwin. *New York Review of Books*, 60(2):26–28.

O'Toole, G. (2020). The brain is merely a meat machine. *Quote Investigator*, https://quoteinvestigator .com/2020/05/06/machines/, 6 May.

O'Toole, G. (2021a). Computer science is not about computers, any more than astronomy is about telescopes. *Quote Investigator*, https://quoteinvestigator.com/2021/04/02/computer-science/, 2 April.

O'Toole, G. (2021b). The search for truth is more precious than its possession. *Quote Investigator*, https://quoteinvestigator.com/2021/11/19/search-truth/, 19 November.

Pandya, H. (2013). Shakuntala Devi, 'human computer' who bested the machines, dies at 83. *New York Times*. 23 April.

Papert, S. (1980). *Mindstorms: Children, Computers, and Powerful Ideas*. Basic Books, New York.

Parlante, N. (2005). What is computer science? *Inroads—The SIGSCE Bulletin*, 37(2):24–25.

Parnas, D. L. (1998). Software engineering programmes are not computer science programmes. *Annals of Software Engineering*, 6:19–37. Page references to pre-print at http://www.cas.mcmaster.ca/serg/ papers/crl361.pdf; a paper with the same title (but American spelling as 'program') appeared in *IEEE Software* 16(6) (1990): 19–30.

Parnas, D. L. (2017). The real risks of artificial intelligence. *Communications of the ACM*, 60(10):27–31.

Parsons, K. M. (2015). What is the public value of philosophy? *Huffington Post*. 8 April.

Parsons, T. W. (1989). More on verification (letter to the editor). *Communications of the ACM*, 32(7):790–791.

Pattis, R. E., Roberts, J., and Stehlik, M. (1995). *Karel the Robot: A Gentle Introduction to the Art of Programming*, Second Edition. John Wiley & Sons, New York.

Paulson, L., Cohen, A., and Gordon, M. (1989). The very idea (letter to the editor). *Communications of the ACM*, 32(3):375.

Pawley, A. L. (2009). Universalized narratives: Patterns in how faculty members define 'engineering'. *Journal of Engineering Education*, 98(4):309–319.

Peacocke, C. (1995). Content, computation and externalism. *Philosophical Issues*, 6:227–264.

Peacocke, C. (1999). Computation as involving content: A response to Egan. *Mind & Language*, 14(2):195–202.

Pearl, J. (2000). *Causality: Models, Reasoning, and Inference*. Cambridge University Press,Cambridge, UK.

Pelletier, F. J. (1999). A history of natural deduction and elementary logic textbooks. *History and Philosophy of Logic*, 20:1–31.

Pelletier, F. J. and Hazen, A. (2021). Natural deduction systems in logic. In Zalta, E. N., editor, *The Stanford Encyclopedia of Philosophy*. Metaphysics Research Lab, Stanford University, Winter 2021 edition.

Perlis, A. (1962). The computer in the university. In Greenberger, M., editor, *Management and the Computer of the Future*, pages 181–217. MIT Press, Cambridge, MA.

Perovic, K. (2017). Bradley's regress. In Zalta, E. N., editor, *The Stanford Encyclopedia of Philosophy*. Metaphysics Research Lab, Stanford University, Winter 2017 edition.

Perruchet, P. and Vinter, A. (2002). The self-organizing consciousness. *Behavioral and Brain Sciences*, 25(3):297–388.

Perry, Jr., W. G. (1970). *Forms of Intellectual and Ethical Development in the College Years: A Scheme*. Holt, Rinehart and Winston, New York.

Perry, Jr., W. G. (1981). Cognitive and ethical growth: The making of meaning. In Chickering, A. and Associates, editors, *The Modern American College*, pages 76–116. Jossey-Bass, San Francisco.

Pessin, A. (2009). The 60-Second Philosopher. Oneworld, London.

Petersen, S. (2007). The ethics of robot servitude. *Journal of Experimental and Theoretical Artificial Intelligence*, 19(1):43–54.

Petroski, H. (2003). Early education. *American Scientist*, 91:206–209. May-June.

Petroski, H. (2005). Technology and the humanities. *American Scientist*, 93:304–307. July-August.

Petroski, H. (2008). Scientists as inventors. *American Scientist*, 96(5):368–371. September-October.

Petzold, C. (2008). *The Annotated Turing: A Guided Tour through Alan Turing's Historic Paper on Computability and the Turing Machine*. Wiley, Indianapolis.

Piccinini, G. (2004). Functionalism, computationalism, and mental contents. *Canadian Journal of Philosophy*, 34(3):375–410.

Piccinini, G. (2006). Computation without representation. *Philosophical Studies*, 137(2):204–241.

Piccinini, G. (2007a). Computational explanation and mechanistic explanation of mind. In Marraffa, M., Caro, M. D., and Ferretti, F., editors, *Cartographies of the Mind: Philosophy and Psychology in Intersection*, pages 23–36. Springer, Dordrecht, The Netherlands.

Piccinini, G. (2007b). Computational modelling vs. computational explanation: Is everything a Turing machine, and does it matter to the philosophy of mind? *Australasian Journal of Philosophy*, 85(1):93–115.

Piccinini, G. (2007c). Computing mechanisms. *Philosophy of Science*, 74(4):501–526.

Piccinini, G. (2008). Computers. *Pacific Philosophical Quarterly*, 89:32–73.

Piccinini, G. (2010). The mind as neural software? Understanding functionalism, computationalism, and computational functionalism. *Philosophy and Phenomenological Research*, 81(2):269–311.

Piccinini, G. (2011). The physical Church-Turing thesis: Modest or bold? *British Journal for the Philosophy of Science*, 62:733–769.

Piccinini, G. (2015). *Physical Computation: A Mechanistic Account*. Oxford University Press, Oxford.

Piccinini, G. (2018). Computation and representation in cognitive neuroscience. *Minds and Machines*, 28(1):1–6.

Piccinini, G. (2020a). The computational theory of cognition. In Müller, V., editor, *Fundamental Issues of Artificial Intelligence*. Springer, Berlin.

Piccinini, G. (2020b). *Neurocognitive Mechanisms: Explaining Biological Cognition*. Oxford University Press, Oxford.

Piccinini, G. and Anderson, N. G. (2020). Ontic pancomputationalism. In Cuffaro, M. and Fletcher, S., editors, *Physical Perspectives on Computation, Computational Perspectives on Physics*. Cambridge University Press, Cambridge, UK.

Piccinini, G. and Bahar, S. (2013). Neural computation and the computational theory of cognition. *Cognitive Science*, 34:453–488.

Piccinini, G. and Maley, C. (2021). Computation in physical systems. In Zalta, E. N., editor, *The Stanford Encyclopedia of Philosophy*. Metaphysics Research Lab, Stanford University, Summer 2021 edition.

Pincock, C. (2011). Philosophy of mathematics. In Saatsi, J. and French, S., editors, *Companion to the Philosophy of Science*, pages 314–333. Continuum.

Plaice, J. (1995). Computer science is an experimental science. *ACM Computing Surveys*, 27(1):33.

Plato (1961a). Phaedrus. In Hamilton, E. and Cairns, H., editors, *The Collected Dialogues of Plato, including the Letters*. Princeton University Press, Princeton, NJ.

Plato (1961b). Republic. In Hamilton, E. and Cairns, H., editors, *The Collected Dialogues of Plato, including the Letters*, pages 575–844. Princeton University Press, Princeton, NJ.

Pleasant, J. C. (1989). The very idea (letter to the editor). *Communications of the ACM*, 32(3):374–375.

Pnueli, A. (2002). Analysis of reactive systems. https://web.archive.org/web/20160505160636/http://cs .nyu.edu/courses/fall02/G22.3033-004/, Fall.

"PolR" (2009). An explanation of computation theory for lawyers. 11 November; http://www.groklaw .net/article.php?story=20091111151305785 and http://www.groklaw.net/pdf2/ ComputationalTheoryforLawyers.pdf.

Popova, M. (2012). What is philosophy? An omnibus of definitions from prominent philosophers. http://www.brainpickings.org/index.php/2012/04/09/what-is-philosophy/.

Popper, K. (1953). *Conjectures and Refutations: The Growth of Scientific Knowledge*. Harper & Row, New York.

Popper, K. (1959). *The Logic of Scientific Discovery*. Harper & Row, New York.

Popper, K. (1972). *Objective Knowledge: An Evolutionary Approach*. Oxford University Press, Oxford.

Popper, K. (1978). Three worlds. https://tannerlectures.utah.edu/_resources/documents/a-to-z/p/popper80.pdf.

Post, E. L. (1941). Absolutely unsolvable problems and relatively undecidable propositions: Account of an anticipation. In Davis, M., editor, *Solvability, Provability, Definability: The Collected Works of Emil L. Post*, pages 375–441. Birkhaäuser, Boston.

Post, E. L. (1943). Formal reductions of the general combinatorial decision problem. *American Journal of Mathematics*, 65:197–215.

Prasse, M. and Rittgen, P. (1998). Why Church's thesis still holds: Some notes on Peter Wegner's tracts on interaction and computability. *The Computer Journal*, 41(6):357–362.

Preston, B. (2013). *A Philosophy of Material Culture: Action, Function, and Mind*. Routledge, New York.

Preston, J. (2020). Paul Feyerabend. In Zalta, E. N., editor, *The Stanford Encyclopedia of Philosophy*. Metaphysics Research Lab, Stanford University, Fall 2020 edition.

Proudfoot, D. and Copeland, B. J. (2012). Artificial Intelligence. In Margolis, E., Samuels, R., and Stich, S. P., editors, *The Oxford Handbook of Philosophy of Cognitive Science*. Oxford University Press.

Putnam, H. (1960). Minds and machines. In Hook, S., editor, *Dimensions of Mind: A Symposium*, pages 148–179. New York University Press, New York.

Putnam, H. (1965). Trial and error predicates and the solution to a problem of Mostowski. *Journal of Symbolic Logic*, 30(1):49–57.

Putnam, H. (1975). The meaning of 'meaning'. In Gunderson, K., editor, *Minnesota Studies in the Philosophy of Science, Vol. 7: Language, Mind, and Knowledge*, pages 131–193. University of Minnesota Press, Minneapolis.

Putnam, H. (1987). *The Many Faces of Realism: The Paul Carus Lectures*. Open Court, LaSalle, IL.

Putnam, H. (2015). Rational reconstruction. *Sardonic Comment* blog, 18 February, http://putnamphil.blogspot.com/2015/02/rational-reconstuction-in-1976-when.html.

Pylyshyn, Z. W. (1984). *Computation and Cognition: Towards a Foundation for Cognitive Science*. MIT Press, Cambridge, MA.

Pylyshyn, Z. W. (1992). Computers and the symbolization of knowledge. In Morelli, R., Brown, W. M., Anselmi, D., Haberlandt, K., and Lloyd, D., editors, *Minds, Brains & Computers: Perspectives in Cognitive Science and Artificial Intelligence*, pages 82–94. Ablex, Norwood, NJ. Page references to http://ruccs.rutgers.edu/images/personal-zenon-pylyshyn/docs/suffolk.pdf.

qFiasco, F. (2018). Book review [of Kasparov & Greengard, *Deep Thinking*]. *Artificial Intelligence*, 260:36–41.

Quillian, M. R. (1994). A content-independent explanation of science's effectiveness. *Philosophy of Science*, 61(3):429–448.

Quine, W. V. O. (1951). Two dogmas of empiricism. *Philosophical Review*, 60:20–43.

Quine, W. V. O. (1976). Whither physical objects? In Cohen, R., Feyerabend, P., and Wartofsky, M., editors, *Essays in Memory of Imre Lakatos*, pages 497–504. D. Reidel, Dordrecht, Holland.

Quine, W. V. O. (1987). *Quiddities: An Intermittently Philosophical Dictionary*. Harvard University Press, Cambridge, MA.

Quine, W. V. O. (1988). An unpublished letter from Quine to Hookway. In Hilary Putnam's *Sardonic Comment* blog, 4 May 2015, http://putnamphil.blogspot.com/2015/05/an-unpublished-letter-from-quine-to.html.

Quine, W. V. O. (1990). *Pursuit of Truth*. Harvard University Press, Cambridge, MA.

Radó, T. (1962). On non-computable functions. *The Bell System Technical Journal*, pages 877–884. May, http://alcatel-lucent.com/bstj/vol41-1962/articles/bstj41-3-877.pdf.

Ralston, A. (1971). *Introduction to Programming and Computer Science*. McGraw-Hill, New York.

Ralston, A. (1999). Let's abolish pencil-and-paper arithmetic. *Journal of Computers in Mathematics and Science Teaching*, 18(2):173–194.

Ramakrishna, P. (2019). 'There's just no doubt that it will change the world': David Chalmers on V.R. and A.I. *New York Times/The Stone*. 18 June.

Ramsey, F. P. (1929). Theories. In Mellor, D., editor, *Foundations: Essays in Philosophy, Logic, Mathematics and Economics; revised edition*, pages 101–125. Humanities Press, 1978, Atlantic Highlands, NJ.

Rapaport, W. J. (1982). Unsolvable problems and philosophical progress. *American Philosophical Quarterly*, 19:289–298.

Rapaport, W. J. (1984). Critical thinking and cognitive development. *American Philosophical Association Newsletter on Pre-College Instruction in Philosophy*, 1:4–5. Reprinted, *Proc. and Addr. Amer. Philosophical Assn* 57(5) (May 1984): 610–615.

Rapaport, W. J. (1986). Searle's experiments with thought. *Philosophy of Science*, 53:271–279.

Rapaport, W. J. (1988a). Syntactic semantics: Foundations of computational natural-language understanding. In Fetzer, J. H., editor, *Aspects of Artificial Intelligence*, pages 81–131. Kluwer Academic Publishers, Dordrecht, The Netherlands.

Rapaport, W. J. (1988b). To think or not to think: Review of Searle's *Minds, Brains and Science. Noûs*, 22(4):585–609.

Rapaport, W. J. (1992a). Logic, predicate. In Shapiro, S. C., editor, *Encyclopedia of Artificial Intelligence, 2nd edition*, pages 866–873. John Wiley, New York.

Rapaport, W. J. (1992b). Logic, propositional. In Shapiro, S. C., editor, *Encyclopedia of Artificial Intelligence, 2nd edition*, pages 891–897. John Wiley, New York.

Rapaport, W. J. (1995). Understanding understanding: Syntactic semantics and computational cognition. In Tomberlin, J. E., editor, *AI, Connectionism, and Philosophical Psychology (Philosophical Perspectives, Vol. 9)*, pages 49–88. Ridgeview, Atascadero, CA.

Rapaport, W. J. (1998). How minds can be computational systems. *Journal of Experimental and Theoretical Artificial Intelligence*, 10:403–419.

Rapaport, W. J. (1999). Implementation is semantic interpretation. *The Monist*, 82:109–130.

Rapaport, W. J. (2000). How to pass a Turing test: Syntactic semantics, natural-language understanding, and first-person cognition. *Journal of Logic, Language, and Information*, 9(4):467–490.

Rapaport, W. J. (2005a). Castañeda, Hector-Neri. In Shook, J. R., editor, *The Dictionary of Modern American Philosophers, 1860–1960*, pages 452–412. Thoemmes Press, Bristol, UK.

Rapaport, W. J. (2005b). Implementation is semantic interpretation: Further thoughts. *Journal of Experimental and Theoretical Artificial Intelligence*, 17(4):385–417.

Rapaport, W. J. (2005c). Philosophy of computer science: An introductory course. *Teaching Philosophy*, 28(4):319–341.

Rapaport, W. J. (2011). A triage theory of grading: The good, the bad, and the middling. *Teaching Philosophy*, 34(4):347–372.

Rapaport, W. J. (2012a). Intensionality vs. intentionality. http://www.cse.buffalo.edu/~rapaport/intensional.html.

Rapaport, W. J. (2012b). Semiotic systems, computers, and the mind: How cognition could be computing. *International Journal of Signs and Semiotic Systems*, 2(1):32–71.

Rapaport, W. J. (2013). How to write. http://www.cse.buffalo.edu/~rapaport/howtowrite.html.

Rapaport, W. J. (2017a). On the relation of computing to the world. In Powers, T. M., editor, *Philosophy and Computing: Essays in Epistemology, Philosophy of Mind, Logic, and Ethics*, pages 29–64. Springer, Cham, Switzerland.

Rapaport, W. J. (2017b). Semantics as syntax. *American Philosophical Association Newsletter on Philosophy and Computers*, 17(1):2–11.

Rapaport, W. J. (2017c). What is computer science? *American Philosophical Association Newsletter on Philosophy and Computers*, 16(2):2–22.

Rapaport, W. J. (2018). What is a computer? A survey. *Minds and Machines*, 28(3):385–426.

Rapaport, W. J. (2021). Yes, AI *Can* match human intelligence. Draft of book chapter in progress, https://cse.buffalo.edu/~rapaport/Papers/aidebate.pdf.

Reese, H. (2014a). The joy of teaching computer science in the age of Facebook. *The Atlantic*. 18 February.

Reese, H. (2014b). Why study philosophy? 'To challenge your own point of view'. *The Atlantic*. 27 February.

Rescorla, M. (2007). Church's thesis and the conceptual analysis of computability. *Notre Dame Journal of Formal Logic*, 48(2):253–280.

Rescorla, M. (2012a). Are computational transitions sensitive to semantics? *Australian Journal of Philosophy*, 90(4):703–721.

Rescorla, M. (2012b). How to integrate representation into computational modeling, and why we should. *Journal of Cognitive Science (South Korea)*, 13(1):1–38.

Rescorla, M. (2013). Against structuralist theories of computational implementation. *British Journal for the Philosophy of Science*, 64(4):681–707.

Rescorla, M. (2014a). The causal relevance of content to computation. *Philosophy and Phenomenological Research*, 88(1):173–208.

Rescorla, M. (2014b). A theory of computational implementation. *Synthese*, 191:1277–1307.

Rescorla, M. (2015). The representational foundations of computation. *Philosophia Mathematica*, 23(3):338–366.

Rescorla, M. (2020). The computational theory of mind. In Zalta, E. N., editor, *The Stanford Encyclopedia of Philosophy*. Metaphysics Research Lab, Stanford University, Fall 2020 edition.

Rey, G. (2012). The Turing thesis vs. the Turing test. *The Philosopher's Magazine*, 57:84–89.

Richards, R. J. (2009). The descent of man. *American Scientist*, 97(5):415–417.

Riskin, J. (2020). Just use your thinking pump! *New York Review of Books*, 67(11):48–50. 2 July.

Roberts, S. (2018). The Yoda of Silicon Valley. *New York Times*. 17 December.

Robertson, D. S. (2003). *Phase Change: The Computer Revolution in Science and Mathematics*. Oxford University Press.

Robertson, J. I. (1979). How to do arithmetic. *American Mathematical Monthly*, 86:431–439.

Robertson Ishii, T. and Atkins, P. (2020). Essential vs. accidental properties. In Zalta, E. N., editor, *The Stanford Encyclopedia of Philosophy*. Metaphysics Research Lab, Stanford University, Winter 2020 edition.

Robinson, J. A. (1994). Logic, computers, Turing, and von Neumann. In Furukawa, K., Michie, D., and Muggleton, S., editors, *Machine Intelligence 13: Machine Intelligence and Inductive Learning*, pages 1–35. Clarendon Press, Oxford.

Roelofs, L. and Buchanan, J. (2019). Panpsychism, intuitions, and the great chain of being. *Philosophical Studies*, 176:2991–3017.

Rogers, Jr., H. (1959). The present theory of Turing machine computability. *Journal of the Society for Industrial and Applied Mathematics*, 7(1):114–130.

Rohlf, M. (2020). Immanuel Kant. In Zalta, E. N., editor, *The Stanford Encyclopedia of Philosophy*. Metaphysics Research Lab, Stanford University, Fall 2020 edition.

Rosenberg, A. (1994). If economics isn't science, what is it? In Hausman, D. M., editor, *The Philosophy of Economics: An Anthology, Second Edition*, pages 376–394. Cambridge University Press, New York.

Rosenblueth, A. and Wiener, N. (1945). The role of models in science. *Philosophy of Science*, 12:316–321.

Rosser, J. B. (1939). An informal exposition of proofs of Gödel's theorems and Church's theorem. *Journal of Symbolic Logic*, 4(2):53–60.

Rosser, J. B. (1978). *Logic for Mathematicians: Second Edition*. Dover Publications, Mineola, NY. First edition (1953) at https://archive.org/details/logicformathemat00ross.

Rothman, J. (2021). Thinking it through: How much can rationality do for us? *The New Yorker*, pages 24–29. 23 August.

Rowlands, M., Lau, J., and Deutsch, M. (2020). Externalism about the mind. In Zalta, E. N., editor, *The Stanford Encyclopedia of Philosophy*. Metaphysics Research Lab, Stanford University, Winter 2020 edition.

Royce, J. (1900). *The World and the Individual*. Macmillan, London.

Russell, B. (1912). *The Problems of Philosophy*. Oxford University Press (1959), London.

Russell, B. (1927). *The Analysis of Matter*. Spokesman (2007), Nottingham, UK.

Russell, B. (1936). The limits of empiricism. *Proceedings of the Aristotelian Society, New Series*, 36:131–150.

Russo, J. (1986). Saturn's rings: What GM's Saturn project is really about. *Cornell University Labor Research Review*, 1(9).

Rutenberg, J. (2019). The tabloid myths of Jennifer Aniston and Donald Trump. *New York Times*. 27 January.

Ryle, G. (1945). Knowing how and knowing that. *Proceedings of the Aristotelian Society, New Series*, 46:1–16.

Samet, J. and Zaitchik, D. (2017). Innateness and contemporary theories of cognition. In Zalta, E. N., editor, *The Stanford Encyclopedia of Philosophy*. Metaphysics Research Lab, Stanford University, Fall 2017 edition.

Samuel, A. L. (1953). Computing bit by bit, or digital computers made easy. *Proceedings of the IRE*, 41(10):1223–1230.

Samuelson, P., Davis, R., Kapor, M. D., and Reichman, J. (1994). A manifesto concerning the legal protection of computer programs. *Columbia Law Review*, 94(8):2308–2431.

Sayre, K. M. (1986). Intentionality and information processing: An alternative model for cognitive science. *Behavioral and Brain Sciences*, 9(1):121–165.

Schagrin, M. L., Rapaport, W. J., and Dipert, R. R. (1985). *Logic: A Computer Approach*. McGraw-Hill, New York.

Scherlis, W. L. and Scott, D. S. (1983). First steps towards inferential programming. In Mason, R., editor, *Information Processing 83*, pages 199–212. JFIP and Elsevier North-Holland. Reprinted in Colburn et al., 1993, pp. 99–133.

Schweizer, P. (2017). Cognitive computation *sans* representation. In Powers, T., editor, *Philosophy and Computing: Essays in Epistemology, Philosophy of Mind, Logic, and Ethics,*. Springer.

Schwitzgebel, E. (2021). Belief. In Zalta, E. N., editor, *The Stanford Encyclopedia of Philosophy*. Metaphysics Research Lab, Stanford University, Winter 2021 edition.

Schwitzgebel, E. (9 January 2012). For all x, there's philosophy of x. http://schwitzsplinters.blogspot .com/2012/01/for-all-x-theres-philosophy-of-x.html.

Seabrook, J. (2019). The next word. *The New Yorker*, pages 52–63. 14 October. See also follow-up letters at https://www.newyorker.com/magazine/2019/11/11/letters-from-the-november-11-2019-issue.

Searle, J. R. (1969). *Speech Acts: An Essay in the Philosophy of Language*. Cambridge University Press, Cambridge, UK.

Searle, J. R. (1980). Minds, brains, and programs. *Behavioral and Brain Sciences*, 3:417–457.

Searle, J. R. (1982). The myth of the computer. *New York Review of Books*, pages 3–6. See correspondence, same journal, 24 June 1982, pp. 56–57.

Searle, J. R. (1983). *Intentionality: An Essay in the Philosophy of Mind*. Cambridge University Press, Cambridge, UK.

Searle, J. R. (1984). *Minds, Brains and Science*. Harvard University Press, Cambridge, MA.

Searle, J. R. (1990). Is the brain a digital computer? *Proceedings and Addresses of the American Philosophical Association*, 64(3):21–37.

Searle, J. R. (1992). *The Rediscovery of the Mind*. MIT Press, Cambridge, MA.

Sellars, W. (1963). Philosophy and the scientific image of man. In *Science, Perception and Reality*, pages 1–40. Routledge & Kegan Paul, London.

Shagrir, O. (1997). Two dogmas of computationalism. *Minds and Machines*, 7:321–344.

Shagrir, O. (2001). Content, computation and externalism. *Mind*, 110(438):369–400.

Shagrir, O. (2002). Effective computation by humans and machines. *Minds and Machines*, 12(2):221–240.

Shagrir, O. (2020). In defense of the semantic view of computation. *Synthese*, 197:4083–4108.

Shagrir, O. (2022). *The Nature of Physical Computation*. Oxford University Press, New York.

Shagrir, O. and Bechtel, W. (2015). Marr's computational-level theories and delineating phenomena. In Kaplan, D., editor, *Integrating Psychology and Neuroscience: Prospects and Problems*. Oxford University Press, Oxford.

Shanker, S. (1987). Wittgenstein versus Turing on the nature of Church's thesis. *Notre Dame Journal of Formal Logic*, 28(4):615–649.

Shannon, C. E. (1937). A symbolic analysis of relay and switching circuits. Technical report, MIT Department of Electrical Engineering, Cambridge, MA. MS thesis, 1940; http://hdl.handle.net/1721 .1/11173.

Shannon, C. E. (1948). A mathematical theory of communication. *The Bell System Technical Journal*, 27:379–423, 623–656. July and October.

Shannon, C. E. (1950). Programming a computer for playing chess. *Philosophical Magazine, Ser. 7*, 41(314).

Shannon, C. E. (1953). Computers and automata. *Proceedings of the Institute of Radio Engineers*, 41(10):1234–1241.

Shapin, S. (2019). A theorist of (not quite) everything. *New York Review of Books*, 66(15):29–31.

Shapiro, F. R. (2000). Origin of the term *software*: Evidence from the JSTOR electronic journal archive. *IEEE Annals of the History of Computing*, 22(2):69–71.

Shapiro, S., Snyder, E., and Samuels, R. (2022). Computability, notation, and *de re* knowledge of numbers. *Philosophies*, 7(1).

Shapiro, S. C. (1989). The Cassie projects: An approach to natural language competence. In Martins, J. and Morgado, E., editors, *EPIA 89: 4th Portuguese Conference on Artificial Intelligence Proceedings*, pages 362–380. Springer-Verlag Lecture Notes in Artificial Intelligence 390, Berlin.

Shapiro, S. C. (1992a). Artificial Intelligence. In Shapiro, S. C., editor, *Encyclopedia of Artificial Intelligence, 2nd edition*, pages 54–57. John Wiley & Sons, New York.

Shapiro, S. C. (1992b). *Common Lisp: An Interactive Approach*. W.H. Freeman, New York. https://www .cse.buffalo.edu/~shapiro/Commonlisp/.

Shapiro, S. C. (2001). Computer science: The study of procedures. Technical report, Department of Computer Science and Engineering, University at Buffalo, Buffalo, NY. http://www.cse.buffalo.edu/ ~shapiro/Papers/whatiscs.pdf.

Shapiro, S. C. and Rapaport, W. J. (1987). SNePS considered as a fully intensional propositional semantic network. In Cercone, N. and McCalla, G., editors, *The Knowledge Frontier: Essays in the Representation of Knowledge*, pages 262–315. Springer-Verlag, New York.

Shapiro, S. C. and Rapaport, W. J. (1991). Models and minds: Knowledge representation for natural-language competence. In Cummins, R. and Pollock, J., editors, *Philosophy and AI: Essays at the Interface*, pages 215–259. MIT Press, Cambridge, MA.

Shapiro, S. C. and Rapaport, W. J. (1995). An introduction to a computational reader of narratives. In Duchan, J. F., Bruder, G. A., and Hewitt, L. E., editors, *Deixis in Narrative: A Cognitive Science Perspective*, pages 79–105. Lawrence Erlbaum Associates, Hillsdale, NJ.

Shapiro, S. C. and Wand, M. (1976). The relevance of relevance. Technical Report 46, Indiana University Computer Science Department, Bloomington, IN. https://legacy.cs.indiana.edu/ftp/techreports/TR46.pdf.

Shelley, M. W. (1818). *Frankenstein; or, the Modern Prometheus*. http://www.literature.org/authors/shelley-mary/frankenstein/.

Shepherdson, J. and Sturgis, H. (1963). Computability of recursive functions. *Journal of the ACM*, 10(2):217–255.

Sheraton, M. (1981). The elusive art of writing precise recipes. *New York Times*. 2 May.

Shieber, S. M., editor (2004). *The Turing Test: Verbal Behavior as the Hallmark of Intelligence*. MIT Press, Cambridge, MA.

Shieber, S. M. (2007). The Turing Test as interactive proof. *Noûs*, 41(4):686–713.

Shieh, D. and Turkle, S. (2009). The trouble with computer simulations: Linked in with Sherry Turkle. *Chronicle of Higher Education*, page A14. 27 March.

Shladover, S. E. (2016). What "self-driving" cars will really look like. *Scientific American*. June.

Sieg, W. (1994). Mechanical procedures and mathematical experience. In George, A., editor, *Mathematics and Mind*, pages 71–117. Oxford University Press, New York.

Sieg, W. (2000). Calculations by man and machine: Conceptual analysis. Technical Report CMU-PHIL-104, Carnegie-Mellon University Department of Philosophy, Pittsburgh, PA. http://repository.cmu.edu/philosophy/178.

Sieg, W. (2006). Gödel on computability. *Philosophia Mathematica*, 14:189–207. Page references to http://repository.cmu.edu/cgi/viewcontent.cgi?article=1112&context=philosophy.

Sieg, W. (2007). On mind & Turing's machines. *Natural Computing*, 6(2):187–205.

Sieg, W. (2008). On computability. In Irvine, A., editor, *Philosophy of Mathematics*, pages 525–621. Elsevier, Oxford. https://www.cmu.edu/dietrich/philosophy/docs/seig/On%20Computability.pdf.

Sieg, W. and Byrnes, J. (1999). An abstract model for parallel computation: Gandy's thesis. *The Monist*, 82(1):150–164.

Siegelmann, H. T. (1995). Computation beyond the Turing limit. *Science*, 268(5210):545–548.

Silver, D. et al. (2016). Mastering the game of Go with deep neural networks and tree search. *Nature*, 539:484–489. 28 January.

Silver, D. et al. (2018). A general reinforcement learning algorithm that masters chess, shogi, and Go through self-play. *Science*, 362(6419):1140–1144.

Simon, H. A. (1962). The architecture of complexity. *Proceedings of the American Philosophical Society*, 106(6):467–482. Reprinted as Simon, 1996a, Ch. 8.

Simon, H. A. (1977). What computers mean for man and society. *Science*, 195(4283):1186–1191.

Simon, H. A. (1996a). Computational theories of cognition. In O'Donohue, W. and Kitchener, R. F., editors, *The Philosophy of Psychology*, pages 160–172. SAGE Publications, London.

Simon, H. A. (1996b). *The Sciences of the Artificial, Third Edition*. MIT Press, Cambridge, MA.

Simon, H. A. and Newell, A. (1956). Models: Their uses and limitations. In White, L. D., editor, *The State of the Social Sciences*, pages 66–83. University of Chicago Press, Chicago.

Simon, H. A. and Newell, A. (1958). Heuristic problem solving: The next advance in operations research. *Operations Research*, 6(1):1–10.

Simon, H. A. and Newell, A. (1962). Simulation of human thinking. In Greenberger, M., editor, *Computers and the World of the Future*, pages 94–114. MIT Press, Cambridge, MA.

Skidelsky, R. (2014). The programmed prospect before us. *New York Review of Books*, 61(6):35–37. 3 April.

Skinner, D. (2006). The age of female computers. *The New Atlantis*, pages 96–103. Spring.

Slocum, J. (1985). A survey of machine translation: Its history, current status, and future prospects. *Computational Linguistics*, 11(1):1–17.

Sloman, A. (1978). *The Computer Revolution in Philosophy: Philosophy, Science and Models of Mind*. Humanities Press, Atlantic Highlands, NJ.

Sloman, A. (1998). Supervenience and implementation: Virtual and physical machines. http://www.cs .bham.ac.uk/research/projects/cogaff/Sloman.supervenience.and.implementation.pdf.

Sloman, A. (2002). The irrelevance of Turing machines to AI. In Scheutz, M., editor, *Computationalism: New Directions*, pages 87–127. MIT Press, Cambridge, MA. Page references to http://www.cs.bham .ac.uk/research/projects/cogaff/sloman.turing.irrelevant.pdf.

Sloman, A. (2020). How to trisect an angle (using P-Geometry). https://www.cs.bham.ac.uk/research/ projects/cogaff/misc/trisect.html.

Smith, A. (1776). *Wealth of Nations*.

Smith, B. C. (1985). Limits of correctness in computers. *ACM SIGCAS Computers and Society*, 14–15(1–4):18–26. Also, *Technical Report* CSLI-85-36 (Stanford, CA: Center for the Study of Language & Information); reprinted in Charles Dunlop & Rob Kling (eds.), *Computerization and Controversy* (San Diego: Academic Press, 1991): 632–646; and in Colburn et al., 1993, pp. 275–293.

Smith, B. C. (1987). The correspondence continuum. Technical Report CSLI-87-71, Center for the Study of Language & Information, Stanford, CA.

Smith, B. C. (2002). The foundations of computing. In Scheutz, M., editor, *Computationalism: New Directions*, pages 23–58. MIT Press, Cambridge, MA.

Smith, B. C. (2019). *The Promise of Artificial Intelligence: Reckoning and Judgment*. MIT Press, Cambridge, MA.

Smith, R. D. (2000). Simulation. In Ralston, A., Reilly, E. D., and Hemmendinger, D., editors, *Encyclopedia of Computer Science, 4th Edition*, pages 1578–1587. Nature Publishing Group, London.

Soames, S. (2016). Philosophy's true home. *New York Times*. 7 March.

Soare, R. I. (1999). The history and concept of computability. In Griffor, E., editor, *Handbook of Computability Theory*, pages 3–36. North-Holland, Amsterdam. Page references to http://www .people.cs.uchicago.edu/~soare/History/handbook.pdf.

Soare, R. I. (2009). Turing oracle machines, online computing, and three displacements in computability theory. *Annals of Pure and Applied Logic*, 160:368–399.

Soare, R. I. (2012). Formalism and intuition in computability. *Philosophical Transactions of the Royal Society A*, 370:3277–3304.

Soare, R. I. (2016). *Turing Computability: Theory and Applications*. Springer, Berlin.

Soni, J. and Goodman, R. (2017). A man in a hurry: Claude Shannon's New York years. *IEEE Spectrum*. 12 July.

Sparrow, R. (2007). Killer robots. *Journal of Applied Philosophy*, 24(1):62–77.

Sprevak, M. (2010). Computation, individuation, and the received view on representation. *Studies in History and Philosophy of Science*, 41(3):260–270.

Sprevak, M. (2018). Triviality arguments about computational implementation. In Sprevak, M. and Matteo, C., editors, *The Routledge Handbook of the Computational Mind*, pages 175–191. Routledge, London.

Srihari, S. N. (2010). Beyond C.S.I.: The rise of computational forensics. *IEEE Spectrum*. 29 November.

Stairs, A. (2014). Response to question about the definition of 'magic'. http://www.askphilosophers .org/question/5735.

Staples, M. (2015). Critical rationalism and engineering: Methodology. *Synthese*, 192(1):337–362.

Steed, S. (2013). Harnessing human intellect for computing. *Computing Research News*, 25(2).

Stein, D. K. (1984). Lady Lovelace's notes: Technical text and cultural context. *Victorian Studies*, 28(1):33–67.

Stepney, S., Braunstein, S. L., Clark, J. A., Tyrrell, A., Adamatzky, A., Smith, R. E., Addis, T., Johnson, C., Timmis, J., Welch, P., Milner, R., and Partridge, D. (2005). Journeys in non-classical computation I: A grand challenge for computing research. *International Journal of Parallel, Emergent and Distributed Systems*, 20(1):5–19.

Stevens, Jr., P. (1996). Magic. In Levinson, D. and Ember, M., editors, *Encyclopedia of Cultural Anthropology*, pages 721–726. Henry Holt, New York.

Stewart, I. (2001). Easter is a quasicrystal. *Scientific American*, pages 80, 82–83. March, http://www.whydomath.org/Reading_Room_Material/ian_stewart/2000_03.html.

Stewart, N. (1995). Science and computer science. *ACM Computing Surveys*, 27(1):39–41.

Stoll, C. (2006). When slide rules ruled. *Scientific American*, 294(5):80–87. May.

Stoppard, T. (2015). *The Hard Problem*. Grove Press, New York.

Strasser, C. and Antonelli, G. A. (2019). Non-monotonic Logic. In Zalta, E. N., editor, *The Stanford Encyclopedia of Philosophy*. Metaphysics Research Lab, Stanford University, Summer 2019 edition.

Strawson, G. (2012). Real naturalism. *Proceedings and Addresses of the American Philosophical Association*, 86(2):125–154.

Strevens, M. (2019). The substantiality of philosophical analysis. *The Brains Blog*. 29 March, http://philosophyofbrains.com/2019/03/28/the-substantiality-of-philosophical-analysis.aspx.

Suber, P. (1988). What is software? *Journal of Speculative Philosophy*, 2(2):89–119. Revised version at https://dash.harvard.edu/bitstream/handle/1/3715472/suber_software.html.

Suber, P. (1997a). Formal systems and machines: An isomorphism. https://legacy.earlham.edu/~peters/courses/logsys/machines.htm.

Suber, P. (1997b). The Löwenheim-Skolem theorem. https://legacy.earlham.edu/~peters/courses/logsys/low-skol.htm.

Swade, D. D. (1993). Redeeming Charles Babbage's mechanical computer. *Scientific American*, pages 86–91. February.

Swoyer, C. (1991). Structural representation and surrogative reasoning. *Synthese*, 87(3):449–508.

Tanenbaum, A. S. (2006). *Structured Computer Organization, Fifth Edition*. Pearson Prentice Hall, Upper Saddle River, NJ.

Tedre, M. (2008). What should be automated? *ACM Interactions*, 15(5):47–49.

Tedre, M. (2015). *The Science of Computing: Shaping a Discipline*. CRC Press/Taylor & Francis, Boca Raton, FL.

Tedre, M. and Sutinen, E. (2008). Three traditions of computing: What educators should know. *Computer Science Education*, 18(3):153–170.

Tenenbaum, A. M. and Augenstein, M. J. (1981). *Data Structures using Pascal*. Prentice-Hall, Englewood Cliffs, NJ.

Teuscher, C. and Sipper, M. (2002). Hypercomputation: Hype or computation? *Communications of the ACM*, 45(8):23–30.

Thagard, P. (1984). Computer programs as psychological theories. In Neumaier, O., editor, *Mind, Language and Society*, pages 77–84. Conceptus-Studien, Vienna.

Tharp, L. H. (1975). Which logic is the right logic? *Synthese*, 31:1–21.

The Economist (2013). Unreliable research: Trouble at the lab. *The Economist*. 19 October.

Thomason, R. H. (2003). Dynamic contextual intensional logic: Logical foundations and an application. In Blackburn, P., editor, *CONTEXT 2003: Lecture Notes in Artificial Intelligence 2680*, pages 328–341. Springer-Verlag, Berlin.

Thompson, C. (2019). The secret history of women in coding. *New York Times Magazine*. 13 February.

Thornton, S. P. (2004). Solipsism and the problem of other minds. *Internet Encyclopedia of Philosophy*. https://iep.utm.edu/solipsis/.

Tingley, K. (2013). The body electric. *The New Yorker*, pages 78–86. 25 November.

Toussaint, G. (1993). A new look at Euclid's second proposition. *Mathematical Intelligencer*, 15(3):12–23.

Tukey, J. W. (1958). The teaching of concrete mathematics. *American Mathematical Monthly*, 65(1):1–9.

Turing, A. M. (1936). On computable numbers, with an application to the *Entscheidungsproblem*. *Proceedings of the London Mathematical Society, Ser. 2*, Vol. 42:230–265. Reprinted with corrections in Davis, 1965, pp. 116–154.

Turing, A. M. (1937). Computability and λ-definability. *Journal of Symbolic Logic*, 2(4):153–163.

Turing, A. M. (1938). On computable numbers, with an application to the *Entscheidungsproblem*: A correction. *Proceedings of the London Mathematical Society, Ser. 2*, 43(1):544–546. https://www .wolframscience.com/prizes/tm23/images/Turing2.pdf.

Turing, A. M. (1939). Systems of logic based on ordinals. *Proceedings of the London Mathematical Society*, S2-45(1):161–228.

Turing, A. M. (1947). Lecture to the London Mathematical Society on 20 February 1947. In Copeland, B. J., editor, *The Essential Turing*, pages 378–394. Oxford University Press, Oxford (2004).

Turing, A. M. (1948). Intelligent machinery. In Copeland, B. J., editor, *The Essential Turing*, pages 410–432. Oxford University Press (2004), Oxford.

Turing, A. M. (1949). Checking a large routine. In *Report of a Conference on High Speed Automatic Calculating Machines*, pages 67–69. University Mathematics Lab, Cambridge, UK. Reprinted in Morris and Jones, 1984.

Turing, A. M. (1950). Computing machinery and intelligence. *Mind*, 59(236):433–460.

Turing, A. M. (1951a). Can digital computers think? In Copeland, B. J., editor, *The Essential Turing*, pages 476–486. Oxford University Press (2004), Oxford. Also in Shieber, 2004, Ch. 6.

Turing, A. M. (1951b). Intelligent machinery, a heretical theory. *Philosophia Mathematica*, 4(3):256–260. Reprinted in Copeland, 2004, Ch. 12 and Shieber, 2004, Ch. 5.

Turing, A. M. (1953). Chess. In Copeland, B. J., editor, *The Essential Turing*, pages 562–575. Oxford University Press (2004), Oxford.

Turing, A. M., Braithwaite, R., Jefferson, G., and Newman, M. (1952). Can automatic calculating machines be said to think? In Copeland, B. J., editor, *The Essential Turing*, pages 487–506. Oxford University Press (2004), Oxford. Also in Shieber, 2004, Ch. 7.

Turner, R. (2018). *Computational Artifacts: Towards a Philosophy of Computer Science*. Springer, Berlin.

Tye, M. (2017). *Tense Bees and Shell-Shocked Crabs: Are Animals Conscious?* Oxford University Press, New York.

Uebel, T. (2021). Vienna Circle. In Zalta, E. N., editor, *The Stanford Encyclopedia of Philosophy*. Metaphysics Research Lab, Stanford University, Fall 2021 edition.

Uglow, J. (2010). The other side of science. *New York Review of Books*, pages 30–31, 34. 24 June.

Vahid, F. (2003). The softening of hardware. *IEEE Computer*, 36(4):27–34.

van Fraassen, B. C. (1989). *Laws and Symmetry*. Clarendon Press, Oxford.

van Leeuwen, J. and Wiedermann, J. (2013). The computational power of Turing's non-terminating circular *a*-machines. In Cooper, S. B. and van Leeuwen, J., editors, *Alan Turing: His Work and Impact*, pages 80–85. Elsevier, Amsterdam.

Vardi, M. Y. (2012). What is an algorithm? *Communications of the ACM*, 55(3):5.

Vardi, M. Y. (2013). Who begat computing? *Communications of the ACM*, 56(1):5.

Vardi, M. Y. (2016). The moral imperative of Artificial Intelligence. *Communications of the ACM*, 59(5):5.

Vardi, M. Y. (2017). Would Turing have won the Turing Award? *Communications of the ACM*, 60(11):7.

Veblen, O. (1904). A system of axioms for geometry. *Transactions of the American Mathematical Society*, 5(3):343–384.

Veblen, T. (1908). The evolution of the scientific point of view. *The University of California Chronicle: An Official Record*, 10(4):395–416.

Vera, A. (2018). Social animals. *The New Yorker*, page 3. 30 April. Letter to the editor.

von Neumann, J. (1945). First draft report on the EDVAC. *IEEE Annals of the History of Computing*, 15(4 (1993)):27–75. Page references to http://web.mit.edu/STS.035/www/PDFs/edvac.pdf, Michael D. Godfrey (ed.).

von Neumann, J. (1948). The general and logical theory of automata. In Jeffress, L., editor, *Cerebral Mechanisms in Behavior: The Hixon Symposium*, pages 1–41. John Wiley & Sons (1951), New York.

von Neumann, J. (1966). *Theory of Self-Reproducing Automata*. University of Illinois Press, Urbana, IL. Arthur W. Burks (ed.).

Wadler, P. (1997). How to declare an imperative. *ACM Computing Surveys*, 29(3):240–263.

Wagner, A. R. and Arkin, R. C. (2011). Acting deceptively: Providing robots with the capacity for deception. *International Journal of Social Robotics*, 3(1):5–26.

Wainer, H. (2012). The survival of the fittists. *American Scientist*, 100:358–361. September-October.

Wallach, W. and Allen, C. (2009). *Moral Machines: Teaching Robots Right from Wrong*. Oxford University Press, New York.

Wallas, G. (1926). *The Art of Thought*. Solis Press (2014), Kent, UK.

Wallich, P. (1997). Cracking the U.S. code. *Scientific American*, page 42. April.

Wang, H. (1957). A variant to Turing's theory of computing machines. *Journal of the ACM*, 4(1):63–92.

Wang, Z., Busemeyer, J. R., Atmanspacher, H., and Pothos, E. M., editors (2013). *Topics in Cognitive Science, Vol. 5, No. 4): The Potential of Using Quantum Theory to Build Models of Cognition*. Cognitive Science Society.

Wangsness, T. and Franklin, J. (1966). "Algorithm" and "formula". *Communications of the ACM*, 9(4):243.

Wartofsky, M. W. (1966). The model muddle: Proposals for an immodest realism. In Wartofsky, M. W., editor, *Models: Representation and the Scientific Theory of Understanding*, pages 1–11. D. Reidel, 1979, Dordrecht, The Netherlands.

Wartofsky, M. W. (1979). Introduction. In Wartofsky, M. W., editor, *Models: Representation and the Scientific Theory of Understanding*, pages xiii–xxvi. D. Reidel, Dordrecht, The Netherlands.

Webb, J. (1980). *Mechanism, Mentalism, and Metamathematics*. D. Reidel, Dordrecht, The Netherlands.

Wegner, P. (1976). Research paradigms in computer science. In *ICSE '76 Proceedings of the 2nd International Conference on Software Engineering*, pages 322–330. IEEE Computer Society Press, Los Alamitos, CA.

Wegner, P. (1995). Interaction as a basis for empirical computer science. *ACM Computing Surveys*, 27(1):45–48.

Wegner, P. (1997). Why interaction is more powerful than algorithms. *Communications of the ACM*, 40(5):80–91.

Wegner, P. (2010). What is computation? The evolution of computation. *Ubiquity*, 2010. November, Article 2.

Wegner, P. and Goldin, D. (2006). Principles of problem solving. *Communications of the ACM*, 49(7):27–29.

Weinberg, S. (2002). Is the universe a computer? *New York Review of Books*, 49(16).

Weiner, J. (2017). The magician who wants to break magic. *New York Times Magazine*. 15 March.

Weizenbaum, J. (1972). On the impact of the computer on society: How does one insult a machine? *Science*, 176(4035):609–614.

Weizenbaum, J. (1976). *Computer Power and Human Reason*. W.H. Freeman, New York.

Welch, P. D. (2007). Turing unbound: Transfinite computation. In Cooper, S., Löwe, B., and Sorbi, A., editors, *CiE 2007*, pages 768–780. Springer-Verlag Lecture Notes in Computer Science 4497, Berlin.

Wheeler, J. A. (1989). Information, physics, quantum: The search for links. In *Proceedings III International Symposium on Foundations of Quantum Mechanics*, pages 354–358. Tokyo. https://philpapers.org/archive/WHEIPQ.pdf.

White, N. P. (1974). What numbers are. *Synthese*, 27:111–124.

White, T. I. (2007). *In Defense of Dolphins: The New Moral Frontier*. Blackwell, Oxford.

White, T. I. (2013). A primer on nonhuman personhood, cetacean rights and 'flourishing'. http://indefenseofdolphins.com/wp-content/uploads/2013/07/primer.pdf.

Wiener, N. (1964). *God and Golem, Inc.: A Comment on Certain Points Where Cybernetics Impinges on Religion*. MIT Press, Cambridge, MA.

Wiesner, J. (1958). Communication sciences in a university environment. *IBM Journal of Research and Development*, 2(4):268–275.

Wigner, E. (1960). The unreasonable effectiveness of mathematics in the natural sciences. *Communications in Pure and Applied Mathematics*, 13(1).

Williams, B. (1998). The end of explanation? *The New York Review of Books*, 45(18):40–44.

Williamson, T. (2007). *The Philosophy of Philosophy*. Blackwell, Oxford.

Williamson, T. (2011). What is naturalism? *New York Times Opinionator: The Stone*. 4 September.

Williamson, T. (2020). *Philosophical Method: A Very Short Introduction*. Oxford University Press, Oxford.

Wilson, D. G. and Papadopoulos, J. (2004). *Bicycling Science, Third Edition*. MIT Press, Cambridge, MA.

Wing, J. M. (2006). Computational thinking. *Communications of the ACM*, 49(3):33–35.

Wing, J. M. (2008a). Computational thinking and thinking about computing. *Philosophical Transactions of the Royal Society A*, 366:3717–3725.

Wing, J. M. (2008b). Five deep questions in computing. *Communications of the ACM*, 51(1):58–60.

Wing, J. M. (2010). Computational thinking: What and why? *The Link*, Carnegie-Mellon University, 17 November, https://www.cs.cmu.edu/~CompThink/resources/TheLinkWing.pdf.

Winograd, T. and Flores, F. (1987). *Understanding Computers and Cognition: A New Foundation for Design*. Addison-Wesley, Reading, MA.

Winston, P. H. (1977). *Artificial Intelligence*. Addison-Wesley, Reading, MA.

Wirth, N. (1971). Program development by stepwise refinement. *Communications of the ACM*, 14(4):221–227.

Wittgenstein, L. (1921). *Tractatus Logico-Philosophicus, 2nd edition*. Humanities Press, 1972, New York.

Wittgenstein, L. (1958). *Philosophical Investigations, Third Edition*. Macmillan, New York.

Wittgenstein, L. (1980). *Remarks on the Philosophy of Psychology, Vol. I*. University of Chicago Press, Chicago.

Wolfram, S. (2002). *A New Kind of Science*. Wolfram Media. http://www.wolframscience.com/nks/.

Wulf, W. (1995). Are we scientists or engineers? *ACM Computing Surveys*, 27(1):55–57.

Zeigler, B. P. (1976). *Theory of Modeling and Simulation*. Wiley Interscience, New York.

Zemanek, H. (1971). Was ist Informatik? (What is informatics?). *Elektronische Rechenanlagen (Electronic Computing Systems)*, 13(4):157–171.

Zimmer, C. (2016). In brain map, gears of mind get rare look. *New York Times*, page A1. 21 July.

Zobrist, A. L. (2000). Computer games: Traditional. In Ralston, A., Reilly, E. D., and Hemmendinger, D., editors, *Encyclopedia of Computer Science, 4th edition*, pages 364–368. Grove's Dictionaries, New York.

Zupko, J. (2018). John Buridan. In Zalta, E. N., editor, *The Stanford Encyclopedia of Philosophy*. Metaphysics Research Lab, Stanford University, Fall 2018 edition.

Index

Philosophy of Computer Science: An Introduction to the Issues and the Literature, First Edition. William J. Rapaport.
© 2023 John Wiley & Sons, Inc. Published 2023 by John Wiley & Sons, Inc.

Printed and bound by CPI Group (UK) Ltd, Croydon, CR0 4YY

21/12/2025

14796378-0005